# Proteins

## Energy, Heat and Signal Flow

# COMPUTATION IN CHEMISTRY

**Series Editor**
Professor Peter Gill
*Research School of Chemistry*
*Australian National University*
*pgill@rsc.anu.edu.au*

Proteins: Energy, Heat and Signal Flow, *edited by David M. Leitner and John E. Straub*

# Proteins

## Energy, Heat and Signal Flow

Edited by
## David M. Leitner
## John E. Straub

**CRC Press**
Taylor & Francis Group
Boca Raton London New York

CRC Press is an imprint of the
Taylor & Francis Group, an **informa** business

CRC Press
Taylor & Francis Group
6000 Broken Sound Parkway NW, Suite 300
Boca Raton, FL 33487-2742

First issued in paperback 2019

© 2010 by Taylor and Francis Group, LLC
CRC Press is an imprint of Taylor & Francis Group, an Informa business

No claim to original U.S. Government works

ISBN-13: 978-1-4200-8703-1 (hbk)
ISBN-13: 978-0-367-38514-9 (pbk)

**Library of Congress Cataloging-in-Publication Data**

Proteins : energy, heat and signal flow / editors, David M. Leitner, John Edward Straub.
    p. ; cm. -- (Computation in chemistry ; 1)
    Includes bibliographical references and index.
    ISBN 978-1-4200-8703-1 (hardcover : alk. paper)
    1. Proteins--Mathematical models. 2. Molecular dynamics. 3. Physical biochemistry.
I. Leitner, David M., 1963- II. Straub, John Edward. III. Title. IV. Series: Computation in chemistry ; 1.
    [DNLM: 1. Proteins--physiology. 2. Computer Simulation. 3. Energy Transfer. 4. Models, Molecular. 5. Thermodynamics. QU 55 P9692 2010]

QP551.5.P763 2010
612.3'98--dc22                                                     2009029151

Visit the Taylor & Francis Web site at
http://www.taylorandfrancis.com

and the CRC Press Web site at
http://www.crcpress.com

# Contents

## PART I    Energy Transduction in Molecular Motors

## PART II    Vibrational Energy Flow in Proteins: Molecular Dynamics-Based Methods

## PART III   Vibrational Energy Flow in Proteins and Nanostructures: Normal Mode-Based Methods

## PART IV   Conformational Transitions and Reaction Path Searches in Proteins

# Series Preface

Computational chemistry is highly interdisciplinary, nestling in the fertile region where chemistry meets mathematics, physics, biology, and computer science. Its goal is the prediction of chemical structures, bondings, reactivities, and properties through calculations *in silico*, rather than experiments *in vitro* or *in vivo*. In recent years, it has established a secure place in the undergraduate curriculum and modern graduates are increasingly familiar with the theory and practice of this subject. In the 21st century, as the prices of chemicals increase, governments enact ever-stricter safety legislations, and the performance/price ratios of computers increase, it is certain that computational chemistry will become an increasingly attractive and viable partner of experiment.

However, the relatively recent and sudden arrival of this subject has not been unproblematic. As the technical vocabulary of computational chemistry has grown and evolved, a serious language barrier has developed between those who prepare new methods and those who use them to tackle real chemical problems. There are only a few good textbooks; the subject continues to advance at a prodigious pace and it is clear that the daily practice of the community as a whole lags many years behind the state of the art. The field continues to advance and many topics that require detailed development are unsuitable for publication in a journal because of space limitations. Recent advances are available within complicated software programs but the average practitioner struggles to find helpful guidance through the growing maze of such packages.

This has prompted us to develop a series of books entitled *Computation in Chemistry* that aims to address these pressing issues, presenting specific topics in computational chemistry for a wide audience. The scope of this series is broad, and encompasses all the important topics that constitute "computational chemistry" as generally understood by chemists. The books' authors are leading scientists from around the world, chosen on the basis of their acknowledged expertise and their communication skills. Where topics overlap with fraternal disciplines—for example, quantum mechanics (physics) or computer-based drug design (pharmacology)—the treatment aims primarily to be accessible to, and serve the needs of, chemists.

This book is the first of the series. No previous text has offered a comprehensive treatment of the computational study of energy flow in proteins, and it exemplifies our approach: These will not be dusty technical monographs but, rather, books that will sit on every practitioner's desk.

# Preface

Proteins serve as molecular machines of the cell, and computational modeling has provided a wealth of insight into the role of energy flow in protein function. Prominent examples of the role of energy transduction in protein function include the chemomechanical cycle in the transport of cargo across the cell, the communication between reactive sites to regulate binding kinetics, charge transport, and the transport of heat to maintain a range of structures important for function. In the last few years, there has been a rapid evolution of computational approaches to study these processes, and their application has provided a detailed picture of energy transport in molecular machines and a guide to experimental work. The range of computational tools developed to study energy flow in proteins and other molecular machines includes direct molecular simulations of protein dynamics, hybrid quantum–classical modeling, harmonic analysis, coarse-graining strategies for the study of large systems over long periods of time, and algorithms for locating pathways on a complex energy landscape along which conformational change and energy flow occur. The contributions in this book detail state-of-the-art computational approaches and describe the major challenges that lie ahead, and provide an updated and complete resource for researchers interested in applications or further development.

This book is divided into four parts. Part I addresses the transport of energy in molecular motors, which function by a combination of chemically driven large-scale conformational changes or charge transport. Parts II and III are dedicated to vibrational energy flow in proteins and nanostructures, and are separated into approaches based on molecular dynamics simulations and those that mainly build on a harmonic picture. Part IV covers the flow of free energy in proteins giving rise to conformational changes involved in allosteric transitions and the role of coupled protein–solvent dynamics in conformational changes, and presents computational approaches developed to locate pathways between protein structures.

Part I focuses on energy conversion in molecular motors, and computational tools developed to study these systems. Full atomistic modeling is generally not feasible, and coarse-graining is needed for at least part of the problem to capture the length and timescales of interest. When quantum mechanical effects play a role, such as in electron or proton transfer, hybrid quantum mechanics/molecular mechanics (QM/MM) methods can be applied, where the QM part consists of an *ab initio* calculation involving a relatively small but relevant part of the protein, and molecular mechanics is used to treat the rest of the protein and solvent. In Chapter 1, Hyeon and Onuchic address the chemomechanical cycle of kinesin-1, which moves unidirectionally along microtubules with the energy of ATP hydrolysis, using a minimal model that preserves only the most important structural details required to capture the chemomechanical cycle and overall dynamics of kinesin. In Chapter 2, Yu and coworkers focus on myosin II, a motor protein involved in muscle contraction and cell division. Several complementary computational approaches are put forth and applied to the myosin II recovery stroke to locate transition paths, identify energetic coupling between remote regions of the protein, and locate residues particularly important in free energy transduction. In Chapter 3, Gao and Shao describe a thermal ratchet model for the study of kinesin and dyein, which yields insights into energy flow resulting from coupling between substrate binding, hydrolysis, product release, and the mechanical motion of the protein motor. In Chapter 4, the last chapter of this part, Stuchebrukhov reviews the mechanism of proton pumping in cytochrome $c$ oxidase, and describes a QM/MM method developed to address the coupled electron transfer and proton translocation steps carried out by this protein.

Parts II and III cover fast energy flow, both computational and experimental study of vibrational energy transport in proteins and nanostructures, which occurs typically on a 10 ps timescale. The four chapters in Part II detail methods based on molecular dynamics simulations for computing vibrational energy flow in proteins. The four chapters in Part III provide approaches that build mainly on a normal mode picture to describe vibrational energy and heat transport in proteins, as well as theoretical approaches that can be applied to model nanostructures, providing insights into the control of thermal transport on the nanoscale.

In the first chapter in Part II (Chapter 5), Kidera and coworkers describe the calculation of vibrational energy transfer rates and pathways in proteins using molecular dynamics simulations. In Chapter 6, Yamato reviews a method for locating vibrational energy transfer channels in proteins based on an analysis of molecular dynamics simulations and illustrates the method with the identification of energy transport pathways and local energy conductivities in photoactive yellow protein. In Chapter 7, Nguyen, Hamm, and Stock present nonequilibrium molecular dynamics approaches to simulate energy flow in peptides and proteins. The simulated energy flow in peptides is compared with experimental measurements, which highlight the important roles that quantum mechanical effects and solvents play in energy transport. In Chapter 8, Nagaoka, Yu, and Tabayanagi describe a method for sequentially time-resolving the protein vibrations and energy transfer from a molecular dynamics simulation. The important role that solvent and hydration structures play in energy transfer is discussed in Chapters 8 and 9.

Chapter 9, the first chapter in Part III, begins the bridge from molecular dynamics to normal mode-based methods. Zhang and Straub discuss directed energy flow in heme proteins computed by molecular dynamics simulations, and also introduce calculations of vibrational energy transfer beginning with a representation in terms of normal modes. Quantum mechanical effects on energy flow can be handled in a relatively straightforward manner in terms of a harmonic basis with the addition of anharmonic coupling. Coarse-graining procedures have been developed to extend harmonic analysis to very large biomolecules. In Chapter 10, Lu and Ma discuss recent developments in coarse-grained network approaches for computing normal modes. Application of normal modes with anharmonic corrections can be extended to the calculation of thermal transport in proteins. In Chapter 11, Leitner discusses heat flow in proteins and the calculation of frequency-resolved local diffusivities in proteins, which reveal networks of energy transport channels, many of which depend sensitively on vibrational frequency. In Chapter 12, Segal describes theoretical approaches to study heat transport in other nanoscale systems, offering a broad range of methods that may be brought to bear on the study of fast energy transport in molecular machines.

Part IV addresses the slower energy transduction associated with conformational changes in proteins, and the search for pathways along which conformational changes occur. The latter is the topic of the first two chapters in this part. The times over which energy associated with conformational changes flows lie beyond those accessible to conventional molecular simulation techniques. In Chapter 13, Elber, Kuczera, and Jas describe a coarse-graining technique that they developed called milestoning to compute kinetics and reaction dynamics of large molecules, and apply the approach to the folding of a helical peptide. In Chapter 14, Wales and coworkers discuss methods for computing important transition networks on a complex energy landscape that yield pathways and rates for structural transformations of peptides and proteins. In Chapter 15, Hilser and Whitten discuss free energy transduction in allostery, exploiting the fact that proteins exist as ensembles of interconverting conformational states to form the basis of a statistical, coarse-grained model that provides insights into rules governing communication in macromolecules, and how binding energy is distributed throughout the structure. In Chapter 16, Tobias, Sengupta, and Tarek present computational approaches for elucidating coupled protein–solvent dynamics, which give rise to conformational changes in proteins that would otherwise not take place in the absence of a solvent.

In general, the contributions in this book provide an updated overview of state-of-the-art computational methods developed to address fundamental questions related to molecular signaling and energy flow in proteins. The integrated presentation is designed to emphasize the interrelations between these disparate approaches that have contributed to our understanding of energy flow in proteins and its role in protein function. It is our hope that by defining the forefront of research in a variety of fields, these contributions frame the current challenges and opportunities for the development of methods and novel applications in the future in the evolving study of energy flow in molecular machines and nanomaterials.

**David M. Leitner**
**John E. Straub**

# Editors

**David M. Leitner, PhD,** was born in Chicago, Illinois. He studied chemistry and chemical engineering at Cornell University, and received his BS in both subjects in 1985. He then returned to Chicago for his graduate studies, receiving his PhD in 1989 at the University of Chicago in chemical physics while working with Steve Berry on dynamics and thermodynamics of small clusters. He did postdoctoral work with Jim Doll at Brown University; with Lorenz Cederbaum at the University of Heidelberg, Germany, as an NSF postdoctoral fellow; and with Peter Wolynes at the University of Illinois at Urbana-Champaign, with whom he carried out theoretical work on quantum mechanical energy flow in large molecules. He held a research faculty position at the University of California, San Diego, and was a frequent guest at Bilkent University in Turkey from 1998 to 2000. Since 2000 he has been at the University of Nevada, Reno, where he is a professor of chemistry, apart from a year as a visiting faculty at Ruhr University, Bochum, Germany, in 2006–2007. His major research interests involve theoretical and computational studies of energy flow in complex systems such as large molecules and glasses, vibrational dynamics and spectroscopy, thermal conduction on the nanoscale, and studies exploring how quantum energy flow in molecules mediates chemical reaction kinetics.

**John E. Straub, PhD,** was born in Denver, Colorado. He attended the University of Maryland at College Park where he carried out research on quantum scattering theory with Millard Alexander. He received his BS in chemistry in 1982. At Columbia University, he carried out research on chemical reaction dynamics with Bruce Berne and received his PhD in chemical physics in 1987. As an NIH postdoctoral fellow at Harvard University, he carried out research with Martin Karplus on dynamics and energy flow in proteins. In 1990, he joined the chemistry department at Boston University where he is currently professor and chair. He has enjoyed visiting professor appointments at the Institute for Advanced Studies at Hebrew University in Jerusalem, Israel, in 1998, and in the Department of Chemistry and Biochemistry at Montana State University in Bozeman, Montana, in 2006. He has served as the chair of the Theoretical Chemistry Subdivision of the American Chemical Society and as the president of the Telluride Science Research Center in Telluride, Colorado. His research interests in the field of theoretical and computational chemistry and biophysics include the development of computational methods for enhanced sampling in molecular dynamics and Monte Carlo simulations, the development of coarse-grained models for complex systems, the study of dynamics and thermodynamics of protein folding and aggregation, and the study of classical and quantum dynamics describing energy flow in liquids and biomolecules.

# Contributors

**Joanne M. Carr**
University Chemical Laboratories
Cambridge, United Kingdom

and

Computational Science
    and Engineering Department
STFC Daresbury Laboratory
Cheshire, United Kingdom

**Qiang Cui**
Department of Chemistry and
    Theoretical Chemistry Institute
and
Graduate Program in Biophysics
University of Wisconsin–Madison
Madison, Wisconsin

**Ron Elber**
Department of Chemistry and
    Biochemistry
Institute for Computational
    Engineering and Sciences
The University of Texas at Austin
Austin, Texas

**Hiroshi Fujisaki**
Integrated Simulation of
    Living Matter Group
Computational Science
    Research Program
RIKEN Brain Science Institute
Wako, Japan

**Yi Qin Gao**
Department of Chemistry
Texas A&M University
College Station, Texas

**Peter Hamm**
Physikalisch Chemisches Institut
Universität Zürich
Zurich, Switzerland

**Vincent J. Hilser**
Department of Biochemistry
    and Molecular Biology
Sealy Center for Structural
    Biology and Molecular
    Biophysics
The University of Texas Medical
    Branch–Galveston
Galveston, Texas

**Changbong Hyeon**
Department of Chemistry
Chung-Ang University
Seoul, South Korea

**Gouri S. Jas**
Department of Chemistry and
    Biochemistry
Baylor University
Waco, Texas

**Mey Khalili**
MITRE Corporation
McLean, Virginia

**Akinori Kidera**
Department of Supramolecular Biology
Yokohama City University
Yokohama, Japan

and

Integrated Simulation of
    Living Matter Group
Computational Science
    Research Program
RIKEN Brain Science Institute
Wako, Japan

**Krzysztof Kuczera**
Departments of Chemistry and
    Molecular Biosciences
University of Kansas
Lawrence, Kansas

**David M. Leitner**
Department of Chemistry
University of Nevada–Reno
Reno, Nevada

**Mingyang Lu**
Department of Biochemistry
    and Molecular Biology
Baylor College of Medicine
Houston, Texas

**Jianpeng Ma**
Department of Biochemistry
    and Molecular Biology
Baylor College of Medicine
Houston, Texas

and

Department of Bioengineering
Rice University
Houston, Texas

**Liang Ma**
Graduate Program in Biophysics
University of Wisconsin–Madison
Madison, Wisconsin

**Yasuhiro Matsunaga**
Integrated Simulation of
    Living Matter Group
Computational Science
    Research Program
RIKEN Brain Science Institute
Wako, Japan

**Kei Moritsugu**
Integrated Simulation of
    Living Matter Group
Computational Science
    Research Program
RIKEN Brain Science Institute
Wako, Japan

**Masataka Nagaoka**
Department of Complex Systems
    Science
Graduate School of Information
    Science
Nagoya University
Nagoya, Japan

**Phuong H. Nguyen**
Institute of Physical and
    Theoretical Chemistry
Goethe University
Frankfurt, Germany

**José N. Onuchic**
Department of Physics
Center for Theoretical
    Biological Physics
University of California
    San Diego
La Jolla, California

**Dvira Segal**
Department of Chemistry
University of Toronto
Toronto, Ontario, Canada

**Neelanjana Sengupta**
Physical Chemistry Division
National Chemical Laboratory
Pune, India

**Qiang Shao**
Department of Chemistry
Texas A&M University
College Station, Texas

**Vanessa K. de Souza**
School of Chemistry
University of Sydney
Sydney, Australia

**Gerhard Stock**
Institute of Physical and Theoretical
    Chemistry
Goethe University
Frankfurt, Germany

**John E. Straub**
Department of Chemistry
Boston University
Boston, Massachusetts

**Birgit Strodel**
University Chemical Laboratories
Cambridge, United Kingdom

**Alexei Stuchebrukhov**
Department of Chemistry
University of California–Davis
Davis, California

**Masayoshi Takayanagi**
Department of Complex Systems
    Science
Graduate School of Information
    Science
Nagoya University
Nagoya, Japan

**Mounir Tarek**
Equipe de Dynamique des
    Assemblages Membranaires
Nancy Université
Vandoeuvre-lès-Nancy, France

**Douglas J. Tobias**
Department of Chemistry
University of California–Irvine
Irvine, California

**David J. Wales**
University Chemical Laboratories
Cambridge, United Kingdom

**Steven T. Whitten**
Department of Biochemistry
    and Molecular Biology
The University of Texas Medical
    Branch–Galveston
Galveston, Texas

**Chris S. Whittleston**
Department of Chemistry
University of Cambridge
Cambridge, United Kingdom

**Takahisa Yamato**
Department of Molecular
    and Human Genetics
Baylor College of Medicine
Houston, Texas

**Yang Yang**
Department of Chemistry and
    Theoretical Chemistry Institute
University of Wisconsin–Madison
Madison, Wisconsin

**Haibo Yu**
Gordon Center for Integrative Science
University of Chicago
Chicago, Illinois

**Isseki Yu**
Department of Chemistry
    and Biological Science
College of Science and Engineering
Aoyama Gakuin University
Sagamihara, Japan

**Yong Zhang**
Department of Chemistry
Boston University
Boston, Massachusetts

# Part I

*Energy Transduction
in Molecular Motors*

# 1 Energy Balance and Dynamics of Kinesin Motors

*Changbong Hyeon and José N. Onuchic*

## CONTENTS

## 1.1 INTRODUCTION

Dynamics of biological systems occur with their characteristic length and energy scales. Although Newtonian mechanics can be used for describing events of any scale, our intuition based on Newtonian mechanics for macroscopic energy and length scales often fails to capture the essence of dynamics in biological systems. The length scale of basic biomaterials such as proteins, RNA, DNA, and cytoskeletal filaments ranges from nanometers to micrometers; the energy scale associated with their interaction, leading to the self-assembly and mutual interaction, is on the order of the thermal energy, $\sim 1$–$20 k_B T$. Since the energy scale of thermal fluctuations ($= k_B T$) of environment is comparable to the interaction energy, the entire molecule as well as the individual noncovalent bond jiggles and wiggles incessantly; hence in biological systems, there is neither smooth trajectory nor rigid body motion as observed in the macroscopic world. In addition to these length and energy scales, the dynamics of bioworld is characterized by a low Reynolds number hydrodynamics (Reynolds number $Re = f_{inertial} / f_{friction} \sim 10^{-5}$ where $f_{inertial}$ and $f_{friction}$ are the inertial force and the frictional force, respectively), which results from the large viscosity value of water environment (~1 cP). Consequently, the dynamics of biological systems vastly differ from the typical phenomena observed in human life. For instance, a "molecular ball" of nanometer size readily loses the memory of its original velocity in less than a picosecond; hence playing football is impossible in the bioworld. All these factors (size, energy, viscosity) make the dynamics in biological systems unique in the sense that biological nanomachines employ a different strategy from human-made macroscopic machines to operate in such a severe environment. In this chapter, we discuss one of the extensively studied biological nanomachines, the conventional kinesin, which the evolutionary processes have tailored over billions of years for its current biological functions, such as organelle transports and force generations.

As one of the tiniest molecular nanomachines, the unidirectional movement of kinesin-1 along microtubules (MTs) has fascinated the communities of various disciplines. The discovery in 1985 of the connection between the adenosine triphosphate (ATP) consumption and the transport of axoplasmic organelles [10,64] suggested an ATP-dependent force-generating molecule designated kinesin. Since then, efforts to understand the physical principle governing the mechanism of kinesin motility have been made through various methods including structural studies [24,25,34,35,37,60], biochemical analysis [17,18,20–22,43,48], single-molecule (SM) experiments [1–3,7,8, 12,26,33,45,46,57,58,63,65], and theoretical studies [4,11,15,16,29,30,32,36,38,52,62]. It has been proposed that the chemical processes and mechanics of kinesin motors are closely interwoven through structural changes [19,33,61,63]. During the chemical processes associated with the change of nucleotide state from ATP to adenosine diphosphate (ADP), kinesins go through a series of conformational changes, resulting in unidirectional stepwise movement along a MT filament. The microscopic rate constants associated with each step of the conformational changes, induced by the nucleotide chemistry at the catalytic site, have been measured from ensemble experiments [17,18,20,22,43,48] and have also been extracted from the comparison between single molecule time traces and the theory based on master equations [15,16,42,62]. Distributions of dwell times between the steps or alternatively the velocity and randomness vary with the ATP concentrations and/or external loads [2,58,65], providing a clue to decipher the energy landscape of kinesin motors [15,29,30,36,47,51]. Many theoretical studies have been devoted to interpret the one-dimensional time traces of SM experiments and to extract the maximal possible information.

While detailed studies on the kinesin dynamics are important, integrating the existing experimental data should also be useful to gain a global perspective on the mechanism of kinesin's motility. The first part of this chapter summarizes the existing experimental data in the literature and investigates the energy balance along the kinesin's reaction cycle. Quantifying how the energy is balanced at each thermodynamic state throughout the reaction cycle is not only a basic step to understand the overall kinesin dynamics, but also provides a further insight into an unresolved issue such as kinesin's backsteps [12].

After reviewing the thermodynamics and kinetics data for kinesins, we present our efforts to understand the detailed dynamics at some steps of the whole reaction cycle. The advantage of understanding physical phenomena at the microscopic level is clear as suggested by R. P. Feynman (from the lecture titled "There's Plenty of Room at the Bottom"): "The problems of chemistry and biology can be greatly helped if our ability to see what we are doing, and to do things on an atomic level, is ultimately developed—a development which I think cannot be avoided." Many of the problems in molecular biology can, in principle, be resolved by increasing the resolution of the instruments. Although current technologies do not allow us to watch the details of kinesin dynamics, a sound computation model can provide clear insights beyond the current experimental findings. For some class of the problems, especially for the problems in biopolymers whose biological functions are encoded in their native topology, investigating the dynamics emanating from the native topology has provided a great deal of insight. Some of the conundrums in the kinesin dynamics can be semiquantitatively understood simply by examining the molecular topology with sound theoretical considerations. We review our recent progress in understanding the kinesin dynamics at the subnanometer level using the computational model based on the molecular structure.

## 1.2  MECHANOCHEMICAL CYCLE OF KINESIN MOTORS

The mechanical work of molecular motors is generated in an open system environment where a nonequilibrium steady state is maintained. Nonzero flux for the fuel supply and disposal of the by-products is essential for molecular motors to be operative; otherwise biological systems degrade into an equilibrium state, leading to death. The dynamics of kinesin motors can be viewed as a cyclic reaction in the nonequilibrium steady state. Although the hydrolysis of a single ATP molecule—the net chemical reaction for the cycle—appears to be the driving force for kinesin's 8 nm steps, mechanistic details behind this net reaction turn out to be far more complicated. The knowledge about the interplay between at least three biological constituents—kinesin, nucleotide, and MT—and their structures is essential to understand the working principle of the kinesin motility.

1. The structure of the kinesin monomer is classified into the three subdomains—head, neck-linker, and tail (see Figure 1.1A). The head domain (residue 1–323) contains a nucleotide-binding pocket, a catalytic site, which controls the conformational state of the neck-linker (residue 324–338) made of 15 amino acids. The neck-helix domain (residues from 339 to the C-terminus), extended from the neck-linker, forms an alpha-helical structure; dimeric kinesins are made via coiled-coil interactions between the neck-helices from two monomers (see Figure 1.1A).

2. As mentioned earlier, the chemical state of the nucleotide changes from the ATP and ADP · P$_i$ to ADP state. Thus, the kinesin head domain undergoes at least four different states from the K · φ (φ denotes the empty binding site) to the K · ADP as summarized in Table 1.1.

3. MTs are made of a head-to-tail alignment of α- and β-tubulin dimers, each of which provides a single binding site for kinesins. The 8 nm step size of the kinesin motor matches with the periodicity of the tubulin dimers along the protofilament (see Figure 1.1B).

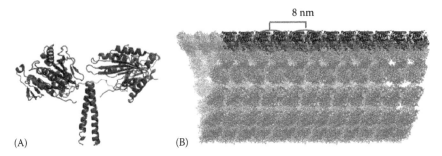

**FIGURE 1.1** (See color insert following page 172.) Kinesin and microtubule. (A) Conventional kinesins are homodimer, each of the monomer is made of head, neck-linker, and neck-helix domain. The neck-linkers of two heads are colored in green and yellow, respectively. The neck-helices from the two monomers associate the two subunits. (B) A microtubule with 13 protofilaments, each of which is made of an 8 nm periodic head-to-tail alignment of the tubulin dimer subunits. A single protofilament that is used as a track for kinesin is colored in red. Kinesins take steps hand-over-hand along the protofilament.

---

**TABLE 1.1**

**Binding Affinity between the Monomeric Kinesin and MT, and the Conformation of Neck-Linker with Varying Nucleotide State**

|  | Binding Affinity to the MT | Neck-Linker State |
|---|---|---|
| $K \cdot \phi$ | Strong | Unzippered |
| $K \cdot ATP$ | Strong | Zippered |
| $K \cdot ADP \cdot P_i$ | Strong | Zippered |
| $K \cdot ADP$ | Weak | Unzippered |

---

The interplay between the kinesin and nucleotide affects both the neck-linker conformation and the binding affinity of head domain to the MT binding site (see Table 1.1). Using cryo-EM (electron microscopy), fluorescence resonance energy transfer (FRET), and electron paramagnetic resonance (EPR) experiments Rice et al. have shown that the chemical state of a nucleotide at the catalytic site determines whether the neck-linker conformation is in a zippered (ordered) state or an unzippered (disordered) state [55]. Upon ATP binding to the empty catalytic site, the neck-linker changes from the unzippered to zippered state. $ADP \cdot P_i$ at the catalytic site maintains the neck-linker in the zippered state. After the release of $P_i$, the neck-linker returns to the unzippered state. This finding proposes the "neck-linker zipper model" as the origin of the force generation responsible for the stepping motion of kinesin motors. It is noteworthy that, instead of the common notion that the ATP hydrolysis is the energy source for biological nanomachines, the power stroke of the motor is produced when ATP binds to the catalytic site. In addition to the neck-linker conformation, the nucleotide-dependent binding affinity of the head domain with respect to the tubulin binding site is another important factor for the proper function of kinesins. The binding between the kinesin and MT is greatly weakened in the ADP state, leading to the dissociation of the kinesin's trailing head from the MT.

Even with a full knowledge about the monomeric kinesins with the varying ligand states, it is still difficult to resolve the mechanism for the coordinated dynamics for dimeric kinesins. However, a number of experimental measurements for the last two decades propose a diagram of the mechanochemical cycle for the coordinated dynamics of dimeric kinesin-1 (see Figures 1.2A or 1.3): (i) Starting from D – ϕ (abbreviated notation for the kinesin dimer state, K · D – K · ϕ), the ATP (=T) binding induces (ii) the conformational changes in the neck-linker of the leading empty head (K · ϕ or ϕ for simplicity), which result in the stepping mechanics (D · T → T · D). (iii) ADP (=D) is removed from the catalytic site of the tethered monomer when the tethered head binds to the next MT binding site. During the reaction steps associated with (iv) and (v), both heads remain strongly bound to the MT, and the ATP hydrolysis occurs in the trailing head. (vi) Once the $P_i$ is released from the trailing head after the hydrolysis, the trailing head containing ADP dissociates from the MT, and the half-cycle is completed.

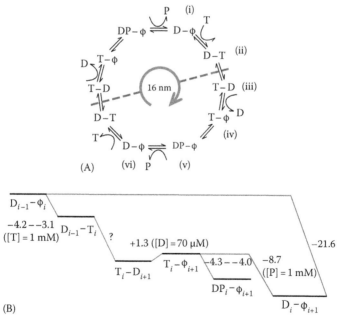

**FIGURE 1.2** (A) Mechanochemical cycle of kinesin motors. Along the reaction cycle two kinesin monomers are depicted in different color. The symbol ϕ represents a kinesin monomer with empty catalytic site, and the symbols T, DP, D represent a kinesin monomer with ATP, ADP · $P_i$, and ADP in its catalytic site, respectively. (B) Free energy diagram along the half-cycle from the state (i) to (vi) in A is calculated using the kinetic data from ensemble experiments summarized in Table 1.2.

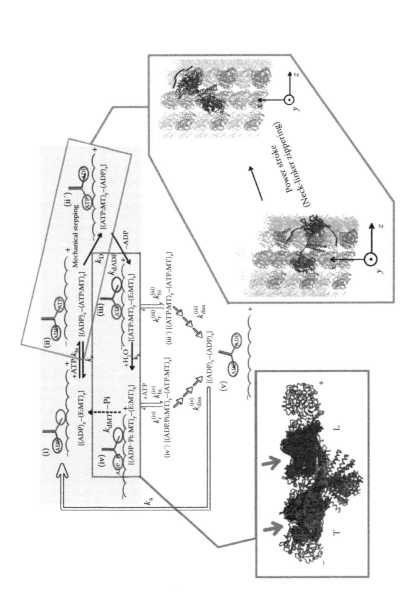

**FIGURE 1.3** (See color insert following page 172.) Mechanochemical cycle of conventional kinesin (kinesin-1) and the summary of our theoretical study. (A) During the kinetic step within the blue box, ATP binding to the leading head is inhibited, which leads to the high level of processivity of the kinesin motor. This is explained by the mechanochemistry due to the asymmetric strain-induced regulation mechanism between the two motor domains on the MT. The thermal ensemble of structures from the simulations shows that the nucleotide binding pocket of the leading head is more disordered than that of the trailing head, both of which are indicated by the green arrows. The perturbed configuration of the catalytic site of the leading head is maintained as long as the trailing head remains bound to the MT. The tension built on the neck-linker of the leading head disrupts the ATP binding pocket from its native-like pocket topology. (B) The kinetic step from (ii) to (ii′) enclosed in the green box denotes the stepping dynamics of the kinesin motor, which is explained by the combined processes of power stroke and diffusional search of the next binding site. Because of the multiplicity of MT binding sites, the pattern of time traces involving stepping dynamics is affected by the rate of power stroke (see Figure 1.5 and text for more details).

Note again that the kinesin dynamics coupled to the state of nucleotide is envisioned as a cyclic reaction, whose material flux is maintained at a nonzero value. As long as the concentrations of ATP, ADP, and $P_i$ are kept far from their equilibrium value and the kinesin motor itself is not fatigued, the kinesin can continuously progress toward the plus end of the MT filament. One can corroborate this notion using the terminology of the equilibrium chemical thermodynamics. The chemical potential ($\Delta\mu$), material flux ($\Delta J = J_+ - J_-$ where $J_+$ is for the forward and $J_-$ is for the backward flux), equilibrium constant ($K$), and reaction rate constant at each step with nucleotide concentrations are related as follows [53,54]:

$$\Delta\mu = -k_B T \log K = -k_B T \log\left(\frac{J_+}{J_-}\right)$$

$$= -k_B T \log\left(\frac{k_T k_S k_{-D} k_h k_{-P}}{k_{-T} k_{BS} k_D k_{-h} k_P}\right) = -k_B T \log\left(\frac{k_T^o [T] k_S k_{-D} k_h k_{-P}}{k_{-T} k_{BS} k_D^o [D] k_{-h} k_P^o [P]}\right)$$

$$= -k_B T \log\left(\frac{[D]_{eq}[P]_{eq}}{[T]_{eq}}\right) + k_B T \log\left(\frac{[D][P]}{[T]}\right)$$

$$= \Delta\mu^o + k_B T \log\left(\frac{[D][P]}{[T]}\right),$$

where the rate constants in the above equations are for each reaction step along the reaction cycle (Figure 1.2). $k_T(= k_T^o[T])$ and $k_{-T}$ are the forward and reverse rate constants for the ATP binding; $k_S$ and $k_{BS}$ are the rate for stepping and backstepping; $k_{-D}$ and $k_D$ are for the ADP release and binding to the catalytic site; $k_h$ and $k_{-h}$ are for the ATP hydrolysis and synthesis; $k_{-P}$ and $k_P(= k_P^o[P])$ are for the dissociation of inorganic phosphate ($P_i$) from the ADP · $P_i$ state. At equilibrium ($K = 1$ or $\Delta\mu = 0$), zero flux ($\Delta J = 0$) is maintained throughout the reaction cycle, namely, the microscopic reversibility for the full cycle is established. For kinesins to move on MTs, a bias of the chemical potential ($\Delta\mu < 0$ or $\Delta J > 0$) supplied by external agents is required. A physiologically maintained nonequilibrium steady-state condition produces such a bias. Since the formation free energy of ATP at the standard condition (25°C, 1 atm) is ~$12k_B T$ ($\Delta\mu^o = -12k_B T$), the chemical potential acting on the kinesin cycle at the physiological condition ([T] = [T]$_{cell}$ = 1 mM, [D] = [D]$_{cell}$ = 70 µM, [P] = [P]$_{cell}$ = 1 mM) is $\Delta\mu \approx -21.6k_B T$. Every cycle of the kinesin dynamics is realized by this free energy bias.

The global potential bias for the cycle is decomposed into the more details by using the kinetic data from the ensemble experiments employing the stopped flow and isotope exchange techniques (see Table 1.2). From this data, we calculate the free energy difference at each step of the kinesin cycle, and build the free energy diagram for the cycle (see Figure 1.2B). The equilibrium constant and the forward and reverse rate constants for each step of reversible chemical reactions are employed to build a "one-dimensional" free energy profile whose coordinate is not necessarily

## TABLE 1.2
## Rate and Equilibrium Constants for the Microsopic Steps along the Kinesin Cycle

| | Rate Constant | Equilibrium Constant |
|---|---|---|
| $k_T$ | $k_T^0 = 2.0 \pm 0.8\,\mu M^{-1}\,s^{-1}$ [48] | $1/K_T = k_{-T}/k_T^0 = 35\,\mu M$ |
| | $k_T = k_T^0[T]_{cell} \approx 2 \times 10^3\,s^{-1}$ [48] | $(K_m \sim 50\,\mu M$ for MT-activated ATPase) |
| $k_{-T}$ | $k_{-T} = 71 \pm 9\,s^{-1}$ [48] | |
| $k_s$ | $k_T = 20\,\mu s$ | $K_s \sim ?$ |
| $k_{BS}$ | $k_{BS} = ?$ | |
| $k_{-D}$ | $k_{-D} = 75\text{--}100\,s^{-1}$ (L), $1\,s^{-1}$ (T) [43] | $K_D = k_D^0/k_{-D} = 5 \times 10^4\,M^{-1}$ (L) |
| $k_D$ | $k_{-D} = 300\,s^{-1}$ [48] (depends strongly on ionic strength) | $K_D = k_D^0/k_{-D} = 5 \times 10^6\,M^{-1}$ (T) [43] |
| | No data | |
| $k_h$ | $k_h > 100 \pm 30\,s^{-1}$ [17] | $K_h^{K \cdot MT} < 39$ |
| $k_{-h}$ | $k_{-h} = 1.3\,s^{-1}$ | |
| $k_{-P}$ | $k_{-P} = 50\,s^{-1}$ | $(k_h/k_{-h} \times k_{-P}/k_P) \approx 6000$ |
| $k_P$ | | |
| $k_{h,-P}$ | $k_{h,-P} = 100\text{--}300\,s^{-1}$ | $K_{h,-P} = 200\,s^{-1}/34\,M^{-1}s^{-1} = 6\,M$ |
| | $k_{P,-h} = 34\,M^{-1}s^{-1}$ [22] | |

parallel to the kinesin's moving direction. The ATP binding free energy to the leading head in vivo ($[T]_{cell} = 1\,mM$) is estimated as $\Delta G = -4.2 \sim -3.1 k_B T$ [$D_{i-1} - \phi_i \leftrightarrow D_{i-1} - T_i$], the dissociation free energy of ADP from the leading head is $+1.3 k_B T$ [$T_i - D_{i+1} \leftrightarrow T_i - \phi_{i+1}$], and the net free energy associated with the two successive reaction steps—the ATP hydrolysis in the kinesin's trailing head and the subsequent $P_i$ release—is $-8.7 k_B T$ [$T_i - \phi_{i+1} \leftrightarrow DP_i - \phi_{i+1} \leftrightarrow D_i - \phi_{i+1}$]. By equating the microscopic free energy changes for kinesin complex with the net chemical free energy (ATP hydrolysis), we find that the free energy difference associated with the stepping ($D_{i-1} - T_i \leftrightarrow T_i - D_{i+1}$) is $\Delta G_{step} = -11.1 \sim -10.0 k_B T$. Interestingly, the value of $\Delta G_{step}$ coincides with the work value one can calculate using the stall force ($f_{stall} \approx 5\text{--}7\,pN$) from the SM experiments and the step size ($\delta x \approx 8\,nm$), i.e., $W_{mech} = f_{stall} \cdot \delta x \approx 9.7 - 13.5 k_B T$. Thus, 50%–60% of the ATP hydrolysis energy is used for mechanically shifting the position of kinesin motors; the rest of the free energy is for making a series of internal conformational changes of the kinesin. The conformational changes, orthogonal to the moving direction along the MT track, do not directly bring the kinesin into motion but play a crucial role in discarding the waste (ADP or $P_i$) of the hydrolysis and in helping the kinesin accept a new ATP molecule required for the next stepping motion. It is noteworthy that the dissociation of ADP from the leading head [$T_i - D_{i+1} \leftrightarrow T_i - \phi_{i+1}$] is involved with a slightly positive free energy change ($+1.3 k_B T$), where the reaction step becomes locally nonspontaneous.

The thermodynamic analysis for each step of mechanochemical cycle, summarized in the free energy diagram (Figure 1.2B), allows us to assess the kinesin dynamics more easily, providing a clear picture of how the net ATP hydrolysis free energy ($\Delta G_{\text{hyd}} = -21.6 k_B T$) is partitioned into each reaction step. As mentioned above, the free energy change due to the nucleotide chemistry in the catalytic site is partitioned into two components: parallel and orthogonal to the kinesin's moving direction. It will be of interest to analyze the thermodynamics of other biological motors, such as myosin and $F_0 F_1$ rotary motor that use ATP as fuel molecules with different "thermodynamic efficiencies," and to study how the ATP hydrolysis free energy is balanced into each part of the reaction cycle.

## 1.3   DECIPHERING THE KINESIN DYNAMICS USING THE STRUCTURE-BASED MINIMAL MODEL

The interactions of a kinesin with the MT and the nucleotide at the catalytic site constantly modify the structure of the kinesin, enabling the kinesin to progress along the MT. This section summarizes our efforts to answer some of the main questions about the structural details of the kinesin dynamics associated with the mechano-chemistry (see Figure 1.3 and the caption for the summary). Using a native topology-based minimal model, we study (a) how the two kinesin head domains connected via neck-linkers are deformed from their native states to regulate the chemistry at each head, (b) which part of the MT surface is being sampled when the kinesin steps, and (c) how the structure of kinesin is internally deformed when the binding dynamics to the MT binding site takes place.

### 1.3.1   Internal Strain Regulation as an Underlying Mechanism for a High Processivity

Kinesins are processive motors that can continuously walk a hundred steps without dissociating from MTs. Compared with other molecular motors, the processivity of kinesins is remarkably high, so that people have conjectured that, without any specific and direct evidence based on the structures, the origin of the high processivity is due to the regulation between the two head domains [19,23,63]. In principle, with a finite probability (~1% per the half-cycle), kinesins can fall off from MTs at any chemical step of the mechanochemical cycle. Thus, the mechanochemical cycle in Figure 1.2A should be modified to include dissociation pathways of the kinesin from the MT. The branching reaction pathways from the cycle are plausible especially for the $T_i - \phi_{i+1}$ and $(DP)_i - \phi_{i+1}$ states, where ATP can bind to the empty leading head ($\phi_{i+1}$). As long as the nucleotide in the trailing head is in the ATP or in the ADP $\cdot$ $P_i$ state, both heads remain bound to the MT with a high binding affinity (see Table 1.1). It is of interest to know how exactly the kinesin structure is perturbed when both heads are bound to the binding sites of the MT surface. A straightforward answer to this type of question requires specific knowledge about the structure at the subnano-meter scale.

We take advantage of a simple computational model [13,27–31] exploiting the $C_\alpha$ carbon coordinates of the kinesin homodimer in its native state. For simplicity, we introduce Go-like interactions between the residues, so that the native structure takes the minimum energy conformation. Our study based on this simple model finds that the thermal ensemble of kinesins, with two heads being simultaneously constrained to the binding sites, adapts the symmetric native conformation into the asymmetric conformations whose neck-linkers becomes zippered in the trailing head and unzippered in the leading head (see structures in blue box in Figure 1.3). The neck-linker of the leading head, topologically strained from its native configuration, perturbs the ATP binding pocket into a nonnative-like environment [29]. With an assumption that nucleotide binding to the binding pocket is the most favorable if the binding pocket is in native-like structures, we expect that the ATP binding is not favored to the topologically strained leading head in the T – φ or DP – φ state. When both heads are bound to the MT, the internal tension built on the neck-linker strains the leading head, particularly the catalytic site. Unless the trailing head dissociates from the MT, alleviating the strain on the leading head, the ATP binding pocket of the leading head remains perturbed away from the native configuration. To recapitulate, (a) the ATP binding to the leading head is inhibited as long as the two heads are strongly bound to the MT. (b) ADP can more easily dissociate from the leading head than from the trailing head because of the neck-linker position [63]. In this way, the high level of processivity unique to the kinesin-1 is achieved through the communication between the two motor domains using the internal tension, which supports the Guydosh and Block's experimental proposal [19] of the rearward strain regulation mechanism between the two motor heads.

Since the extension of the neck-linker made of 15-amino acids can be obtained from the simulation (contour length $L \approx 5.7$ nm, extension $x = 3.1 \pm 0.8$ nm), with a proper range of persistence length for the polypeptide chain ($l_p \approx 0.4$–$1.0$ nm) [40,41,56] one can estimate the internal tension on the neck-linker as $f = 7$–$15$ pN by using the force–extension ($f - x$) relation of the worm-like chain model [44]:

$$f = \frac{k_B T}{l_p} \left( \frac{1}{4(1 - x/L)^2} - \frac{1}{4} + \frac{x}{L} \right)$$

Our conclusion on the tension-induced regulation mechanism for the nucleotide chemistry to the catalytic site is also in agreement with the straightforward experiment by Hackney et al. in which the processivity of the kinesin was measured by varying the length of neck-linker. When a single amino-acid was inserted to increase the length of the neck-linker (or decrease the tension on the neck-linker), the processivity of the kinesin was reduced by twofold.

Internal strain is one of the clever regulation mechanisms adopted by molecular motor systems. Our simple model unambiguously shows the differences in the structures between the leading and trailing heads, elucidating the role of internal tension in the regulation mechanism of the kinesin.

### 1.3.2 STEPPING DYNAMICS OF KINESINS

The stepping motion involving the reaction step $D_{i-1} - T_i \rightarrow T_i - D_{i+1}$ (Figure 1.2) is unarguably the most spectacular motion along the kinesin's mechanochemical cycle. By precisely probing this dynamics, one can understand an issue like how kinesins explore the MT surface and track parallel to the MT axis [6,14]. However, neither the temporal nor the spatial resolution of the current SM techniques is sufficient to measure the details of the stepping dynamics. The resolution problem has even brought up a controversy among experimentalists about the existence of substepping [6,12,50]. Again, the problem arises due to the lack of resolution in the current instruments.

To reveal the microstructures associated with the stepping dynamics, we extend our structure-based model of kinesin dimers. As mentioned earlier in discussing the mechanochemical cycle, the stepping dynamics results from the neck-linker zippering dynamics (power stroke) that is induced by the ATP binding to the catalytic site of an empty leading head. According to Table 1.1, the neck-linker of an empty kinesin head is in the unzippered state and an ATP binding induces the zippering of the neck-linker. Thus, we mimic this dynamics by adapting the neck-linker zipper contacts (the contacts inside the green and yellow circles in Figure 1.4) of the two heads (see Figure 1.4). In the previous section, we retained the neck-linker zipper contacts for both heads to study the effect of the internal tension. To study the stepping dynamics, however, we take into account the fact that the tethered head is in the ADP state and that the MT-bound head changes its state from an empty to ATP state. The neck-linker zipper contacts of the MT-bound head (the contacts inside the green circles in Figure 1.4) are turned on and off depending on the nucleotide state; the neck-linker zipper contacts of the tethered head (the contacts inside the yellow circles in Figure 1.4) is always kept in an off-state since the tethered head contains an ADP during the stepping. The attractive neck-linker zipper contacts should induce the gradual zippering of the neck-linker to the neck-linker binding motif ($\beta 10$) (for the nomenclature of the structure motif of kinesin, see [37]) in the timescale of $\tau_p$.

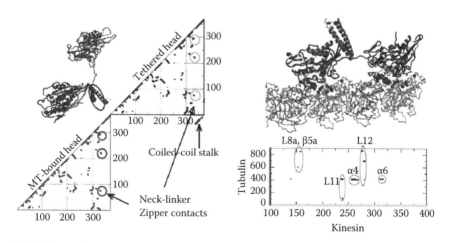

**FIGURE 1.4** **(See color insert following page 172.)** Native contact maps for kinesin dimer (left) and the interface between the kinesin and tubulin binding site (right). The neck-linker zipper contacts that play several important roles for the kinesin function are enclosed in the circles (the green circles for the MT-bound head, and the yellow circles for the tethered head).

While the conformational transition of neck-linker occurs with the timescale of $\tau_p$, the tethered head in ADP state undergoes diffusion on the MT surface. The conformational transition of neck-linker in the MT-bound head affects the diffusional motion of the tethered head that searches for the next tubulin binding site in a time-dependent fashion. The diffusion of the tethered head is rectified by the conformational change of the neck-linker of the MT-bound head.

While the above procedure for simulating the stepping dynamics is straightforward, a brute force simulation of stepping dynamics using the entire construct is still computationally expensive because of the need of incorporating the diffusion tensor to naturally gain the correct translational diffusion coefficient for the kinesin molecule. For this reason, we avoid an alternative method to simulate the stepping dynamics:

1. We construct the two extreme cases of three-dimensional free energy surface (or potentials of mean force [PMF]) felt by the tethered head by exhaustively sampling the kinesin configurations on the MT surface by modeling the neck-linker of the MT-bound head either in the ordered (zippered) or in the disordered (unzippered) state.
2. We mimic the power stroke of the kinesin motor by switching the PMF from the one with the disordered neck-linker to the other with the ordered neck-linker in a finite switching time ($\tau_p$). Thus, the stepping dynamics of the kinesin tethered head is simulated using a diffusional motion of a quasiparticle on the time-varying PMF.

Note again that the kinesin, as a microscopic machine, must adopt the "diffusion" with a rectification of the underlying free energy surface as a basic strategy to navigate in a low Reynolds number environment. In our computational model, the ATP binding transforms the underlying free energy landscape from the one with unzippered neck-linker to the other with the zippered neck-linker. The tethered head of the kinesin searches for the next binding site via the diffusion on the dynamically changing free energy surface. The above line of thought is summarized in the following equation with Figure 1.5A:

$$\vec{r}(t + \Delta t) = \vec{r}(t) - D_K^{\text{eff}} \vec{\nabla} F(x, y, z, t) \Delta t / k_B T + \vec{R}(t)$$

with

$$\langle R_\alpha(t) R_\beta(t + \Delta t) \rangle = 2D \Delta t \delta_{\alpha\beta}$$

and

$$F(x, y, z, t) = \begin{cases} -k_B T \log\left[\left(1 - \dfrac{t}{\tau_P}\right) e^{-F_0(x,y,z)/k_B T} + \dfrac{t}{\tau_P} e^{-F_1(x,y,z)/k_B T}\right] & (0 < t \le \tau_P) \\ F_1(x, y, z) & (t > \tau_P) \end{cases}.$$

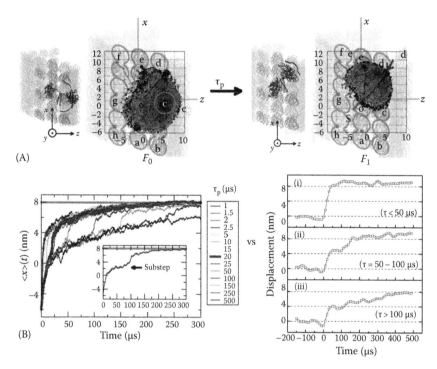

**FIGURE 1.5** **(See color insert following page 172.)** Stepping dynamics of kinesin motors. (A) Free energy surfaces, projected on the MT surface, are obtained by sampling the position of tethered head while the neck-linker zipper contacts in the MT-bound head are switched off ($F_0(x,y,z)$) or switched on ($F_1(x,y,z)$). The variation from $F_0(x,y,z)$ to $F_1(x,y,z)$ can be interpreted as a power stroke (neck-linker zippering) dynamics of the kinesin motor, which leads to the stepping dynamics when combined with the diffusional motion of the tethered head. (B) The average time traces from the BD simulation of a quasiparticle with varying $\tau_p$ (the panel on the left), and the average time traces measured using optical tweezers by Higuchi and coworkers (the three panels on the right) [50]. Higuchi and coworkers divided the individual SM time traces into three groups depending on the stepping time ((i) $t < 50\,\mu s$, (ii) $t = 50 - 100\,\mu s$, (iii) $t > 100\,\mu s$) and averaged over each ensemble. (Adapted from Nishiyama, M. et al., *Nat. Cell Biol.*, 3, 425, 2001.)

The diffusion of a quasiparticle representing the centroid of the tethered head $\vec{r}(t)$ is simulated using Brownian dynamics (BD) simulations on the continuously varying time-dependent free energy surface $F(x, y, z, t)$. We calculate $F(x, y, z, t)$ using the two extreme free energy surfaces $F_0(x, y, z)$ and $F_1(x, y, z)$ [5,28], each of which is obtained by thermodynamically sampling the position of the tethered head with and without the neck-linker zipper contacts on the MT-bound head, respectively. For the BD simulation of a quasiparticle, we choose an effective diffusion constant $D_K^{eff} = 2\,\mu m^2/s$. Although $D_K^{eff} = 2\,\mu m^2/s$ is rather small compared to the diffusion constant for a spherical object of $a \approx 4\,nm$ size in the water of $1\,cP$ ($D = k_B T/6\pi\eta a \approx 55\,\mu m^2/s$ from Stokes–Einstein relation), we justify the value $D_K^{eff} = 2\,\mu m^2/s$ by reasoning that the experimental data we attempt to fit was obtained by probing, instead of the kinesin molecule, the position of a $0.5\,\mu\text{m}$ microbead connected to the kinesin through a long tail domain [50], and that the effective viscosity near the MT, where a number of counterions is condensed, is likely to be larger than 1 cP (see Supporting Information of [30] for the details).

With the selected value of $D_K^{\text{eff}}$, the fastest average time trace in Higuchi and coworkers experiment [50] is reproduced at $\tau_p = 20\,\mu\text{s}$ (see Figure 1.5B). Because of the geometry of the MT surface characterized by a multiple potential binding sites for the kinesin, if the timescale of the power stroke (switching time from $F_0(x, y, z)$ to $F_1(x, y, z)$) is greater than $20\,\mu\text{s}$, an intermediate substep, corresponding to the slow time traces in Higuchi and coworker experiment, emerges in the average time traces (Figure 1.5B on the left). The substep is due to the molecule transiently trapped to the sideway binding site (c in Figure 1.5A) at the adjacent protofilament. For a complex system like kinesins on MTs, a detailed look at a kinetic ensemble reveals multiple timescales for the parallel routes for the dynamics.

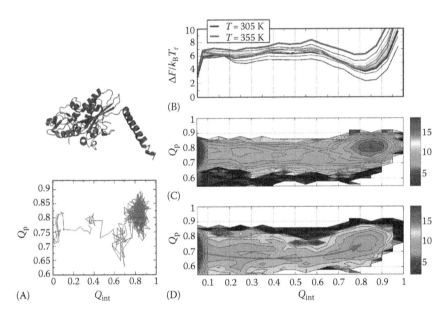

**FIGURE 1.6** (See color insert following page 172.) PMF between the tethered kinesin head and the MT binding site "e" as a function of $Q_p$ and $Q_{\text{int}}$. $Q_p$ is the fraction of native contacts of the MT binding motifs whose structures are colored in orange in A. $Q_{\text{int}}$ is the fraction of native contacts for the binding interface between the kinesin and MT (the contact map for the interface is shown in Figure 1.4). (A) The MT binding motifs of the kinesin head domain are colored in orange (top). An exemplary binding trajectory as a function of $Q_p$ and $Q_{\text{int}}$ (panel on the bottom). Note that $Q_p$ value reaches the minimum value prior to the binding. (B) One-dimensional projection of free energy profile as a function of $Q_{\text{int}}$ at varying temperatures. (C, D) Two-dimensional free energy surface for the binding event as a function of $Q_p$ and $Q_{\text{int}}$. The partial unfolding along the binding process is more clearly manifested at a higher temperature. An average binding route suggested from the free energy is drawn in (D).

### 1.3.3 PARTIAL UNFOLDING OF STRUCTURE FACILITATES THE BINDING PROCESS

Unlike macroscopic machines, biomolecules constantly experience substantial fluctuations in their structures. Compared to the thermal energy, an individual noncovalent bond is only marginally stable. Thus, biomolecules can partially unfold and refold if this dynamics is more effectively exploited to achieve their functional goal. Kinesins also adopt this strategy to effectively march along the MT.

Indeed, the partial disruption of the internal structure facilitates the binding dynamics of kinesin motors. We probe the binding events by monitoring the fractional native contacts of the kinesin's MT binding motifs made of $\alpha 4$, $\alpha 6$, $\alpha 5$, L12, and $\beta 5$ ($Q_p$) and of the interface between the kinesin and MT ($Q_{int}$) (see Figure 1.6). The two-dimensional free energy surface (or an exemplary binding trajectory) calculated as a function of $Q_p$ and $Q_{int}$ (Figure 1.6A) shows that the transition state ensemble prior to the complete binding has the lowest $Q_p$ value throughout the binding dynamics. The transient partial unfolding helps the kinesin detour around the high free energy barrier, which is also called a fly-casting mechanism in protein–protein or protein–DNA association processes [39,59].

## 1.4 FUTURE WORK AND CONCLUDING REMARKS

All the reaction steps in the mechanochemical cycle are, in principle, reversible. This notion becomes very important especially when mechanical forces are externally applied to kinesin motors. The stall force can be estimated, in good agreement with the SM experiments, by setting the equilibrium constant to one [16]. The force larger than the stall force can eventually bias the gradient of the chemistry in the reverse direction. Recent SM experiments by Carter and Cross observed processive backsteps under a resisting load of 14 pN [12]. However, several issues remain to be answered concerning the physics of the backstep; Is the backstep due to the reverse reaction of the cycle? In other words, is ATP synthesized, as in $F_1F_0$-ATP synthase [9], while kinesins take backsteps? The SM version of the equilibrium constant ($K_{SM}$) [12], calculated using the ratio between the number of forward and backward steps, with varying external forces ($f$), $K_{SM} = K_{SM}^o \exp(-f\delta/k_BT) = 802 \times \exp(-0.95 \times f)$, does not provide a straightforward interpretation. The suggested value of $\delta(= 0.95 \times k_BT) \approx 4$ nm does not match with the spacing between the tubulin dimers. The equilibrium constant under zero tension, $K_{SM}^o = 802$, corresponding to the $\Delta\mu = -6.8k_BT$, neither match with the hydrolysis free energy ($\Delta\mu = -21.6k_BT$) nor with our estimate of the free energy change for the mechanical part of stepping dynamics ($\Delta G_{step} = -11.1 \sim -10.0k_BT$). Is the cyclic form of the kinesin reaction still valid under a large resisting force? It is plausible to think about going beyond the single cycle [42] and to propose the backsteps induced by the hydrolysis of ATP [49] and/or the existence of a futile cycle [58]. Combining our structure-based model with the master equations describing the "expanded mechanochemical cycle" shown in Figure 1.7 will provide a clear understanding of the kinesin dynamics under a large external load.

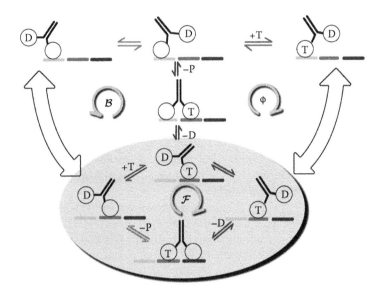

**FIGURE 1.7** An expanded mechanochemical cycle of kinesin motors that includes the normal cycle ($\mathcal{F}$ stands for forward cycle. A reversal of $\mathcal{F}$ cycle would be the backstep induced by an ATP synthesis) and the possibility of (A) backstep induced by ATP hydrolysis ($\mathcal{B}$) and (B) futile cycle ($\phi$). This type of modification of reaction path is plausible to explain the mechanism of kinesin's backstepping.

## ACKNOWLEDGMENTS

We are grateful to Dr. Stefan Klumpp and Professor Olga Dudko for many useful comments and discussions. This work was supported by the Center for Theoretical Biological Physics sponsored by the NSF (Grant PHY-0822283) with additional support from NSF-MCB-0543906. C.H. is supported by the Korea Science and Engineering Foundation (KOSEF) grant (No. R01-2008-000-10920-0) and the Korea Research Foundation Grant funded by the Korean Government (KRF-C00142 and KRF-C00180).

## REFERENCES

1. Alonso MC, Drummond DR, Kain S, Hoeng J, Amos L, Cross RA. 2007. An ATP gate controls tubulin binding by the tethered head of kinesin-1. *Science* 316: 120.
2. Asbury CL, Fehr AN, Block SM. 2003. Kinesin moves by an asymmetric hand-over-hand mechanism. *Science* 302: 2130.
3. Ashkin A. 1997. Optical trapping and manipulation of neutral particles using lasers. *Proceedings of the National Academy of Sciences of the USA* 94: 4853.
4. Astumian RD, Bier M. 1994. Fluctuation driven ratchets—Molecular motors. *Physical Review Letters* 72: 1766.
5. Best RB, Chen YG, Hummer G. 2005. Slow protein conformational dynamics from multiple experimental structures: The helix/sheet transition of arc repressor. *Structure* 13: 1755.

6. Block SM. 2007. Kinesin motor mechanics: Binding, stepping, tracking, gating, and limping. *Biophysical Journal* 92: 2986.

7. Block SM, Asbury CL, Shaevitz JW, Lang MJ. 2003. Probing the kinesin reaction cycle with a 2D optical force clamp. *Proceedings of the National Academy of Sciences of USA* 100: 2351.

8. Block SM, Goldstein LSB, Schnapp BJ. 1990. Bead movement by single kinesin molecules studied with optical tweezers. *Nature* 348: 348.

9. Boyer PD. 1997. The ATP synthase—A splendid molecular machine. *Annual Review of Biochemistry* 66: 717.

10. Brady ST. 1985. A novel brain ATPase with properties expected for the fast axonal-transport motor. *Nature* 317: 73.

11. Bustamante C, Keller D, Oster G. 2001. The physics of molecular motors. *Accounts of Chemical Research* 34: 412.

12. Carter NJ, Cross RA. 2005. Mechanics of the kinesin step. *Nature* 435: 308.

13. Clementi C, Nymeyer H, Onuchic JN. 2000. Topological and energetic factors: What determines the structural details of the transition state ensemble and "en-route" intermediates for protein folding? An investigation for small globular proteins. *Journal of Molecular Biology* 298: 937.

14. Fehr AN, Asbury CL, Block SM. 2008. Kinesin steps do not alternate in size. *Biophysical Journal* 94: L20.

15. Fisher ME, Kim YC. 2005. Kinesin crouches to sprint but resists pushing. *Proceedings of the National Academy of Sciences of USA* 102: 16209.

16. Fisher ME, Kolomeisky AB. 2001. Simple mechanochemistry describes the dynamics of kinesin molecules. *Proceedings of the National Academy of Sciences of USA* 98: 7748.

17. Gilbert SP, Johnson KA. 1994. Pre-steady-state kinetics of the microtubule-center-dot-kinesin ATPase. *Biochemistry* 33: 1951.

18. Gilbert SP, Moyer ML, Johnson KA. 1998. Alternating site mechanism of the kinesin ATPase. *Biochemistry* 37: 792.

19. Guydosh NR, Block SM. 2006. Backsteps induced by nucleotide analogs suggest the front head of kinesin is gated by strain. *Proceedings of the National Academy of Sciences of USA* 103: 8054.

20. Hackney DD. 1988. Kinesin ATPase—Rate-limiting ADP release. *Proceedings of the National Academy of Sciences of USA* 85: 6314.

21. Hackney DD. 1994. Evidence for alternating head catalysis by kinesin during microtubule-stimulated ATP hydrolysis. *Proceedings of the National Academy of Sciences of USA* 91: 6865.

22. Hackney DD. 2005. The tethered motor domain of a kinesin–microtubule complex catalyzes reversible synthesis of bound ATP. *Proceedings of the National Academy of Sciences of USA* 102: 18338.

23. Hancock WO, Howard J. 1999. Kinesin's processivity results from mechanical and chemical coordination between the ATP hydrolysis cycles of the two motor domains. *Proceedings of the National Academy of Sciences of USA* 96: 13147.

24. Hirose K, Lockhart A, Cross RA, Amos LA. 1995. Nucleotide-dependent angular change in kinesin motor domain bound to tubulin. *Nature* 376: 277.

25. Hirose K, Lockhart A, Cross RA, Amos LA. 1996. Three-dimensional cryoelectron microscopy of dimeric kinesin and ncd motor domains on microtubules. *Proceedings of the National Academy of Sciences of USA* 93: 9539.

26. Hunt AJ, Gittes F, Howard J. 1994. The force exerted by a single kinesin molecule against a viscous load. *Biophysical Journal* 67: 766.

27. Hyeon C, Dima RI, Thirumalai D. 2006. Pathways and kinetic barriers in mechanical unfolding and refolding of RNA and proteins. *Structure* 14: 1633.

28. Hyeon C, Lorimer GH, Thirumalai D. 2006. Dynamics of allosteric transitions in GroEL. *Proceedings of the National Academy of Sciences of USA* 103: 18939.
29. Hyeon C, Onuchic JN. 2007. Internal strain regulates the nucleotide binding site of the kinesin leading head. *Proceedings of the National Academy of Sciences of USA* 104: 2175.
30. Hyeon C, Onuchic JN. 2007. Mechanical control of the directional stepping dynamics of the kinesin motor. *Proceedings of the National Academy of Sciences of USA* 104: 17382.
31. Hyeon C, Thirumalai D. 2007. Mechanical unfolding of RNA: From hairpins to structures with internal multiloops. *Biophysics Journal* 92: 731.
32. Julicher F, Ajdari A, Prost J. 1997. Modeling molecular motors. *Reviews of Modern Physics* 69: 1269.
33. Kawaguchi K, Ishiwata S. 2001. Nucleotide-dependent single- to double-headed binding of kinesin. *Science* 291: 667.
34. Kikkawa M, Okada Y, Hirokawa N. 2000. 15 angstrom resolution model of the monomeric kinesin motor, KIF1A. *Cell* 100: 241.
35. Kikkawa M, Sablin EP, Okada Y, Yajima H, Fletterick RJ, Hirokawa N. 2001. Switch-based mechanism of kinesin motors. *Nature* 411: 439.
36. Kolomeisky AB, Fisher ME. 2007. Molecular motors: A theorist's perspective. *Annual Review of Physical Chemistry* 58: 675.
37. Kozielski F, Sack S, Marx A, Thormahlen M, Schonbrunn E, et al. 1997. The crystal structure of dimeric kinesin and implications for microtubule-dependent motility. *Cell* 91: 985.
38. Leibler S, Huse DA. 1993. Porters versus Rowers—A unified stochastic-model of motor proteins. *Journal of Cell Biology* 121: 1357.
39. Levy Y, Onuchic JN, Wolynes PG. 2007. Fly-casting in protein-DNA binding: Frustration between protein folding and electrostatics facilitates target recognition. *Journal of the American Chemical Society* 129: 738.
40. Li HB, Linke WA, Oberhauser AF, Carrion-Vazquez M, Kerkviliet JG, et al. 2002. Reverse engineering of the giant muscle protein titin. *Nature* 418: 998.
41. Li HB, Oberhauser AF, Redick SD, Carrion-Vazquez M, Erickson HP, Fernandez JM. 2001. Multiple conformations of PEVK proteins detected by single-molecule techniques. *Proceedings of the National Academy of Sciences of USA* 98: 10682.
42. Liepelt S, Lipowsky R. 2007. Kinesin's network of chemomechanical motor cycles. *Physical Review Letters* 98: 258102.
43. Ma YZ, Taylor EW. 1997. Interacting head mechanism of microtubule–kinesin ATPase. *Journal of Biological Chemistry* 272: 724.
44. Marko JF, Siggia ED. 1995. Stretching DNA. *Macromolecules* 28: 8759.
45. Mehta AD, Rief M, Spudich JA, Smith DA, Simmons RM. 1999. Single-molecule biomechanics with optical methods. *Science* 283: 1689.
46. Meyhofer E, Howard J. 1995. The force generated by a single kinesin molecule against an elastic load. *Proceedings of the National Academy of Sciences of USA* 92: 574.
47. Miyashita O, Onuchic JN, Wolynes PG. 2003. Nonlinear elasticity, proteinquakes, and the energy landscapes of functional transitions in proteins. *Proceedings of the National Academy of Sciences of USA* 100: 12570.
48. Moyer ML, Gilbert SP, Johnson KA. 1998. Pathway of ATP hydrolysis by monomeric and dimeric kinesin. *Biochemistry* 37: 800.
49. Nishiyama M, Higuchi H, Ishii Y, Taniguchi Y, Yanagida T. 2003. Single molecule processes on the stepwise movement of ATP-driven molecular motors. *Biosystems* 71: 145.
50. Nishiyama M, Muto E, Inoue Y, Yanagida T, Higuchi H. 2001. Substeps within the 8-nm step of the ATPase cycle of single kinesin molecules. *Nature Cell Biology* 3: 425.

51. Okazaki K, Koga N, Takada S, Onuchic JN, Wolynes PG. 2006. Multiple-basin energy landscapes for large-amplitude conformational motions of proteins: Structure-based molecular dynamics simulations. *Proceedings of the National Academy of Sciences of USA* 103: 11844.

52. Qian H. 2000. A simple theory of motor protein kinetics and energetics. II. *Biophysical Chemistry* 83: 35.

53. Qian H. 2006. Open-system nonequilibrium steady state: Statistical thermodynamics, fluctuations, and chemical oscillations. *Journal of Physical Chemistry B* 110: 15063.

54. Qian H. 2007. Phosphorylation energy hypothesis: Open chemical systems and their biological functions. *Annual Review of Physical Chemistry* 58: 113.

55. Rice S, Lin AW, Safer D, Hart CL, Naber N, et al. 1999. A structural change in the kinesin motor protein that drives motility. *Nature* 402: 778.

56. Rief M, Gautel M, Oesterhelt F, Fernandez JM, Gaub HE. 1997. Reversible unfolding of individual titin immunoglobulin domains by AFM. *Science* 276: 1109.

57. Rosenfeld SS, Fordyce PM, Jefferson GM, King PH, Block SM. 2003. Stepping and stretching—How kinesin uses internal strain to walk processively. *Journal of Biological Chemistry* 278: 18550.

58. Schnitzer MJ, Visscher K, Block SM. 2000. Force production by single kinesin motors. *Nature Cell Biology* 2: 718.

59. Shoemaker BA, Portman JJ, Wolynes PG. 2000. Speeding molecular recognition by using the folding funnel: The fly-casting mechanism. *Proceedings of the National Academy of Sciences of USA* 97: 8868.

60. Sindelar CV, Budny MJ, Rice S, Naber N, Fletterick R, Cooke R. 2002. Two conformations in the human kinesin power stroke defined by X-ray crystallography and EPR spectroscopy. *Nature Structural Biology* 9: 844.

61. Terada TP, Sasai M, Yomo T. 2002. Conformational change of the actomyosin complex drives the multiple stepping movement. *Proceedings of the National Academy of Sciences of USA* 99: 9202.

62. Tsygankov D, Fisher ME. 2007. Mechanoenzymes under superstall and large assisting loads reveal structural features. *Proceedings of the National Academy of Sciences of USA* 104: 19321.

63. Uemura S, Ishiwata S. 2003. Loading direction regulates the affinity of ADP for kinesin. *Nature Structural Biology* 10: 308.

64. Vale RD, Reese TS, Sheetz MP. 1985. Identification of a novel force-generating protein, kinesin, involved in microtubule-based motility. *Cell* 42: 39.

65. Visscher K, Schnitzer MJ, Block SM. 1999. Single kinesin molecules studied with a molecular force clamp. *Nature* 400: 184.

# 2 Mechanochemical Coupling in Molecular Motors: Insights from Molecular Simulations of the Myosin Motor Domain

*Haibo Yu, Yang Yang, Liang Ma, and Qiang Cui*

## CONTENTS

## 2.1    INTRODUCTION

In the context of energy flow in biomolecules, molecular motors[1,2] are arguably the most fitting topic for in-depth investigations. The hallmark feature of molecular motors is the high efficiency of energy transduction. Defined by the amount of work done (e.g., pulling the cargo forward) divided by the amount of energy input (e.g., ATP hydrolysis) per functional cycle, the efficiency of molecular motors typically ranges from 0.25 to almost 1.0,[1] which is incredibly high compared to all artificial machines. The efficiency is even more remarkable considering that the sites where the chemistry and mechanical work are carried out are typically far apart (few nanometers, see below), making allostery another feature of molecular motors.[3] Not surprisingly, revealing the operating mechanism of these natural "nanomachines" has been one of the most enticing challenges in biophysics for many decades.[4] As a matter of fact, a significant body of work[5] has been done already in the 1970s on the thermodynamic and kinetic principles that govern the efficiency of free energy transduction in molecular motors and molecular pumps. Being phenomenological in nature, however, these studies do not incorporate molecular-scale descriptions and were limited by the resolution of experimental data available at the time.

In recent years, breakthroughs in structural biology and other biophysical techniques, particularly single molecule spectroscopy, have provided a wealth of information at unprecedented spatial and temporal resolutions on the function of many molecular motors.[3,6-8] The availability of these experimental data has provided the opportunity to develop models that describe the energy transduction process in molecular motors with molecular level of detail; this is an important challenge to tackle because models with molecular details are required for making sufficiently specific predictions that can be tested experimentally. In this regard, in addition to very specific mechanistic details, there are also principle-level questions that need to be addressed. For example, a few general questions that have been driving our own research are: to what degree the thermodynamic and kinetic constraints proposed in the pioneering phenomenological models[5,9] are satisfied in molecular motors? If these constraints are indeed important, how are they "implemented" in structural and energetic terms in specific motor systems? Compared to other biomolecules that use the similar chemical substrate (e.g., ATP/GTP) but do not carry out energy transduction (e.g., signaling proteins or phosphoryl transferase), do molecular motors have unique structural and energetic features?

The specific system that we use to explore these questions is the conventional myosin, also termed myosin II, which plays key physiological functions in muscle contraction and cell division.[10] It is an ideal system for in-depth theoretical and computational analysis because its structural and kinetic properties have been characterized by a large body of diverse experimental techniques.[11,12] For example, at the time our research was initiated, myosin II was one of the few motor systems for which high-resolution x-ray structures are available for multiple functional states[13,14]; since then, multiple high-resolution x-ray structures have also been obtained for myosin V and VI, two other widely studied members of the myosin superfamily that are more processive in nature compared to myosin II.[15] The functional cycle of myosin II is best described by the celebrated Lymn–Taylor scheme (Figure 2.1a),[11,16] in which

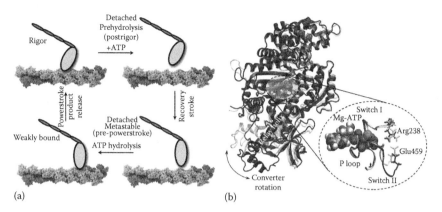

**FIGURE 2.1** (See color insert following page 172.) Function and structure of conventional myosin. (a) The Lymn–Taylor kinetic scheme for the functional cycle[11,16,57]; only a myosin monomer is shown for simplicity. The process of interest here is the recovery stroke. (b) Structural differences between two x-ray conformations (postrigor,[14] in blue, and pre-powerstroke,[13] in green) of the *Dictyostelium discoideum* myosin motor domain. The most visible transition is the converter rotation, although there are also notable changes in the nucleotide binding site and the relay helix that connects the converter and the nucleotide binding site. With ADP·VO$_4^-$ bound, the nucleotide binding site of the pre-powerstroke state has a closed configuration (in yellow); with ATP bound, the nucleotide binding site in the postrigor state is open (in blue) due to displacement of the Switch II loop.

mechanical work is done in the "powerstroke" step, during which the myosin motor domain transits from a weak actin-binding mode to a strong actin-binding mode and thereby pushes the actin polymer in a specific direction. An equally important step is the "recovery stroke," in which the motor domain swings relative to the lever arm and becomes primed for the subsequent powerstroke. Although the recovery stroke does not produce any work because it occurs while the motor domain is detached from actin following ATP binding (Figure 2.1a), it is particularly interesting for studying mechanochemical coupling since ATP hydrolysis is known to be tightly coupled to the recovery stroke.[11] Indeed, if ATP hydrolysis is weakly coupled to the recovery stroke, futile hydrolysis may occur, which compromises the efficiency of the motor.

The complexity and fascinating aspects of the recovery stroke are better illustrated in Figure 2.1b, in which two x-ray structures of the myosin II motor domain are superimposed. One structure was obtained with ATP[14] (the structure is very similar when the active site is occupied by an ATP analogue such as AMP·PNP[17], with a root-mean-square deviation in the Cα of 0.16 Å, and the only major difference in the active site involves a rotation of Asn233 sidechain) as the ligand and is generally interpreted as the postrigor conformation. The other structure was solved with ADP·VO$_4^-$ as the ligand[13]; since ADP·VO$_4^-$ is considered a transition state analog for ATP hydrolysis, the corresponding x-ray structure is believed to reflect the conformation of the pre-powerstroke kinetic state. The two structures are highly similar for most regions except for the C-terminal converter subdomain, which undergoes a major rotation of ~60° that presumably correlates with the lever arm rotation; in the following, we will use "converter rotation" and "lever arm rotation" interchangeably. The nucleotide-binding site also shows subtle but distinct transitions: it is closed tightly in the pre-powerstroke state but opened up in the postrigor state due to the displacement of the Switch II (SwII) motif. Finally, the relay helix and SH1 helix that connect the nucleotide-binding site and the converter also undergo notable structural changes; for example, the relay helix is straight in the postrigor conformation but develops a kink in the pre-powerstroke conformation.

As discussed above, the most interesting feature of the recovery stroke is that ATP hydrolysis and converter rotation are tightly coupled. This is not only important from the functional perspective but also supported by fluorescence experiments[18]; a single dominant orientation was observed for the lever arm with nonhydrolyzable ATP analog, while multiple orientations were observed in the presence of ATP hydrolysis. The tight coupling between the two processes of very different physical nature is most remarkable given the observation that the nucleotide-binding site is more than $40\text{\AA}$ away from the converter, a feature in fact quite common among molecular motors.[3] The simplest "explanation" for this allosteric mechanochemical coupling, which has been rather well accepted in the literature, is based on two premises: (1) numerous x-ray structures[12] with different nucleotides indicate a clear correlation between SwII displacement and converter rotation, thus the two sites must be structurally coupled, presumably through the intervening motifs such as the relay helix; (2) with the open configuration of the nucleotide binding site (SwII), ATP cannot be hydrolyzed due to the lack of proper water structure and perhaps other hydrogen-bonding interactions.[13,19] In other words, ATP hydrolysis is tightly coupled to converter rotation because hydrolysis cannot occur until SwII is closed, which in turn is coupled structurally to the distant converter rotation; the role of the slightly exothermic hydrolysis is to pull the equilibrium toward the pre-powerstroke conformation.[20]

Although this description is logical, there are several questions that remain unanswered, and here we focus on two of the most pressing ones. First, as a typical allostery issue,[21] the molecular mechanism for the observed structural coupling between SwII and converter needs to be defined; in particular, the residues and interactions that mediate the coupling need to be identified and their roles explained. Second, the causality between the hydrolysis process and the various structural transitions needs to be established. Although it is known[18] that hydrolysis is required for the lever arm swing (see above), the precise mechanism for how hydrolysis "drives/ stabilizes" the lever arm swing has not been defined at a molecular level; even if we accept that hydrolysis occurs following the entire recovery stroke,[20] question remains how the hydrolysis product preferentially stabilizes the pre-powerstroke orientation of the converter. In this regard, it is worth stressing that in many molecular motors, large-scale conformational transitions are believed to be coupled to the binding rather than the hydrolysis of ATP, which makes intuitive sense considering that the stereochemical difference between ATP and $ADP \cdot P_i$ is much more subtle than that brought about by the binding of the highly charged $Mg^{2+} \cdot ATP$. For myosin II, the binding energy of ATP is spent to detach the motor domain from actin (Figure 2.1a), which is presumably the reason for the role of hydrolysis in the recovery stroke becoming more apparent.

The presence of multiple lengths- and time-scales as well as the involvement of both major structural transitions and chemistry present great challenges to both experimental and theoretical studies. In the following, we summarize results from our recent computational studies[22–26] to provide insights with regard to these two questions. One important goal here is to illustrate how different simulation techniques can be combined to address challenging mechanistic issues in such complex biomolecular systems as molecular motors. At the end, we discuss findings from our simulations in the context of the more general questions outlined for molecular motors in the beginning of Section 2.1. We believe that the mechanistic insights, simulation protocols, and the framework of analysis established from these recent studies can be useful for the investigation of molecular motors beyond the specific myosin system. Finally, a few remarks regarding future directions are given.

## 2.2 ALLOSTERIC COUPLING AT THE LOCAL AND GLOBAL SCALES[22,23,26]

Since the actual timescale for the recovery stroke is ~10 ms,[11] it is unrealistic to directly simulate the cascade of conformational transitions that couple the SwII displacement and converter rotation. Therefore, we chose to combine a set of computational methods to explore common features that emerge from these analyses; since the methods are based on different principles and have different underlying assumptions, insights from these analyses are complementary and sometimes reinforce each other.

### 2.2.1 NORMAL MODE ANALYSIS AND STATISTICAL COUPLING ANALYSIS: INTRINSIC STRUCTURAL FLEXIBILITY AND "HOTSPOT" RESIDUES

Since large-scale conformational transitions are essential to function, it is expected that molecular motors have intrinsic structural flexibility that facilitates these transitions. This is best illustrated through normal mode analysis (NMA),[27] which was carried out using our implementation[28] of the block normal mode approach.[29] The results shown in Figure 2.2 are qualitatively similar to the observations made in many large biomolecular assemblies,[30–33] i.e., there is a large degree of correlation between the motions in the first few low-frequency modes and the structural difference between two functional states. This is shown more quantitatively with two commonly used descriptors, the involvement coefficients ($I_k$) and the cumulative involvement coefficients ($CI_n$),

$$I_k = \frac{\mathbf{X}_1 - \mathbf{X}_2}{|\mathbf{X}_1 - \mathbf{X}_2|} \cdot \mathbf{L}_k \qquad (2.1)$$

$$CI_n = \sum_{k=1}^{n} I_k^2 \qquad (2.2)$$

where

$\mathbf{X}_1 - \mathbf{X}_2$ is the displacement vector between two conformations $(\mathbf{X}_1, \mathbf{X}_2)$

$\mathbf{L}_k$ is the $k$th eigenvector.

As shown in Figure 2.2a, using less than 20 lowest frequency modes, more than 50% of the displacement can be accounted for, indicating that the motor domain has

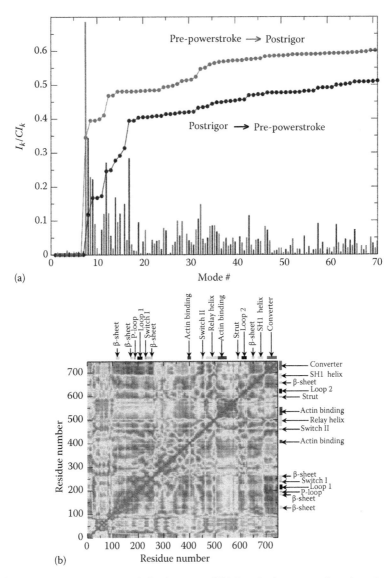

(a)

(b)

**FIGURE 2.2** **(See color insert following page 172.)** Results from normal mode and statistical coupling analyses.[22,26] (a) Involvement and cumulative involvement coefficients (Equations 2.1 and 2.2) for the recovery stroke using normal modes from the postrigor (black) and prepowerstroke (red) structures. (b) Covariance matrix the Cα atoms calculated using normal modes for the postrigor structure. Red (blue) indicates positive (negative) correlation.

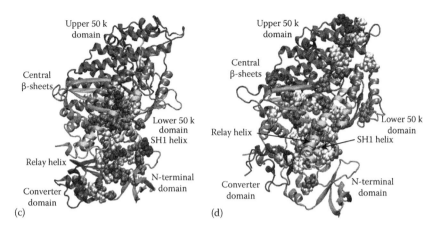

**FIGURE 2.2 (continued)** **(See color insert following page 172.)** (c) Hinge and bending residues in the postrigor structure identified by DynDom[34] based on using normal modes with involvement coefficients larger than 0.10. (d) Mapping of the 52 strongly coupled core residues (in van der Waals form) identified by SCA to the pre-powerstroke structure. In (c) and (d), residues are colored based on residue type; blue: basic, orange: acidic, ice-blue: polar, white: nonpolar.

inherent flexibility in the specific direction of the recovery stroke. Comparatively, with the same number of modes, the $CI_n$ for the pre-powerstroke state is notably higher than that for the postrigor state, which can be interpreted to suggest that the former is more flexible in the direction of the functional transition. This in fact is consistent with the FRET study of Spudich and coworkers,[18] who found that the orientation of the converter (lever arm) is relatively rigid in the postrigor state but spans a broader range of angles in the pre-powerstroke state.

It is also useful to examine the covariance matrix between different structural motifs; NMA is arguably more useful than atomistic molecular dynamics (MD) for this purpose because NMA is not limited by timescale as atomistic MD is. A number of features in the correlation map (Figure 2.2b) is immediately evident. Residues within each structural element are found to exhibit cooperative motions, which is consistent with the fact that the recovery stroke is dominated by motions of relatively rigid domains (Figure 2.1b). The motions of active site motifs (P-loop, SwI, and SwII) are highly correlated with each other and with the relay helix, the relay loop (466–518) and the SH1 helix (669–690). A negative correlation is found between the active site residues and the strut motif (590–593). The converter domain has significant anticorrelation with the central β-sheets and loop 1. Finally, by examining the covariance in detail, it is found that the correlations between Phe482 and Gly680, Ile499 and Phe692 are among the highest around the kink region in the relay helix, which highlights the importance of interactions between these residue pairs in maintaining the mechanochemical coupling (also see below).

Another valuable piece of information from NMA pertains hinge and bending residues, which likely play a major role in structural transitions. They can be identified via, for example, the DynDom algorithm,[34] given the eigenvectors for modes with high $I_k$ values. Compared to the similar analysis using only the two x-ray structures, hinge analysis based on important low-frequency modes revealed a richer amount of information regarding "hotspot" residues potentially important for conformational transition. For example, based on the postrigor and pre-powerstroke x-ray structures only, a handful of hinge residues in the relay helix and near the converter domain were located. Using the normal modes with $I_k$ larger than 0.10, the distribution of the bending/hinge residues was found to depend on the initial structure and the nature of the conformational transition. In the postrigor→pre-powerstroke transition (Figure 2.2c), the hinge residues in the six dominant modes occur mainly in the N-terminal domain, the P-loop, Switch II, as well as the relay helix. In the reverse transition that starts with the pre-powerstroke state (data not shown), the bending/hinge residues were mainly found in the relay helix, SH1 helix, and the converter domain; two hinges (Val124, Ala125) were also found in the central β-sheet region. Among all the hinges identified, a small but significant fraction (39) are highly conserved (>80% across all species), which supports their functional importance.

Although the hinge analysis provides targets for experimental mutation studies, one limitation is that the number of hinge residues is generally large. Therefore, it is interesting to supplement the hinge analysis with other types of information based on, for example, approximate coupling energies and sequence conservation[35]. In our study, we used the statistical coupling analysis (SCA) approach of Lockless and Ranganathan,[36] who proposed to define coupling network in an allosteric system via the analysis of coevolved residues in homologous sequences. The SCA approach is particularly interesting because residues revealed in the SCA are, by definition, at most moderately conserved (~40%–80% in our analysis) and therefore may not have caught the attention of mutagenesis studies. Application of SCA to 709 myosin sequences identified 52 significantly coupled residues in the myosin motor domain, which distribute globally in the motor domain and occupy sites in the intersubdomain linkers such as Switch II, relay helix, SH1 helix, strut, and the central β-sheet. Mapping the 52 coupled residues onto the pre-powerstroke structure (Figure 2.2d) shows more clearly that a cluster of coupled residues appear in the intersection of four subdomains and extends from the upper 50kDa domain, nucleotide-binding pocket, lower 50kDa domain to the converter domain. Most of them are located in the key regions including intersubdomain linkers (yellow), the central β-sheet (cyan) and the actin-binding interface. Overall, despite the fact that the hinge analysis and SCA are based on very different principles with the hinge analysis being more sensitive to which structural transition is being analyzed, many common residues emerge. Indeed, 21 out of the 52 SCA identified residues are hinge or bending residues, while most of the rest were found to engage in important interactions in targeted MD (TMD) simulations (see below). Since the SCA algorithm works with sequence information only, the identified residues are not guaranteed to be involved in allostery and might instead play a role in, for example, cooperative folding. Therefore, combining SCA and physically motivated computational analysis (NMA and TMD) helps narrow down the list of important residues and assign potential functional roles to these residues.

## 2.2.2 Targeted Molecular Dynamics: Approximate Transition Paths

To further explore residues/interactions that dictate the coupling between converter rotation and SwII displacement, the approximate transition path for the recovery stroke was studied using TMD simulations[37] with an implicit solvent model,[38] as an alternative to minimum energy path analysis.[39] The main goal was to observe the formation of *transient* interactions that are not present in either end-states and therefore difficult to identify using the static x-ray structures. Analysis of the results[22] indicates that different types of interactions (polar vs. hydrophobic) along the relay helix play an important role during the recovery stroke. Around halfway in the relay helix, the hydrophobic cluster provides stabilization to the kink of the relay helix, while at the joint between the relay helix and the relay loop region, strong polar interactions facilitate cooperative changes in the relay helix, the SH1 helix, and the converter domain.

Overall, the TMD results deviate substantially from the minimum energy path (MEP) results of Fischer and coworkers.[39] The MEP study produced a highly ordered transition path that initiates from the active site and propagates sequentially through the relay helix to the converter domain. The TMD results, however, indicate that most rotation of the converter domain occurs in the first stage of the transition (Figure 2.3a) while structural changes in the relay helix and complete closure of the active site occur at a later stage to stabilize the converter conformation via a series of hydrogen bonding interactions as well as hydrophobic contacts. For example, the "unwinding" of the relay helix happens much later in the TMD simulations compared to the MEP description.[39] In contrast to the MEP results, which suggest that the kink and unwinding in the relay helix are induced by SwII closure via a single hydrogen-bonding interaction involving Asn475, the TMD simulations tend to suggest that the converter rotation, via strong polar interactions to the relay helix and the relay loop, induces the formation of the hydrophobic cluster halfway in the relay helix as well as some polar interactions (e.g., Asn483–Glu683) to produce the kink in the relay helix. The interactions between SwII and the relay helix stabilize, in return, the new conformation of the relay helix.

Which description is closer to reality is debatable because both methods have limitations. The MEP approach searches for saddle points and the steepest descent paths on the potential energy surface[40]; the method does not require the specification of a predetermined "reaction coordinate" although the result likely depends on the initial assumptions in the calculation due to the existence of large number of degenerate paths. More importantly, the MEPs do not include the effect of thermal fluctuation and therefore calculations tend to produce highly ordered sequence of events. The TMD approach,[37] on the other hand, relies on a predetermined reaction coordinate (e.g., the root-mean-square-deviation relative to a target structure) and enforces a monotonic decrease in the value of the reaction coordinate during the finite-temperature simulation through a holonomic constraint, thereby tends to favor large-scale structural changes in the initial stage (which, in fact, is a feature found in many recent discussions of allostery[21,41,42]). The encouraging aspect, on the other hand, is that both TMD and MEP studies point to the importance of a consistent set of hydrophobic interactions (e.g., between Phe482, Phe487, Phe503, Phe506, and Phe652) and hydrogen bonding interactions (e.g., Glu497–Lys743) between the relay helix, active site, and the converter domain (Figure 2.3b).

In addition to identifying transient interactions, TMD simulations can also be used to examine the behavior of "reporter residues" during the transition and help examine basic assumptions made in experimental studies that monitored spectroscopic

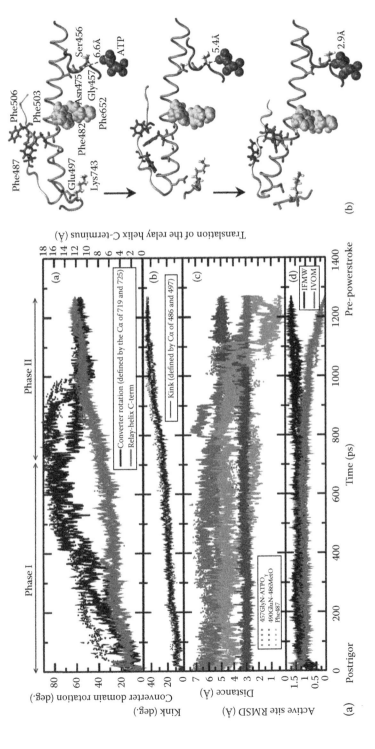

**FIGURE 2.3 (See color insert following page 172.)** Results from targeted molecular dynamics[22] and potential of mean force simulations for the recovery stroke. (a) Variation of critical structural parameters along the three TMD trajectories for illustrating the sequence of events along the approximation transition paths. (b) Three snapshots (at 0.0 s, 630.0 and 1270.0 ps) from one of the three TMD simulations, in the same format as Figure 2 in Ref.[39] for comparison, to illustrate the proposed coupling between the small motion of SwII with the large translation of the relay helix C-terminus.

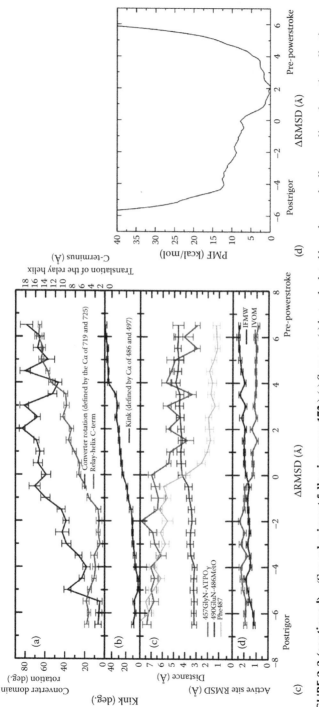

**FIGURE 2.3 (continued)** **(See color insert following page 172.)** (c) Same as (a) but calculated based on umbrella sampling along the collective coordinate, ΔRMSD between the postrigor and pre-powerstroke states; (d) Calculated PMF for the recovery stroke along ΔRMSD.

signature of these residues. In myosin, for example, a commonly studied residue is Trp501, whose fluorescence has been used to characterize structural transitions in myosin.[20,43] We monitored the solvent accessible surface area (SASA) of Trp501 during several independent TMD trajectories, which yielded rather diverse behaviors; Trp501 can be fully exposed to solvent or almost fully buried by other groups during the transition, although its SASA at the end of TMD simulations is similar to values from equilibrium simulations. This observation supports the proposal that the accessibility of water should not contribute much to ATP-induced fluorescence enhancement[43]; rather, the fluorescence enhancement of Trp501 during the recovery stroke is likely due to changes in the distances between Trp501 and nearby charged/polar residues. In fact, the observed structural changes in TMD trajectories are consistent with the conclusion drawn from fluorescence experiment that Trp501 has at least three different microenvironments,[43] which highlights the complexity inherent in the interpretation of such experiments.

### 2.2.3 POTENTIAL OF MEAN FORCE: CHARACTERIZE THE ENERGETICS OF COUPLING

Another approach for characterizing the coupling between remote structural motifs is to explore the energetics of key conformational transitions. We have carried out this type of studies at two different scales.

First, PMF calculations were attempted for the entire recovery stroke process, going between the postrigor and the pre-powerstroke x-ray structures with ATP as the ligand. A collective reaction coordinate is defined as the differential root-mean-square-difference ($\Delta$RMSD) of a conformation relative to the two x-ray structures.[44] Umbrella sampling along this $\Delta$RMSD coordinate was carried out using the generalized Born solvent model,[38] which allowed totally ~50 ns simulations. Analysis of key geometrical properties in different windows found very similar trends as in the TMD simulations (compare Figure 2.3a and c), indicating no major mechanistic change between the nonequilibrium TMD simulations and the umbrella sampling simulations that are closer to equilibrium. Overall, the calculated PMF (Figure 2.3d) is largely downhill in nature and reveals a rather broad basin around the pre-powerstroke state. The broad feature is qualitatively consistent with results from NMA as well as experimental observation[18] that the pre-powerstroke state has a significant degree of flexibility in the lever arm/converter. The downhill nature is qualitatively similar to the finding of a simulation study using myosin II from a different organism and explicit solvent,[45] although the degree of exothermicity is significantly smaller in our result. The quantitative nature of these PMF calculations is unclear given the scale of the structural transition and various approximations inherent in this type of analysis. Nevertheless, the results suggest that the recovery stroke is largely diffusive in nature and does not involve any major energetic bottleneck; these features appear to be consistent with a recent study of Elber and coworkers using the stochastic path approach (R. Elber, private communication). In other words, it is not valid to use simple transition state theory to connect the timescale of the recovery stroke and the underlying free energy profile (either computationally or experimentally), and it is likely that a minimal model requires understanding the dependence of the diffusive property of the system on the value of the reaction coordinate.[46]

A more "manageable" set of PMF calculations concerns the local open/close transition of the active site in the two x-ray structures. The ΔRMSD coordinate is again used as the collective reaction coordinate, although the RMSDs were defined only in terms of the atoms in the SwI and SwII motifs. Since only local structural transitions were explored, a finite-sphere simulation protocol based on explicit solvent and the generalized solvent boundary potential (GSBP)[47] has been adopted. The results (Figure 2.4a) were most revealing, although expected to some degree. In the postrigor state, the open/close transition was found to be largely thermoneutral with a small barrier of ~5 kcal/mol. In the pre-powerstroke state, by contrast, the active site strongly prefers the closed configuration and the open configuration is higher in free energy by at least 8 kcal/mol. The quantitative connection between these results to experimental data[20,43] is not straightforward, partly because the precise nature of the structural transition investigated by the fluorescence experiment was not clear. For example, fluorescence studies indicated that the open/close transition in the postrigor state is fast and nearly thermoneutral, although it is likely that the process studied involves both active site closure (or SwII displacement) and converter rotation. Nevertheless, the qualitative difference between the open/close PMFs in the postrigor and pre-powerstroke states clearly indicates that converter rotation is accompanied by structural transitions that propagate to the surrounding region of the active site such that the energetics of SwII displacement are significantly perturbed. In fact, data from the PMF simulations can also be used to construct the $(\phi, \psi)$-free energy profile of residues near the active site. The results (not shown here)[23] clearly indicate that the motion of these residues becomes substantially restricted in the prepowerstroke state. As discussed below, since the ATP hydrolysis activity is very sensitive to the active site configuration (including the position of water molecules), this tight coordination between converter orientation and active site stability suggests that the converter rotation is also likely coupled to ATP hydrolysis (i.e., "mechanochemical coupling").

## 2.3 REGULATION OF ATP HYDROLYSIS BY LOCAL AND DISTANT STRUCTURAL CHANGES[24,25]

To firmly establish the causality between chemistry (ATP hydrolysis) and various structural transitions in myosin, it is important to explicitly characterize the energetics of the chemical reaction in different conformations. Technically, this requires well-defined ways to generate relevant conformations in addition to available x-ray structures and a sufficiently reliable combined quantum mechanical/molecular mechanical (QM/MM) method. For the generation of relevant conformations, the extensive classical simulations discussed above have provided crucial clues (see Section 2.3.2). Regarding a proper QM/MM method, it is a major challenge since it is well-known that phosphate chemistry has complex mechanisms and a meaningful analysis requires a sufficiently accurate and flexible QM method[48,49]; at the same time, the very nature of the mechanochemical coupling analysis requires sufficient conformational sampling, which calls for computational efficiency. The development of a QM/MM method that balances accuracy and efficiency has been an important topic in our group, which is outside the scope of this chapter. We refer interested

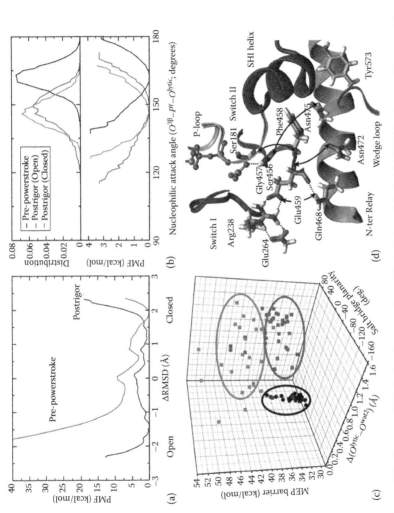

**FIGURE 2.4 (See color insert following page 172.)** Active site properties and hydrolysis activity.[23,25] (a) PMF for the open/close transition of the active site in the postrigor (black) and pre-powerstroke (red) structures; (b) Comparison of the distribution and corresponding PMF for the nucleophilic attack angle based on equilibrium simulation for the pre-powerstroke state and two postrigor structures. (c) Minimum energy path (MEP) barriers for the first step of ATP hydrolysis calculated starting from snapshots collected from equilibrium simulations of the pre-powerstroke state and a closed postrigor structure (see text). The barriers are plotted against the Arg238–Glu459 salt bridge planarity and the differential distance between the lytic water and Wat2 in the reactant and transition state. The black dots indicate data for the pre-powerstroke state; the blue, green, and red set indicate data from the closed postrigor simulations with different behaviors of Wat2.[25] (d) Key hydrogen-bonding interactions in the active site region of the closed postrigor structure. The arrows indicate interactions that are broken when SwII is displaced to close the active site in the postrigor state; i.e., rearrangements in the N-terminus of the relay helix and wedge loop are required to form the stable active site as in the pre-powerstroke state.

readers to recent reviews and articles for details.[50,51] In the following, we discuss results using a semiempirical-based DFT approach, SCC-DFTB,[52] which has been specifically parameterized recently to study phosphate hydrolysis chemistry[53]; the specific protocol is referred to as SCC-DFTBPR/MM-GSBP,[54] where the GSBP approach[47] is used to treat long-range electrostatics.

## 2.3.1 HYDROLYSIS IN DIFFERENT X-RAY STRUCTURES: ROLE OF ACTIVE SITE RESIDUES AND WATER MOLECULES

As discussed in Section 2.1, it is generally believed that the pre-powerstroke structure corresponds to the active ATPase conformation while the postrigor structure reflects a conformation incapable of hydrolyzing ATP. It is useful to verify these well-accepted speculations as an important validation to the computational protocol. Moreover, the precise reason that the postrigor state is incapable of hydrolyzing ATP is not as clear as commonly thought. The most quoted reason[13,19] refers to the position of active-site water molecules identified in the postrigor x-ray structure; they are not in the "in-line" attack orientation as discussed in the phosphoryl transfer literature, and therefore believed to be incapable of efficiently attacking the γ phosphate of ATP. However, the energy cost for changing the orientation of water is expected to be too small to effectively regulate the hydrolysis activity, thus other factors should be analyzed systematically.

For the pre-powerstroke state, we focused on a nominal associative hydrolysis mechanism[55] in which the Ser236 sidechain serves as a "relay group" for the proton transfer from the lytic water to the γ phosphate during the initialstep of the hydrolysis; involvement of a proton relay is generally thought as necessary based on stereochemical considerations.[19,56] A subsequent rotation of a hydroxyl group in the "pentavalent" intermediate helps completely break the covalent interaction between the ADP moiety and the inorganic phosphate, leading to the final hydrolysis product. The efficiency of the SCC-DFTBPR/MM-GSBP protocol makes it possible to generate the two-dimensional PMF that corresponds to this mechanism. The first step of the reaction has a barrier about 16 kcal/mol; the nominal pentavalent intermediate is 12 kcal/mol higher than the reactant state, and the P–O bond between the γ and β phosphate groups is largely broken (the $P^\gamma$–$O^{3\beta}$ distance is 2.60 Å). In fact, the $P^\gamma$–$O^{3\beta}$ bond is already significantly weakened in the transition state, thus the actual mechanism has partial dissociative character. For the second step of the reaction, the PMF result indicates a very small barrier from the intermediate followed by an energetically downhill proton transfer from the γ phosphate to $O^{3\beta}$ in ADP. Summing the energetics from the two PMFs, the ATP hydrolysis in the pre-powerstroke state is calculated to be nearly thermoneutral and the rate-limiting barrier is 16 kcal/mol. These are consistent with experimentally measured equilibrium constant and rate constant,[57,58] and this is the first time that the experimental data are computationally reproduced for ATP hydrolysis in myosin. Such close agreements indicate that the SCC-DFTBPR/MM approach is at least semiquantitatively reliable for our purpose of exploring mechanochemical coupling.

It is worth noting that several previous calculations,[59] including one of ours,[24] used MEP calculations to study the hydrolysis. In all cases, the rate-limiting barrier is significantly higher than the PMF barrier and experimental value. Analysis of structural features in the active site from PMF calculations revealed a possible reason. Superposition of the transition state structures from the PMF and MEP calculations indicates that the key residues, including water molecules in the active site, have very similar positions except for the orientation of a nearby serine (Ser181) sidechain and the water that hydrogen bonds to it (Wat3). Evidently, as proton transfer from the lytic water to $O^{2\gamma}$ proceeds, it is energetically more favorable for Ser181 to break hydrogen bonding to $O^{2\gamma}$ and seek alternative interactions with nearby groups. Since such an isomerization also involves changing the orientation of Wat3, the process is difficult to sample in local MEP calculations; perturbation analysis of the MEP results indicates that removing the hydrogen bonding interaction between Ser181 and ATP lowers the barrier by ~8 kcal/mol, which is a significant portion of the difference between MEP and PMF barriers. Nevertheless, similarity in the average position of most key active site residues suggests that the MEP calculations, if compared carefully, are appropriate for analyzing how the hydrolysis energetics depend on the conformational state of the motor domain.

For the postrigor state, we carried out extensive MEP calculations using different levels of QM/MM methods. These calculations consistently point to significantly higher (by more than 14 kcal/mol) barriers compared to the values found for the pre-powerstroke conformation (consistently comparing MEP barriers in both cases). Moreover, the hydrolysis reaction is significantly endothermic rather than nearly thermoneutral. Therefore, the calculations support the well-accepted postulate that the postrigor conformation is hydrolysis incompetent. The question, however, is why. As shown in Figure 2.4b, there is indeed significant difference between the distribution of the key "nucleophilic attack angle" ($O^{3\beta}$–$P^{\gamma}$–$O^{lytic}$) based on equilibrium simulations of the two conformations. For the pre-powerstroke state, the distribution peaks around 165°, while it is ~150° for the postrigor state. Since the average nucleophillic attack angle in the transition state is ~169° according to both MEP and PMF calculations, the free energy penalty associated with properly aligning the lytic water in the postrigor state (i.e., the difference in the angular potential of mean force between 150° and 169° in the closed postrigor state, see Figure 2.4b) is on the order of 2–3 kcal/mol. Although this is not a small contribution in kinetic terms (corresponds to 30- to 150-fold change in the rate constant at 300 K according to transition state theory), it is clear that the nucleophilic attack angle is not as dominating a factor as commonly suggested[13,19] for dictating the hydrolysis activity.

Based on the structures from the MEP simulations, we carried out perturbative analysis to explore contributions from amino acids to the hydrolysis energetics. These calculations revealed that a number of polar and charged amino acids, such as Arg238 and Ser456, have notably different contributions to the hydrolysis barrier in the two active site conformations due to change in positions; e.g., Arg238 is substantially closer to ATP in the postrigor conformation and therefore can stabilize the reactant state more effectively in the postrigor conformation than the pre-powerstroke conformation. Therefore, like in the discussion of many enzyme cases,[60,61] electrostatic interactions seem to be more important than stereoelectronic factors for controlling the energetics of the chemical process.

## 2.3.2 HYDROLYSIS IN THE CLOSED POSTRIGOR STATE: ROLE OF STRUCTURAL CHANGES BEYOND THE ACTIVE SITE

To clearly illustrate the causality between ATP hydrolysis and various structural transitions implicated in the recovery stroke, we should examine the ATP hydrolysis activity at different stages of the recovery stroke. As the first step, we ask whether it is sufficient to turn on efficient ATP hydrolysis by simply closing the active site via the much-discussed SwII displacement. Since chemistry is generally controlled by local interactions, it is possible that a local structural change in the postrigor state is sufficient to switch on the ATPase activity. For this purpose, we generated a closed postrigor configuration by displacing only SwII in the postrigor x-ray structure; it is meaningful to displace SwII itself because, as discussed above, our PMF simulations have shown that the open/close transition of SwII in the postrigor state is a low-barrier and nearly thermoneutral process[23]; one might argue whether this "intermediate" has significant population in reality, although the answer depends on whether the recovery stroke is highly cooperative or largely diffusive in nature, which is an interesting issue worth further analysis. This structure was carefully equilibrated with extensive MD simulations, and a large set of QM/MM MEP calculations was then carried out for the first step of the hydrolysis reaction starting from snapshots sampled in the MD trajectories. MEP instead of PMF calculations were employed because it is easier to correlate structural features of the active site and energetics of the hydrolysis in the MEP framework.

Remarkably, unlike the small variations in the MEP barrier found for configurations sampled in the pre-powerstroke state ($31.3 \pm 1.8$ kcal/mol), the MEP barrier fluctuates significantly (ranging from 32.0 to 53.0 kcal/mol with a standard deviation of 4.4 kcal/mol) and is almost always substantially higher in the closed postrigor state (Figure 2.4c). This is rather unexpected given that the active site remains essentially closed during the simulations of both states. Analysis of the "nucleophilic attack angle" reveals a similar picture as discussed above for the open postrigor state; i.e., there is indeed a notable (2–3 kcal/mol) contribution but it is not sufficient to explain the difference in barriers.

Further combined structural analysis and perturbative analysis of residual contributions to the MEP barriers indicate that another factor important to efficient hydrolysis is the structural stability of the active site. As shown in Figure 2.4d, displacing SwII *alone* to close the active site in the postrigor state leaves many crucial interactions unformed or even breaks existing interactions. For example, although Gly457 forms a stable hydrogen bonding interaction with the γ phosphate of ATP upon SwII displacement, the interaction between Ser456 main chain and Asn475 in the relay helix is broken; the interaction between Gly457 and γ phosphate also becomes weaker in the later segment of the simulations. Similarly, although Glu459 forms a salt-bridge interaction with Arg238 when SwII is displaced, the interactions between the main chain of Glu459 and Asn472 in the relay helix are lost; instead, the carbonyl of Glu459 forms a hydrogen bond with the sidechain of Gln468. As a result, the extensive hydrogen-bonding network that involves Arg238, Glu459, Glu264, and Gln468 observed in the pre-powerstroke state is not present in the closed postrigor state, which explains the higher rotational flexibility of Glu459 in the latter case. Finally, since Tyr573 in the "wedge loop" remains far from SwII in the closed postrigor state, there is ample space for Phe458 to sample multiple rotameric states and its main chain interaction with Ser181 remains unformed, in contrast to the situation in the pre-powerstroke state. Although this additional structural flexibility seems fairly subtle (certainly compared to the striking converter rotation), perturbative analysis indicates that these status have a significant impact on the hydrolysis barrier. For example, as Glu459 rotates out of the salt-bridge plane, the second active-site water (Wat2) tightly associated with its sidechain forms a hydrogen-bonding interaction with the lytic water only in the reactant state but not the transition state of hydrolysis. Accordingly, Wat2 makes an *unfavorable* contribution (~6 kcal/mol) to the hydrolysis barrier, which is consistent with the higher hydrolysis barriers found for configurations collected from this part of the trajectory. Evidently, the emerging picture is that the transition from the postrigor state to a structurally stable closed active site (which apparently is critical to efficient ATP hydrolysis) relies on not only the displacement of SwII but also more extensive structural rearrangements in the nearby region, such as the N-terminus of the relay helix and the wedge loop. Experimentally,[62] it was recently found that perturbing the conformational equilibrium in the converter region indeed had a notable impact on the apparent ATPase activity, although the impact on the hydrolysis rate was not clear.

## 2.4  CONCLUDING DISCUSSIONS AND FUTURE PERSPECTIVES

The tight coupling between ATPase activity and large-scale conformational transition of remote domains is the hallmark feature of many biomolecular motors[3] and crucial for their high thermodynamic efficiency.[9] While the process of large-scale structural transition induced by ATP binding has been visualized and discussed in several systems,[63–65] the underlying mechanism for the allosteric coordination of ATP hydrolysis and large-scale conformational rearrangements remains elusive. In the past few years, we have been combining a battery of computational techniques to study the mechanochemical coupling in a prototypical molecular motor, myosin II.[22–26] These studies allowed us to both answer specific mechanistic issues and gain insights into some general principles behind the motor function.

### 2.4.1  STRUCTURAL BASIS FOR THERMODYNAMIC AND KINETIC CONSTRAINTS IN MOTOR FUNCTION

As pointed out elegantly by Hill and coworkers in their pioneering analysis of coupled biological processes,[5,9,66] biological systems need to meet specific thermodynamic and kinetic constraints to acquire the corresponding characteristics necessitated by their biological functions. Specifically for myosin, the constraints based on functional considerations are[66]: (1) processes that occur in the detached states have small free energy drops; (2) in particular, the ATP hydrolysis step in the motor domain (i.e., not the entire ATPase cycle, see below for further clarification) is close to be thermoneutral; and (3) the ATP hydrolysis step is tightly coupled to the reorientation of the lever arm (i.e., recovery stroke requires the hydrolysis of one and only one ATP). The first two constraints arise because no work can be done in the detached states (no force can be exerted on the actin); considering that the sum of free energy changes through a functional cycle is strictly a constant (the hydrolysis free energy of ATP in solution), processes in the detached states should have small free energy drops. In principle, large free energy increases and drops may cancel to still leave a large free energy drop for the powerstroke, but the large free energy increases for certain steps may compromise the speed of the motor. The third constraint minimizes the amount of futile ATP hydrolysis because otherwise a loose coupling may lead to rebinding of the motor to actin with a lever arm orientation incapable of powerstroke.

These "idealized" constraints are derived by considering a motor with the optimal efficiency and most biological motors do not need to have the perfect thermodynamic efficiency for their biological function.[1] Nevertheless, they provide an interesting set of guidelines to design molecular simulations to probe the "design principles" of molecular motors at the molecular scale. Molecular features of myosin in connection to constraint 3 are discussed in detail in the Section 2.4.2. Constraint (1) is consistent with the findings from NMA and PMF calculations that myosin has intrinsic structural flexibility at both the domain level and local scale (SwII displacement in postrigor). In addition, a series of hydrophobic and polar interactions are observed in TMD simulations[22] to facilitate the large-scale structural rearrangements in the relay helix and converter domain and help maintain a low energy cost. For other processes other than the powerstroke, although not explicitly studied by us, it is natural to envision how they can be nearly thermoneutral.[11,12] The detachment of the motor domain from actin is compensated by the binding of ATP; the rebinding of the motor domain to actin following ATP hydrolysis is known to be only weak association; the weak→strong actin-binding transition is compensated with dissociation of hydrolysis products and may partially be coupled to the powerstroke.

For constraint (2), which is consistent with the measured equilibrium constant,[57,58] QM/MM simulations suggest that the small free energy drop is likely controlled by well-orchestrated electrostatic environment of the active site. Both favorable and unfavorable contributions from residues/water in the closed active site have been identified,[24,25] which together result in a small hydrolysis free energy. In this context, we emphasize that one has to distinguish the free energy drop for ATP hydrolysis in the motor domain, which is small in magnitude, from that for the entire functional cycle; the latter is strictly equivalent to the hydrolysis free energy of ATP in solution and the ultimate thermodynamic driving force for the vectorial motion of the motor.

### 2.4.2 MECHANISTIC INSIGHTS INTO MECHANOCHEMICAL COUPLING (CONSTRAINT 3)

As discussed in a recent review on allostery,[21] the key challenge for understanding a specific system is to provide a molecular level description regarding how events of different spatial scales (domain vs. sidechain) and physicochemical nature (hydrolysis vs. structural isomerization) are tightly coupled to achieve the desired biological function. For example, it is interesting to ask whether there is distinct causality among the different processes, and if so, what is the specific sequence of event; it is entirely possible that multiple pathways are present and contribute to different degrees.[67] Ultimately, answering this question requires a complete characterization of the free energy landscape of the system, which is extremely challenging to both experimental and computational studies. It is rarely possible to even construct projection of the free energy landscape in reduced dimensions[68] because identifying the most relevant variables to span the reduced dimensions is, by itself, an intense subject of research.

Given these difficulties, the best approach at this stage appears to be combining different techniques and scrutinizing the results collectively. This is illustrated here by the combination of NMA, TMD, and PMF simulations, and SCA, in the pursuit of a molecular mechanism for the structural aspect of the recovery stroke. The NMA results emphasized that the intrinsic structural flexibility and correlated motions already present in the equilibrium fluctuations of the myosin motor domain have a high degree of overlap with the direction of functional conformational transition. The TMD results, on the other hand, revealed potential transient interactions that facilitate the nonequilibrium process of conformational transition, which contain both hydrophobic and hydrophilic contacts. Combining NMA, TMD, and SCA seems to be a productive way for identifying "hotspot" residues that play important roles in the transition, which occupy sites in the intersubdomain linkers such as SwII, relay helix, SH1 helix, strut and the central β-sheet. These observations support the idea that "nanomachines" like molecular motors consist of semirigid domains interacting via flexible regions (e.g., hinges), such that local events (e.g., SwII displacement) can propagate over a long distance to affect activities elsewhere (e.g., converter rotation).[21]

Regarding the role of ATP hydrolysis, our QM/MM analysis[24] and classical PMF simulations for the active site[23] indicated clearly that the hydrolysis product preferentially stabilizes the closed configuration of the active site. Since the active site closure is structurally coupled to converter rotation, the hydrolysis product indirectly stabilizes the pre-powerstroke orientation of the converter/lever arm. The preference of the closed configuration by the hydrolysis product also helps prevent premature release of $P_i$, which is believed to be the driving force for powerstroke.[12]

Turning to the sequence of events during the coupled recovery stroke and hydrolysis processes, it is common to assume that hydrolysis is strictly controlled by the immediate environment and the large-scale structural changes are the consequence of the local structural changes associated with the hydrolysis process; i.e., the hydrolysis product preferentially stabilizes a specific configuration of the active site, which in turn propagates into major structural changes in remote regions.[23,39] In the specific case of myosin, fluorescence experiments indicated that a conformational change step precedes hydrolysis,[20] although whether the entire recovery stroke has to finish before hydrolysis is not clear due to the limited structural resolution in such study. Our most recent analysis[25] demonstrated that turning on the ATP hydrolysis activity in myosin requires rather extensive conformational rearrangements beyond the immediate environment of ATP; this observation does not prove that the converter rotation has to finish prior to hydrolysis, although the result highlights the complexity of mechanochemistry in motor systems. Specifically, we illustrate that efficient ATP hydrolysis relies directly on the proper positioning of not only the first coordination sphere of the γ-phosphate, such as the commonly discussed lytic water,[19] but also second-coordination sphere motifs. In fact, the second-sphere electrostatic stabilization is found to be more significant than aligning the lytic water into the in-line attack orientation.[19] Importantly, we demonstrate that the ability of these second-coordination sphere residues to make their crucial electrostatic stabilization relies on the configuration of residues in the nearby N-terminus of the relay helix and the "wedge loop." Without the structural support from those motifs, residues in a closed active site in the postrigor motor domain undergo subtle structural variations that lead to consistently higher calculated ATP hydrolysis barriers than in the pre-powerstroke state. In other words, structural transitions remote from the active site can play a very active role in regulating the hydrolysis of ATP, rather than merely responding to structural changes in the active site in a passive fashion. We note that the analyses were made possible by generating a "hybrid" structure between the available x-ray structures, which highlights the unique value of computational studies in dissecting the causal relationship between different processes. Whether this "hybrid" structure has significant population in reality is an interesting issue worth additional analysis.

### 2.4.3 Future Directions

Based on the above discussions, our current understanding in how different events are coupled in the myosin recovery stroke is summarized in Figure 2.5, which represents a qualitative characterization of the complex free energy landscape underlying the process. Further studies at the atomic or near-atomic scale are needed to make the landscape characterization and therefore our understanding more complete. For example, more sophisticated methods[69–71] are required to go beyond the MEP and TMD models to characterize the structural transition pathways; the most interesting questions of fundamental importance are whether there exist multiple pathways of transition with very different sequences of events and whether certain "nonnative" interactions indeed play crucial roles in dictating the kinetics of the transition, as found in the folding of some proteins.[72] Given the complexity of the problem, it is likely that effective coarse-grained models beyond simple Go models need to be developed, which is a fascinating subject of research.

In addition to myosin, it is of interest to use the similar techniques to analyze other molecular motors in a comparative fashion. Of particular interest are kinesin and helicases, in which the chemical step occurs when the motor is bound on its tracks; therefore, it is possible that somewhat different sets of constraints and structural features can be found in these motors compared to myosin. Once mechanochemical coupling is well understood in several representative systems, it might be possible to design computational protocols that allow one to predict conformational transitions coupled to ATP binding and/or hydrolysis based on, for example, combination of experimental data at different resolutions (e.g., x-ray and electron microscopy [EM], fluorescence resonance energy transfer [FRET]); meeting this challenge can have a

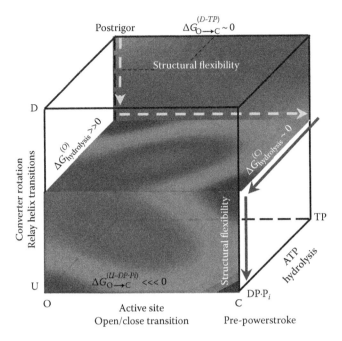

**FIGURE 2.5**  **(See color insert following page 172.)** A schematic view for the coupling between three major events (converter rotation/relay helix kinking, active site closure and ATP hydrolysis) in the recovery stroke based on results from simulation studies. The scheme highlights the intrinsic structural flexibility of the converter,[22,26] the coupling between open/close transition of the active site to both ATP hydrolysis and converter rotation,[22–24] and that turning on ATP hydrolysis requires extensive structural changes beyond active site closure (SwII displacement).[25]

major impact on our ability to regulate and design signaling proteins and molecular motors in general.

Finally, other than studying molecular motors as isolated allosteric enzymes, it is tempting to start building molecular models to study the interaction between motors and their tracks, the effect of external force and cooperativity between motors so that more direct connections can be made to state-of-the-art single molecule studies.[3,6–8] Although large-scale atomistic simulations are helpful to provide guidance in this context, development of effective coarse-grained models at both particle[73,74] and continuum levels[75,76] is likely required and remains the central challenge.

## ACKNOWLEDGMENTS

Q.C. acknowledges other group members and collaborators who made significant contributions to the work discussed here. The studies reviewed have been generously supported from the National Institutes of Health (R01-GM071428). Q.C. also acknowledges the Alfred P. Sloan Research Fellowship. Computational resources from the National Center for Supercomputing Applications at the University of Illinois are greatly appreciated.

## REFERENCES

1. J. Howard, *Mechanics of Motor Proteins and the Cytoskeleton* (Sinauer Associates, Inc., Sunderland, MA, 2001).
2. M. Schliwa, ed., *Molecular Motors* (Wiley-VCH, Weinheim, 2002).
3. R. D. Vale and R. A. Milligan, *Science* **288**, 88 (2000).
4. C. Bustamante, D. Keller, and G. Oster, *Acc. Chem. Res.* **34**, 412 (2001).
5. T. L. Hill, *Free Energy Transduction in Biology* (Academic Press, New York, 1977).
6. A. D. Mehta, R. S. Rock, M. Rief, J. A. Spudich, M. S. Mooseker, and R. E. Cheney, *Nature* **400**, 590 (1999).
7. S. M. Block, *Biophys. J.* **92**, 2986 (2007).
8. J. R. Moffitt, Y. R. Chemla, S. B. Smith, and C. Bustamante, *Annu. Rev. Biochem.* **77**, 205 (2008).
9. E. Eisenberg and T. L. Hill, *Science* **227**, 999 (1985).
10. B. Alberts, D. Bray, J. Lewis, M. Raff, K. Roberts, and J. D. Watson, *Molecular Biology of the Cell* (Garland Publishing, Inc., New York, 1994).
11. M. A. Geeves and K. C. Holmes, *Annu. Rev. Biochem.* **68**, 687 (1999).
12. M. A. Geeves and K. C. Holmes, *Adv. Prot. Chem.* **71**, 161 (2005).
13. C. A. Smith and I. Rayment, *Biochemistry* **35**, 5404 (1996).
14. C. B. Bauer, H. M. Holden, J. B. Thoden, R. Smith, and I. Rayment, *J. Biol. Chem.* **275**, 38494 (2000).
15. J. Menetrey, A. Bahloul, A. L. Wells, C. M. Yengo, C. A. Morris, H. L. Sweeney, and A. Houdusse, *Nature* **435**, 779 (2005).
16. R. W. Lymn and E. W. Taylor, *Biochemistry* **10**, 4617 (1971).
17. A. M. Gulick, C. B. Bauer, J. B. Thoden, and I. Rayment, *Biochemistry* **36**, 11619 (1997).
18. W. M. Shih, Z. Gryczynski, J. R. Lakowicz, and J. A. Spudich, *Cell* **102**, 683 (2000).
19. W. W. Cleland and A. C. Hengge, *Chem. Rev.* **106**, 3252 (2006).
20. A. Málnási-Csizmadia, R. J. Woolley, and C. R. Bagshaw, *Biochemistry* **39**, 16135 (2000).
21. Q. Cui and M. Karplus, *Prot. Sci.* **17**, 1295 (2008).
22. H. Yu, L. Ma, Y. Yang, and Q. Cui, *PLoS Comput. Biol.* **3**, 0214 (2007).
23. H. Yu, L. Ma, Y. Yang, and Q. Cui, *PLoS Comput. Biol.* **3**, 0199 (2007).
24. G. H. Li and Q. Cui, *J. Phys. Chem. B* **108**, 3342 (2004).
25. Y. Yang, H. Yu, and Q. Cui, *J. Mol. Biol.* **381**, 1407 (2008).
26. G. H. Li and Q. Cui, *Biophys. J.* **86**, 743 (2004).
27. Q. Cui and I. Bahar, eds., *Normal Mode Analysis: Theory and Applications to Biological and Chemical Systems*, Mathematical and Computational Biology Series (Chapman and Hall/CRC, New York, 2005).
28. G. H. Li and Q. Cui, *Biophys. J.* **83**, 2457 (2002).
29. F. Tama, F. X. Gadea, O. Marques, and Y. H. Sanejouand, *Proteins* **41**, 1 (2000).
30. Q. Cui, G. H. Li, J. P. Ma, and M. Karplus, *J. Mol. Biol.* **340**, 345 (2004).
31. J. P. Ma, *Structure* **13**, 373 (2005).
32. I. Bahar and A. J. Rader, *Curr. Opin. Struct. Biol.* **15**, 586 (2005).
33. F. Tama and C. L. Brooks III, *Annu. Rev. Biophys. Biomol. Struct.* 115–134 (2006).
34. S. Hayward and H. J. C. Berendsen, *Proteins* **30**, 144 (1998).
35. W. J. Zheng, B. R. Brooks, and D. Thirumalai, *Proc. Natl. Acad. Sci. U. S. A.* **103**, 7664 (2006).
36. S. W. Lockless and R. Ranganathan, *Science* **286**, 295 (1999).
37. J. Schlitter, M. Engels, P. Krüger, E. Jacoby, and A. Wollmer, *Mol. Simu.* **10**, 291 (1993).
38. W. Im, M. Lee, and C. L. Brooks, *J. Comput. Chem.* **24**, 1691 (2003).
39. S. Fischer, B. Windshugel, D. Horak, K. C. Holmes, and J. C. Smith, *Proc. Natl. Acad. Sci. U. S. A.* **102**, 6873 (2005).

40. D. J. Wales, *Energy Landscapes* (Cambridge University Press, Cambridge, U.K. 2003).
41. D. Kern and E. R. P. Zuiderweg, *Curr. Opin. Struct. Biol.* **13**, 748 (2003).
42. O. Miyashita, J. N. Onuchic, and P. G. Wolynes, *Proc. Natl. Acad. Sci. U. S. A.* **100**, 12570 (2003).
43. A. Málnási-Csizmadia, M. Kovács, R. J. Woolley, S. W. Botchway, and C. R. Bagshaw, *J. Biol. Chem.* **276**, 19483 (2001).
44. N. K. Banavali and B. Roux, *Structure* **13**, 1715 (2005).
45. H. J. Woo, *Biophys. Chem.* **125**, 127 (2007).
46. J. Chahine, R. J. Oliveira, V. B. P. Leite, and J. Wang, *Proc. Natl. Acad. Sci. U. S. A.* **104**, 14646 (2007).
47. W. Im, S. Berneche, and B. Roux, *J. Chem. Phys.* **114**, 2924 (2001).
48. J. Aqvist, K. Kolmodin, J. Florian, and A. Warshel, *Chem. Biol.* **6**, R71 (1999).
49. S. Admiraal and D. Herschlag, *Chem. Biol.* **2**, 729 (1995).
50. D. Riccardi, P. Schaefer, Y. Yang, H. B. Yu, N. Ghosh, X. Prat-Resina, P. König, G. H. Li, D. G. Xu, H. Guo, et al., *J. Phys. Chem. B* **110**, 6458 (2006).
51. Y. Yang, H. Yu, D. York, Q. Cui, and M. Elstner, *J. Phys. Chem. A* **111**, 10861 (2007).
52. M. Elstner, D. Porezag, G. Jungnickel, J. Elsner, M. Haugk, T. Frauenheim, S. Suhai, and G. Seifert, *Phys. Rev. B* **58**, 7260 (1998).
53. Y. Yang, H. Yu, D. York, M. Elstner, and Q. Cui, *J. Chem. Theo. Comp.* **4**, 2067 (2008).
54. P. Schaefer, D. Riccardi, and Q. Cui, *J. Chem. Phys.* **123**, Art. No. 014905 (2005).
55. A. Hassett, W. Blättler, and J. R. Knowles, *Biochemistry.* **21**, 6335 (1982).
56. B. Gerratana, G. A. Sowa, and W. W. Cleland, *J. Am. Chem. Soc.* **122**, 12615 (2000).
57. D. R. Trentham, J. F. Eccleston, and C. R. Bagshaw, *Q. Rev. Biophys.* **9**, 217 (1976).
58. A. Málnási-Csizmadia, D. S. Pearson, M. Kovács, R. J. Woolley, M. A. Geeves, and C. R. Bagshaw, *Biochemistry* **40**, 12727 (2001).
59. S. M. Schwarzl, J. C. Smith, and S. Fischer, *Biochemistry* **45**, 5830 (2006).
60. Q. Cui and M. Karplus, *Adv. Prot. Chem.* **66**, 315 (2003).
61. A. Warshel, *Annu. Rev. Biophys. Biomol. Struct.* **32**, 425 (2003).
62. A. Malnasi-Csizmadia, J. Toth, D. S. Pearson, C. Hetenyi, L. Nyitray, M. A. Geeves, C. R. Bagshaw, and M. Kovacs, *J. Biol. Chem.* **282**, 17658 (2007).
63. H. Y. Wang and G. Oster, *Nature* **396**, 279 (1998).
64. Y. Gao and M. Karplus, *Curr. Opin. Struct. Biol.* **14**, 250 (2004).
65. P. H. von Hippel and E. Delagoutte, *Cell* **104**, 177 (2001).
66. T. L. Hill and E. Eisenberg, *Q. Rev. Biophys.* **14**, 463 (1981).
67. Q. Lu and J. Wang, *J. Am. Chem. Soc.* **130**, 4772 (2008).
68. L. Ma and Q. Cui, *J. Am. Chem. Soc.* **129**, 10261 (2007).
69. W. C. Swope, J. W. Pitera, and F. Suits, *J. Phys. Chem. B* **108**, 6571 (2004).
70. S. Yang and B. Roux, *PLoS Comput. Biol.* **4**, e1000047 (2008).
71. A. K. Faradjian and R. Elber, *J. Chem. Phys.* **120**, 10880 (2004).
72. A. Zarrine-Afsar, S. Wallin, A. Neculai, P. Neudecker, P. Howell, A. Davidson, and H. Chan, *Proc. Natl. Acad. Sci. U. S. A.* **105**, 9999 (2008).
73. C. Hyeon and J. N. Onuchic, *Proc. Natl. Acad. Sci. U. S. A.* **104**, 2175 (2007).
74. C. Clementi, *Curr. Opin. Struct. Biol.* **18**, 10 (2008).
75. G. H. Lan and S. X. Sun, *Biophys. J.* **88**, 999 (2005).
76. Y. Tang, G. Cao, X. Chen, J. Yoo, A. Yethiraj, and Q. Cui, *Biophys. J.* **91**, 1248 (2006).

# 3 The Chemomechanical Coupling Mechanisms of Kinesin and Dynein

*Yi Qin Gao and Qiang Shao*

## CONTENTS

Dynein, kinesin, and myosin form the three families of molecular motors that are responsible for the motility of eukaryotic cells [1–3]. In recent years, numerous studies have been performed to characterize the structure, function, and mobility of these protein motors. In particular, the use of single molecule spectroscopic techniques have made possible the direct observation and quantitative measurement of the motions and mechanical responses of these motor proteins. The understanding of the chemomechanical coupling mechanism, through which chemical energy of adenosine triphosphate (ATP) hydrolysis is used to perform mechanical work, has been the focus of most research studies. Generally speaking, these protein motors all follow a similar strategy [4,5] in which the motor core binds and hydrolyzes ATP and coordinates the chemical energy release with the conformational changes that lead to physical motions, although much of the molecular mechanism is still not known. Chemical transitions including ATP binding and hydrolysis and/or adenosine diphosphate (ADP) and inorganic phosphate ($P_i$) release in the motor domain regulate the binding of a motor protein to its track, such as the microtubule for kinesin and dynein and the actin filament for myosin, and induce the directional or oscillatory motions of the motor proteins along the tracks. The proper functions of these motor proteins are essential for the survival and reproduction of cells. Their malfunctions are known to be linked to a large number of diseases.

Although our understanding on how exactly the chemical energy is converted into mechanical energy is still at a very early stage of research, the large body of experimental data available is making a detailed picture of molecular mechanism more approachable than ever. Experiments have showed that both traditional kinesin [6] and myosin (in particular myosin V) [7] walk by a hand-over-hand mechanism (in which the two motor domains alternate their leading positions alternately during walking) and there exists a tight coupling between the chemical and mechanical processes. For example, every ATP hydrolyzed generates an 8 nm movement of kinesin along the microtubule in a large range of external load [8–13]. On the other hand, dramatically different behaviors of different motor proteins have also been reported. For example, although dynein processes by 8 nm steps when it walks against a heavy load [14–16] and a single-headed dynein is not able to move along the microtubule [16], similar to conventional kinesin, recent single molecule experiments [17–22] showed that the chemomechanical properties of dynein are very different from kinesin.

Theoretical modeling has proven to be useful in understanding the experimental results and is shedding light on the chemomechanical coupling mechanisms of motor proteins. Atomic detailed and coarse-grained molecular dynamics simulations and normal mode analysis have been used to understand the motions and conformational changes of motor proteins. In addition, both kinetic modeling [23,24] and master equation approaches [25–27] using thermal ratchet-type models have also been used to study their kinetic and mechanical properties. A full account of these studies is beyond the scope of this chapter. Instead, we focus on our recent effort in the modeling of the external force and ATP concentration dependence in the walking mechanism of kinesin and dynein, under the theoretical framework of the Brownian ratchet model. Through these studies, we try to build, based on structural and biochemical experimental data, a physical model to account for the directional motion of both kinesin and dynein, in particular, the coupling between ATP hydrolysis and walking. In the following, we first provide a summary on the structural and kinetic data of both kinesin and dynein.

Kinesins are microtubule-based linear motors and exist with a large population in eukaryotic cells. They have a large variety of different structures and have been found to function as monomers, dimers, or tetramers in cells [28,29]. Based on structure and function, a number of their subspecies have been classified; for example, 45 different kinesin species are found in humans [29]. These microtubule-based motors are involved in many biological functions, including cargo transportation, mitosis, control of microtubule dynamics, as well as in signal transduction [28–32]. Among the different family members, the only conserved domain is the catalytic core [32]. In a more general way, they can be classified into three categories: N-terminal kinesins, C-terminal kinesins, and M kinesins, and among them, the N-terminal kinesins are the majority kinesins in humans [28]. The motions of the kinesins on microtubules are directional: N-terminal kinesins move to the plus end and C-terminal kinesins move toward the minus end of microtubule [28,29].

Each kinesin monomer has an N-terminal motor head, which is responsible for the binding of nucleotides as well as microtubule, a neck-linker, a long coiled-coil involved in dimerization, and a globular cargo-binding tail domain formed by a light chain [28,29,32]. The neck-linker is an extension from the motor head and is thought to serve as a lever-arm in force generation [29–31], consistent with its observed nucleotide-dependent conformational change [29,30,33]. It is believed to be coupled with force transduction or generation during kinesin walking. The comparison of the kinesin 1 cryo-EM (electron microscopy) structures in different nucleotide states indicated that ATP binding to a microtubule-associated kinesin head causes a conformational change involving a tilt of the stalk in the forward walking (+) direction [34]. In addition, ADP release likely induces redocking of the neck-linker and thus stepping of kinesin [35,36]. Two distinct states (open and closed) exist for the switch region (switch I), which flanks the active site [31]. Only the closed state is active in ATP hydrolysis, indicating that this switch region is an important element in coupling the conformational change of kinesin to ATP hydrolysis, also in accordance with the structure difference between the 5′-adenylyl-$\beta,\gamma$-imidodiphosphate (AMPPNP)- and ATP-bound motor domain in fluorescence microscopy studies [37].

An impressive amount of kinetics data have also been collected for kinesin. It has been determined that ATP binds to kinesin with a rate constant of ~4 $\mu M^{-1}$ $s^{-1}$ [31] and the dissociation constant, $K_d$, is about 75 $\mu M$ [31]. (Therefore, ATP dissociates from its binding site with a rate constant of ~150 $s^{-1}$.) In the absence of microtubule, the rate constant for ATP hydrolysis catalyzed by kinesin is ~6 $s^{-1}$ [4] and ADP releases from a kinesin with a rate constant of ~0.002 $s^{-1}$ [38]. Both ATP hydrolysis [39] and ADP release [40] increase to 100–300 $s^{-1}$ and ~20 $s^{-1}$, respectively, when kinesin binds microtubule. It was also found that the chemical processes at the two heads are cooperative: the ADP release from one head is further accelerated (60–300 $s^{-1}$) when ATP is bound at the other head. However, it is believed that ADP release remains the rate-limiting step [31,38]. Regardless of the experimental conditions, $P_i$ releases at a rate of >100 $s^{-1}$ [31,41]. The kinetic measurements also showed that an ADP occupied kinesin head only binds the microtubule weakly ($K_d$ ~ 10–20 $\mu M$ [4]) compared to an empty head [31,42].

Detachment force measurement in single molecule experiments confirmed that an ADP-occupied kinesin head binds microtubule much more weakly than a nucleotide-free or AMPPNP-occupied kinesin head [43–45]. The experimentally determined unbinding force (the force applied to detach kinesin from microtubule) for the ADP state is below 4 pN [43,45], whereas the unbinding forces for the empty and AMPPNP states are both greater than 6 pN [45]. It was also observed [45] that for all three states, a larger force is needed to detach the kinesin head when it is pulled in the minus direction (the direction opposite to the motion of kinesin) than when it is pulled in the plus direction. The unbinding force of the ADP state is 3.3–3.4 pN with a plus-end load and becomes 3.6–3.9 pN under a minus-end load. The unbinding force of the AMPPNP and empty states is 6.1–6.9 pN in the plus direction and 9.1–10 pN in the minus direction. The unbinding force is similar for a monomeric and dimeric kinesin if the heads are at the ADP or empty states, indicating that under these conditions, only one of the two heads of a dimer binds to the microtubule. The observation that when both kinesin heads are occupied by

ADP only one head binds microtubule is consistent with earlier x-ray structural studies [46]. On the other hand, in the presence of AMPPNP, the detaching force is much larger for a dimer than for a monomer, suggesting that under these conditions, both heads are attached to microtubule. It is likely that one of the two heads is occupied by AMPPNP and the other is empty [45].

Most single molecular mobility experiments were performed on the conventional kinesin, which is a homodimer and each of the monomer contains a heavy chain of ~120 kDa [29]. Conventional kinesin is a processive motor and its step size is around 8 nm, the axial distance between two adjacent kinesin binding sites on microtubule. This step size was shown to be invariant in a large range of ATP concentration and external load [47,48]. The stepping of kinesin is tightly coupled to ATP turnover, in the presence of both low and high external load [47]. One step of kinesin requires only one ATP molecule, unless the load is high. Different models have been proposed to account for the processive stepping of kinesin [29,31]. In most existing models, ATP binding to an empty kinesin head causes the other head with a bound ADP to move forward, which subsequently reattaches to microtubule and releases ADP, although the details of these models differ. There have been suggestions of an inchworm model (in which one of the two heads always keeps the leading position), but recent single molecule experiments convincingly showed that kinesin walks by a hand-over-hand mechanism, in which the two heads alternately take the leading position. Different research groups have studied the external load as well as ATP concentration dependence of the kinesin motion. In spite of extensive studies, the molecular basis of the chemomechanical coupling mechanism is still obscure. In fact, some experimental observations still contradict each other. Both force-dependent [49] and force-independent [50] Michaelis–Menten constants (the ATP concentration at which the speed of kinesin is one-half of its maximum value obtained at saturating ATP concentrations) have been observed. It has also been observed that kinesin takes not only forward (toward the plus-end of microtubule) but also backward steps [50,51], with the latter occurring more often with the increase of external load [50,51]. In a recent experiment in which a large range of external load is applied in both assisting and hindering directions [51], Carter and Cross observed sustained backward steps of an 8 nm step size at large hindering external forces. The backward steps, similar to the forward steps, are ATP-dependent. Both forward and backward stepping occurs very fast, on the microsecond time scale without detectable substeps, in contrast to some earlier experiments [52,53]. To account for the observation of both forward and backward steps, Carter and Cross proposed a model in the prestroke state of which only one head (presumably at an empty state) binds to microtubule and the other head with an ADP is detached. The latter head takes a position between two microtubule binding sites along the microtubule axial. ATP binding moves the detached head forward, which subsequently binds to microtubule and releases ADP, while at the same time the former attached head hydrolyzes ATP and detaches from the microtubule. However, the model is more consistent with a stepping pattern with both steps of a kinesin head being intermediate (between 0 and 17 nm and close to 8 nm) and it is in contradiction to the observation [54] that each of the two heads alternately takes steps of ~17 and ~0 nm.

Yildiz et al. [54] showed that each kinesin head takes alternative ~17 and ~0 nm steps during its walking, with an average step size of the dimer of 8.3 nm. Combined with an earlier observation that kinesin steps are taken without a stalk rotation [55], these results suggest that kinesin walks by an asymmetric hand-over-hand mechanism. The asymmetry of kinesin walking is also consistent with experiments on truncated [56] or mutant homodimeric kinesins [57], which "limp" along the microtubule. The difference in the motions of the two structurally identical monomers suggests that the stepping of kinesin is intrinsically asymmetric, although this asymmetry is not directly observable for a wild-type kinesin [56]. The asymmetry in the hand-over-hand mechanism was confirmed more recently by several novel experiments in which homodimeric kinesin also shows "limping" under certain conditions. A single amino acid mutation in the P-loop (nucleotide-binding domain) of Drosophila kinesin causes ADP release to be about 3.6-fold faster and the gliding velocity to be 3.3-fold slower. At low forces and/or low ATP concentrations, successive 8 nm steps are observed. However, at high load and high ATP concentrations, 16 nm steps are observed. A careful analysis showed that the 16 nm steps are actually rapid, double 8 nm steps. The 16 nm steps are thus due to alternating long and short dwell times. In another experiment, wild-type kinesin with truncated stalk also showed significant limping behavior in the presence of external force. It was shown in this experiment that the extent of limping increases with shortening of the stalk. Since both experiments mentioned above were performed with homodimeric constructions, the observed limping of kinesin suggests that in the stepping of kinesin, the two heads play different roles (which may not be evident for a wild-type kinesin walking under low external load), and the walking is intrinsically asymmetric.

In another experiment [58], the further depletion of the ATP hydrolysis activity of one of the two heads through mutation, so that the ATP hydrolysis by the mutated kinesin head is at least 700 times slower than that by the wild-type, does not demolish kinesin walking entirely. The speed of kinesin was reduced by a factor of ~9 as a result of this mutation. In addition, when the solution of nucleotides contains both ATP and AMPPNP, which does not hydrolyze, kinesin maintains processive walking. This observation is surprising and poses serious challenge to the current understanding of the hand-over-hand mechanism. One of the possible explanations of the above observation is that only one of the two heads play dominating roles in driving the motion and the other only plays assisting roles. In this picture, the two heads are intrinsically nonequivalent and it is inconsistent with the experimental observations that the two heads are equivalent. The other possibility is that during stepping, there exists some kind of rescue mechanism due to the cooperation between the two heads, namely, the mutant head regains some ATP hydrolysis capability when fused with another wild-type kinesin head. The third possibility is that in the walking of a wild-type kinesin, both heads generate power strokes for the motion of kinesin regardless of whether they are in the front or rear position. An earlier experiment showed that a single-headed kinesin fused with a different protein, which does not have ATP hydrolysis activity but does bind to microtubule, also walks processively along microtubule, although at a slower speed than the wild-type kinesin. This experimental observation may suggest that without a possible rescue mechanism, a kinesin with a single motor domain can still function.

As mentioned above, dynein is another family of motor proteins. Although dynein is a well-known force-generating element in cells and there has been significant progress in structural and mechanistic studies [18,59–62], the lack of atomic detailed structural information makes our understanding of its cargo transport mechanism lag behind that of kinesin and myosin. This relative slow advance is partly due to the large and complex structure of dynein. To maintain a complete molecular activity, the required size of this motor protein is about 380 kDa [60]; its total molecular weight is about 1.2 MDa [60]. Not surprisingly, most structural information of dynein comes from EM studies. Dynein has a motor domain formed by seven different globular subdomains, aggregating together to result in a ring-like architecture surrounding a central cavity [60–62]. Moreover, the dynein motor domain contains multiple nucleotide-binding sites, unlike that of myosin or kinesin. Six out of the seven globular domains are AAA-like (ATPase associated with diverse cellular activities) domains and are named AAA1 to AAA6. Four of the six AAA domains, AAA1 to AAA4, bind ATP with different binding affinities [61,63,64]. AAA1 appears to contain the active site for ATP hydrolysis, although AAA2 to AAA4 also have some catalytic activity [63,65], and experiments have indicated that AAA3 is essential for the dynein-induced microtubule gliding activity [60,66].

The other essential structural elements of dynein include the microtubule-binding stalk (MT stalk), an unusually long (13 nm) structure projected out of the ring-like motor domain, and the stem, also an elongated projection [60–62,65]. The MT stalk contains a small microtubule-binding domain and the stem regulates the binding of dynein with cargoes. MT stalk and the stem are considered potential lever arms [61,62]. In the EM image, the MT stalk and the stem point to different directions [61,62], depending on the power-stroke state. In both axonemal and cytoplasmic dynein, the ATP hydrolysis and product release processes are associated with a rotation of the ring of the motor domain around the motor–stem junction [4,60,65], which is considered to be an essential element in the power stroke of dynein. This motion is estimated to produce a movement of the motor domain along the microtubule of 15 nm [65], which is expected to be larger when the movement of the MT stalk is also taken into account [65].

The main chemomechanical difference between kinesin and dynein lies in their different responses to the external load and to the change of ambient ATP concentration. Firstly, the kinesin stall force has a much weaker dependence on the ATP concentration [11,13] than that of dynein [17]. With an ATP concentration change of 5 M to 1 mM, the stall force increases only by ~20% for kinesin, and the dynein stall force increases more than 4 times when ATP concentration changes from 0.1 to 1 mM. Secondly, the maximal stall force in the presence of saturating concentrations of ATP is significantly larger for kinesin (>8 pN) [11,13] than for dynein (about 1.1 pN) [17]. Thirdly, while kinesin always takes 8 nm steps under all forces smaller than the stall force, dynein's step size varies with the external load [5,17]. When the external force is large, its step size is 8 nm [14,15]. This step size increases with decreasing external force: When the external force vanishes, the step size of dynein is most likely to be 24 or 32 nm [17].

## 3.1  THEORETICAL MODEL

### 3.1.1  DESCRIPTION OF THE MODEL

In this section, we first summarize the basic models used [67–69] for the description of kinesin and dynein motions driven by ATP hydrolysis. For both kinesin and dynein, two-dimensional master equations are used to describe the motion of the protein motors, although the variables represented by the two dimensions are different. In the case of kinesin, these two variables are used to describe the positions of the two motor domains and, for dynein, to be able to treat the variable chemomechanical coupling, one variable is used to describe its physical motion along the microtubule and the other used to describe its chemical coordinate.

### 3.1.2  THE KINESIN MODEL

The step-by-step model for the hand-over-hand walking mechanism and the definition of various parameters, such as the neck-linker length and the variation of the step size during stepping, which leads to asymmetric walking, are given in Scheme 3.1. In

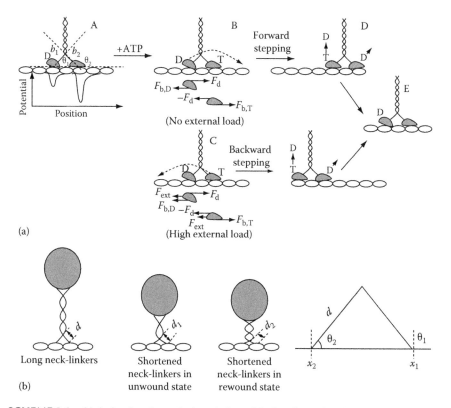

**SCHEME 3.1**  (a) A simple schematic description of the hand-over-hand mechanism of kinesin, and (b) the schematic depiction of the effects of the neck-linker length on the "length." (From Shao, Q. and Gao, Y.Q., *Proc. Natl. Acad. Sci. U. S. A.*, 103, 8072, 2006; Shao, Q. and Gao, Y.Q. *Biochemistry*, 46, 9098, 2007. With permission.)

short, the stepping of kinesin is described by the motion of its head positions on the microtubule. The two heads are connected through the neck-linker, which goes through unwinding and rewinding cycles when kinesin walks. The binding between the kinesin heads and the microtubule is modeled as a function of their chemical states, which determine the configuration of the head domains with respect to the neck-linkers. As a result, the binding of both kinesin heads are coupled. In this model, kinesin binds the microtubule with a nucleotide-state-dependent affinity in the following order: posthydrolytic ADP/P$_i$ state > the ATP state ≈ the empty state > the ADP state, consistent with the detachment force measurement [43,45]. The model assumes that the ATP (or AMPPNP) binding induces a conformational change of the front kinesin head and thus provides a driving force in the plus direction. It further assumes that the hydrolysis of ATP generates additional driving forces. Fluorescence polarization experiment suggested that the posthydrolysis state tends to be more rigid than the prehydrolysis state [70,71]. The prehydrolytic state, in accordance with experiments [70,71], is assumed to bind microtubule more weakly than the posthydrolytic state. To take into account the two-step force-generating mechanism by ATP binding and hydrolysis, four possible chemical states are included for each kinesin head (ATP, ADP/P$_i$, ADP, and empty). (In Ref. [68] on the wild-type kinesin, since ATP hydrolysis occurs fast, only three states were considered.) The ATP state is taken into account explicitly in the case of mutants where there is absence of hydrolysis of AMPPNP or ATP. The asymmetry in kinesin walking is speculated to be due to the unwinding and rewinding of the lower part of the central coiled-coil stalk during the stepping of the kinesin, leading to a change in the length of the neck-linkers. The change of the neck-linker length is included in the simple model by the change of the "leg length" parameter $d$ between 5.2 nm in the unwound and 4.7 nm in the rewound state of the coiled-coil of a wild-type kinesin [68]; resulting in a change of the lengths of the two neck-linkers summing up to ~1 nm, which is the approximate length of one repeat of the coiled-coil [72]. A similar model is used for a truncated kinesin. Its only difference from the wild-type kinesin is its shortened coiled-coil stalk and thus a shorter neck-linker [72]. In the model for the truncated kinesin, the value of $d$ (for both unwound and rewound states) is smaller than that for the wild-type kinesin (see Table 3.1 for the values of $d$).

Once the model is constructed, simple potential energy functions [68,69] are used to describe the interaction between each of kinesin heads and the microtubule binding sites as well as that between the two kinesin heads. The binding strengths used in our model are 9 kcal·mol$^{-1}$ for the ADP/P$_i$ state, 7 kcal·mol$^{-1}$ for the ATP or empty state, and 4 kcal·mol$^{-1}$ for the ADP state. These values were obtained by fitting to the measurement of detaching forces [43,45]. Given the chemical states of the two kinesin heads, the positions of the kinesin heads are then calculated using a master equation approach, including both diffusions of the motor heads along the microtubule and the chemical transitions of each head [68,69]:

$$x(t + \Delta t) - x(t) = (F_{x,i}/\zeta_{x,i})\Delta t + \Delta x(t). \tag{3.1}$$

## TABLE 3.1
### Parameter Values for the Kinesin Model

| Neck-length $d_1/d_2$ (nm) | Conventional kinesin | 5.2/4.7 (unwinding/ rewinding state) |
|---|---|---|
| | Kinesin with shorter neck-linkers | 4.6/4.1 (unwinding/ rewinding state) |
| Kinesin–microtubule binding | ATP-bound $V_{s,0}^{ATP}$ | 7.0 |
| affinity (kcal · mol$^{-1}$) | Empty $V_{s,0}^{E}$ | 7.0 |
| | ADP/P$_i$-bound $V_{s,0}^{ADP/P_i}$ | 9.0 |
| | ADP-bound $V_{s,0}^{ADP}$ | 4.0 |
| Chemical transition rate constant | ADP release (s$^{-1}$) | 260/2.6 (front/rear head) |
| | ATP binding ($\mu$M$^{-1}$ s$^{-1}$) | 3.0/0.3 (front/rear head) |
| | ATP hydrolysis (s$^{-1}$) | 8.0/800.0 (front/rear head) |
| | ATP release (s$^{-1}$) | 10.0 |
| The width of the binding potential $\alpha$ (nm) (Equation 3.1) | | 0.45/0.20 (forward/backward) |
| Parameters in Equation 3.2 | Internal force constant $K_b(k_BT/nm^2)$ | 0.047/0.04/0.031 (ADP · P$_i$/ ATP/empty state) |
| | $\Delta x_1(\Delta x_2)$ (nm) | −8.1 (ADP/P$_i$ and ATP state) 8.1 (empty state) |
| Overstretching potential $V_0$ (Equation 3.3) | | $10\ k_BT$ |
| Chemical transition constraint | Chemically allowed range within binding site (nm) | 0.16 |
| | Optimal neck-linker angle | 40° |
| | Chemically allowed angle deviation for ATP binding | ±5° |
| Diffusion constant, $D$ (nm$^2$ · s$^{-1}$) | | $2.0 \times 10^4$ |

The exact value of the stalk length is not used in this study. However, the stalk length change is the inverse of the neck-linker length change during the kinesin walking: the longer the neck-linker becomes, the shorter the stalk is.

In Equation 3.1, $\zeta_{x,i}$ is the friction constant for the motion along $x$. $\Delta x(t)$ is the random displacement due to the stochastic force and in the presence of white noise, it can be written as

$$\Delta x(t) = \sqrt{2D_{x,i}\Delta t}\,R_1$$

where the diffusion coefficient, $D = \zeta/k_BT$. In this equation, $k_B$ is the Boltzmann constant and $T$ is the temperature. $R_1$ is a random number with a normal Gaussian distribution. In Equation 3.1, $F_{x,i}$ is the force acting along the coordinate $x$. These forces can be further written as $F_{x,i} = -\partial V/\partial x - F_{ext}$. $V$ is the two-dimensional potential of mean force acting on the motor when it is at chemical state $i$ and $F_{ext}$ is the external load (e.g., the force applied through an optical trap). The parameters used for the wild-type and truncated kinesins are given in Table 3.1.

### 3.1.3   THE DYNEIN MODEL

To account for the variations in the chemomechanical coupling ratio, a loose chemomechanical coupling model was introduced for dynein [67], in which two reaction coordinates were employed to describe the configuration change and translocation of dynein, respectively. The first coordinate was used to describe the conformation of the dynein that controls the ATP hydrolysis reaction and the second coordinate simply represents the location of a dynein on a microtubule. The first coordinate was called the chemical reaction coordinate ($\alpha$) and the second the physical reaction coordinate ($x$). The details of the model include

1. One ATP molecule is hydrolyzed per cycle of reaction; a series of chemical states are defined according to the occupation of the active ATP hydrolysis site. For simplicity, only three chemical states are considered explicitly in the simple model: the ATP (ADP·P$_i$), ADP, and empty states. The binding of ATP to a motor domain detaches the stalk domain from microtubule [4,73]. ATP hydrolysis provides the energy for the conformational change in $\alpha$. Each chemical state prefers a different conformation of the nucleotide binding site including the P-loop region [64]. The chemical transitions occur in the following order: ATP binding and hydrolysis → P$_i$ release → ADP release. During this process of chemical energy release, dynein walks in the minus end along the microtubule. During the translocation of dynein, conformational changes leading ADP and P$_i$ release occur in the motor domain. After P$_i$ and ADP are released, dynein returns to its poststroke (ATP-free) state and rebinds microtubule. As a consequence, through this cycle of dynein conformational change (along the $\alpha$-coordinate), one ATP molecule is hydrolyzed and the dynein has displaced from its original position on the microtubule.

2. ATP hydrolysis is activated by dynein binding to microtubule, in accordance with the experimental observation that the presence of microtubule stimulates the ATPase activity of dynein [64]. The rates of other chemical transitions (binding of ATP and release of ADP·P$_i$) depend explicitly only on $\alpha$, so that the chemical reaction is not tightly coupled to the translocation of dynein. The transitions between chemical states are random processes and the probabilities for the occurrence of these transitions are determined by the rate constants of the transitions.

3. The binding sites of dynein on microtubule are distributed according to the microtubule lattice structure with a separation distance of 8.1 nm. The chemical transitions of dynein (e.g., binding of ATP) induce the conformational changes along the $\alpha$-coordinate, and the latter changes the binding affinity between the dynein and the microtubule. When dynein is in a poststroke state (e.g., the empty state), it binds preferably to one of the binding sites on the microtubule. The binding of ATP and subsequent ATP hydrolysis at the active site of dynein bound to microtubule, which transforms dynein into a prestroke state, induce the conformational change of the motor domain (the $\alpha$-coordinate). This change of dynein conformation

reduces its binding affinity with the microtubule and a power stroke is generated for dynein to displace in the minus direction of the microtubule (the comparison of structures obtained in the presence of ADP + vanadate and in the absence of nucleotide suggests that a power stroke may be produced when ADP and $P_i$ are released in the ATP hydrolysis cycle [1,4,62]). This driving force is represented by a harmonic term. The motions of dynein, in the direction of translocation along the microtubule and in conformational changes associated with chemical transitions, are again described by a two-dimensional Euler equation as given by Equation 3.1, but with one of the $x$ being $\alpha$.

## 3.2 CHEMOMECHANICAL COUPLING OF KINESIN AND DYNEIN

### 3.2.1 KINESIN

#### 3.2.1.1 Hand-Over-Hand Walking of Kinesin

Calculated trajectories for kinesin moving along microtubule in the absence of external forces are given in Figure 3.1, which show consecutive forward steps with two heads switching the leading position: Kinesin walks by a hand-over-hand mechanism. The calculated maximum speed of kinesin at saturating ATP concentrations is about 400 nm s$^{-1}$, with a Michaelis–Menten constant of 29 µM, in reasonable agreement with the experimental data. In the presence of a hindering force, kinesin is more likely to take backward steps (see Figure 3.1c for +10 pN) and finally sustained backward stepping with an 8 nm step size is observed at large hindering forces (>10 pN). The backward motion, which is very slow compared to the forward motion in the absence of external forces, also follows a hand-over-hand mechanism (Figure 3.1d).

The speed of kinesin calculated as a function of the external force showed that applied forces influence the kinesin motion in a rather complicated way (Figure 3.2, left panel): The speed of kinesin shows a maximum at an intermediate external load (~5 pN) and even larger external load slows down the kinesin walking speed, consistent with the experimental results of Carter and Cross [51]. A hindering force slows down the speed of kinesin in the force range of 0 to about 8 pN (for [ATP] = 1 mM), without changing the direction of the net motion and the stall force appears to be insensitive to the ATP concentration, which decreases from about 8 pN at 5 mM of ATP to 7 pN at 5 µM of ATP.

The ratio between forward and backward step numbers was calculated for external forces of different magnitude. The change of ATP concentration has a small influence on the calculated ratio, which decreases monotonically with the external force. At the stall force, the ratio becomes unity, resulting in a zero speed of kinesin. These results are in good agreement with the experiments in that the forward/backward ratio is largely linearly dependent on the external force. Calculations were also performed to study the force dependence of the dwell time between individual steps for both forward and backward motions at ATP concentrations of 1 mM and 10 µM (Figure 3.3). In general, the dwell time is smaller for 1 mM of ATP than for 10 µM of ATP. At the lower concentration, ATP binding is rate-limiting.

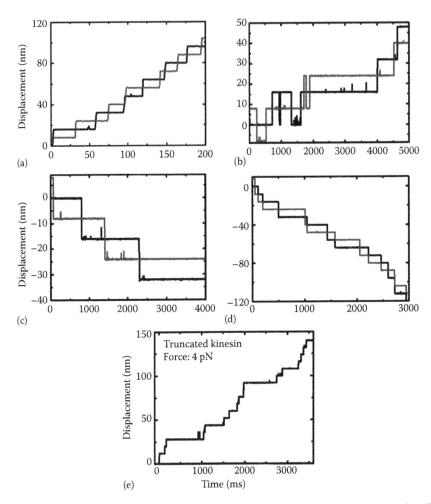

**FIGURE 3.1** Representative trajectories for the wild-type kinesin in the presence of (a) 0; (b) 7; (c) 10 and (d) 18 pN of external forces, and (e) for the truncated kinesin in the presence of 4 pN. (From Shao, Q. and Gao, Y.Q., *Proc. Natl. Acad. Sci. U. S. A.,* 103, 8072, 2006; Shao, Q. and Gao, Y.Q., *Biochemistry,* 46, 9098, 2007. With permission.)

In the force range of −15 to 5 pN, the dwell time is insensitive to the external force. When the external force increases from 5 to 10 pN, a sharp increase of the dwell time from ~0.05 to ~1 s for [ATP] = 1 mM and from ~0.1 to ~1 s for [ATP] = 10 μM was observed, in agreement with experimental data [51].

## 3.2.2 LIMPING OF KINESIN

From Figure 3.1a through d, it is observed that the wild-type kinesin walks in a symmetric way, in that the dwell time is invariant with respect to switching the leading positions of the two heads. In contrast, the mutant kinesin was shown

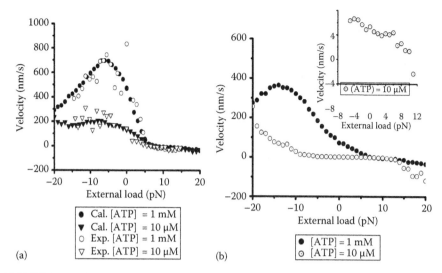

**FIGURE 3.2**    The force dependence of the speed of (a) wild type and (b) truncated kinesins. (From Shao, Q. and Gao, Y.Q., *Proc. Natl. Acad. Sci. U. S. A.* 103, 8072, 2006; Shao, Q. and Gao, Y.Q., *Biochemistry,* 46, 9098, 2007. With permission.)

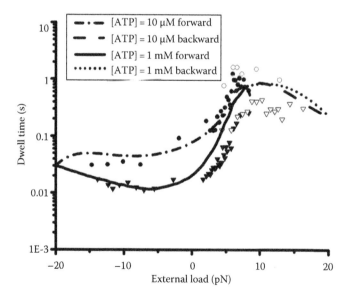

**FIGURE 3.3**    The forward/backward dwell times as a function of external load. (From Shao, Q. and Gao, Y.Q., *Proc. Natl. Acad. Sci. U. S. A.*, 103, 8072, 2006. With permission.)

to limp, showing alternative long and short dwell times during kinesin stepping, although processive motion is also observed (Figure 3.1e). The limping factor $L$ is defined as the ratio between the successive long and short dwell times [72]. In the present model, the mutant kinesin is modeled by a shortened neck-linker. The calculations showed that the limping factor increases largely with the shortening of the "leg (neck-linker) length," with mean long dwell time increasing but the mean short dwell time remaining invariant, consistent with experimental observations.

Asymmetry in kinesin walking is also shown by a mutant kinesin of which only one head has hydrolysis activity. Our calculations showed the maximum speed of such a mutant, walking by a hand-over-hand fashion, is about nearly sixfold smaller than that of wild-type. Limping is obvious for such a mutant (Figure 3.4a). A similar limping mechanism is observed for the walking of wild-type kinesin modeled in the presence of both ATP and AMPPNP. In the presence of the hindering force (Figure 3.4b), kinesin frequently takes long pauses due to the binding of AMPPNP at the rear head. These long pauses are normally terminated by a quick backward step, which is followed by either another segment of long pause or by forward steps. The average waiting time before those backward steps is ~3 s. On the contrary, the long pause during the processive motion of the kinesin appears rarely at small external forces (Figure 3.4c). These calculated results are again consistent with experimental results [58].

### 3.2.3 MOTION OF DYNEIN AND EFFECTS OF THE EXTERNAL FORCE

Figure 3.5a shows a typical trajectory for a dynein moving along the microtubule track in the absence of an external load. The speed of dynein as a function of ATP

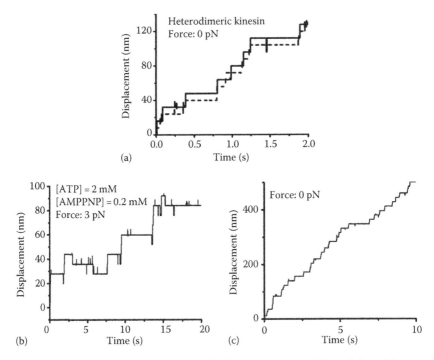

**FIGURE 3.4** Representative trajectories for hetero-constructed (a) and for wild type of kinesin in the mixture of ATP and AMPPNP at 3 pN (b) and 0 pN (c) of external forces. (From Shao, Q. and Gao, Y.Q., *Biochemistry*. 46, 9098, 2007. With permission.)

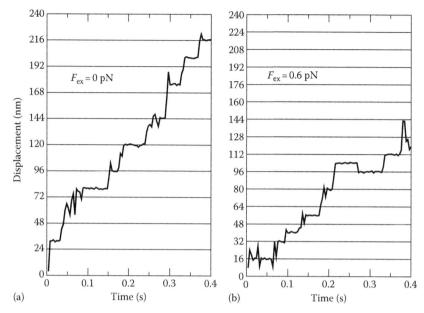

**FIGURE 3.5** Calculated traces for the translocation of dynein along microtubule in the presence of constant external load. [ATP]= 1 mM, and (a) $F_{ex} = 0$ pN, (b) $F_{ex} = 0.6$ pN. (From Gao, Y.Q., *Biophys. J.*, 90, 811, 2006. With permission.)

concentration was calculated and fitted to a simple Michaelis–Menten kinetics, which yielded a maximum speed, $V_{cat}$, of 0.51 $\mu$m·s$^{-1}$ and a $K_M$ of 38 $\mu$M. Both values are in reasonable agreement with the experimental results obtained for both cytoplasm dynein [64,74] and the monomeric dynein a [75]. It is also seen from the trajectory that in the absence of external force, dynein takes large step sizes (16, 24, or 32 nm), consistent with experimental observations.

The effects of the external load on the translocation of dynein were studied using two types of calculations. In the first one, a force term, $-kx$ (where $k$ is a force constant), is added to $F_{x,i}$ in Equation 3.1 to mimic the experimental forces applied through an optical trap. These simulations again showed that dynein takes large step sizes (24 or 32 nm) when the displacement is small (which corresponds to a small external load). With dynein displacing further away (increasing $x$), the force increases and step size decreases. When the external force is large enough, dynein stops moving forward. Occasional backward steps also occur under these conditions. It is interesting to point out that the calculated stall force (the force at which dynein stops moving forward) in the presence of 1 mM ATP is about 1 pN (displacement of ~80 nm) and it decreases with decreasing ATP concentration, again in agreement with experimental observations [17]. In a second type of simulation, a constant external force was applied. Not surprisingly, the walking speed of dynein decreases with increasing external load. The ATP concentration dependence of stall forces was also calculated by extrapolating the force dependence of walking speed. These simulations yielded a stall force 1.1 pN at high concentrations of ATP (5 mM), consistent with the first-type simulations, and about 0.4 pN at low concentrations of ATP (~0.1 mM). The stall force decreases approximately linearly with ATP concentration in the range of 0.1–1 mM, consistent with the experimental results, although as mentioned above, the experiments were not performed with a constant force [17].

The most interesting characteristic in the force dependence of dynein motion is the change of step size with the external force. As seen from the exemplary trajectories shown in Figure 3.5, the increase of the external force from 0 to 0.6 pN not only decreases the speed of motion, but also decreases the sizes of the steps from ~32 to ~16 nm. This dependence of step size on external force is more clearly seen in the pair wise distribution functions given in Figure 3.6. In this figure, when the external load vanishes, the highest peaks occur in multiples of 24 nm, indicating that the most abundant steps at low external loads are of the size of 24 nm. When the external force increases, the distribution shifts toward smaller values. When the external force is 0.6 pN, the highest peaks occur in multiples of 16 nm, whereas the count of 24 nm steps becomes significantly less, indicating that the 16 nm steps replace 24 nm steps as the most abundant ones. When the external force is large (close to the maximum stall force of 1.1 pN), dynein mainly takes 8 nm steps. These observations are consistent with experiments. These results can be easily understood [67] by considering the nature of the loose coupling model. In the presence of a constant external force $F_{ext}$, the assumed harmonic potential energy profile for the translocation along microtubule becomes

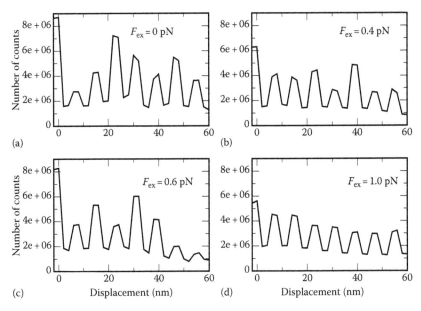

**FIGURE 3.6**  Pair wise distribution functions calculated in the presence of an external load of the magnitude of (a) 0 pN, (b) 0.4 pN, (c) 0.6 pN, and (d) 1.0 pN. The ATP concentration is 1 mM for (a), (b) and (c), and is 5 mM for (d). (From Gao, Y.Q., *Biophys. J.*, 90, 811, 2006. With permission.)

$$V_{x,\text{ATP}}(x) = \frac{K_{\text{PS}}}{2}(x - x_{\text{d}} - d_{\text{max}})^2 + F_{\text{ext}}x. \quad (3.2)$$

The minima of such a potential is located at $x_{\text{min}}$, where $K_{\text{PS}}$ is a force constant, $x$ is the position of the center of mass of dynein, $x_{\text{d}}$ is the coordinate of the dynein binding sites along microtubule, and $d_{\text{max}}$ is the maximum step length (30 nm) that a dynein can achieve.

$$x_{\text{min}} = d_{\text{max}} + x_{\text{d}} - \frac{F_{\text{ext}}}{K_{\text{PS}}}. \quad (3.3)$$

The average step size under this condition is then

$$x_{\text{min}} - x_{\text{d}} = d_{\text{max}} - \frac{F_{\text{ext}}}{K_{\text{PS}}}, \quad (3.4)$$

which decreases linearly with the applied force. Due to the periodicity of the dynein–microtubule interaction, the step size of dynein takes values of multiples of 8.1 nm. Using $d_{\text{max}} \approx 30$ nm and $K_{\text{PS}} \approx 0.04$ pN nm$^{-1}$, the average step size can be calculated: the dominant step size is 32 nm when the external force is between 0 and 0.08 pN and becomes 24 nm for $F_{\text{ext}}$ being 0.08 to 0.4 pN, and further decreases to 16 nm when 0.4 pN $< F_{\text{ext}} <$ 0.72 pN. It finally reduces to 8 nm when $F_{\text{ext}} > 0.72$ pN. This simple estimation is in excellent agreement with experiments [17].

## 3.3 DISCUSSION

The thermal ratchet model has been used in a number of studies for motor proteins. (The essential elements of the thermal ratchet model modes include driving forces provided by ATP hydrolysis and the thermal motions induced by the environment and we do not distinguish between thermal ratchet and power stroke models here. As pointed out in Ref. [76], the difference between these two models is not unambiguously defined.) These studies have shown that with a careful choice of parameters and by incorporating the experimentally determined structural and kinetics information, the general mechanical properties of various motor proteins can be calculated. In most cases, these calculations allow direct and favorable comparison with single molecule mechanical measurements. In this chapter, we have summarized our recent efforts in making use of the thermal ratchet model to study the external load dependence of the stepping of kinesin and dynein. The two motors behave drastically differently in their responses to the external force: whereas kinesin exhibits a tight chemomechanical coupling with a constant coupling ratio, dynein changes gears in response to the external load.

The average number of ATP hydrolyzed for every kinesin step (backward or forward) was calculated as a function of external force and the results showed the chemomechanical coupling ratio to be close to unity in a large range of external forces. The calculations also showed that the range of the forces that the tight coupling sustains increases with ATP concentration (−15 to 6 pN for [ATP] = 1 mM and −6 to 4 pN for [ATP] = 10 μM). On the other hand, a large negative force may cause kinesin to take more than one step per ATP hydrolyzed when the ATP concentration is low, corresponding to a forced sliding. The force dependence of chemomechanical coupling ratio at large positive forces show a more complicated behavior in the region of forward/backward stepping transition: Near the stall force, the backward and forward stepping become loosely coupled to ATP hydrolysis, with more than one ATP consumed for a successful step. Taking into account that the net motion is close to zero in this range of force, this loose coupling is due to the balance between the force produced by kinesin and external load, which leads to a quick forward/backward of one head without moving the other (see Scheme 3.1a). However, when the force further increases, the chemomechanical coupling ratio approaches unity again, indicating the recovery of a tight coupling during the consecutive backward motion in the range of external load. This ratio becomes less than 1 at even larger hindering forces (>15 pN), and slippage to the minus-end of microtubule occurs due to the large force.

Very different from what was observed for kinesin, dynein shows a variable step size during its translocation along the microtubule. The calculations are consistent with a scenario that the consumption of one ATP molecule allows the motor protein to translocate a distance that is dependent on the external load. The essential feature of the dynein model is loose coupling between the chemical reaction and the physical translocation. This loose coupling is modeled by invoking two different reaction coordinates; one coordinate is used to describe the displacement of dynein on the microtubule and the other coordinate represents the conformational change controlling the chemical transition rates. The chemical energy release through ATP hydrolysis is achieved through several chemical transitions including ATP binding, hydrolysis, and ADP, $P_i$ release. The chemical energy drives the conformational changes of dynein in both coordinates mentioned above. The two reaction coordinates are largely uncoupled. However, the conformation of dynein motor domain does have an effect on the binding affinity of dynein to microtubule. Since the force applied against the motion of dynein changes the potential energy surface and slows down the translocation of dynein along the microtubule but does not directly affect the conformational change, the relative speed of the two types of motions vary with the external load. When the external force is small, dynein moves a long distance on microtubule before its conformational change permits it to rebind microtubule. On the other hand, a large external load slows down its translocation on the microtubule. Therefore, a smaller step size over one cycle of conformational change (as well as the hydrolysis of one ATP) is observed. Another possible mechanism for the control of the loose or tight chemomechanical coupling could also arise from the coupling between the heads of the protein motors. It was shown in recent publications [77,78] that the strain between the two legs of kinesin is important for the successful movement. In our simple kinetic model, the strain is not considered explicitly but through the tilting angle dependence of the ATP hydrolysis rate.

A detailed molecular and mode analysis such as that performed for the myosin motors [79–81] is expected to provide further information on the coupling and strain between the two motor heads.

A variable chemomechanical coupling mechanism of dynein may optimize its thermodynamic efficiency in its cargo transportation in cells [5,17,67], which was analyzed in Ref. [67] and is summarized in the following. The thermodynamic efficiency of a dynein can be calculated as

$$\eta = \frac{F_{ext}x}{n\Delta G},$$

where

x is the step length of dynein
n is the average number of ATP molecules hydrolyzed per step taken
$\Delta G$ is the chemical energy available from the hydrolysis of an ATP molecule

Under the condition that the hydrolysis of every ATP leads to a single step, $n = 1$, making use of Equation 3.4, $\eta$ can be expressed as

$$\eta = \frac{1}{\Delta G}\left(F_{ext}d_{max} - \frac{F_{ext}^2}{K_{PS}}\right). \tag{3.5}$$

In contrast, the thermodynamic efficiency of a kinesin can be written as

$$\eta = \frac{F_{ext}d}{\Delta G}$$

and $d = 8.1\,nm$. Since $d_{max} > d$, the thermodynamic efficiency is obviously larger for a dynein than for a kinesin under small external loads. The thermodynamic efficiencies of dynein and kinesin are compared in Figure 3.7c for different values of external loads.

As mentioned above, the noticeable feature of the dynein thermodynamic efficiency is that it is significantly larger than that of kinesin in the presence of smaller external load and is approximately constant when the external force is in the range of 0.2–0.8 pN. Although the maximum thermodynamic efficiency that a dynein can achieve is much smaller than that achievable by a kinesin, the larger steps that dynein takes in the presence of small external load makes dynein more efficient under these conditions. However, a tight coupling mechanism has its own advantage, in particular when motors walk against large external forces. As seen from Figure 3.1a through d, this tight coupling makes kinesin a persistent and tough motor over a large range of forces, even preventing its fast slipping (backward) under loads beyond stall forces. The difference between kinesin and dynein may indicate the different tasks and working conditions in the biological system.

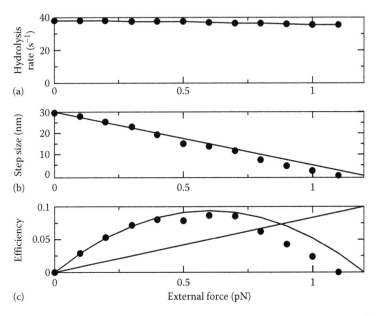

**FIGURE 3.7**    (a) Rate of ATP hydrolysis as a function of external load, (b) Filled circles: Step sizes of dynein calculated as the ratio between the speed of dynein and the ATP hydrolysis rate. Line: Average step size of dynein calculated using Equation 3.4. (c) Filled circles: Calculated thermodynamic efficiency using data points given in (b). Curved line: Calculated thermodynamic efficiency using Equation 3.5. Straight line: Thermodynamic efficiency of a tight coupling motor (such as kinesin) with a step size of 8.1 nm. The thermodynamic curve is only shown in the low external force for kinesin, since the chemomechanical coupling ratio is not determined at high forces, although the forward/backward ratios have been determined. The thermodynamic efficiency should also take a bell shape and vanishes at its stall force (~8 pN). (From Gao, Y.Q., *Biophys. J.*, 90, 811, 2006. With permission.)

## ACKNOWLEDGMENT

The authors acknowledge the support of the ACS-PRF, the Camile and Henry Dreyfus Foundation, and the Searle Scholar Program for the support of this research.

## REFERENCES

1. Burgess, S. A. and Knight, P. J. (2004) Is the dynein motor a winch?, *Curr. Opin. Struct. Biol. 14*, 138–146.
2. Alberts, B. (1998) The cell as a collection of protein machines: Preparing the next generation of molecular biologists, *Cell 92*, 291–294.
3. Vale, R. D. (2003) The molecular motor toolbox for intracellular transport, *Cell 112*, 467–480.
4. Oiwa, K. and Sakakibara, H. (2005) Recent progress in dynein structure and mechanism, *Curr. Opin. Cell Biol. 17*, 98–103.
5. Karplus, M. and Gao, Y. Q. (2004) Biomolecular motors: The F-1-ATPase paradigm, *Curr. Opin. Struct. Biol. 14*, 250–259.

6. Yildiz, A., Tomishige, M., Vale, R. D., and Selvin, P. R. (2004) Kinesin walks hand-over-hand, *Science 303*, 676–678.

7. Yildiz, A., Forkey, J. N., McKinney, S. A., Ha, T., Goldman, Y. E., and Selvin, P. R. (2003) Myosin V walks hand-over-hand: Single fluorophore imaging with 1.5 nm localization, *Science 300*, 2061–2065.

8. Block, S. M., Schnitzer, M. J., and Visscher, K. (2000) Force production by single kinesin motors, *Mol. Biol. Cell 11*, 426a.

9. Schnitzer, M. J., Visscher, K., and Block, S. M. (2000) Force production by single kinesin motors, *Nat. Cell Biol. 2*, 718–723.

10. Visscher, K., Schnitzer, M. J., and Block, S. M. (1999) Mechanochemical coupling in kinesin under load, *Biophys. J. 76*, A43.

11. Visscher, K., Schnitzer, M. J., and Block, S. M. (1999) Single kinesin molecules studied with a molecular force clamp, *Nature 400*, 184–189.

12. Nishiyama, M., Higuchi, H., Ishii, Y., Taniguchi, Y., and Yanagida, T. (2003) Single molecule processes on the stepwise movement of ATP-driven molecular motors, *Biosystems 71*, 145–156.

13. Nishiyama, M., Higuchi, H., and Yanagida, T. (2002) Chemomechanical coupling of the forward and backward steps of single kinesin molecules, *Nat. Cell Biol. 4*, 790–797.

14. Sakakibara, H., Kojima, H., Sakai, Y., Katayama, E., and Oiwa, K. (1999) Inner-arm dynein c of *Chlamydomonas flagella* is a single-headed processive motor, *Nature 400*, 586–590.

15. Hirakawa, E., Higuchi, H., and Toyoshima, Y. Y. (2000) Processive movement of single 22S dynein molecules occurs only at low ATP concentrations, *Proc. Natl. Acad. Sci. U. S. A. 97*, 2533–2537.

16. Iyadurai, S. J., Li, M. G., Gilbert, S. P., and Hays, T. S. (1999) Evidence for cooperative interactions between the two motor domains of cytoplasmic dynein, *Curr. Biol. 9*, 771–774.

17. Mallik, R., Carter, B. C., Lex, S. A., King, S. J., and Gross, S. P. (2004) Cytoplasmic dynein functions as a gear in response to load, *Nature 427*, 649–652.

18. Cross, R. A. (2004) Molecular motors: Dynein's gearbox, *Curr. Biol. 14*, R355–R356.

19. Gennerich, A., Carter, A. P., Reck-Peterson, S. L., and Vale, R. D. (2007) Force-induced bidirectional stepping of cytoplasmic dynein, *Cell 131*, 952–965.

20. Reck-Peterson, S. L., Yildiz, A., Carter, A. P., Gennerich, A., Zhang, N., and Vale, R. D. (2006) Single-molecule analysis of dynein processivity and stepping behavior, *Cell 126*, 335–348.

21. Nan, X., Sims, P. A., and Xie, X. S. (2007) Variable step sizes of dynein in vivo revealed by particle tracking at microsecond time resolution with dark field microscopy, *Biophys. J.*, 202a.

22. Nan, X. L., Sims, P. A., Chen, P., and Xie, X. S. (2005) Observation of individual microtubule motor steps in living cells with endocytosed quantum dots, *J. Phys. Chem. B 109*, 24220–24224.

23. Fisher, M. E. and Kolomeisky, A. B. (2001) Simple mechanochemistry describes the dynamics of kinesin molecules, *Proc. Natl. Acad. Sci. U. S. A. 98*, 7748–7753.

24. Kolomeisky, A. B., Stukalin, E. B., and Popov, A. A. (2005) Understanding mechanochemical coupling in kinesins using first-passage-time processes, *Phys. Rev. E 71*, 031902.

25. Astumian, R. D. (2000) The role of thermal activation in motion and force generation by molecular motors, *Phil. Trans. R. Soc. Lond. Biol. Sci. 355*, 511–522.

26. Peskin, C. S. and Oster, G. (1995) Coordinated hydrolysis explains the mechanical-behavior of kinesin, *Biophys. J. 68*, S202–S211.

27. Thomas, N., Imafuku, Y., Kamiya, T., and Tawada, K. (2002) Kinesin: A molecular motor with a spring in its step, *Proc. R. Soc. Lond. Biol. Sci. 269*, 2363–2371.

28. Hirokawa, N. (1998) Kinesin and dynein superfamily proteins and the mechanism of organelle transport, *Science 279*, 519–526.

29. Yildiz, A. and Selvin, P. L. (2005) Kinesin: Walking, crawling or sliding along? *Trends Cell Biol. 15*, 112–120.

30. Asbury, C. L. (2005) Kinesin: World's tiniest biped, *Curr. Opin. Cell Biol. 17*, 89–97.

31. Cross, R. A. (2004) The kinetic mechanism of kinesin, *Trends Biochem. Sci. 29*, 301–309.

32. Vale, R. D. (2003) The molecular motor toolbox for intracellular transport, *Cell 112*, 467–480.

33. Cross, R. A., Crevel, I., Carter, N. J., Alonso, M. C., Hirose, K., and Amos, L. A. (2000) The conformational cycle of kinesin, *Phil. Trans. R. Soc. Lond. Biol. Sci. 355*, 459–464.

34. Rice, S., Lin, A. W., Safer, D., Hart, C. L., Naber, N., Carragher, B. O., Cain, S. M., Pechatnikova, E., Wilson-Kubalek, E. M., Whittaker, M., Pate, E., Cooke, R., Taylor, E. W., Milligan, R. A., and Vale, R. D. (1999) A structural change in the kinesin motor protein that drives motility, *Nature 402*, 778–784.

35. Hirose, K., Lockhart, A., Cross, R. A., and Amos, L. A. (1995) Nucleotide-dependent angular change in kinesin motor domain bound to tubulin, *Nature 376*, 277–279.

36. Yun, M., Bronner, C. E., and Park, C. G. (2003) Rotation of the stalk/neck and one head in a new crystal structure of the kinesin motor protein, Ncd, *EMBO J. 22*, 5382–5389.

37. Asenjo, A. B., Krohn, N., and Sosa, H. (2003) Configuration of the two kinesin motor domains during ATP hydrolysis, *Nat. Struct. Biol. 10*, 836–842.

38. Kull, F. J. and Endow, S. A. (2002) Kinesin: Switch I & II and the motor mechanism, *J. Cell Sci. 115*, 15–23.

39. Hackney, D. D. (1988) Kinesin ATPase—rate-limiting ADP release, *Proc. Natl. Acad. Sci. U. S. A. 85*, 6314–6318.

40. Ma, Y. Z. and Taylor, E. W. (1997) Kinetic mechanism of a monomeric kinesin construct, *J. Biol. Chem. 272*, 717–723.

41. Lockhart, A., Cross, R. A., and McKillop, D. F. (1995) ADP release is the rate-limiting step of the MT activated ATPase of non-claret disjunctional and kinesin, *FEBS Lett. 368*, 531–535.

42. Xing, J., Wriggers, W., Jefferson, G. M., Stein, R., Cheung, H. C., and Rosenfeld, S. S. (2000) Kinesin has three nucleotide-dependent conformations—Implications for strain-dependent release, *J. Biol. Chem. 275*, 25413–25423.

43. Kawaguchi, I. and Ishiwata, S. (2001) Nucleotide-dependent single- to double-headed binding of kinesin, *Science 291*, 667–669.

44. Kawaguchi, K., Uemura, S., and Ishiwata, S. (2003) Equilibrium and transition between single- and double-headed binding of kinesin as revealed by single-molecule mechanics, *Biophys. J. 84*, 1103–1113.

45. Uemura, S., Kawaguchi, K., Yajima, J., Edamatsu, M., Toyoshima, Y. Y., and Ishiwata, S. (2002) Kinesin-microtubule binding depends on both nucleotide state and loading direction, *Proc. Natl. Acad. Sci. U. S. A. 99*, 5977–5981.

46. Kozielski, F., Sack, S., Marx, A., Thormahlen, M., Schonbrunn, E., Biou, V., Thompson, A., Mandelkow, E. M., and Mandelkow, E. (1997) The crystal structure of dimeric kinesin and implications for microtubule-dependent motility, *Cell 91*, 985–994.

47. Coy, D. L., Wagenbach, M., and Howard, J. (1999) Kinesin takes one 8 nm step for each ATP that it hydrolyzes, *J. Biol. Chem. 274*, 3667–3671.

48. Schnitzer, M. J. and Block, S. M. (1997) Kinesin hydrolyses one ATP per 8 nm step, *Nature 388*, 386–390.

49. Schnitzer, M. J., Visscher, K., and Block, S. M. (2000) Force production by single kinesin motors, *Nat. Cell Biol. 2*, 718–723.

50. Nishiyama, M., Higuchi, H., and Yanagida, T. (2002) Chemomechanical coupling of the forward and backward steps of single kinesin molecules, *Nat. Cell Biol. 4*, 790–797.
51. Carter, N. J. and Cross, R. A. (2005) Mechanics of the kinesin step, *Nature 435*, 308–312.
52. Coppin, C. M., Finer, J. T., Spudich, J. A., and Vale, R. D. (1996) Detection of sub-8-nm movements of kinesin by high-resolution optical-trap microscopy, *Proc. Natl. Acad. Sci. U. S. A. 93*, 1913–1917.
53. Nishiyama, M., Muto, E., Inoue, Y., Yanagida, T., and Higuchi, H. (2001) Substeps within the 8-nm step of the ATPase cycle of single kinesin molecules, *Nat. Cell Biol. 3*, 425–428.
54. Yildiz, A., Tomishige, M., Vale, R. D., and Selvin, P. R. (2004) Kinesin walks hand-over-hand, *Science 303*, 676–678.
55. Hua, W., Chung, J., and Gelles, J. (2002) Distinguishing inchworm and hand-over-hand processive kinesin movement by neck rotation measurements, *Science 295*, 844–848.
56. Asbury, C. L., Fehr, A. N., and Block, S. M. (2003) Kinesin moves by an asymmetric hand-over-hand mechanism, *Science 302*, 2130–2134.
57. Higuchi, H., Bronner, C. E., Park, H. W., and Endow, S. A. (2004) Rapid double 8-nm steps by a kinesin mutant, *EMBO J. 23*, 2993–2999.
58. Guydosh, N. R. and Block, S. M. (2006) Backsteps induced by nucleotide analogs suggest the front head of kinesin is gated by strain, *Proc. Natl. Acad. Sci. U. S. A. 103*, 8054–8059.
59. Koonce, M. P. and Samso, M. (2004) Of rings and levers: The dynein motor comes of age, *Trends Cell Biol. 14*, 612–619.
60. Samso, M. and Koonce, M. P. (2004) 25 angstrom resolution structure of a cytoplasmic dynein motor reveals a seven-member planar ring, *J. Mol. Biol. 340*, 1059–1072.
61. Burgess, S. A., Walker, M. L., Sakakibara, H., Knight, P. J., and Oiwa, K. (2003) Dynein structure and power stroke, *Nature 421*, 715–718.
62. Burgess, S. A., Walker, M. L., Sakakibara, H., Oiwa, K., and Knight, P. J. (2004) The structure of dynein-c by negative stain electron microscopy, *J. Struct. Biol. 146*, 205–216.
63. Takahashi, Y., Edamatsu, M., and Toyoshima, Y. Y. (2004) Multiple ATP-hydrolyzing sites that potentially function in cytoplasmic dynein, *Proc. Natl. Acad. Sci. U. S. A. 101*, 12865–12869.
64. Kon, T., Nishiura, M., Ohkura, R., Toyoshima, Y. Y., and Sutoh, K. (2004) Distinct functions of nucleotide-binding/hydrolysis sites in the four AAA modules of cytoplasmic dynein, *Biochemistry 43*, 11266–11274.
65. Mallik, R. and Gross, S. P. (2004) Molecular motors: Strategies to get along, *Curr. Biol. 14*, R971–R982.
66. Kikushima, K., Yagi, T., and Kamiya, R. (2004) Slow ADP-dependent acceleration of microtubule translocation produced by an axonemal dynein, *FEBS Lett. 563*, 119–122.
67. Gao, Y. Q. (2006) Simple theoretical model explains dynein's response to load, *Biophys. J. 90*, 811–821.
68. Shao, Q. and Gao, Y. Q. (2006) On the hand-over-hand mechanism of kinesin, *Proc. Natl. Acad. Sci. U. S. A. 103*, 8072–8077.
69. Shao, Q. and Gao, Y. Q. (2007) Asymmetry in kinesin walking, *Biochemistry 46*, 9098–9106.
70. Asenjo, A. B., Krohn, N., and Sosa, H. (2003) Configuration of the two kinesin motor domains during ATP hydrolysis, *Nat. Struct. Biol. 10*, 836–842.
71. Asenjo, A. B., Weinberg, Y., and Sosa, H. (2006) Nucleotide binding and hydrolysis induces a disorder-order transition in the kinesin neck-linker region, *Nat. Struct. Mol. Biol. 13*, 648–654.

72. Asbury, C. L., Fehr, A. N., and Block, S. M. (2003) Kinesin moves by an asymmetric hand-over-hand mechanism, *Science 302*, 2130–2134.

73. Inaba, K. (1995) Atp-dependent conformational changes of dynein—Evidence for changes in the interaction of dynein heavy-chain with the intermediate chain-1, *J. Biochem. Tokyo 117*, 903–907.

74. Toba, S. and Toyoshima, Y. Y. (2004) Dissociation of double-headed cytoplasmic dynein into single-headed species and its motile properties, *Cell Motil. Cytoskel. 58*, 281–289.

75. Shiroguchi, K. and Toyoshima, Y. Y. (2001) Regulation of monomeric dynein activity by ATP and ADP concentrations, *Cell Motil. Cytoskel. 49*, 189–199.

76. Block, S. M. (2007) Kinesin motor mechanics: Binding, stepping, tracking, gating, and limping, *Biophys. J. 92*, 2986–2995.

77. Hyeon, C. and Onuchic, J. N. (2007) Mechanical control of the directional stepping dynamics of the kinesin motor, *Proc. Natl. Acad. Sci. U. S. A. 104*, 17382–17387.

78. Hyeon, C. and Onuchic, J. N. (2007) Internal strain regulates the nucleotide binding site of the kinesin leading head, *Proc. Natl. Acad. Sci. U. S. A. 104*, 2175–2180.

79. Yu, H. B., Ma, L., Yang, Y., and Cui, Q. (2007) Mechanochemical coupling in the myosin motor domain. I. Insights from equilibrium active-site simulations, *Plos Comput. Biol. 3*, 199–213.

80. Yu, H., Ma, L., Yang, Y., and Cui, Q. (2007) Mechanochemical coupling in the myosin motor domain. II. Analysis of critical residues, *Plos Comput. Biol. 3*, 214–230.

81. Li, G. H. and Cui, Q. (2004) Analysis of functional motions in Brownian molecular machines with an efficient block normal mode approach: Myosin-II and $Ca^{2+}$-ATPase, *Biophys. J. 86*, 743–763.

# 4 Electron Transfer Reactions Coupled with Proton Translocation: Cytochrome Oxidase, Proton Pumps, and Biological Energy Transduction

*Alexei Stuchebrukhov*

## CONTENTS

The correlated transport of electrons and protons in cytochrome *c* oxidase (CcO), the terminal enzyme in the respiratory electron transport chain of aerobic organisms, is discussed. This enzyme catalyzes the reduction of atmospheric oxygen to water, and converts the free energy of oxygen reduction into the membrane proton gradient by pumping protons across the membrane. The proton gradient subsequently drives the synthesis of ATP. This enzyme is a quantum mechanical energy-converting machine, for the electron transport in it is solely due to quantum tunneling. Computer simulations and theoretical modeling point to a possible molecular mechanism of this redox-driven proton pump. Recent progress in this area is discussed.

## 4.1   ENERGY TRANSFORMATIONS IN AEROBIC CELLS

Mitochondria generate most of the energy in animal cells. This occurs primarily through oxidative phosphorylation, a process in which electrons are passed along a series of redox enzymes located in the inner mitochondrial membrane toward molecular oxygen, which is reduced with the addition of electrons and protons to water. The electron transport chain consists of three major enzyme complexes: NADH dehydrogenase, bc1, and CcO. Each of the complexes contains several metal redox sites, which form a continuous chain along which electrons eventually get to molecular oxygen. The final complex in the chain is CcO. Each of these complexes works as a proton pump, which utilizes the free energy of oxygen reduction to transfer protons from the mitochondrion matrix surrounded by the inner mitochondrion membrane to cytosol. The free energy stored in proton gradient created in the process is later utilized by ATP synthase to make ATP (Figure 4.1). The same basic principles work in some bacteria and plant mitochondria [1,2] and all other aerobic organisms.

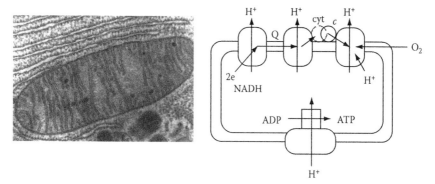

**FIGURE 4.1** Schematics of the mitochondrion. Electron transport chain, proton pumps, and ATP synthase are components of oxidative phosphorylation. Major complexes of electron transport chain are NADH dehydrogenase, bc1 complex, and cytochrome $c$ oxidase. The quinone, Q, and cytochrome $c$ are intermediates that shuttle electrons between the complexes. Each of the three complexes is a proton pump. Cytochrome $c$ oxidase is the terminal complex of the electron transport chain.

The dysfunction of electron transport chain is associated with a number of human mitochondrial diseases [3]. Also, an incorrect or incomplete reduction of oxygen that occasionally happens in the process of oxidative phosphorylation gives rise to the production of oxygen free radicals or reactive oxygen species (ROS), which have been recently linked to cell aging, apoptosis, and a number of degenerative diseases of heart and brain in humans [4]. These findings underscore the importance of the research efforts to characterize quantitatively the whole electron transport chain and to gain an understanding of the mechanisms of proton pumping, oxygen reduction, and ATP production. The knowledge gained in these studies may open up in future the ways to improve the efficiency of energy-generating machinery in cells and to keep production of ROS under control.

In this chapter, we will be concerned with the principles and molecular details of proton pumps. These molecular devices are energy-transducing machines. A remarkable feature of these machines is that they are essentially based on quantum principles: electron transport along the chain of redox centers of electron transport chain occurs via quantum mechanical tunneling. Without this fundamental process, which is due to wave properties of matter, the overall machinery of energy transduction would not be possible—at least in the form that we know it. Although most of biology can be described in terms of classical physics, this area certainly justifies the notion of "quantum biology" and since electrons and protons are involved, one can also think about it as "elementary particle biophysics."

The structures of CcO, bc1 complex, and part of NADH dehydrogenase are now known. The mechanism of pumping in bc1 is mostly understood and is based on the so-called quinone Q-cycle postulated by Peter Mitchell [5]. The molecular mechanisms of proton pumping by complex I is completely unknown, and there has been significant progress in understanding complex IV, CcO, the subject of detailed discussion of the remainder of this chapter.

In the following sections, we will give a brief review of CcO (Section 4.2) and will discuss some key computational methods used in the research of this enzyme (Sections 4.3 through 4.7). Section 4.3 highlights most important computational issues involved in CcO mechanism; Section 4.4 discusses long-distance electron tunneling aspect of the problem; the following Section 4.5 discusses electrostatic calculations involved in understanding the membrane potential generated by the enzyme; Section 4.6 focuses on the $pK_a$ calculations problem in proteins and reviews a method that combines density functional theory (DFT) and electrostatic calculations; finally, Section 4.7 discusses calculations of proton transfer rates and a related interesting problem of proton gating in CcO. Section 4.8 concludes the chapter; it reiterates some unresolved computational problems of CcO, and discusses possible ways to tackle these problems in the future.

## 4.2 CYTOCHROME c OXIDASE: A REVIEW

Cytochrome $c$ oxidase, an enzyme, is the last element in the electron transport chain. In 1978, Wikström discovered that CcO is an unusual "non-Mitchellian" proton pump [6], in the sense that there are no physical carriers of protons in the enzyme, such as quinines, as in bc1 complex, and protons move through the enzyme "by themselves." The proton translocation against the electrochemical gradient across the membrane appears to be entirely due to electrostatic interaction with the electrons passing through the system. The structure of CcO is now available at high resolution (2.3 Å; Figure 4.2) [7], however, the details of its molecular mechanism remain the subject of intense debate [8–12]. In a diverse range of organisms, the central processing unit (CPU) of the enzyme is almost identical, suggesting that all aerobic life uses similar mechanism of oxygen reduction and proton pumping.

In the electron transport chain, CcO receives electrons from cytochrome $c$, a water-soluble heme protein, on the cytoplasmic side of the membrane, and transfers them through a series of electron transfer steps to the active site, which contains a heme iron and a copper, where the electrons are used to reduce the molecular oxygen. The protons needed for this reaction are taken from the mitochondrion matrix side through two proton-conducting channels. In addition to these chemical protons, four more protons, per every oxygen molecule reduced, are translocated across the membrane. The overall enzymatic reaction of CcO is

$$4e\ (\text{cyt } c) + 4H^+\ (\text{in}) + 4H^+\ (\text{in}) + O_2 \rightarrow 2H_2O + 4H^+\ (\text{out}),$$

where "in" and "out" correspond to the inside and outside mitochondrion inner membrane (Figure 4.2). Four protons thus are consumed to make two water molecules and additional four are pumped across the mitochondrial membrane. Thus, four electrons and eight protons are involved in one cycle of the enzyme. Both chemical and pumped protons contribute to the creation of an electrochemical membrane gradient of about 200 mV. The studies of CcO are driven by two related key questions: (1) How does the enzyme catalyze oxygen reduction and water formation, and (2) what is the mechanism of proton pumping?

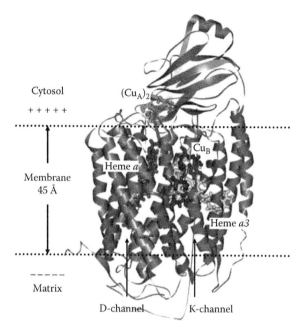

Cytosol   $(Cu_A)_2$

+ + + +

Membrane
45 Å

Matrix

Heme a

$Cu_B$

Heme a3

D-channel          K-channel

**FIGURE 4.2** The structure of a membrane-bound cytochrome oxidase from bovine heart [13]. The enzyme consists of 13 subunits, total molecular mass is 204 kDa. In subunits I and II (shown), there are four redox centers: binuclear $Cu_A$ (in subunit II), low-spin heme a, high-spin heme a3, and $Cu_B$ (all in subunit I). The oxygen-binding site is between Fe ion of heme a3 and $Cu_B$. Although structure of CcO is known, it proved to be extremely difficult to understand the details of its proton translocation machinery. The key for understanding are coupled electron and proton transfer reactions occurring in the enzyme.

## 4.2.1   THE ENZYME CYCLE

The CcO cycle involves six distinct steps (Figure 4.3), assigned by the state of the catalytic site: fully reduced, R($Fe^{2+}/Cu^{1+}$) → oxygen bound, A → peroxy, P (oxygen bond is broken) → ferryl, F ($Fe^{4+}=O^{2-}$) → O ($Fe^{3+}/Cu^{2+}$)→E ($Fe^{3+}/Cu^{1+}$) to R. The kinetics of the formation of each of these steps is now well resolved [14–16].

Most of the steps involve correlated electron and proton transfer. Two electrons are present in the R state. Acceptance of the third electron from the heme a results in the F state and water formation; the fourth electron brings the system to the O state with release of an additional water molecule. The current data indicate two (or possibly three) protons are pumped in R to O transitions and two (or one) during O to R. The timescales of the reactions vary in the range of 10 μs to 1 ms. The wild-type enzyme pumps about 1000 protons per second, with stoichiometry 4H+/4e.

Recently, Wikström and coworkers have proposed [18] that the cycle of a working enzyme may actually involve the so-called high-energy intermediates: $O_H$ and $E_H$—nonequilibrium states in which somehow additional energy for proton pumping is stored, and in which the redox potential of the binuclear center that drives the

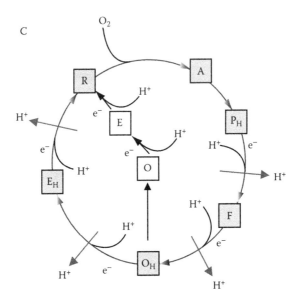

**FIGURE 4.3**   The enzyme cycle and the states involved according to Belevich et al. [17].

electron and proton transport is much higher than that in the equilibrium O and E. The true nature of these states is unknown, however.

### 4.2.2  ELECTRON TRANSFER

The binuclear $Cu_A$ center is the first site to receive electrons from cytochrome $c$ (Figure 4.4). These electrons are transferred to the low-spin heme $a$ and then to the binuclear center of heme $a3$ and $Cu_B$. At almost all steps, electron transfer between the redox centers is coupled with proton transfer. For theory of rates of electron transfer (ET) reactions coupled with proton translocation, such as in CcO, the reader is referred to Ref. [19].

### 4.2.3  WATER IN THE PROTEIN

Water plays a critical role in the function of the enzyme, as it provides the "medium" for proton translocation [12]. Recent high-resolution structural data indicate that there is plenty of water inside the protein. The problem is that most important dynamic water molecules that carry protons are not seen in the x-ray structures. Here, computer simulations play the critical role.

The simulations (molecular dynamics [MD] and free energy calculations) have identified two important water chains in the enzyme (Figure 4.5), which connect Glu242, a known proton donor, both for chemical and pumped protons, to the binuclear Fe–Cu center (BNC), and to the region "above" heme $a3$, via its Prop D. The exact number of water molecules in this region, however, is not known.

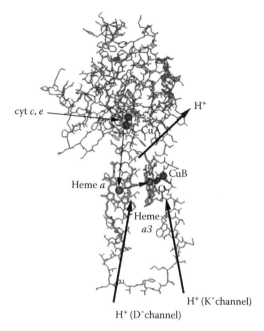

cyt $c, e$

$H^+$

CuA

CuB

Heme $a$

$O_2$

Heme $a3$

$H^+$ (K^channel)

$H^+$ (D^channel)

**FIGURE 4.4** Redox sites, electron transfer paths, and proton channels of CcO. Electrons are derived from cyt $c$, and delivered to $O_2$ bound in the binuclear center heme $a3$/Cu$_B$. Protons for oxygen chemistry are supplied via two channels: K and D. Putative exit channel for protons, schematically indicated on the diagram, have been identified in recent work. (From Popovic D.M. and Stuchebrukhov A.A., *J. Phys. Chem. B*, 109, 1999, 2005. With permission.)

This finding immediately suggested that one chain can be used to deliver chemical protons for oxygen chemistry in BNC, and another to translocate pumped protons. The majority of currently discussed models of CcO are based on this idea. However, it is not clear yet how the enzyme controls the gating of protons at Glu242 site to be delivered for chemistry and for pumping.

### 4.2.4 POSSIBLE PROTONATION SITES

Which ionizable amino acids are participating in the translocation of protons (water chains are carrying protons between such sites) is one of the key questions. Several such side chains have been identified in experimental studies, in particular in the K- and D- proton conducting channels, which are the entry points of the protons. These are the sites roughly below the level of heme $a$/heme $a3$ (see Figure 4.2). The identity of such sites above the line heme $a$/heme $a3$ is mainly unknown at present.

The calculations performed in our group recently have provided some leads in directing experimental research of this subject. In particular, we predicted a possible involvement of His291 [22], one of the ligands of Cu$_B$ center, and a possible proton exit site Lys173B/Aps171B [20] on the positive side of the enzyme. Recent experiments (still preliminary at present) by two independent groups indicate that mutation

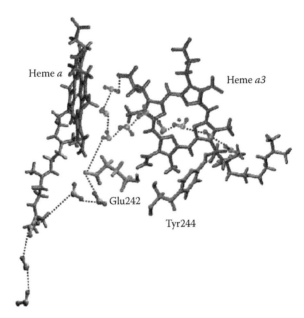

**FIGURE 4.5**  Computer simulations have revealed two bifurcated chains of water molecules in the catalytic site of CcO [21]. One branch connects Glu242 to the binuclear center, and another connects Glu242 to Prop *D* of heme *a3*, indicating that pumped protons can be channeled using one branch, while chemical protons can use the other. Most of the current models of CcO pump are essentially based on this structure.

of Lis173B/Aps171B significantly impairs the turnover of the enzyme: down to 3% of weight, while the mutation of a nearby threonine renders enzyme incapable of pumping but retaining the turnover activity, as the modeling suggests.

### 4.2.5  PUMPING MECHANISM

In the past 30 years, there have been many speculations as to how the enzyme pumps protons [6]. Until very recently, two major proposals have been mainly discussed. One proposal, due to Wikström [23], was based on the so-called histidine cycle. This mechanism postulated that one of the His ligands of $Cu_B$ acts as a molecular "arm" that carries the pumped protons and releases them in pairs. The physical principles of energy coupling to redox chemistry in this scheme are not clear. The other proposal, due to Michel [24], was based on the assumption that heme *a* is the pump element. A somewhat similar proposal is due to Yoshikawa [25], who suggests that there is a third H-channel for proton translocation.

Recent breakthrough experiments have led to significant changes in the field. Currently a le/1H+ pumping ratio is favored by most researches in the field, but whether protons are pumped in the E to R transition, or R to E in the cycle of a working enzyme with a load is not entirely clear yet. Based on the findings of two chains of water molecules and (still putative) redox-dependent protonation state of His291, we proposed a detailed le/1H+ pumping model [26], shown in Figures 4.6 and 4.7.

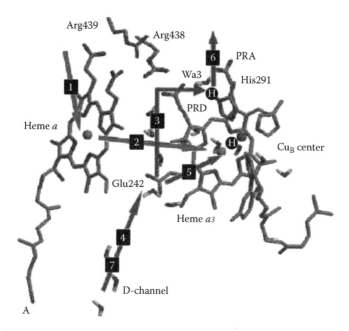

**FIGURE 4.6** The model of CcO pump [26]. Upon electron transfer (1,2) two PT reactions occur: PT Glu242 to His291 (3), followed by, after re-protonation (4), PT Glu242 to BNC (5); due to repulsion of the two protons, the proton from His291 is expelled to the positive side of the membrane (6).

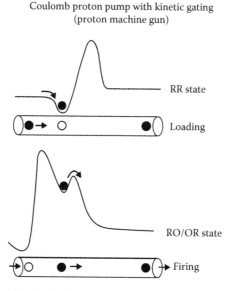

**FIGURE 4.7** Basic principles of the proposed pumping mechanism of CcO. Schematically shown are the energy profiles for proton translocation along the proton conducting channel in RR (loading) and RO/OR (firing) states. The pump works as a "proton machine gun." The energy profiles provide kinetic gates, which ensure unidirectionality of proton translocation.

This model suggests that upon the electron transfer between heme $a$ and heme $a3$, two proton transfers (PTs) occur. The first proton transfer occurs between Glu242 and His291, which is followed by the second proton transfer between (reprotonated) Glu242 and the binuclear center. The calculations suggest that proton–proton repulsion between the two protons on His291 and in the binuclear center can give rise to the expulsion of the proton from His291 to the upper side of the membrane, which will result in the pumping event.

The recent experiments mainly support the above proposal. Most detailed data come from ingenious experiments in which the membrane potential generated by the enzyme upon single electron injection is measured with microsecond resolution. For example, in the F to O transition, the data show one short protonic phase followed by another slow and large in amplitude protonic phase [27]. These two kinetic phases correspond to two proton transfer reactions of the above model [28].

The basic physical principle involved in such a scheme is shown in Figure 4.7. The enzyme has a proton conducting channel whose energy profile depends on the redox state of the catalytic center. Upon arrival of an electron, the RR (both metal sites in the binuclear center are reduced) state is formed and the energy profile along the proton conducting channel has a high barrier from the P-side (positive) of the membrane, as shown in the figure; a proton from the N-side of the membrane arrives to the proton loading site (PLS) of the enzyme. The barrier provides the kinetic gating for the process, so that the loading occurs with a low potential proton from the N-side of the membrane. Upon arrival of the chemical proton to the catalytic center, the RO or OR redox state is formed, which has a different energy profile; now the barrier is higher toward the N-side of the membrane and therefore the proton expulsion from the PLS occurs toward the P-side of the membrane. Again, kinetic gating is involved in the directionality of the process. According to the model, the enzyme works as a "Coulomb machine gun": in the RR state the proton is loaded, while in RO/OR state, the proton is fired. The enzyme cycles between the loading and firing states, as electrons flow through the system. The driving force for electrons (and for the chemical proton that expels the loaded proton) is oxygen reduction chemistry occurring in the catalytic center of the enzyme. Clearly a constant flux of electrons (i.e., supply of energy) is needed to maintain the machine in the nonequilibrium state and to pump protons from the low- to high-chemical potential side of the membrane.

## 4.2.6 ENZYME CYCLE

A detailed catalytic cycle that corresponds to the proposed model is shown in Figure 4.8. We are not going discuss the details of chemistry involved in the catalysis of oxygen reduction; the interested reader is referred to the original papers, Refs. [22,26].

Recently, Wikström and coauthors have proposed a new pumping scheme [29], which is almost identical to our model in general, yet there is one significant difference: the loading stage in our model corresponds to a P (or R) level of the reduction of the binuclear center, whereas the new proposal involves one additional electron, as in $P_R$ state. This difference emerges as a focal point of the discussion of two possible types of models as they correspond to two different charging states of the binuclear catalytic center of the enzyme in the pumping stroke of the cycle.

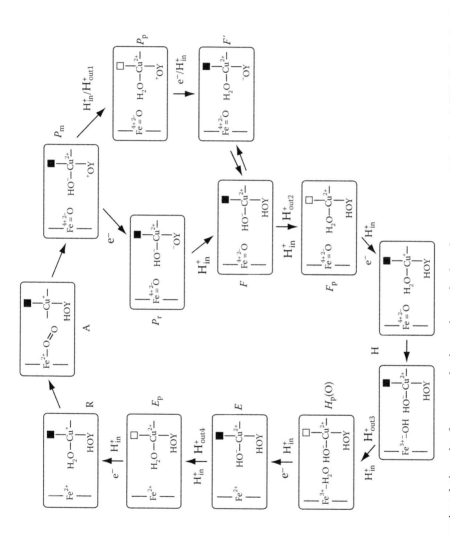

**FIGURE 4.8** Proposed catalytic cycle of oxygen reduction and pumping based on our computational work [26]. The filled and open squares are for His291 ligand in the protonated and deprotonated states, respectively. The pumping step is the protonation of OH− group in the catalytic center, which is coupled with deprotonation of His291, as shown in Figure 4.6.

Some other theoretical possibilities have been discussed by Siegbahn et al. [30], Olsson and Warshel [31], and Xu and Voth [32] recently. Siegbahn et al. proposed that the PLS is not His291, but rather a nearby Prop A of heme *a3*. The energetics of their model raises questions however; they report that the chemical proton has no driving force left after the first proton transfer to Prop A has happened. Our most recent calculations indicate that there is a remaining driving force for the second proton. Still, the identity of the PLS remains unknown.

Given remarkable progress made in the field in the past few years (due to both experiment and simulation), the pace of discovery, and the degree of atomistic detail at which possible models are now discussed, it is likely that significant breakthroughs will occur in the field in the near future. In such developments, simulations will play a crucial role.

In the remainder of this chapter, we will summarize the major computational results on the mechanism of CcO obtained so far, and discuss a few selected computational topics involved in the studies of CcO.

## 4.3 COMPUTATIONAL STUDIES OF PROTON PUMPING MECHANISM IN CYTOCHROME OXIDASE

In this section, we summarize the major computational results on the mechanism of CcO. Naturally, we will mainly focus on the studies done by the author's research group.

### 4.3.1 WATER CHAINS IN CcO

The disordered water molecules are not seen in x-ray structure, yet they are critical for proton transport in the enzyme. In MD simulations, we have discovered [21] that water molecules in the catalytic cavity in CcO form two branching chains (Figures 4.5 and 4.6) that connect an experimentally known proton donor Glu242, the Fe–Cu binuclear center, and the putative PLS His291, from which protons can be pumped using our mechanism of proton–proton repulsion. The possibility of such connectivity has been discussed; however, they were never observed through computer simulations before.

This finding immediately suggested that one branch of the structure can be used to supply protons to the binuclear center for chemistry, and another branch for pumping protons. Following this proposal, Wikström and coworkers suggested [33] that water orientation in these chains maybe used for gating pumped and chemical protons by the "water-gate" mechanism.

### 4.3.2 PROTON LOADING SITE OF THE PUMP

The second important finding was identification of a potential PLS of the pump as one of the $Cu_B$ ligands His291 [22]. In various calculations, which currently include a mixture of DFT and continuum electrostatics, we showed that in "RR" state of the binuclear $Fe_{a3}$–$Cu_B$ center of the enzyme, delta-Nitrogen of His291 is protonated, where as in "RO" or "OR" states, $pK_a$ is low enough to suggest a deprotonated state. Together with the two chains of water molecules, this finding led us to propose a model [26] based on electrostatic repulsion between protons and kinetic gating principle (Figures 4.6 and 4.7). According to this model, an electron transfer to BNC is coupled with two proton transfers—one is from Glu242 to His291 and the second one is to BNC. Due to proton–proton repulsion, the first proton is expelled to the upper side of the membrane, which is a pumping event. The sequence of two proton transfer reactions is determined kinetically. The major features of model have strong experimental support. In recent experiments on the membrane potential kinetics [34], one fast and one slow protonic phase are observed; the current view is that these two phases correspond to two proton transfer reactions that are present in our model.

### 4.3.3 KINETIC GATING OF PROTON CONDUCTING CHANNELS

The entry points for proton translocation in CcO are well known; these are the entries of D- and K-channels, which are utilized for delivery of chemical and pumped protons. However, nothing is known about the proton exit channels in CcO. In fact, the pathways of proton translocation beyond Glu242 and the binuclear center remain poorly understood. Based on our calculations, we recently proposed [20] that Lys173B/Aps171B groups are possible proton exit sites on the positive side of the enzyme and a prediction was made that the pumping activity of the enzyme should be significantly affected upon mutation of these groups. Two recent independent experimental reports from Gennis and Brzezinski laboratories indicates that upon mutation of Lys171B or nearby Threonine either reduce activity to only 3% of wild-type (Gennis data), or renders protein unable to pump protons at all while maintaining oxygen reduction turnover (P. Brzezinski, private communication). Energetics of the proton conducting channels in CcO suggests their possible gating functional role [20], which can explain why protons do not leak through the enzyme in the "wrong" direction, across the membrane.

### 4.3.4 PROTON COLLECTING ANTENNA OF THE PUMP

The supply of protons to the mouth of proton pumps has stringent requirements, given the efficiency and rate of pumping, and was suggested to occur via a coupled diffusion of protons along the surface of the protein (and membrane) and in the bulk. Theoretical analysis indicates that this type of reduced-dimensionality coupled surface-bulk diffusion has many interesting and unusual properties, which give rise to the "antenna" effect by which protons are collected at the entrance of the pump with a surprising efficiency [35,36].

### 4.3.5 MEMBRANE POTENTIAL GENERATED BY CcO

Time-resolved measurement of the membrane potential generated by the enzyme upon single-electron injection has been one of the most fruitful techniques in the studies of CcO [17]. However, the quantitative interpretation of such experiments requires a computational input. We have recently started developing a theory and the necessary computational tools for interpretation of potentiometry experiments [28,37]. It appears that not only geometry of the enzyme determines the amplitudes of the potentials observed, but also the rates of the processes themselves. Using this theory, we have shown that upon injection of the electron into the enzyme the first proton is transferred not to the binuclear center, but to a group "above" BNC. The identity of the group, which is according to our model is a PLS, could not be determined with certainty from such an analysis; however, the proposed His291 site turned out to be in the group of most likely candidates for PLS.

### 4.3.6  QM/MM Calculations of p$K_a$'s in Proteins

The central computational problem of redox enzymes, which couple electrons and protons, is to be able to correctly evaluate protonation state of different groups (including water molecules) for different redox states of the enzyme—in other words, to calculate redox-dependent p$K_a$ values in proteins [38–43]. These are difficult calculations, mainly because of the uncertainties involved in modeling of the dielectric properties of the proteins. Similar problem exists for calculations of redox potentials of protein redox groups.

We have recently started to explore a type of calculations in which DFT treatment of the quantum mechanical (QM) site is combined with either continuum electrostatics treatment of the protein, or with microscopic molecular mechanics/dynamics treatment of the protein, or with a combined molecular mechanics and continuum electrostatics treatment of the protein in a truly multiscale type of calculations. All these calculations have a spirit of QM/MM (quantum mechanics combined with molecular mechanics) method, which is currently in wide use in protein calculations. The DFT and the solvation energy calculations are performed in a self-consistent way. The work aims at both improving the QM part of p$K_a$ calculations and the MM or electrostatic part, in which of the protein dielectric properties are involved. In these studies, an efficient procedure has been developed for incorporating inhomogeneous dielectric models of the proteins into self-consistent DFT calculations, in which the polarization field of the protein is efficiently represented in the region of the QM system by using spherical harmonics and singular value decomposition techniques [41,42].

### 4.3.7  Theory of Electron Transfer Reactions Coupled with Proton Translocation

Theory of this type of electron transfer reactions has been recently reviewed by the author in Ref. [19,44]; the reader is also referred to a related Ref. [45] where concerted ET/PT reactions are discussed.

In the next sections, we discuss four computational areas that are at the core of CcO studies.

## 4.4  LONG-DISTANCE ELECTRON TUNNELING IN CcO

To study electron tunneling in the enzyme, the tunneling currents method [44,46] developed by the author can be used. This method is now implemented at semiempirical (ZINDO) level and other ab initio methods, and can be used for mapping electron transfer tunneling paths, and calculation of electron transfer rates in CcO.

The mapping of electron transfer pathways has been described in Refs. [44,46,47]. The general idea of the approach is to examine dynamics of charge redistribution in the system "clamped" at the transition state of electron transfer reaction. Briefly, the essence of the method is as follows. The tunneling dynamics of a many-electron system is described by following wave function:

$$| \Psi(t) > = \cos\left(\frac{T_{DA}}{\hbar}t\right)|D> - i\sin\left(\frac{T_{DA}}{\hbar}t\right)|A>, \qquad (4.1)$$

where $|D>$ and $|A>$ are diabatic states corresponding to localization of the tunneling electron on the donor (heme $a$) and on the acceptor complexes respectively, and $T_{DA}$ is the transfer matrix element. The states $|D>$ and $|A>$ can be calculated at Hartree–Fock (HF) [48] or DFT level [49]. The extension of the method beyond a one-determinant description has been proposed as well, however, not implemented yet.

The redistribution of charge in the system during the tunneling transition is described in terms of current density $\vec{J}(\vec{r}, t)$ and its spatial distribution $\vec{J}(\vec{r})$, which is given by the matrix element between states A and D of the current density operator (this is the so-called transition current):

$$\vec{j}(\vec{r},t) = -\vec{J}(\vec{r})\sin\frac{2T_{DA}t}{\hbar}, \quad \vec{J}(\vec{r}) = -i < A \,|\, \hat{\vec{j}}(\vec{r})\,|\, D >, \qquad (4.2)$$

or, in terms of interatomic currents $J_{ab}$, which are introduced as follows:

$$\frac{dP_a}{dt} = \sum_b j_{ab}, \quad j_{ab} = -J_{ab}\sin\frac{2T_{DA}t}{\hbar}, \qquad (4.3)$$

where $P_a$ are atomic populations. The total current through an atom is proportional to probability that the tunneling electron will pass through this atom during the tunneling jump. The streamlines of the current $\vec{J}(\vec{r})$ represent the whole manifold of the so-called Bohmian trajectories [50]. Both interatomic currents $J_{ab}$ and current density $\vec{J}(\vec{r})$ provide full information about the tunneling process and, in particular, about the distribution of the tunneling current in space, i.e., about the tunneling pathway.

Both $\vec{J}(\vec{r})$ and $J_{ab}$ are related to the tunneling matrix element.

$$T_{DA} = -\hbar \sum_{a\in\Omega_D, b\notin\Omega_D} J_{ab} = -\hbar \int_{\partial\Omega_D} (d\vec{s} \cdot \vec{J}) \qquad (4.4)$$

and eventually to the rate of electron transfer reaction.

An example of this type of calculations for the tunneling transition $Cu_A \rightarrow$ heme $a$ in CcO, is shown in Figure 4.9 [51]. The tunneling current method has been used to get insights into electronic communication of all four redox centers in CcO.

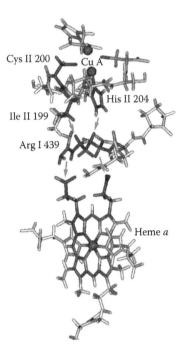

**FIGURE 4.9** **(See color insert following page 172.)** $Cu_A$ to heme $a$ ET transition in CcO [51]. Electron transfer occurs via two interfering paths running from $Cu_A$, via Arg439, and to heme $a$ group, and includes a jump between subunits I and II. Water at the interface between the subunits facilitates electron tunneling between the subunits. The color coding is used to show atomic probabilities for tunneling electron to pass a given atom.

The complete analysis of electron transfer is typically performed in two steps. First the electronic coupling is calculated for a given nuclear configuration of the protein and then the dynamics is included and the averaging and other dynamics effects are incorporated into calculations of electron transfer rates. Various theories that combine pure electronic and dynamic effects are then utilized to calculate the rate of electron transfer [44,52,53]. In recent years, a number of theories and computational methods for electron tunneling in proteins have been developed and reviewed extensively. Of particular interest is the work of the Beratan and Onuchic groups, who pioneered the area of tunneling pathways [54].

A quick estimate of the electron transfer rate can be obtained with the empirical Moser–Dutton relation [55–58]. This model neglects all structural aspects of the protein medium and treats it as an effective dielectric continuum. It has been reported that this empirical theory works rather well in predicting electron transfer rates in some heme proteins. The reason for a success of a model that neglects the details of the of electron tunneling pathways is likely due to a significant averaging occurring due to thermal fluctuation of the protein. The various protein dynamic effects on the rate of electron transfer has been recently actively discussed in the literature [44,59,60]; in particular, the dynamic effects were suggested to play a key role in a surprisingly fast (nanosecond) electron transfer between heme $a$ and heme $a3$ in CcO [60–62].

## 4.5 MEMBRANE POTENTIAL GENERATED BY CYTOCHROME *c* OXIDASE

The principal obstacle for detailed experimental characterization of proton translocation by CcO is that it is difficult to monitor where the protons are in the molecule at any given time. Unlike electrons, which are transferred between four redox centers—$Cu_A$, heme *a*, heme *a3*, and $Cu_B$—and can be monitored by time-resolved spectroscopy of the metal centers, the protons are notoriously difficult to detect, in particular in real time.

### 4.5.1 POTENTIOMETRY

The time-resolved measurement of the membrane potential generated by the enzyme upon single electron injection has been one of the most fruitful approaches in the studies of CcO. This ingenious experimental technique has been adopted in the past few years by all major research groups in the area [17,27,63,64]. The schematics and the idea of such measurements are shown below. The enzyme is incorporated into artificial liposomes, which are adsorbed on the measuring membrane. A laser pulse is used to inject an electron into the enzyme, and the kinetics of the membrane potential developed due to charge movements in the enzyme (and across the liposome membrane) are measured with microsecond resolution in real time (Figure 4.10). The membrane potential reflects charge transfer processes occurring in the enzyme.

The kinetic curves typically show distinctly a fast electron transfer phase (Figure 4.10c), followed by several kinetic phases due to proton translocation. The observed voltage on the measuring membrane is proportional to dielectric distance traveled

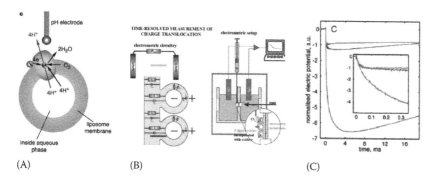

(A)  (B)  (C)

**FIGURE 4.10** (See color insert following page 172.) (A) Schematics of artificial liposome and reconstituted CcO; after injection of an electron, several proton transfer reactions occur that result in a change of membrane potential. (B) The measuring electrochemical cell (Wikström and Konstantinov, private communication). The liposomes are adsorbed on one side of the measuring membrane. The charge transfer across the liposome membrane (± separation) results in potential difference across the measuring membrane; (C) Typical signal observed in such measurements: initial potential drop, green curve, is due to electron injection—ET $Cu_A$ to heme *a* (normal to membrane)—followed by two or more slow kinetic phases due to motion of protons across the membrane. Experimental data are the rates and amplitudes of the signal.

by the charge in the enzyme. Remarkably, the charge transfers over distance of few angstroms can be detected in such electrochemical cells in real time. Different kinetic phases correspond to different charge transfer processes; the amplitude of each phase corresponds to dielectric distance traveled by the charge, and the rate of the kinetic phase corresponds to the rate of the corresponding charge transfer process.

Recent landmark experiment by the Wikström group [17] on CcO from *Paracoccus denitrificans* has revealed four kinetic phases: one is due to electron injection, 10 μs phase, and three phases due to motion of protons in the enzyme with rate constants 150 μs, 800 μs, and 2.4 ms, and corresponding amplitudes 12%, 42%, 30%, and 16% (see Figure 4.11).

Presumably, the first protonic phase (150 μs) observed in the experiment corresponds to the transfer of the pump proton to PLS from the N-side of the membrane. The loading occurs upon electron transfer between heme *a* and heme *a3* as we predicted in our model [26] (see Figure 4.6). The identity of PLS however is unknown.

Although potentiometry provide most valuable data that reflect charge transfer processes in the enzyme, getting molecular insights from the experiments proved to be difficult because of lack of theory for proper quantitative interpretation of the data. Until recently, it has not been possible to directly relate the potentiometric data to a specific molecular mechanism of charge translocation in CcO.

Our recent work has partially resolved the difficulties of interpretation of potentiometric experiments and provided a quantitative method for deciphering potentiometric data. Using this theory, we have performed quantitative analysis of the data of Belevich et al., in which the experimental amplitudes and rates are related to specific residues that exchange electrons and protons, and generate the observed membrane potential. Using this theory, we have tested proposed candidates for the proton pump site of the enzyme against experimental potentiometric data [37].

### 4.5.1.1 Limitations of Potentiometry: Need for Detailed Computer Simulations

The fundamental limitation of potentiometry is that one cannot distinguish between the sites that have sufficiently close potentials. Such sites will generate the same experimental signal. The number of possibilities, however, can be reduced to only a few, e.g., His326/PropA/PropD/Arg474 group shown in Figure 4.6. To get further insights, additional independent data—measurements or calculations—are required. Despite the limitations, there is little doubt that potentiometry, in particular in combination with optical spectroscopy as in Ref. [17] and with computational and theoretical analysis developed in Ref. [37] will be one of the most fruitful techniques for characterization of proton pumps in the future.

## 4.6 CALCULATION OF REDOX-COUPLED PROTONATION STATES OF KEY RESIDUES OF CcO: THE SEARCH FOR THE PUMP ELEMENT

The key unknown in CcO proton pumping mechanism is the identity of the group that plays the role of the pump element. Currently, most of the researches in the area agree that the most likely mechanism of pumping involves an intermediate proton

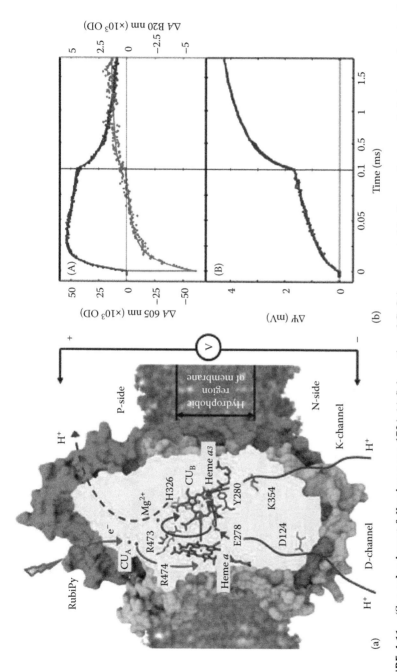

**FIGURE 4.11** (See color insert following page 172.) (a) Schematics of CcO incorporated in the membrane, and electron and proton transfer reactions in the enzyme in the experiment of Belevich et al. [17]. Two key subunits are shown together with the four redox centers Cu$_A$, heme $a$, heme $a3$, and Cu$_B$. Upon laser flash, an electron from Ru–bpy complex is injected into the enzyme, which drives the enzyme from O to E state generating membrane potential. The electron transfer path is indicated in red. Proton transfer is in blue. Protons are delivered via two channels D and K. The pumped proton is released on the P-side of the membrane from an yet unknown PLS group above heme $a3$. (b) Upper panel is kinetics of absorption changes of heme $a$ and Cu$_A$ centers; lower panel—kinetics of the membrane potential. The kinetic curves reveal four kinetic phases with rate constants: 10 µs, 150 µs, 800 µs, and 2.4 ms, and corresponding amplitudes: A1 = 12%, A2 = 42%, A3 = 30%, A4 = 16%.

transfer from Glu242 to yet unidentified group called proton loading site (PLS) in the region "above" heme *a3* (see Figure 4.6) [11,30] to which the pumping proton is loaded, and then the next "chemical" proton transferred to the binuclear center displaces the pumping proton from PLS to the positive side of the membrane, as we predicted in our model [26]. Although the overall scheme is supported by recent potentiometric experiments, the identity of the PLS site remains unknown. Our analysis of potentiometry data provides important clues as to where the pump element is located; however, the ultimate atomistic picture of the pump can emerge only from direct computer simulations of redox-coupled proton transfer reactions in the enzyme, and from the direct experimental probes of the predicted sites that exchange protons.

The key computational step in the analysis of proton coupled electron transfer reactions is the evaluation of the energy of protonation ($pK_a$) of different groups of the enzyme in different redox states of its metal centers, as well as calculations of redox potentials of the binuclear metal center of CcO for different protonation states of its ligands (Figure 4.12). Thus the first step in the analysis concerns the energetics of different redox and protonation states of CcO; the kinetics issues can be addressed after that.

While the treatment of charges in homogeneous solvents (including water) is a relatively well developed computational procedure, the evaluation of energetics of

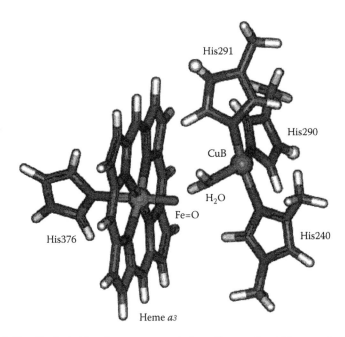

**FIGURE 4.12**   Catalytic binuclear center consisting of heme *a3* and $Cu_B$, together with their ligands is treated quantum mechanically (QM). The protonation state of the ligands of metal centers Fe and $Cu_B$, as well as that of His291 in different redox states of the enzyme is one of the issues of the proposed study. The cavity around this center separates the QM system, and the rest of the protein, which is treated using continuum electrostatics, or microscopic MM or MD modeling, or their combination.

charges in an inhomogeneous and dynamic protein environment is still one of the greatest challenges of computational biophysics [65,66]. The evaluation of the energy of charging (electron or proton) involves the evaluation of three types of energies: the quantum "self-energy" of the group involved, including its polarization by the protein environment, the interaction of the group with immediate environment (the interface) of the protein (hydrogen bonds, bonding and nonbonding interactions), and the long-range electrostatic energy of interaction (including protein relaxation, polarization, and interaction with internal polarizable water). The critical issue is to correctly describe dielectric response of the inhomogeneous medium of the protein which includes cavities, dynamic water molecules, membrane and the movable ions of the solutions, and to describe the "boundary" between the quantum subsystem and the rest of the protein [66].

The methods and programs for $pK_a$ calculations in proteins have been steadily improving in the past years [65,67,68], in part due to efforts of this group [38–40,69,70]. These methods range from fully microscopic (based on MD free energy evaluation, e.g., [71]) to fully phenomenological (continuum electrostatics free energy calculations based on Poisson–Boltzmann equation (e.g., [72,73]) and also include a combination of the two limiting approaches, as in the example of PDLD/S semimicroscopic methodology of Warshel [66].

Most of the previous studies have focused on the $pK_a$ values of standard ionizable amino acids. In our case, most interesting groups are ligands of the metal centers of the protein, such as His291 of the $Fe_{a3}$/$Cu_B$ binuclear center, and therefore the $pK_a$ of the so-called model compound (in water) of these groups is not known, which adds a new dimension to the problem.

Our contribution to the field is a method that combines DFT treatment of the titratable groups involved, and an electrostatic, MD, or MM treatments of the protein medium [40,43]. In such calculations, a group of interest is treated quantum mechanically (QM), while the rest of the protein/membrane/solvent is treated with either continuum electrostatic model or with a combination of MD/MM of the protein surrounding the cavity of the group, with embracing medium described by continuum electrostatic models, in a truly multiscale approach. The caveats of such calculations have been recently discussed in the literature [70]. The important issues concern self-consistent charges of the quantum subsystem, correct geometry of system in both protonation states, correct charge distribution in the protein, which involves a question of protonation state of other titratable groups in the protein, etc. The last issue is extremely important; as we showed, the use of so-called standard protonation state for the surrounding medium can result in a significant error for solvation energy of protonation. But most important is how we treat the polarizable inhomogeneous medium of the protein, including internal water molecules in it.

The combination of DFT and continuum electrostatic calculations has been used successfully in the past to compute the $pK_a$ values and redox potentials in various enzymes and proteins [74–78]. However, the parameterization of the protein (including that of internal water) is not universal [79–82], and needs to be adjusted specifically for a system in question—in our case CcO—if we want to accurately calculate energetics of the system.

The main computational tool here is a combination of DFT and continuum electrostatics [40].

The calculations are typically carried out as follows. The whole system (CcO) is divided into an active site complex (QM system) and the surrounding medium—protein, membrane, and the external aqueous phase. For example, to explore the protonation states of the binuclear center and its ligands, the QM system is defined as shown in Figure 4.12. The protonation of state of His291, one of the ligands of $Cu_B$, at different redox states of the $Fe_{a3}/Cu_B$ binuclear center is one of the sites of our interest; the $H_2O/OH^-$ ligand to $Cu_B$ or Fe $a3$ metals is another such site.

The proton affinity of the complex is calculated with an accurate DFT method, and the dielectric effects of medium are treated with continuum electrostatic or microscopic MD methods.

The dielectric medium involves the protein itself, the membrane, and the surrounding solvent. The protein itself is an inhomogeneous dielectric with many internal cavities, which contain water molecules. These regions should have higher dielectric constant than the regions of "dry" protein. For regions of dry protein, a specific value of dielectric constant 4 is typically assumed, and for the cavities a higher value in the range 10–80 is used. Neither of these values is well-defined; however, the issue can be addressed in direct MD simulations of the protein (see discussion below).

The general scheme of $pK_a$ evaluation is well known. The $pK_a$ value is related to the free energy of deprotonation, which is a sum of two contributions: the free energy of deprotonation in vacuum ($\Delta G_{vac}^{deprot}$), and the solvation energy difference between the deprotonated and protonated forms of the protonatable group ($\Delta\Delta G_{solv}^{deprot}$):

$$pK_a = \frac{1}{kT\ln 10}\left(\Delta G^{deprot}\right) = \frac{1}{kT\ln 10}\left(\Delta G_{vac}^{deprot} + \Delta\Delta G_{solv}^{deprot}\right). \qquad (4.5)$$

To avoid the issue of the proton solvation energy, the $pK_a$ value of the group of interest is calculated by comparing the above terms for a suitable reference compound, for which the $pK_a$ value is experimentally known (for calculation of $pK_a$ of His291, for example, the model compound 4-methylimidazole will be used); then the *shift* of $pK_a$ value is calculated: $pK_a^{site} = pK_a^{model} + \Delta pK_a$. The relative shift of the group's $pK_a$ involves evaluation of $\Delta E_{elec} = \Delta E_{elec}^{A^-} - \Delta E_{elec}^{HA}$—a DFT-calculated electronic energy of deprotonation and $\Delta\Delta E_{elec}$ its shift relative to the model compound (usually this term is neglected, e.g., [66,79,83–86], as the "electronics" of the reference compound is assumed to be the same as that of the group in the protein. For our problems, this is not the case); and $\Delta G_{solv} = \Delta G_{solv}^{A^-} - \Delta G_{solv}^{HA}$— the difference in solvation energy between the deprotonated and protonated forms and $\Delta\Delta G_{solv}$—the shift of the solvation energy relative to the model compound. The total solvation energy is divided into several components: $G_{solv}^x = G_{Born}^x + G_{strain}^x + G_q^x$ ($x$ = deprotonated or protonated form). The Born solvation energy ($G_{Born}^x$) and the energy of interaction with protein charges ($G_q^x$) are two main contributions to the free energy of solvation of the active site complex in the protein environment. The third term ($G_{strain}^x$) is the so-called strain energy; this term reflects the energy associated with the reorganization of electron density of the QM system induced by the reaction field of the polarized medium. The $pK_a$ of the site of interest is then be calculated as

$$pK_a^{site} = pK_a^{model} + \frac{1}{kT\ln 10}\left(\Delta\Delta E_{elec} + \Delta\Delta G_{strain} + \Delta\Delta G_{Born} + \Delta\Delta G_q\right), \quad (4.6)$$

where first two energy terms are evaluated by using quantum chemistry and last two terms are obtained from solvation electrostatic calculations. (Alternatively, one can calculate the absolute value of $pK_a$. [87] We have confirmed that both methods—the absolute and relative $pK_a$ values—give practically identical results.)

An example of the calculation on the binuclear catalytic complex of CcO shown in Figure 4.12, in which we specifically are interested in $pK_a$ of His291 delta nitrogen in different redox states of the enzyme is describe below.

### 4.6.1 DENSITY FUNCTIONAL CALCULATIONS: $\Delta\Delta E_{EL}$, $\Delta G_{STRAIN}$ TERMS

The starting structure of the active site was taken from the x-ray structure of bovine heart CcO obtained at 2.3 Å resolution (PDB code, 2OCC). [13] The QM-model (Figure 4.12) consists of the Cu atom, methylimidazole molecules representing coordinated histidines 240, 290, and 291, a methyl group representing tyrosine 244 (which is cross-linked to His240), and an $H_2O$ or $OH^-$ $Cu_B$-ligand. Energies of each protonation/redox state of interest were calculated using DFT [88,89] and the Jaguar 5.5 quantum chemistry package. [90] Both the B3LYP [91] and PBE0 [92] hybrid density functionals were employed; all calculations use unrestricted wave functions. Typically, the geometry optimization is carried out in vacuum, and therefore without the surrounding charges of the protein; however, the charge effects should be expected. All optimizations were done using the $6-31 + G^*$ (for nonmetal atoms) and LACVP+$^*$ (for Fe and Cu) basis sets, and with both the B3LYP and PBE0 functionals. Separate optimizations were conducted for the singlet, triplet, and quintet spin states to verify that the DFT method predicted the correct ordering of the spin states by comparing our results with the literature values in Refs. [93,94]. Of special interest is the triplet state of the iron(IV)–oxo porphyrin.

Geometry optimization for different redox/protonation states is a critical issue. Which atoms to constrain to keep the right balance between the structure from x-ray and giving the system enough freedom to adjust as quantum mechanics requires? How to include correctly constrains that protein matrix exerts on the QM subsystem? These are open questions that need careful consideration in the calculations.

To calculate $\Delta\Delta E_{el}$, $\Delta G_{strain}$ terms, and to obtain the set of the QM electrostatic potential (ESP) charges for solvation calculations, the complex is surrounded by a continuum dielectric and the self-consistent reaction field (SCRF) method is applied as implemented in Jaguar [95,96] The program used in our calculation did not allow treatment of inhomogeneous dielectrics; we therefore had to modify the program to do such calculations. The new procedure is based on the effective method of reproducing the reaction field of the protein in the cavity of the QM system using spherical harmonics and single value decomposition methods developed by us [41,42].

The ESP atomic charges can be generated by using a modified version of the CHELPG procedure [97]. For the binuclear center shown in Figure 4.12 the computations were performed on five different structures: [Fe=O($H_2O$)$Cu^{2+}$ (HisH)], [Fe=O($H_2O$)$Cu^{2+}$ (His$^-$)], [Fe=O($H_2O$)$Cu^+$ (HisH)], [Fe=O($H_2O$)$Cu^+$ (His$^-$)], and [Fe=O($HO^-$)$Cu^{2+}$ (HisH)], which represent different oxidation states of $Cu_B$ center, different protonation states of His291 site, and $H_2O$/$OH^-$ ligand of $Cu_B$. Similar calculations were performed on the isolated ionizable groups which are not ligands to metal centers.

### 4.6.2   CONTINUUM ELECTROSTATIC CALCULATIONS: $\Delta\Delta G^{BORN}$, $\Delta\Delta G_Q$ TERMS

Once the charges of the QM system have been calculated, its geometry optimized, and the equilibrium charge distribution of the protein has been determined, the solvation energy of the QM system in the protein can be calculated. For solvation, the MEAD program [98] is typically used. The solvation energy in the protein consists of the Born solvation energy and the energy of interaction with the protein charges. To find these contributions, the Poisson–Boltzmann equation (or PBE) is solved numerically. Various dielectric models are used to probe sensitivity of the predictions on the model. [99–102]. Of particular interest is the dependence of the results of calculations on the value of the dielectric constant associate with water filled cavities inside of the protein. This subtle issue has also been explored in direct microscopic simulations of the protein solvation energy (see below). In these calculations, the ESP fitted DFT charges are used for the QM-system, while partial charges of the protein atoms are taken from the CHARMM22 parameter set, [103] modified so as to reflect the equilibrium protonation state of the enzyme [22]. The protein charges [104] and radii [103] that we use in this application have been used successfully before to calculate the electrostatic energies, protonation probabilities of titratable groups, and redox potentials of cofactors in different proteins [23,104–107].

### 4.6.3   MICROSCOPIC MODELING OF POLARIZABLE PROTEIN MEDIUM:
### FIXING EMPIRICAL PARAMETERS FOR CONTINUUM CALCULATIONS

Continuum electrostatic models [72,108–113] are presently most developed and commonly used for the evaluation of the solvation energies in proteins; however, they carry a number of limitations and uncertainties, which cannot be avoided unless the microscopic interactions of the quantum subsystem and the protein are taken into account [114]. For example, it is not clear which dielectric constant of the polarizable water cavities one should use in such calculations; even the usually assumed dielectric "constant" of a dry protein (typically assumed as 4 [99,115,116]) is not that well defined—many studies indicate that the "effective" dielectric of the protein is much higher [114,117–119]—primarily due to internal water [120], and partially due to protein (nonlinear) charge relaxation. Proteins are also inhomogeneous media. It is understood that only microscopic simulations should eventually provide a correct picture and remove the inherent uncertainty of phenomenological approach [71,114,115,121–132]. Despite the drawbacks, the continuum models provide most computationally efficient approach for the treatment of the protein electrostatics, which make possible large-scale investigation of the enzyme properties, such as CcO.

The microscopic approach is challenging. Part of the problem of a full-scale MD simulation is the convergence issue: the long-range nature of electrostatic interactions enforces the inclusion of too many atoms into the simulation box, as a result one cannot (yet) run long-enough trajectories to have sufficient statistics, to capture realistic timescales of protein relaxation, which can be in the microseconds to milliseconds range. Here one can use an intermediate approach: part of the protein (outside QM part) which include all redox sites and internal water is treated explicitly with MD, while remaining part of CcO, membrane, and external water is treated with continuum electrostatics PB model. The accuracy of the force fields used, and in particular the issues of electronic polarization, are the key issues in this approach. As new efficient polarizable models are being developed, one can use a model that combines the usual nonpolarizable force-field MD with continuous treatment of electronic polarization, as described below.

The procedure of fixing continuum model parameters of a protein is typically based on the comparison of the free energy of charging of a selected group (His291 ligand to $Cu_B$, for example) determined by MD simulations and by continuum simulations with adjustable parameters. This approach has been used in the past to study dielectric relaxation in several proteins [115,122,127,128,130,133,134]. Dielectric constants of the protein $\varepsilon_p$ and water cavities $\varepsilon_{cav}$ can be determined in such calculations as adjustable parameters that best reproduce free energy terms obtained in microscopic simulations. The parameters found in such calculation can be further used in the continuum (more efficient) calculations of redox potentials and protonation energetics of other groups in CcO [135].

### 4.6.4 CONTRIBUTION OF THE ELECTRONIC POLARIZATION TO SOLVATION ENERGY: MDEC METHOD

One of the critical issues of microscopic simulations is the contribution of the electronic polarization to total solvation energy. Unlike in highly polar media, such as water, in protein the electronic part of the solvation energy needs to be taken into account explicitly [130,136,137]. One approach is to use fully polarizable force field. While such force fields are still being developed, one can use a semiempirical approach, developed recently [138], where the electronic polarizability is taken into account using the electronic continuum model, while the nuclear polarization part is calculated using nonpolarizable force-field MD. In this model, total solvation energy is a sum of the nuclear part and electronic part; the former is obtained from MD simulations, and the latter from the continuum model (solution of PB equation outside the QM system with dielectric constant around 2). The model states that the total Born solvation energy, the part of energy related to dielectric properties of the protein, can be calculated as follows:

$$\Delta G_{tot}^{Born} = \frac{1}{\varepsilon_{el}} \Delta G_{MD}^{Born} + \Delta G_{el}^{Born}, \qquad (4.7)$$

where

$\Delta G_{MD}^{Born}$ is the energy obtained with standard nonpolarizable MD simulations

$\Delta G_{el}^{Born}$ is the electronic part of the reaction-field energy

The latter is estimated by using phenomenological electronic continuum (EC) model. In this model, the QM system is represented by a set of point charges located inside the QM cavity; inside the cavity the dielectric constant $\varepsilon$ is set to be unity, while outside $\varepsilon = \varepsilon_{el}$. The charging energy $\Delta G_{el}^{Born}$ is found by solving the Poisson equation. The factor $1/\varepsilon_{el}$ in Equation 4.7 reflects the electronic screening of interaction of the charges on the QM system that undergoes charging and the charges of the surrounding protein medium.

In high-dielectric media, such as water, the above expression produces results which are very close to those of the standard approach; the latter does not consider electronic polarization energy at all, and assumes $\Delta G_{tot}^{Born} = \Delta G_{MD}^{Born}$. For example, for a spherical ion of radius $R$, the total charging energy in high-dielectric medium is approximately $Q^2/2R$. It is this value that the standard MD simulation is supposed to reproduce, therefore, $\Delta G_{MD}^{rf} = Q^2/2R$. The electronic energy, on the other hand, is $\Delta G_{el}^{rf} = (1 - 1/\varepsilon_{el})Q^2/2R$. Thus in this case Equation 4.7 predicts $\Delta G_{tot}^{rf} \simeq \Delta G_{MD}^{rf}$, as in the standard approach.

In the low-dielectric media, however, the solvation energy is mainly due to electronic polarization; the standard nonpolarizable MD will produce in this case a negligible contribution $\Delta G_{MD}^{Born}$. Equation 4.7 predicts in this case $\Delta G_{tot}^{Born} \simeq \Delta G_{el}^{Born}$, as expected. Equation 4.7 therefore can be considered as interpolation between high- and low-dielectric cases.

Using the above methodology, one can evaluate the polarization energy contributions of the protein and internal water to the total solvation energy of the system. On the other hand, the same energy can be calculated explicitly, using continuum electrostatics methods with variable adjustable phenomenological parameters for dielectric of dry protein and that of water cavities. Using this method one can resolve the difficult issue of uncertainty of dielectric parameters $pK_a$ calculations [30,31,43]. (For *all* groups of our interest, the full-scale MD free-energy $pK_a$ simulation is unrealistic.)

In summary, the general methodology of $pK_a$ calculations is well developed, however, the accurate calculations in proteins is still a huge challenge. The key problem is the correct description of the dielectric properties of the protein medium. Using better parameters derived from the microscopic simulations, the redox potentials and $pK_a$ values of protonatable ligands of the catalytic centers, as well as other the candidates for proton pump site, will have to be reevaluated in the future work.

## 4.7 KINETIC GATING MECHANISM FOR CHEMICAL AND PUMPED PROTONS IN CcO

One of the key questions of proton pumping in CcO concerns the nature of the mechanism of gating of protons donated by Glu242: all experimental evidence points out that one proton from this group is transferred to a yet unknown PLS above heme $a3$, and the second proton (after re-protonation of Glu242) is transferred to the binuclear center—Steps 3, 4, and 5 in Figure 4.6. How does the protein know where the proton

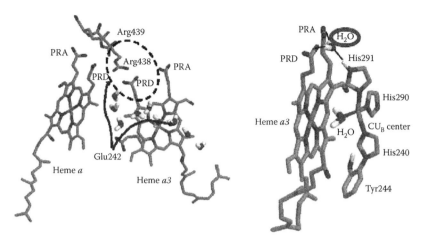

**FIGURE 4.13** Two water chains beyond Glu242 residue in the D-channel of CcO identified in our simulations, left. Protons from Glu242 can either be transferred to PropD/Arg438/His291 site, or to the binuclear center heme $a3$ (OH$^-$)Cu$_B$, right. Water between PRA and PRD of heme $a3$ is part of proton transfer path to His291.

from Glu242 should go, and how does it distinguish between the "pumped" and "chemical" protons? This puzzle of CcO still awaits its solution.

Earlier in our simulations, we found two bifurcated water chains in the catalytic cavity of CcO that connect Glu242, an experimentally known donor of both chemical and pumped protons, to the binuclear Fe$_{a3}$–Cu$_B$ catalytic center, and to PropD/Arg438/His291 groups [21] (Figure 4.13).

This finding suggests a possible mechanism of proton transfer in this region: one branch of the water cluster can be used by pumped protons the other by chemical protons (Figure 4.14). There is driving force for both of these reactions [39]. The mechanism of gating of chemical and pumped protons, however, is not clear.

Two water chains leading to the catalytic site

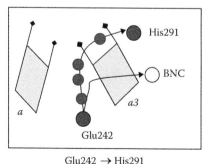

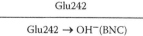

**FIGURE 4.14** Two channels of proton transfer: from Glu242 to BNC (Fe$_{a3}$–Cu$_B$ center) and from Glu242 to PropD/Arg438/His291. The challenge is to understand the gating mechanism of protons in these two channels.

Wikström and coworkers have recently proposed that the gating involves orientation of water molecules in the chains [33]. Our data indicate instead that the mechanism of gating is more likely of kinetic nature. Namely, the structure of hydrogen bonding in the chain leading to Prop D of heme *a3* appears to be more tight than that in the chain connecting Glu242 and the binuclear center, suggesting different energy barriers for proton transfer along these two chains [26].

One possibility is that the barrier for proton transfer of the pumped proton is lower than that for the chemical proton; hence, the rate of pumped proton transfer is faster than for chemical proton. In order to evaluate this proposal, one needs to study the energy profile along the proton transfer path.

The energetics of proton transfer along the water chains can be evaluated using methods that combine QM for a system, MM (or MD) for the protein (including internal water), and continuum electrostatics for external solvent and the membrane. (The latter can treated via generalized solvent boundary potential (GSBP) [139]). The methods of such calculations are currently well developed [32,140, 141–151], and have been reviewed recently [152]. At the time of this writing, such

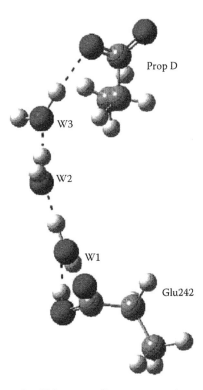

**FIGURE 4.15** Proton transfer QM system: Proton water wire, proton donor Glu242, and proton acceptor Prop D. The boundary of the QM/MM system is at beta-carbons of Glu242 and Prop D, where the dangling bonds are substituted with H-atoms. This system interacts with the medium via LJ (walls of the QM cavity) and Coulomb forces screened by the external solvent and the membrane.

calculations are under way in the research group of the author. The QM system for one reaction of our interest (proton transfer from Glu242 to Prop D) is detailed in Figure 4.15.

The most interesting aspects of this calculation concern the proton transfer reaction itself: (1) Does proton transfer occur via "excess" proton, $H_3O^+$, or $OH^-$, "proton hole" [153]? The energetics can resolve this question. (2) Does the charge transfer occur via localized hopping mechanism (as assumed by Warshel [65]) or via a Grotthuss- or soliton-like [19] delocalized concerted mechanism? The energy minimization should provide the answer to this question as well. The relaxation of the protein around the QM system in response to charge advancement along the chain of water molecules is the key here and needs to be accurately evaluated. (3) How do the timescales of protein dielectric relaxation/reorganization compare to that of proton transfer? i.e., will the protein dielectric reorganization adiabatically follow proton transfer or the relaxation will be incomplete? This novel aspect of the dynamic nature of protein dielectric relaxation needs to be carefully examined. Finally, what are the factors that determine the rate of proton transfer in this case? Does tunneling play a role and will water chain formation a limiting factor here? It is hoped that the analysis of these basic questions will help us to get insights into the gating mechanism of proton translocation in CcO.

## 4.8 CONCLUSIONS

In this chapter, we have reviewed recent progress in the computational studies of CcO. It is clear that a number of outstanding problems still remain to be solved in order to address the molecular details of the proton pumping mechanism of this enzyme. At present, perhaps the most pressing issue is to develop a detailed computational model for the analysis of potentiometric data; this is the only experimental technique that allows monitoring in real time the charge translocation within the enzyme. So far, only continuum model has been developed; a detailed microscopic simulation of the system, together with the membrane, will be critical for precise determination of the nature of the groups that exchange protons with the enzyme. The precise calculation of redox-dependent $pK_a$ values of the groups inside the protein is another fundamental issue for computational research. This area involves a whole complex of difficult problems such as an accurate description of dielectric properties of proteins. The latter issue, in turn, is related to relaxation properties of proteins (micro- and millisecond timescale) in response to an introduced charge and the amount of water inside the protein. The dynamics of water is clearly plays key roles in both the kinetics of proton transfer reactions and thermodynamics of protonatable sites that exchange the protons inside the protein. None of the above problems of course can be adequately addressed without accurate force-fields that treat properly the electronic polarizability of the condensed phase materials. The latter issue clearly will be increasingly more important in simulations of proteins.

It is the complexity and interdependence of fundamental issues so characteristic for biophysical research that makes this area both exciting and challenging for present and future computational biophysicists.

## ACKNOWLEDGMENTS

I would like to acknowledge members of my research group who participated in the development of the ideas discussed in this chapter: Dragan Popovic, Jason Quenneville, Xuehe Zheng, Yuri Georgievskii, Emile Medvedev, Ryogo Sugitani, Dmitry Medvedev, Dmitry Makhov and Igor Leontyev. This work has been supported in part by the NSF grant PHY 0646273, and NIH GM054052.

## REFERENCES

1. W.A. Cramer and D.B. Knaff, *Energy Transduction in Biological Membranes*, Springer-Verlag, New York (1990).
2. V.P. Skulachev, *Membrane Bioenergetics*, Springer-Verlag, Berlin (1988).
3. F. Palmeri, *Biochim. Biophys. Acta* 1772 (2008) 564–578.
4. M.T. Lin and M.F. Beal, *Science* 443 (2006) 787.
5. P. Mitchell, *Nature* 191 (1961) 144–148.
6. M. Wikstrom, *Biochim. Biophys. Acta* 1655 (2004) 241–247.
7. T. Tsukihara, K. Shimokata, Y. Katayama, H. Shimada, K. Muramoto, H. Aoyama, M. Mochizuki, K. Shinzawa-Itoh, E. Yamashita, M. Yao, Y. Ishimura, and S. Yoshikawa, *Proc. Natl. Acad. Sci. U. S. A.* 100 (2003) 15304.
8. H. Michel, *Biochemistry* 38 (1999) 15129–15140.
9. D. Zaslavsky and R.B. Gennis, *Biochim. Biophys. Acta* 1458 (2000) 164–179.
10. M. Wikström, *Biochemistry* 39 (2000) 3515–3519.
11. P. Brzezinski and P. Adelroth, *Curr. Opin. Struct. Biol.* 16 (2006) 465–472.
12. J.P. Hosler, S. Ferguson-Miller, and D.A. Mills, *Annu. Rev. Biochem.* 75 (2006) 165–187.
13. S. Yoshikawa, K. Shinzawa-Itoh, R. Nakashima, R. Yaono, E. Yamashita, N. Inoue, M. Yao, M.J. Fei, C.P. Libeu, T. Mizushima, H. Yamaguchi, T. Tomizaki, and T. Tsukihara, *Science* 280 (1998) 1723–1729.
14. T.K. Das, F.L. Tomson, R.B. Gennis, M. Gordon, and D.L. Rousseau, *Biophys. J.* 80 (2001) 2039–2045.
15. J. Han, S. Takahashi, and D. Rousseau, *J. Biol. Chem.* 275 (2000) 1910–1919.
16. I. Szundi, G.-L. Liao, and O. Einarsdottir, *Biochemistry* 40 (2001) 2332–2339.
17. I. Belevich, D.A. Bloch, N. Belevich, M. Wikström, and M.I. Verkhovsky, *Proc. Natl. Acad. Sci. U. S. A.* 104 (2007) 2685–2690.
18. D. Bloch, I. Belevich, A. Jasaitis, A. Puustinen, M.I. Verkhovsky, and M. Wikström, *Proc. Natl. Acad. Sci. U. S. A.* 101 (2004) 529–533.
19. A.A. Stuchebrukhov, *J. Theor. Comput. Chem.* 2 (2003) 91–118.
20. D.M. Popovic and A.A. Stuchebrukhov, *J. Phys. Chem. B* 109 (2005) 1999–2006.
21. X.H. Zheng, D.M. Medvedev, J. Swanson, and A.A. Stuchebrukhov, *Biochim. Biophys. Acta—Bioenergetics* 1557 (2003) 99–107.
22. D.M. Popovic and A.A. Stuchebrukhov, *J. Am. Chem. Soc.* 126 (2004) 1858–1871.
23. M. Wikström, *Biochim. Biophys. Acta* 1458 (2000) 188–198.
24. H. Michel, *Proc. Natl. Acad. Sci. U. S. A.* 95 (1998) 12819–12824.
25. S. Yoshikawa, *Proc. Natl. Acad. Sci.* 100 (2003) 15304–15309.
26. D.M. Popovic and A.A. Stuchebrukhov, *FEBS Lett.* 566 (2004) 126–130.
27. S.A. Siletsky, A.S. Pawate, K. Weiss, R.B. Gennis, and A.A. Konstantinov, *J. Biol. Chem.* 279 (2004) 52558–52565.
28. D.M. Medvedev, E. Medevedev, and A.A. Stuchebrukhov, *Biochim. Biophys. Acta* 1710 (2005) 47–56.
29. I. Belevich, M.I. Verkhovsky, and M. Wikström, *Nature* 440 (2006) 829–832.

30. P.E.M. Siegbahn, M.R.A. Blomberg, and M.L. Blomberg, *J. Phys. Chem. B* 107 (2003) 10946–10955.
31. M.H. Olsson and A. Warshel, *Proc. Natl. Acad. Sci. U. S. A.* 103 (2006) 6500–6505.
32. J. Xu and G.A. Voth, *Proc. Natl. Acad. Sci. U. S. A.* 102 (2005) 6795–6800.
33. M. Wikström, M.I. Verkhovsky, and G. Hummer, *Biochim. Biophys. Acta* 1604 (2003) 61–65.
34. D. Bloch, I. Belevich, A. Jasaitis, C. Ribacka, A. Puustinen, M.I. Verkhovsky, and M. Wikstrom, *Proc. Natl. Acad. Sci. U. S. A.* 101 (2004) 529–533.
35. Y. Georgievskii, E.S. Medvedev, and A.A. Stuchebrukhov, *J. Chem. Phys.* 116 (2002) 1692–1699.
36. Y. Georgievskii, E.S. Medvedev, and A.A. Stuchebrukhov, *Biophys. J.* 82 (2002) 2833–2846.
37. R. Sugitani, E.S. Medvedev, and A.A. Stuchebrokhov, *Biochim. Biophys. Acta* 1777 (2008) 1129–1139.
38. A.A. Stuchebrukhov, Coupled electron and proton transfer reactions in proteins and computational challenges of membrane redox pumps, In Cundari, T. (ed.), *Reviews in Computational Chemistry*, Wiley-VCH, Weinheim (2007) [Invited review article].
39. J. Quenneville, D.M. Popovic, and A.A. Stuchebrukhov, *Biochim. Biophys. Acta* 1757 (2006) 1035–1046.
40. D.M. Popovic, J. Quenneville, and A.A. Stuchebrukhov, *Modern Methods for Theoretical Physical Chemistry of Biopolymers*, Elsevier, Amsterdam, the Netherlands (2006).
41. D. Makhov, D.M. Popovic, and A.A. Stuchebrukhov, *J. Phys. Chem. B* 110 (2006) 12162–12166.
42. D. Makhov and A.A. Stuchebrukhov, *Phys. Stat. Sol. (b)* 242 (2006) 2030–2037.
43. D.M. Popovic, J. Quenneville, and A.A. Stuchebrukhov, *J. Phys. Chem. B* 109 (2005) 3616–3626.
44. A.A. Stuchebrukhov, *Theor. Chem. Acc.* 110 (2003) 291–306.
45. Y. Georgievskii and A.A. Stuchebrukhov, *J. Chem. Phys.* 113 (2000) 10438–10450.
46. A.A. Stuchebrukhov, *Adv. Chem. Phys.*, 118 (2001) 1–44.
47. A.A. Stuchebrukhov, *J. Chem. Phys.* 118 (2003) 7898–7906.
48. A.A. Stuchebrukhov, *Int. J. Quantum Chem.* 77 (2000) 16–26.
49. J. Wang and A.A. Stuchebrukhov, *Int. J. Quantum Chem.* 80 (2000) 591–597.
50. J. Cushing, A. Fine, and S. Goldstein, *Bohmian Mechanics and Quantum Theory: An Appraisal*, Kluwer Academic, Dordrecht the Netherlands (1996).
51. D.M. Medvedev, I. Daizadeh, and A.A. Stuchebrukhov, *J. Am. Chem. Soc.* 122 (2000) 6571–6582.
52. S.S. Skourtis and D.N. Beratan, *Adv. Chem. Phys.* 106 (1999) 377–452.
53. J.J. Regan and J.N. Onuchic, *Adv. Chem. Phys.* 107 (1999) 497–553.
54. D.N. Beratan, J.N. Onuchic, J.R. Winkler, and H.B. Gray, *Science* 258 (1992) 1740–1741.
55. C.C. Page, C.C. Moser, X. Chen, and P.L. Dutton, *Nature* 402 (1999) 47–52.
56. C.C. Moser, J.M. Keske, K. Warncke, R.S. Farid, and L.P. Dutton, *Nature* 355 (1992) 796.
57. C.C. Moser and P.L. Dutton, *Biochim. Biophys. Acta* 1101 (1992) 171–176.
58. C.C. Page, C.C. Moser, and P.L. Dutton, *Curr. Opin. Struct. Biol.* 7 (2003) 551–556.
59. T.R. Prytkova, I.V. Kurnikov, and D.N. Beratan, *Science* 315 (2007) 622–625.
60. M.L. Tan, I. Balabin, and J.N. Onuchic, *Biophys. J.* 86 (2004) 1813–1819.
61. A. Jasaitis, F. Rappaport, E. Pilet, U. Liebl, and M.H. Vos, *Proc. Natl. Acad. Sci. U. S. A.* 102 (2005) 10882–10886.
62. A. Jasaitis, M.P. Johansson, M. Wikström, M.H. Vos, and M.I. Verkhovsky, *Proc. Natl. Acad. Sci. U. S. A.* 104 (2007) 20811–20814.

63. A. Namslauer, A.S. Pawate, R. Gennis, and P. Brzezinski, *Proc. Natl. Acad. Sci. U. S. A.* 100 (2003) 15543–15547.
64. M. Ruitenberd, A. Kannt, E. Bamberg, B. Ludwig, H. Michel, and K. Fendler, *Proc. Natl. Acad. Sci. U. S. A.* 97 (2000) 4632–4636.
65. A. Warshel, *Annu. Rev. Biophys. Biomol. Struct.* 32 (2003) 425–443.
66. A. Warshel, *Proteins* 44 (2001) 400–417.
67. J. Mongan, D.A. Case, and A.J. McCammon, *J. Comput. Chem.* 25 (2004) 2038–2048.
68. M. Feig and C.L. Brooks III, *Curr. Opin. Struct. Biol.* 14 (2004) 217–224.
69. A.A. Stuchebrukhov, *J. Theor. Comp. Chem.* 2 (2003) 91–118.
70. A.A. Stuchebrukhov and D.M. Popovic, *J. Phys. Chem. B* 110 (2006) 17286–17287.
71. T. Simonson, *J. Am. Chem. Soc.* 120 (1998) 4875.
72. D. Bashford and M. Karplus, *Biochemistry* 29 (1990) 10219.
73. M.K. Gilson and B. Honig, *Proteins Struct. Function Genetics* 4 (1988) 7–18.
74. J.M. Mouesca, J.L. Chen, L. Noodleman, D. Bashford, and D.A. Case, *J. Am. Chem. Soc.* 116 (1994) 11898–11914.
75. J. Li, C.L. Fischer, J.L. Chen, D. Bashford, and L. Noodleman, *J. Phys. Chem.* 96 (1996) 2855–2866.
76. J. Li, M.R. Nelson, C.Y. Peng, D. Bashford, and L. Noodleman, *J. Phys. Chem. A* 102 (1998) 6311–6324.
77. J. Li, C.L. Fisher, R. Konecny, D. Bashford, and L. Noodleman, *Inorg. Chem.* 38 (1999) 929–939.
78. G.M. Ullmann, L. Noodleman, and D.A. *J. Biol. Inorg. Chem.* 7 (2002) 632–639.
79. J. Antosiewicz, J.A. McCammon, and M.K. Gilson, *Biochemistry* 35 (1996) 7819.
80. S. Varma and E. Jakobsson, *Biophys. J.* 86 (2004) 690.
81. R. Luo, J. Moult, and M.K. Gilson, *J. Phys. Chem. B* 101 (1997) 11226–11236.
82. E.J. Nielsen and A.J. McCammon, *Protein Sci.* 12 (2003) 313–326.
83. E.G. Alexov and M.R. Gunner, *Biophys. J.* 74 (1997) 2075–2093.
84. Y.Y. Sham, Z.T. Chu, and A. Warshel, *J. Phys. Chem. B* 101 (1997) 4458–4472.
85. Y.Y. Sham, I. Muegge, and A. Warshel, *Biophys. J.* 74 (1998) 1744–1753.
86. D.M. Chipman, *J. Phys. Chem. A* 106 (2002) 7413–7422.
87. J. Quenneville, D.M. Popovic, and A.A. Stuchebrukhov, *J. Phys. Chem. B* 108 (2004) 18383–18389.
88. P. Hohenberg and W. Kohn, *Phys. Rev.* 136 (1964) B864–874.
89. K. Kohn and L.J. Sham, *Phys. Rev.* 140 (1965) A1133.
90. Jaguar 5.5, Schroedinger, Inc., L.L.C., Portland, OR, 1991–2003. ed.
91. A.D. Becke, *J. Chem. Phys.* 98 (1993) 5648.
92. J.P. Perdew and M. Ernzerhof, *J. Chem. Phys.* (1996) 9982.
93. M.T. Green, *J. Am. Chem. Soc.* 122 (2000) 9495.
94. M.T. Green, *J. Am. Chem. Soc.* 123 (2001) 9218.
95. D.J. Tannor, B. Marten, R. Marphy, R.A. Friesner, D. Sitkoff, A. Nicholls, M.G.W.A.I. Ringnalda, and B. Honig, *J. Am. Chem. Soc.* 116 (1994) 11875.
96. B. Marten, K. Kim, C. Cortis, R.A. Friesner, R.B. Murphy, M.N. Ringnalda, D. Sitkoff, and B. Honig, *J. Phys. Chem.* 100 (1996) 11775.
97. C.M. Breneman and K.B. Wiberg, *J. Comput. Chem.* 11 (1990) 361.
98. D. Bashford, in Ishikawa, Y., Oldehoeft, R.R., Reynders, J.V.W., and Tholburn, M. (eds.), *Scientific Computing in Object-Oriented Parallel Environments*, Springer, Berlin (1997), pp. 233–240.
99. K. Sharp and B. Honig, *Ann. Rev. Biophys. Biophys. Chem.* 19 (1990) 301.
100. T. Simonson and D. Perahia, *Proc. Natl. Acad. Sci. U. S. A.* 92 (1995) 1082–1086.
101. T. Simonson and C.L. Brooks, *J. Am. Chem. Soc.* 118 (1996) 8452–8458.
102. A. Warshel and A. Papazyan, *Curr. Opin. Struct. Biol.* 8 (1998) 211–217.

103. A.D. MacKerell Jr., D. Bashford, M. Bellot, R.L. Dunbrack, J.D. Evanseck, M.J. Field, S. Fischer, J. Gao, H. Guo, S. Ha, D. Joseph-McCarthy, L. Kuchnir, K. Kuczera, F.T.K. Lau, C. Mattos, S. Michnick, T. Ngo, D.T. Nguyen, B. Prodholm, I. Reiher, W.E., B. Roux, M. Schlenkrich, J.C. Smith, R. Stote, J. Straub, M. Watanabe, J. Wiórkiewicz-Kuczera, D. Yin, and M. Karplus, *J. Phys. Chem.* 102 (1998) 3586–3616.

104. D.M. Popovic, S.D. Zaric, B. Rabenstein, and E.W. Knapp, *J. Am. Chem. Soc.* 123 (2001) 6040–6053.

105. D.M. Popovic, A. Zmiric, S.D. Zaric, and E.W. Knapp, *J. Am. Chem. Soc.* 124 (2002) 3775–3782.

106. P. Vagedes, B. Rabenstein, J. Åqvist, J. Marelius, and E.W. Knapp, *J. Am. Chem. Soc.* 122 (2000) 12254–12262.

107. B. Rabenstein and E.W. Knapp, *Biophys. J.* 80 (2001) 1141–1150.

108. J. Warwicker and H. Watson, *J. Mol. Biol.* 157 (1982) 671.

109. M.E. Davis and J.A. McCammon, *Chem. Rev.* 90 (1990) 509.

110. M.R. Gunner and B. Honig, *Proc. Natl. Acad. Sci. U. S. A.* 88 (1991) 9151.

111. B. Honig and A. Nicholls, *Science* 268 (1995) 1144.

112. P. Beroza and D.A. Case, *J. Phys. Chem.* 100 (1996) 20156.

113. D.M. Popovic, J. Quenneville and A.A. Stuchebrukhov, *J. Phys. Chem. B* 109 (2005) 3616.

114. C.N. Schutz and A. Warshel, *Proteins* 44 (2001) 400.

115. T. Simonson and D. Perahia, *Proc. Natl. Acad. Sci. U. S. A.* 92 (1995) 1082.

116. T. Simonson and C.L. Brooks, *J. Am. Chem. Soc.* 118 (1996) 8452.

117. J. Antosiewicz, J.A. McCammon, and M.K. Gilson, *J. Mol. Biol.* 238 (1994) 415.

118. O. Miyashita, J.N. Onuchic, and M.Y. Okamura, *Biochemistry* 42 (2003) 11651.

119. I. Muegge, T. Schweins, R. Langen, and A. Warshel, *Structure* 4 (1996) 475.

120. C.A. Fitch, D.A. Karp, K.K. Lee, W.E. Stites, E.E. Lattman, and B.E. Garcia-Moreno, *Biophys. J.* 82 (2002) 3289–3304.

121. A. Warshel, S.T. Russell, and A.K. Churg., *Proc. Natl. Acad. Sci. U. S. A.* 81 (1984) 4785.

122. Y. Sham, I. Muegge, and A. Warshel, *Biophys. J.* 74 (1998) 1744.

123. T. Simonson, *Rep. Prog. Phys.* 66 (2003) 737.

124. P. Smith, R. Brunne, A. Mark, and W.F. van Gunsteren, *J. Phys. Chem. B* 97 (1993) 2009.

125. L. Krishtalik, A. Kuznetsov, and E. Mertz, *Proteins* 28 (1997) 174.

126. H. Nakamura, *Q. Rev. Biophys.* 29 (1996) 1.

127. G. Archontis and T. Simonson, *J. Am. Chem. Soc.* 123 (2001) 11047.

128. T. Simonson, G. Archontis, and M. Karplus, *J. Phys. Chem. B* 103 (1999) 6142.

129. T. Simonson, J. Carlsson, and C.D., *J. Am. Chem. Soc.* 126 (2004) 4167.

130. G. Archontis and T. Simonson, *Biophys. J.* 88 (2005) 3888.

131. T. Simonson, *Int. J. Quantum Chem.* 73 (1999) 45.

132. J. Pitera, M. Falta, and W. Van Gunsteren, *Biophys. J.* 80 (2001) 2546.

133. T. Simonson, D. Perahia, and A.T. Brunger, *Biophys. J.* 59 (1991) 670.

134. G. King, F. Lee, and A. Warshel, *J. Chem. Phys.* 95 (1991) 4366.

135. I.V. Leontyev and A.A. Stuchebrokhov, *J. Chem. Phys.* (2008) (in press).

136. I.V. Leontyev, M.V. Vener, I.V. Rostov, M.V. Basilevsky, and M.D. Newton, *J. Chem. Phys.* 119 (2003) 8024.

137. M.V. Vener, I.V. Leontyev, and M.V. Basilevsky, *J. Chem. Phys.* 119 (2003) 8038.

138. I.V. Leontyev and A.A. Stuchebrokhov, *J. Chem. Phys.* 130 (2009) 085102.

139. W. Im, S. Berneche, and B. Roux, *J. Chem. Phys.* 114 (2001) 2924–2937.

140. Q. Cui and M. Karplus, *J. Phys. Chem. B* 107 (2003) 1071.

141. H. Park and J.K.M. Merz, *J. Am. Chem. Soc.* 127 (2005) 4232–4221.

142. E. Rosta, M. Klahn, and A. Warshel, *J. Phys. Chem, B*, 110 (2006) 2934–2941.

143. N. Reuter, A. Dejaegere, B. Maigret, and M. Karplus, *J. Phys. Chem. A* 104 (2000) 1720–1735.

144. A. Nemukhin, B. Grigorieva, I. Topol, and S.K. Burt, *J. Phys. Chem. B* 107 (2003) 2958–2965.

145. V. Guallar and R.A. Friesner, *J. Am. Chem. Soc.* 126 (2004) 8501–8508.

146. B. Gherman, M. Baik, S. Lippard, and R.A. Friesner, *J. Am. Chem. Soc.* 126 (2004) 2978–2990.

147. B. Gherman, S. Lippard, and R.A. Friezer, *J. Am. Chem. Soc.* 127 (2005) 1025–1037.

148. P.H. Konig, M. Hoffmann, and Q. Cui, *J. Phys. Chem. B* 109 (2005) 9082–9095.

149. J.K.M. Merz, *Encyclopedia Comput. Chem.* 4 (1998) 2330–2343.

150. A.-N. Bondar, S. Fischer, and J. Smith, *J. Am. Chem. Soc.* 126 (2004) 14668–14677.

151. A.-N. Bondar, J.C. Smith, and S. Fischer, *Photochem. Photophys. Sci.* 5 (2006) 547–552.

152. Q. Cui, *J. Phys. Chem. B* 110 (2006) 6458–6469.

153. D. Riccardi, P. Konig, H. Yu, X. Prat-Resina, M. Elstner, T. Frauenheim, and Q. Cui, *J. Am. Chem. Soc.* 128 (2006) 16302–16311.

# Part II

## Vibrational Energy Flow in Proteins: Molecular Dynamics-Based Methods

# 5 Molecular Dynamics Simulation of Proteins: Two Models of Anharmonic Dynamics

*Akinori Kidera, Kei Moritsugu,*
*Yasuhiro Matsunaga, and Hiroshi Fujisaki*

## CONTENTS

## 5.1 INTRODUCTION

The biological function of a protein can be interpreted as a physicochemical process of sequential switching motions and associated chemical reactions occurring as a response to a certain external perturbation. These switching motions, or the structural transitions between the resting and the activated states, are a consequence of the highly anharmonic dynamics on the rugged potential surface of the protein molecule [1]. Thus, understanding such anharmonic dynamics is essential for elucidating the expression and regulation of molecular functions. Molecular dynamics (MD) simulation has made significant contributions to studies of the anharmonic dynamics of proteins. One of the most successful models for analyzing the anharmonic dynamics in MD trajectories is based on the quasiharmonic approximation, where the distribution function is approximated up to the second-order with respect to coordinates [2,3]. The anharmonicity is highlighted by embedding the MD trajectory into a few largest-amplitude modes to depict the landscape of large-scale collective motions [4,5]. Although the quasiharmonic picture has been successfully used for analyses of the equilibrium distribution, it is not necessarily suitable for time-domain analysis of relaxation behaviors.

Many MD studies on the kinetic processes of proteins have revealed the existence of a specific pathway of energy flow [6–10]. In this chapter, for the purpose of studying anharmonic dynamics, we propose two models in which protein motions are assumed to be described by perturbed harmonic oscillators [11–13]. These models are based on the success of the harmonic/quasiharmonic model in the description of the equilibrium properties of proteins. Firstly, we consider vibrational energy transfer between normal modes, based on the Lagrangian [11,12]:

$$L = \sum_{i=1}^{N}\left(\frac{1}{2}\dot{q}_i^2 - \frac{1}{2}\omega_i^2 q_i^2\right) - \sum_{i,j,k=1}^{N}\alpha_{ijk}q_iq_jq_k - \sum_{i,j,k,l=1}^{N}\beta_{ijkl}q_iq_jq_kq_l - \cdots, \quad (5.1)$$

where
  $q_i$ is the mass-weighted coordinate of the $i$th normal mode
  $\omega_i$ is the frequency of the associated normal mode
  $N$ is the number of internal degrees of freedom in the system

The higher-order terms with coefficients $\alpha_{ijk}$, $\beta_{ijkl}$, ... represent the perturbation of the harmonic part in the Lagrangian or the mode couplings. Through these coupling terms, the vibrational energy is transferred among normal modes. In other words, the anharmonic dynamics can be explained by tracing the vibrational energy transfer from one mode to another.

Secondly, the scope is extended to cover a much wider range of the conformational space using the following Lagrangian [13]:

$$L = \frac{1}{2}\sum_{i=1}^{N}\left(\left[\dot{\mathbf{x}}\cdot\mathbf{e}_i(t)\right]^2 - \omega_i(t)^2\left[\left(\mathbf{x}-\mathbf{x}_0(t)\right)\cdot\mathbf{e}_i(t)\right]^2\right), \quad (5.2)$$

where
  $\mathbf{x}$ is an $N$-dimensional vector representing the Cartesian coordinates
  $\omega_i(t)$ and $\mathbf{e}_i(t)$ are the frequency and the mode vector, respectively, for mode $i$ defined instantaneously at time $t$
  $\mathbf{x}_0(t)$ is the origin of vibration at time $t$

If these basis functions, $\omega_i(t)$, $\mathbf{e}_i(t)$, and $\mathbf{x}_0(t)$, are time-independent, then Equation 5.2 is simply a Lagrangian for a set of harmonic oscillators. In a protein system, however, strong anharmonicity makes the basis functions time-dependent, and thus we refer to the model as "moving normal mode coordinates."

## 5.2   VIBRATIONAL ENERGY TRANSFER IN PROTEINS

Since the vibrational energy transfer in a protein molecule occurs in a high dimensional space, an approach that prescribes selecting a small number of modes that participate in the transfer process is not appropriate. Rather, simulation should be performed using a heuristic approach. However, classical MD simulations are not necessarily suitable for simulation of energy transfer, particularly when the transfer process contains high-frequency modes, in which quantum effects are dominant [14]. As shown below, at ordinary temperatures, the off-resonance transfer obstructs observations of the slow process of energy transfer. To overcome this problem, we used a classical MD simulation at near 0 K to investigate the coupling of normal modes through energy transfer. In a protein molecule at 0 K, a specified mode (referred to hereafter as the perturbed mode) was exclusively assigned a small excess of vibrational energy to see how the imposed energy was transferred to other modes. In the simulation, since all other modes were kept in a minimum energy state, we could specifically observe the time course of the energy transfer between the perturbed mode and the resonance modes.

### 5.2.1 Energy Transfer at Near-Zero Temperatures

We will explain energy transfer at near zero temperatures using the classical formulation of a model system composed of only two modes, the perturbed mode $q_1$ and the resonance mode $q_2$. The Lagrangian is

$$L = \tfrac{1}{2}\left(\dot{q}_1^2 + \dot{q}_2^2\right) - \tfrac{1}{2}\left(\omega_1^2 q_1^2 + \omega_2^2 q_2^2\right) - \alpha q_1^2 q_2, \tag{5.3}$$

where the anharmonic term is limited to $q_1^2 q_2$ with the coupling coefficient $\alpha$. When $\alpha$ is sufficiently small, the time evolution of $q_1$ and $q_2$ is given up to the first order in $\alpha$ as [11,15]

$$q_1(t) = A_1 \cos\tau_1 - \alpha A_1 A_2 \left[ \frac{\cos(\tau_1 + \tau_2)}{\omega_1^2 - (\Omega_1 + \Omega_2)^2} + \frac{\cos(\tau_1 - \tau_2)}{\omega_1^2 - (\Omega_1 - \Omega_2)^2} \right] + O(\alpha^2),$$

$$q_2(t) = A_2 \cos\tau_2 - \frac{\alpha A_1^2}{2} \left[ \frac{\cos 2\tau_1}{\omega_2^2 - 4\Omega_1^2} + \frac{1}{\omega_2^2} \right] + O(\alpha^2), \tag{5.4}$$

where
$\Omega_i$ is the modulated frequency [$\Omega_i = \omega_i + O(\alpha)$] due to the perturbation
$\tau_i = \Omega_i t + \theta_i$, with $\theta_i$ being the initial phase
$A_1$ and $A_2$ are determined by the initial condition

In this system, when the resonance condition is satisfied, i.e., $2\omega_1 - \omega_2 \equiv \Delta \sim O(\alpha)$, the coupling term has a significant contribution to the zeroth order of $\alpha$.

$$E_1 \simeq \frac{1}{2} A_1^2 \omega_1^2 - \frac{\alpha}{\Delta} \frac{A_1^2 A_2 \omega_1}{2} \cos(2\tau_1 - \tau_2) + O(\alpha)$$

$$E_2 \simeq \frac{1}{2} A_2^2 \omega_2^2 + \frac{\alpha}{\Delta} \frac{A_1^2 A_2 \omega_2}{4} \cos(2\tau_1 - \tau_2) + O(\alpha). \tag{5.5}$$

Equation 5.5 clearly shows that the resonance condition is satisfied in the energy transfer. It also indicates that energy conservation is satisfied at the zeroth order of $\alpha$, i.e., $E_1 + E_2 = $ const., and that energy transfer requires a period, $\sim 2\pi/(2\omega_1 - \omega_2) = 2\pi/\Delta$, which is much longer than the frequencies of the two modes. This formula can easily be extended to a system with higher order coupling and more than three oscillators.

In this study, deoxymyoglobin (PDB entry: 1 mwd) was chosen as the simulation system. The trajectory was obtained by MD simulations using an initial minimum-energy structure at 0 K perturbed by kinetic energy of 1 kcal/mol (corresponding to 0.136 K after equilibration) imposed on mode 2861 with a frequency of 827.2 cm⁻¹. Figure 5.1 shows the time course of total energy after perturbation of myoglobin. This system is weakly anharmonic, as shown by the sum of the harmonic total energy, $\sum_i (\dot{q}_i^2 + \omega_i^2 q_i^2)/2$, which does not drift. This result clearly indicates that energy transfer occurs via Fermi resonance of the third-order coupling terms. The perturbed energy of mode 2861 is transferred to modes, 1589 (362.9 cm⁻¹), 1860 (464.5 cm⁻¹), and 5858 (1654.5 cm⁻¹) through third-order coupling. The amount of energy transferred to the other 7415 modes is negligible, as shown in Figure 5.1, although a small but clear energy transfer through fourth-order resonance can also be observed [12].

Figure 5.1 shows two different types of third-order coupling: Two modes (2861 and 5858) couple in the term, $\alpha q_{2861}^2 q_{5858}$, satisfying the resonance condition, 827.2 cm⁻¹ × 2 ≈ 1654.5 cm⁻¹; and three modes (2861, 1589, and 1860) couple in the term, $\alpha q_{2861} q_{1589} q_{1860}$, the resonance condition of which is 827.2 cm⁻¹ ≈ 362.9 cm⁻¹ + 464.5 cm⁻¹. Furthermore, energy transfer through the term $\alpha q_i q_j q_k$ can be seen to occur earlier than energy transfer through the term $\alpha q_i^2 q_j$. At $t = 0$, all amplitudes besides $q_{2861}$ are zero. Hence, the coupling term $\alpha_1 q_{2861}^2 q_{5858}$ yields a nonzero force,

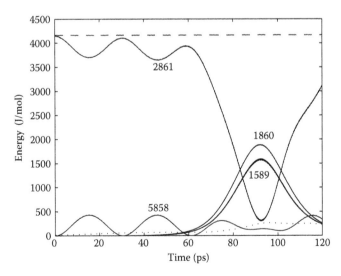

**FIGURE 5.1**  Time course of the total energy of the perturbed mode 2861, and resonance modes 1589, 1860, and 5858. The sum of the total energy for the other modes (dotted curve), and for the whole system (broken curve) are also shown. (From Moritsugu, K. et. al., *J. Phys. Chem. B* 107, 3309, 2003. With permission.)

$-\partial L/\partial q_{5858}$, to excite mode 5858. The other modes require a longer period, ~80 ps, to undergo indirect excitation probably through the low-frequency modes. This can be understood as a kind of the induction phenomena [15].

It has been noted that the frequency rule is a necessary condition but not a sufficient condition for energy transfer. Because of the high density of modes owing to the large number of degrees of freedom in myoglobin, many modes, besides the modes described above, satisfy the resonance condition. Among these many modes, the target mode for energy transfer is selected by coupling coefficients $\alpha$ and $\beta$, as shown in Equation 5.1. As seen in Equation 5.5, the amount of resonance energy is determined by prefactor $\alpha/\Delta$, which includes both the resonance condition $\Delta$ and the coupling coefficient [12,16]. Here, we examine the correlation between the value of $\alpha/\Delta$ and the energy transferred in the MD simulation.

To demonstrate the reason why mode 5858 was selected, values of $\alpha$ were calculated numerically by differentiating the potential energy, i.e., $\partial^3 V/\partial q_{2861}^2 \partial q_i$. This process revealed that $|\alpha/\Delta|$ was largest at $i = 5858$, demonstrating that energy transfer at near-zero temperatures is well described by the resonance scheme of Equation 5.5, i.e., the energy is transferred to modes with small modulated frequency difference ($\Delta$) and large coupling coefficient $\alpha$. Energy transfer through coupling of $\alpha q_i q_j q_k$ was also well described by the factor $|\alpha/\Delta|$; that is, the values of $|\alpha/\Delta|$ (where $\Delta = \omega_{2861} - \omega_i - \omega_j$) were especially large for $i = 1589$ and $j = 1860$ [12].

Finally, we considered the structural implications of mode coupling. We defined the following quantity to measure the geometrical overlap of the three modes $i, j, k$:

$$G_{ijk} = \sum_{p=1}^{n} m_p^{3/2} \left|\mathbf{e}_{ip}\right|\left|\mathbf{e}_{jp}\right|\left|\mathbf{e}_{kp}\right|, \tag{5.6}$$

where

$n$ is the number of atoms in the protein

$\mathbf{e}_{ip}$ is the eigenvector component for mode $i$ and atom $p$

A high correlation was found between the absolute values of the coupling coefficients $|\alpha|$, and their corresponding $G$ values, indicating that the coupling coefficients are determined mostly by the geometric overlap of the corresponding normal modes [11,12].

The possibility of evaluating the coupling coefficients using the analogy of multidimensional infrared (IR) spectroscopy [17] was also considered on the basis of the classical formulation shown in Equation 5.3. In the case of the third-order term $\alpha_2 q_1^2 q_2$, the coupling coefficient $\alpha_2$ corresponds to the off-diagonal peak intensity of the power spectra, $I(\Omega_1, |\Omega_1 - \Omega_2|)$. Here, the power spectra associated with multidimensional IR spectroscopy can be calculated via 2D Fourier transform of the following time correlation function:

$$\left\langle q_1(t_1)q_1(t_2)q_2(0) \right\rangle = \lim_{T \to \infty} \frac{1}{2T} \int_{-T}^{T} dt q_1(t+t_1)q_1(t+t_2)q_2(t). \tag{5.7}$$

Therefore, in principle, it is possible to obtain the coupling coefficient from an MD trajectory. If the system contains only two variables, as in Equation 5.3, the spectral intensity (i.e., the Fourier transform of Equation 5.7) at the off-diagonal peak becomes to the first-order term of $\alpha_2$ as follows.

$$I(\Omega_1, |\Omega_1 - \Omega_2|) \sim \left[ \frac{\alpha_2 A_1 A_2^2}{\omega_1^2 - (\Omega_1 - \Omega_2)^2} \right]^2. \tag{5.8}$$

However, in the case where mode 1 couples with many modes with coupling coefficients $\alpha_i$, the peak intensity becomes

$$I(\Omega_1, |\Omega_1 - \Omega_2|) \sim \left[ \frac{\alpha_2 A_1 A_2^2}{\omega_1^2 - (\Omega_1 - \Omega_2)^2} + \sum_i \frac{\alpha_i A_1 A_i^2}{\omega_1^2 - (\Omega_1 - \Omega_i)^2} \lim_{T \to \infty} \frac{1}{2T} \int_{-T}^{T} dt \cos\left[ (\Omega_2 - \Omega_i) t \right] \right]^2. \tag{5.9}$$

This means that the spectral intensity contains the contributions from not only $\alpha_2$ but also all the other $\alpha_i$ values. In order to separate $\alpha_2$ from the other values, integration of the period $2\pi/(\Omega_2 - \Omega_i)$ is necessary.

## 5.2.2   ENERGY TRANSFER AT FINITE TEMPERATURES

At near-zero temperatures, all modes in the system are almost freezing or have a near-zero amplitude, except those participating in energy transfer. Therefore, it is possible to observe the energy transfer through the resonance mechanism, which requires a long time without disturbance from off-resonance coupling. In contrast, in an environment at a finite temperature where all the modes are oscillating, the perturbation energy will dissipate to the other modes through off-resonance coupling in a much shorter time, before the resonance-type transfer is completed. Actually, it has been found that the energy transfer from modes 2861–5858 at finite temperatures quickly decreases to the level of the thermal fluctuation at ~50 K, indicating that the resonance contribution has almost disappeared [12]. Energy transfer occurs through the resonance scheme only at near-zero temperatures, and the off-resonance scheme starts to dominate the energy transfer when the temperature is raised slightly above zero.

Now we consider off-resonance energy transfer at temperatures over 50 K in order to examine mode coupling in a thermally excited state. At high temperatures, the resonant energy transfer between a pair of modes will be disturbed by thermal conduction. To overcome this problem, we calculated the energy transfer within a short-time period (1 fs) to obtain the average excess energy:

$$\Delta \bar{E}_i^k(T) = \left\langle \left| E_i^k(T; t; m) - E_i(T; t; m) \right| \right\rangle, \tag{5.10}$$

where
   $E_i(T; t; m)$ is the kinetic energy for mode $i$ at time $t$ ($=1$ fs), calculated from the $m$th
      MD trajectory at temperature $T$

$E_i^k(T; t; m)$ is kinetic energy calculated from the same trajectory but with a perturbation involving a certain amount of kinetic energy being added to mode $k$ at $t = 0$

We chose $k = 2861$ as above.

According to Equation 5.3, within a short time period after perturbation, i.e.,
$\Delta t$ [$\ll 2\pi/(2\omega_1 + \omega_2)$], is derived at the first order of $\alpha$ as [12]

$$E_1(\Delta t) - E_1(0) = \frac{\alpha A_1^2 A_2}{2} \omega_1 \Delta t \left[ \sin(2\theta_1 + \theta_2) + \sin(2\theta_1 - \theta_2) \right],$$

$$\tag{5.11}$$

$$E_2(\Delta t) - E_2(0) = \frac{\alpha A_1^2 A_2}{4} \omega_2 \Delta t \left[ \sin(2\theta_1 + \theta_2) - \sin(2\theta_1 - \theta_2) + 2\sin\theta_2 \right].$$

A simulation for 1 fs after perturbation can cover only an infinitesimal area of the
potential surface, unlike simulations for a longer timescale, which can better explore
the rugged potential surface. In this sense, our simulation of energy transfer during 1 fs
after perturbation does not intend to analyze the kinetic process of energy transfer, but
intends to study the shape of the local potential surface around an instantaneous structure formed at a finite temperature. Therefore, it is reasonable to analyze the results of
such simulations from the viewpoint of series expansion of potential energy.

MD simulations were carried out at 13 different temperatures ranging from 5 to
300 K, with a perturbation of 0.1 kcal/mol kinetic energy added to mode 2861, and
with a time step of 0.1 fs to allow the short-time dynamics to be followed. The excess
kinetic energy, $\Delta \bar{E}_i^{2861}(T)$, was averaged over the interval of $t = 0$–1 fs. To cover the
small amount of samples, we repeated the MD simulations 500 times with 500 dif-
ferent initial conditions taken from 500 ps trajectories at each temperature with the
interval of 1 ps. In analyses of the trajectories, we were able to successfully calculate
the average excess energy even at 300 K.

Figure 5.2 plots the sums of the average excess energy, $\sum_{i \neq 2861} \Delta \bar{E}_i^{2861}(T)$, for all
modes other than mode 2861. In this figure, a sudden increase in the gradient can be
seen at ~180 K. The inflection in the average excess energy at ~180 K is reminiscent
of the glass transition-like phenomena observed in protein dynamics [18], which
represents a dynamic transition from harmonic vibration to anharmonic diffusive
motions at around 200 K.

To clarify the relation between the inflection in the transferred energy and the
glass transition, the energy transfer was recalculated by reducing the variety of initial structures for the MD simulations. The average excess energy was calculated
using the results of 100 simulations with 100 different initial structures, which was
taken from the first 100 ps trajectory, instead of the full 500 ps trajectory. This analysis was based on the expectation that within 100 ps there is a large probability that a
myoglobin molecule will remain within the same energy basin of the conformational
substate, which can be well represented by the original set of normal modes in the
quasiharmonic sense. Over the period of 500 ps, however, the myoglobin molecule
may jump into another substate, which should be represented by a new set of normal
modes. When the dynamics in the new conformational substate is represented by the
original set of normal modes, a number of coupling terms should be added to explain

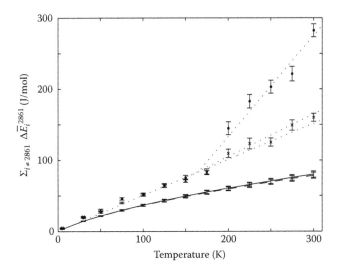

**FIGURE 5.2**  Sum of the absolute values of the average excess energy for all modes, except for mode 2861, averaged over 500 ps trajectories (•) or over five sets of first 100 ps trajectories (×) plotted as a function of temperature. Error bars represent standard deviation. Dotted lines are linear least-squares fittings. The same values were calculated by the simulation model for harmonic oscillators containing up to third-order coupling (solid curve) and up to fourth-order coupling (broken curve). See text for details. (From Moritsugu, K. et. al., *J. Phys. Chem. B* 107, 3309, 2003. With permission.)

the change in the mode space from the original set to the new set. Such coupling terms do not necessarily represent real energy transfer but rather artifacts caused by the change in the mode space. In Figure 5.2, there is no obvious inflection at ~180 K in the average excess energy for the first 100 ps trajectory, implying that the changes in the mode space due to a jump to a new conformational substate occur above the glass transition temperature for the full 500 ps.

We next tried to decompose the values of $\Sigma_{i\neq2861}\Delta\bar{E}_i^{2861}(T)$ into contributions from various coupling terms in Equation 5.1. Toward this end, a model system was constructed comprising $N$ (=7419) harmonic oscillators with the frequencies of the myoglobin normal modes and coupling with mode 2861 through the $(N-1)(N-2)/2$ third-order coupling terms and through part of the fourth-order terms with coefficients $\alpha_{2861ij}$ and $\beta_{2861ijk}$, respectively. Since it is not possible to calculate all of the $N^3$ fourth-order terms, we used the following fourth-order terms in the model: all of the terms of $\beta_3q_{2861}^3q_i$, $\beta_4q_{2861}^2q_i^2$, $\beta_{2861,i^3}q_{2861}q_i^3$ and $\beta_{(2861)^2ij}\,q_{2861}^2q_iq_j$ $(i \neq j)$, and 4,268,893 terms of $\beta_{2861ijk}q_{2861}q_iq_jq_k$ $(i,j,k \neq 2861)$. This model system mimics a myoglobin molecule without coupling terms higher than the fourth order. The MD simulation was repeated 100 times with random changes of the initial condition. As a reference, we carried out the same simulations but included only the third-order coupling terms.

The results are plotted in Figure 5.2. At temperatures below 50 K, this model explains the energy transfer behavior almost perfectly. However, with increasing temperature, the model starts to deviate from the simulation results. At 300 K, the amount of transferred energy explained by the model becomes ~50% of the simulation values, showing that the coupling terms, which are eliminated in the model system, contribute largely. Furthermore, the third-order terms play dominant roles in energy transfer and fourth-order coupling contributes only slightly.

In summary, we have developed a scheme for off-resonance energy transfer at finite temperatures as follows. The characteristic features of the elementary process of energy transfer over a short-time period are well described by a model of coupled harmonic oscillators with lower-order coupling terms, even at room temperature. Over a longer period, energy transfer occurs as an accumulation of the elementary processes or as indirect energy transfer via a number of intermediate modes. Above the glass transition temperature, the protein molecule moves to a new conformational substate, and the associated rotation of the mode space causes apparent mode couplings.

## 5.3   MOVING NORMAL MODES

In the previous section, we showed that motions traversing conformational substates in proteins would induce a change or a transition of the mode space and cause artificial coupling among modes if the protein dynamics were observed using a fixed set of modes. To deal with these larger motions, we extend the scope of the model to describe a wider conformational space than that covered by the perturbation scheme described above. These changes of the mode space may originate from strong anharmonicity in the conformational space spanned by the low-frequency or large-amplitude normal modes. Since it has been argued that biologically significant motions occur frequently in the space of the low-frequency normal modes [19], it is important to develop a model describing the anharmonic dynamics occurring in the low-frequency normal modes.

We first assume that, within a sufficiently short period of time on the trajectory, the system can be considered to approximately behave as a set of harmonic oscillators. Under this assumption, it is possible to define normal mode coordinates characterizing the short-term dynamics at any instant of time on the trajectory. It is, however, expected that due to the influence of anharmonicity, any two sets of coordinate systems, each derived from different portions of the trajectory, would not be identical to each other, but rather would have different origins and orientations. Based on this idea, we introduced the time-dependent Lagrangian $L(t)$ in Equation 5.2, which we refer to as "moving normal mode coordinates."

The moving normal mode coordinates were calculated at each instant in time of the trajectories with a small time-window of size $t_w$ by using two different methods of defining the set of eigenvectors: local principal component analysis (PCA) and multivariate frequency domain analysis (MFDA). There is another way to calculate the normal mode coordinates, known as instantaneous normal mode analysis (INMA) [20], which has been used in studies of vibrational energy relaxation in proteins [21,22]. However, since we focus on low-frequency normal modes that are sensitive to normal modes with negative eigenvalues, we chose local PCA and MFDA, which guarantee positive definiteness.

### 5.3.1   MOVING NORMAL MODES BY LOCAL PCA

The local PCA is the eigenvalue problem for the time-dependent variance–covariance matrix $\mathbf{C}(t)$ of the distribution function, whose elements are defined by $c_{ij}(t) = \langle (x_i(t) - \langle x_i \rangle)(x_j(t) - \langle x_j \rangle) \rangle$, where $x_i$ is the $i$th coordinate and the average is taken between

$t$ and $t + t_w$. As an application of the moving normal mode coordinates, we analyzed 5 ns MD trajectories of myoglobin under several equilibrium conditions in order to evaluate the environmental influences on anharmonic dynamics in the protein [13]. The number of internal degrees of freedom, $N$, in Equation 5.2 is 453 ($=153 \times 3 - 6$), corresponding to the 153 C$\alpha$ atoms in myoglobin.

Here, we will discuss the rotation of large-amplitude moving normal modes $\mathbf{e}_i(t)$. The translational motion of the origin of the moving normal modes, $\mathbf{x}_0(t)$, has been discussed in Ref. [13]. Rotation of the moving normal mode coordinates was first monitored by using the average of the cosine of the rotation angle, $\langle \cos \theta_i(\tau) \rangle = \langle \mathbf{e}_i(t) \cdot \mathbf{e}_i(t+\tau) \rangle$. In calculating this quantity, we encountered a difficulty in deciding whether modes $\mathbf{e}_i(t + \tau)$ originated from $\mathbf{e}_i(t)$ when another mode $j$ had a similar eigenvalue, $\lambda_j \sim \lambda_i$, or the modes $i$ and $j$ were in the resonance condition. The perturbation expansion of $\mathbf{e}_i(t + \Delta t)$ gives the relation for a small value of $\varepsilon (\sim \Delta t)$,

$$\mathbf{e}_i(t + \Delta t) = \mathbf{e}_i(t) + \varepsilon \sum_{j \neq i} \mathbf{e}_j(t) \frac{\Delta c_{ij}}{\lambda_i - \lambda_j} + O(\varepsilon^2), \tag{5.12}$$

where $\varepsilon \Delta c_{ij}$ is the variation in $c_{ij}(t)$ within time interval $\Delta t$. When $\lambda_i \sim \lambda_j$, the second term on the right-hand side becomes large enough to change $\mathbf{e}_i$ significantly within $\Delta t$. In such a resonant case, correspondence of those modes cannot be defined, and rotation loses physical meaning. Since the density of states is extremely high in myoglobin, particularly in the high-frequency region, such resonance inevitably occurs. Therefore, rotation can be defined only for several lowest-frequency modes with sufficiently large gaps between contiguous eigenvalues. This problem will be discussed again later, in Section 5.3.2.

Figure 5.3a and b show the decay curves for $\langle \cos \theta_i(\tau) \rangle$ at 30 and 300 K, respectively, each of which was produced from data for different solvation conditions, and for two sizes of time-window, $t_w = 100$ ps and 1 ns. The data shown in Figure 5.3a and b are the average values calculated for the two to three largest-amplitude modes. Actually, all of the other smaller-amplitude modes suffer from the problem of resonance.

In order to explain the behaviors of the decay curves in Figure 5.3a and b, we adopted the following damped oscillator model:

$$\langle \cos \theta(\tau) \rangle = \exp(-\gamma_R \tau) \left[ \cos(\tilde{\omega} \tau) + \frac{\gamma_R}{\tilde{\omega}} \sin(\tilde{\omega} \tau) \right], \tag{5.13}$$

where $\tilde{\omega} = (\omega_R^2 - \gamma_R^2)^{1/2}$ with the angular velocity $\omega_R$ and the damping coefficient $\gamma_R$. The model of Equation 5.13 corresponds well with the values in Figure 5.3a and b with the adjustable parameters $\omega_R$ and $\gamma_R$. Figure 5.3c and d shows the values of $\omega_R$ and $\gamma_R$ thus obtained. All sets of the parameters $\omega_R$ and $\gamma_R$ indicate near-critical damping, i.e., the ratio $\gamma_R / \omega_R \simeq 0.8$, irrespective of temperature or solvation condition. It was further found that the $t_w$ dependences of $\omega_R$ and $\gamma_R$ are

$$\omega_R (\text{or } \gamma_R) \sim 1/t_w. \tag{5.14}$$

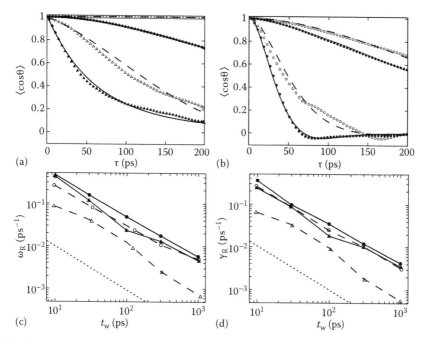

**FIGURE 5.3**  Cosine of the rotation angle of the moving normal mode coordinate, $\langle\cos\theta(\tau)\rangle$, plotted as a function of time $\tau$. (a) Values for trajectories at 30 K in water (filled triangles) and in vacuum (open triangles); (b) values for trajectories at 300 K in water (filled circles) and in vacuum (open circles). Each figure contains data for different time-windows, $t_w = 100$ ps and $t_w = 1$ ns. The best-fit curves of the model function of Equation 5.13 are also shown as a solid curve for data in water and a broken curve for data in vacuum. In (c) and (d), the angular velocity $\omega_R$ and the damping coefficient $\gamma_R$ derived by fitting to the values in (a) and (b) are shown for data at 30 K (c) and at 300 K (d). The symbols are the same as those in (a) and (b), and $\omega_R$ (or $\gamma_R$) = $1/t_w$ with an arbitrary intercept is shown with a dotted line. (From Moritsugu, K. and Kidera, A., *J. Phys. Chem. B* 108, 3890, 2004. With permission.)

Since PCA filters out the vibrational motions with frequencies higher than $2\pi/t_w$, Equation 5.14 means that the rotational velocity of the moving normal mode coordinates is determined entirely by the fastest motions recorded in the trajectory. In other words, the rotation is so quick that the same pattern of motion never appears after one cycle of vibration, or the protein continuously changes its pattern of motion by rotating the mode space.

In the above analyses, we were able to investigate the rotational dynamics of only a limited number of large-amplitude modes due to the problem of resonance. In order to avoid the difficulty in analyzing the rotation of a single mode, we tried to analyze the rotational dynamics of a subset of normal modes by using canonical covariance analysis.

Canonical covariance provides a quantitative measure of how similar two sets of coordinate systems are in the scale of covariance. Here, the method is explained in terms of two sets of variance–covariance matrices, $\mathbf{C}(t)$ and $\mathbf{C}(t+\tau)$. First, $\mathbf{C}$ is decomposed into a matrix $\mathbf{X}$ as, $\mathbf{C} = \mathbf{X}\mathbf{X}^t$. Then, the level of similarity is measured using the eigenvalues of the cross-covariance matrix, $\mathbf{X}(t)^t\mathbf{X}(t+\tau)$. In the analysis, to describe the rotation of the coordinate system spanned by the large-amplitude normal modes, we used the partial matrix $\mathbf{C}_m$ and $\mathbf{X}_m$ ($\mathbf{C}_m = \mathbf{X}_m\mathbf{X}_m^t$) containing only the $m$ largest-amplitude modes, rather than the full matrix of $\mathbf{C}$ and $\mathbf{X}$. Furthermore, the trace was normalized to give the normalized trace of canonical covariance $R_{cc}$,

$$R_{cc}(\tau)=\left\langle \mathrm{Tr}\left[\mathbf{X}_m(t)^t\mathbf{X}_m(t+\tau)\right]^{1/2} \middle/ \left[\sum_{i=1}^{m}\lambda_i(t)\sum_{i=1}^{m}\lambda_i(t+\tau)\right]^{1/2}\right\rangle. \quad (5.15)$$

Figure 5.4a and b shows the decay of the normalized trace of canonical covariance $R_{cc}(\tau)$, which was calculated for the 10 largest-amplitude modes ($m = 10$) explaining as much as 60% of the total variance on average. In all these curves, $R_{cc}$ decreases almost linearly up to $\tau = t_w$, and then becomes almost constant beyond this point ($\tau > t_w$), irrespective of temperature or solvation condition. These features indicate that relaxation of rotation is completed within the time window $t_w$, and no further decay occurs for $\tau > t_w$. Such fast rotation of the set of the mode axes is consistent with rotation of a single mode axis as shown in Figure 5.3.

In Figure 5.4c, the average values of $R_{cc}$ for $\tau > t_w$ are plotted as a function of $t_w$. In order to explain the behavior shown in Figure 5.4c, we propose the following interpretation of the rotation of the coordinate system: The partial matrix of variance–covariance $\mathbf{C}_m(t)$ is a random matrix in a space of a small number of degrees of freedom $N_R$ ($\geq m$). The matrix $\mathbf{C}_m(t + t_w)$ is another entirely new random matrix in the same space of $N_R$ dimension. The dimension $N_R$ of the space in which rotation from $\mathbf{C}_m(t)$ to $\mathbf{C}_m(t + t_w)$ occurs determines the value of $R_{cc}(\tau > t_w)$.

Based on this idea, we obtained the relation between $R_{cc}(\tau > t_w)$ and $N_R$, as shown in Figure 5.4d. The values of $N_R$ were calculated as a function of $R_{cc}(\tau > t_w)$ in the following manner. The random matrix $\mathbf{C}_m(t)$ is given by

$$\mathbf{C}_m(t) = \mathbf{E}_m^t(t)\Lambda_m\mathbf{E}_m(t) \quad \text{with } \mathbf{E}_m(t) = \mathbf{E}_{N_R}\mathbf{R}(t). \quad (5.16)$$

Here, the $m$ eigenvectors $\mathbf{E}_m$ are given as linear combinations of fixed $N_R$ eigenvectors $\mathbf{E}_{N_R}$ with a randomly generated rotation matrix $\mathbf{R}$ (an $N_R \times m$ matrix satisfying the normalization condition, $\mathbf{R}^t\mathbf{R} = \mathbf{I}$). The corresponding eigenvalues $\Lambda_m$ were taken from the average values in the trajectory. In the same way as done for Equation 5.16, the random matrix $\mathbf{C}_m(t + W)$ is calculated. Then, the numerator of $R_{cc}(W)$ in Equation 5.15 is calculated by

$$\mathrm{Tr}\left[\mathbf{X}_m^t(t)\mathbf{X}_m^t(t+W)\right] = \mathrm{Tr}\left[\Lambda_m\mathbf{R}(t)^t\mathbf{R}(t+W)\right]. \quad (5.17)$$

After $10^3$ pairs of random matrices were prepared, the average and the standard deviation of $R_{cc}$ were calculated according to Equation 5.17 for various values of $N_R$ and then plotted in Figure 5.4d.

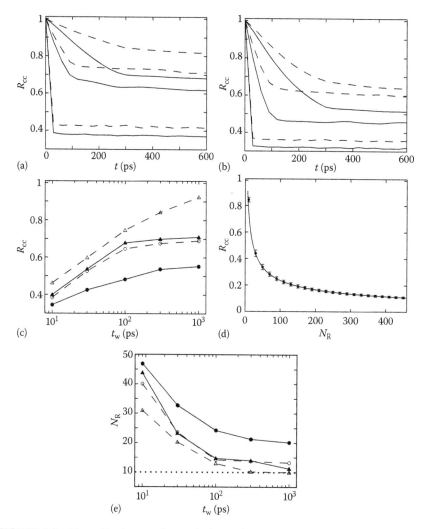

**FIGURE 5.4** Normalized trace of canonical covariance, $R_{cc}(\tau)$ (Equation 5.15), calculated (a) from trajectories at 30 K in water (solid curves) and in vacuum (dashed curves) and (b) from trajectories at 300 K in water (solid curves) and in vacuum (dashed curves). Each figure contains three sets of data, which were calculated with time-windows of three different sizes: $t_w$ = 300, 100, and 10 ps, from top to bottom, respectively. (c) Converged values of $R_{cc}$ or $R_{cc}(\tau > t_w)$ are plotted against $t_w$, calculated from trajectories at 30 K in water (filled triangles) and in vacuum (open triangles), and at 300 K in water (filled circles) and in vacuum (open circles). (d) Correspondence between $R_{cc}(\tau > t_w)$ and the rotational degrees of freedom, $N_R$, derived from the model calculations (see text). The value of $m$ in Equation 5.15 was set to 10. Data presented here are the averages of values calculated from the trajectories at 300 K in water with $t_w$ = 100 ps, together with deviations (error bars). (e) Rotational degrees of freedom, $N_R$, of the configurational space are plotted against the time-window, $t_w$. The dotted line at $N_R$ = 10 indicates the lowest limit of $R_{cc}$. (From Moritsugu, K. and Kidera, A., *J. Phys. Chem. B* 108, 3890, 2004. With permission.)

Using the results shown in Figure 5.4d, the $R_{cc}$ values were converted into the rotational degree of freedom $N_R$ and then plotted against $t_w$ in Figure 5.4e. In Figure 5.4e, it can be clearly seen that the rotation of the space spanned by the 10 mode axes occurs in a small configurational space. For example, at 300 K in water, the rotation of the 10 largest-amplitude modes for $t_w = 1$ ns occurs only in 20-dimensional space (out of 453 possible dimensions).

Finally, we will discuss the dependence of $N_R$ on various conditions. The following behaviors were found to be consistent with the findings described above. Upon solvation, the protein receives more anharmonic perturbations from the solvent molecules and displays anharmonic dynamics to a greater extent. Thus, rotation in water occurs in a larger space than that in vacuum [$N_R$(in water) > $N_R$(in vacuum)]. This influence is enhanced at higher temperature [$N_R$(300 K) > $N_R$(30 K)]. A larger values of $t_w$ makes the potential surface smoother and $N_R$ becomes a monotonically decreasing function of $t_w$.

### 5.3.2 Moving Normal Modes Using MFDA

In the previous section, we utilized local PCA to represent the moving normal modes, which depicted the locally harmonic but globally anharmonic dynamics of proteins. The major difficulty we met within the local PCA was the fact that two principal modes determined in adjacent time-windows, $e_i(t)$ and $e_j(t + \tau)$, did not have a definite correspondence when two or more modes had similar eigenvalues of the variance–covariance matrix $C$, or the quasidegeneracy in $C$. From a statistical viewpoint, this difficulty can be attributed to statistical fluctuation in the estimation of the principal modes due to the small sampling size in the determination of local PCA.

Suppose that we have an MD trajectory with $T$ total sampling steps, all we can do is to *estimate* the *true* principal modes $E = \{e_j\}$ by diagonalization of the sample variance–covariance matrix, $\hat{C}_{ij} = (1/T)\sum_{t=1}^{T} v_i(t)v_j(t)$ where $\hat{}$ denotes a sample quantity. Here we use atomic velocity $v_i$ for the variables instead of the coordinates $\Delta x_i$. If $T$ is large compared with the number of degrees of freedom $N$, there is a well known Gaussian limit theory [23]:

$$\sqrt{T}(\hat{e}_j - e_j) \to \mathcal{N}_N(0, A), \tag{5.18}$$

where $\mathcal{N}_N$ is an $N$-variate Gaussian distribution with mean zero and covariance $A$ given in Ref. [23]. Equation 5.18 implies that the estimator $\hat{e}_j$ is an unbiased estimator and that the standard deviation from the true principal modes, $e_j$, decreases with $T^{1/2}$. On the other hand, when $T$ is comparable to $N$, or the trajectory has a long time correlation comparable to $T$, the performance of the estimator $\hat{e}_j$ becomes radically worse. In this case, the Gaussian limit theory should be replaced with the random matrix theory [24]. According to the random matrix theory, deviation from the true principal modes no longer decreases with $T^{1/2}$ and the estimator $\hat{e}_j$ is biased. In local PCA, this becomes a serious problem because the sampling interval $T$ has to be limited to a relatively small time window.

The main contribution of this section is to show another way to reduce statistical fluctuation and to determine more reliable modes for the moving normal modes of proteins. A key idea is to utilize the Fourier transformation before the PCA is applied to the time series. According to the uncertainty relation in time and frequency spaces, the resolution of the Fourier transformation is proportional to $T^{-1}$, related to the so-called Rayleigh frequency, $f_R = 1/(T\Delta t)$.

Here, we briefly describe MFDA. Considering the usual situations where the trajectory is sampled with discrete time steps, we focus on *discrete* time and *continuous* frequency spaces. For simplicity, we set the sampling time interval of the trajectory data, $\Delta t$, to unity. We define the spectral density matrix $\mathbf{S}(f)$ as the inverse Fourier transform of a lagged variance–covariance matrix $\mathbf{C}(\tau) = \langle \mathbf{v}(t)\mathbf{v}^t(t+\tau)\rangle$,

$$\mathbf{S}(f) = \sum_{\tau=-\infty}^{\infty} \mathbf{C}(\tau)e^{-i2\pi f\tau}. \tag{5.19}$$

where $\mathbf{S}(f)$ is Hermitian and non-negative definite [25]. This is considered to be a multidimensional extension of the celebrated Wiener–Khinchin theorem. A Hermitian matrix can be diagonalized by a unitary operation,

$$\mathbf{S}(f)\mathbf{E}(f) = \mathbf{E}(f)\Lambda(f), \tag{5.20}$$

where
$\mathbf{E}(f)^\dagger\,\mathbf{E}(f) = \mathbf{I}$, $\Lambda$ is the eigenvalue matrix

This decomposition can be interpreted as a PCA on the bandpass-filtered process. An inverse Fourier transform using the first few eigenvectors provides a reduced representation of the original multivariate time series [25]. When a narrowband frequency-domain structure is present in the dynamics, it can be shown that Equation 5.20 provides stronger optimal decomposition of the multivariate time series than does conventional PCA [25].

For the estimator of the spectral density matrix $\mathbf{S}(f)$ under the finite sampling step $T$, we adopt the multitaper method (MTM) [26]. The estimator of MTM is given by averaging $K$ independent estimates,

$$\hat{S}_{ij}^{MTM}(f) = \sum_{k=1}^{K} \hat{S}_{ij}^{k}(f) \text{ with } \hat{S}_{ij}^{k}(f) = \frac{1}{T}\left(\sum_{t=1}^{T} w_t^k v_i e^{-i2\pi ft}\right)\left(\sum_{t=1}^{T} w_t^k v_j e^{-i2\pi ft}\right)^*. \tag{5.21}$$

Here, the asterisk denotes the complex conjugate. Particular taper sequences $w_t^k(k = 1, \ldots, K)$ are known to maximally minimize the bias of the estimator and are known as the discrete prolate spheroidal sequences (DPSS) [26,27], which are defined as the eigenvectors of the following eigenvalue problem:

$$\sum_{t'=1}^{T} \frac{\sin\left[2\pi W(t-t')\right]}{\pi(t-t')} w_{t'}^k = \gamma_k w_t^k. \tag{5.22}$$

Since the DPSS are mutually orthogonal, the MTM provides approximately uncorrelated estimates $\{\hat{S}_{ij}^{k}(f)\}$, which ensures the existence of multiple degrees of freedom within a given narrow frequency band of $2W$ [28]. Since the value of $W$ commonly used is $2/T$ to $3/T$, the Fourier transform in the MTM is able to extract the *local* dynamics characterized by a spectral density in a narrow frequency bin with a width of $4/T$ to $6/T$.

Here, we apply MFDA to the results of MD simulations for bovine pancreatic trypsin inhibitor (BPTI) (PDBentry: 1bpi). The MD simulations for BPTI were carried out at 300 K in two different environments: in vacuum and in explicit water. Both were performed for 1 ns with an integration time step of 0.5 fs. The atomic velocities were saved every $\Delta t = 4$ fs. This is short enough to avoid the aliasing effect in the Fourier transformation. The trajectories of the recorded velocities were subjected to analysis. A time-window with a length of $T\Delta t = 200$ ps was used for the Fourier transformation. For the MTM, we chose $2W = 4/(T\Delta t) \approx 0.67$ cm$^{-1}$, and $K$ (=3) tapers were used. In the process of the MTM, the trajectories were zero-padded to give 131,072 ($= 2^{17}$) steps for efficient calculation of the fast Fourier transform. Thus, the estimate of $\hat{S}(f)$ was obtained at frequencies $\{f_i\}$ ($i = 0, \ldots, 2^{16}-1$, $f_0 = 0$ cm$^{-1}$, and $w \equiv f_{i+1} - f_i \sim 0.13$ cm$^{-1}$). The details are given in Ref. [29]. Here, we focus only on the first eigenvector $\hat{\mathbf{e}}_1$ of $\hat{S}(f)$ at each frequency $f$ as the representative direction of the local dynamics.

As in the previous section, to obtain a quantitative measure of the overlapping of the subspaces composed of the eigenvectors in a frequency bin determined for different time-windows, we calculated canonical correlation coefficients for the two sets of eigenvectors. They are those determined in the time-window of 0–200 and 200–400 ps, that is,

$$\hat{\mathbf{E}}^{0-200}(f) = \left\{\hat{\mathbf{e}}_1^{0-200}(f_i)\right\} \quad \text{and} \quad \hat{\mathbf{E}}^{200-400}(f) = \left\{\hat{\mathbf{e}}_1^{200-400}(f_i)\right\}$$

with

$$f_i \in (f - \Delta f/2, f + \Delta f/2). \tag{5.23}$$

Here, three different frequency bin widths ($\Delta f$), 0.26 ($= 2w$), 1 and 10 cm$^{-1}$, were used. Since we focus only on the eigenvectors, we calculated canonical correlation coefficients instead of canonical covariance used in the previous section. Figure 5.5a and b show the frequency dependence of canonical correlation coefficients calculated for the MD trajectories in vacuum and in water, respectively.

We found that, in the high frequency region of 3000–3700 cm$^{-1}$, the canonical correlation coefficients are large under both solvation conditions. This is reasonable because the dynamics in this frequency region involves bond-stretching motions and the subspaces spanned by the vectors should be stable over time. At frequencies lower than 2000 cm$^{-1}$, the coefficients are much smaller than those at 3000–3700 cm$^{-1}$, indicating that the subspaces overlap less owing to the anharmonic character of the dynamics. The influences of solvation condition are observed most clearly at frequencies of 0–300 cm$^{-1}$, where the coefficients are smaller in water than in vacuum.

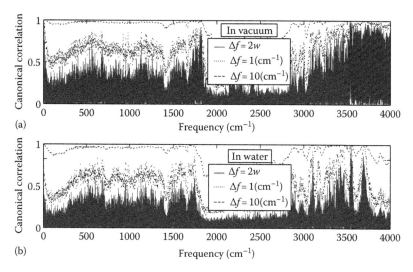

**FIGURE 5.5** Canonical correlation coefficients for two sets of the first eigenvectors of **S**, determined in the time-windows of 0–200 and 200–400 ps, respectively, plotted against frequency. (a) MD trajectory in vacuum and (b) in water.

Recalling the discussion in the previous section, this means that the dimension of the subspace explored by anharmonic dynamics is larger in water than in vacuum; that is, $N_R$(in water) > $N_R$(in vacuum), which is consistent with the results of the local PCA analysis. Thus, we conclude that solvent molecules introduce anharmonic perturbations most clearly to the low frequency region of 0–300 cm$^{-1}$. The small influence of solvation in the frequency region of 300–2000 cm$^{-1}$ suggests that the anharmonicity in this frequency region originates from the protein itself, and not strongly affected by the solvent. Another consequence of MFDA is that canonical correlation becomes almost unity for the bin width $\Delta f = 10$ cm$^{-1}$. This means that the anharmonic dynamics of the protein occurs in rather lower-dimensional spaces as was also concluded in the previous section. In the case of BPTI, a rough estimate of the dimensions spanned by the vectors in $\Delta f = 10$ cm$^{-1}$ is of the order of 10.

In Figure 5.6a through d, we show canonical correlation coefficients for $f \sim 12.2$ cm$^{-1}$ and $f \sim 3546.4$ cm$^{-1}$ for the time-window of 0–200 ps and the subsequent time-windows of 200–400, 400–600, 600–800, and 800–1000 ps. The vectors at $f \sim 3546.4$ retain high correlations, even with the narrow frequency bin $\Delta f = 2w$, irrespective of time-windows and solvation condition, indicating that there is no detectable anharmonicity in this high-frequency region. In Figure 5.7c and d, the vectors at $f \sim 3546.4$ calculated for the vacuum simulation are superimposed on the structure of BPTI. They are localized at Asn24 side-chain of BPTI irrespective of the time-window. The arrows in the figures show the bond-stretching motions involving the hydrogen atoms.

In contrast, the coefficients of vectors around $f \sim 12.2$ cm$^{-1}$ decrease as a function of time and converge to certain values depending on the width of the frequency bin. It is instructive that, in the solvated condition (Figure 5.6b), the relaxation of rotation is complete within 200 ps while it is slower in vacuum. In particular, the vectors in the narrow bin $\Delta f = 2w$ become almost orthogonal after 200 ps in the solvated

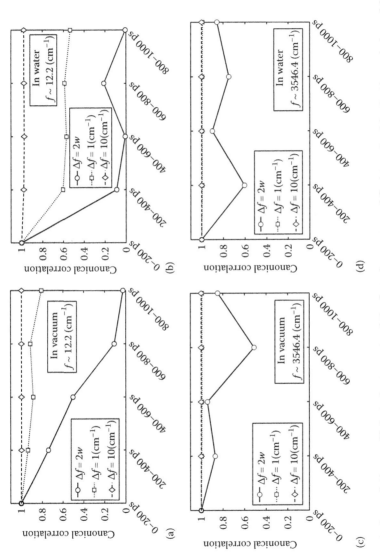

**FIGURE 5.6** Canonical correlation coefficients for two sets of the first eigenvectors of **S**, determined in the time-window of 0–200, 200–400, 400–600, 600–800, or 800–1000 ps. Three different frequency bin widths ($\Delta f$), 0.26, 1, and 10 cm$^{-1}$, were used: $f \sim 12.2$ cm$^{-1}$ in vacuum (a) and in water (b); $f \sim 3546.4$ cm$^{-1}$ in vacuum (c) and in water (d).

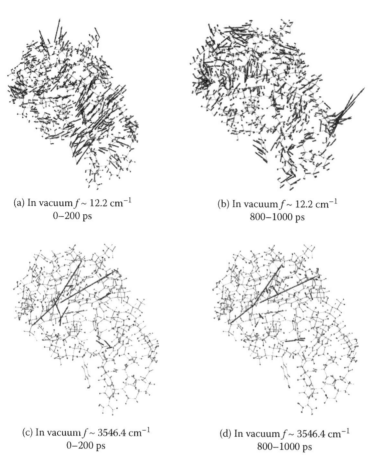

(a) In vacuum $f \sim 12.2$ cm$^{-1}$
0–200 ps

(b) In vacuum $f \sim 12.2$ cm$^{-1}$
800–1000 ps

(c) In vacuum $f \sim 3546.4$ cm$^{-1}$
0–200 ps

(d) In vacuum $f \sim 3546.4$ cm$^{-1}$
800–1000 ps

**FIGURE 5.7**  Eigenvector $\hat{\mathbf{e}}_1$ (12.2 cm$^{-1}$) determined in time-windows of 0–200 ps (a) and 800–1000 ps (b), and eigenvectors $\hat{\mathbf{e}}_1$ (3546.4 cm$^{-1}$) determined in time-windows of 0–200 ps (c) and 800–1000 ps (d). These data were derived from the MD trajectory in vacuum.

condition. This result in MFDA is consistent with that in local PCA. From this we learn that, for the low-frequency dynamics of a solvated protein, it is hard to define a single mode as a harmonic approximation of the dynamics even within a single time-window. To track and characterize the evolution of such anharmonic dynamics, we would have to abandon the notion of a single harmonic mode and adopt *a bundle of modes* defined in a certain frequency bin. The vectors at $f \sim 12.2$ cm$^{-1}$ are plotted in Figure 5.7a and b in the vacuum condition. While the vector in the 0–200 ps time-window shows the movements of the alpha-helices and the beta-strands of BPTI, in the 800–1000 ps time-window, the loop-regions and the terminal residues are prominent. It seems likely that the loop-regions and/or the terminal residues are easily perturbed by solvent molecules and therefore display highly anharmonic behavior in the solvated condition.

## 5.4   CONCLUDING REMARKS

In this chapter, we introduced two models for a collective description of anharmonic protein dynamics, anharmonically coupled oscillators (Equation 5.1) and moving normal modes (Equation 5.2). In Section 5.2, we applied the model of Equation 5.1 to the problem of vibrational energy transfer and showed that the Fermi resonance is a key ingredient in energy flow of a protein at low temperatures. The works based on the similar pictures [14,16,21,22] succeeded to explain the experimental time scales of energy flow [14,30–32]. However, a limitation of the model was evident in the application to energy transfer at finite temperatures. Besides the classical approximation, this is because the coupled oscillator model inevitably fails when a jump among conformational substates occurs. Thus, in Section 5.3, we proposed another model that is suitable for capturing large scale motions of proteins, that is, moving normal modes. We have used two totally different methods to define normal modes for the moving normal modes, local PCA, and MFDA. In spite of the large difference in their definitions, two sets of normal modes gave almost the same conclusion, i.e., in the space spanned by the low-frequency normal modes, the notion of a single harmonic mode does not hold, but a bundle of modes appears to be a correct picture.

## REFERENCES

1. H. Frauenfelder, F. Parak, and R. D. Young, *Annu. Rev. Biophys. Biophys. Chem.* **17**, 451 (1988).
2. A. Amadei, A. B. M. Linssen, and H. J. Berendsen, *Proteins* **17**, 283 (1993).
3. S. Hayward, A. Kitao, and N. Go, *Protein Sci.* **3**, 936 (1994).
4. A. Kitao and N. Go, *Curr. Opin. Struct. Biol.* **9**, 164 (1999).
5. H. J. Berendsen and S. Hayward, *Curr. Opin. Struct. Biol.* **10**, 165 (2000).
6. D. E. Sagnella and J. E. Straub, *J. Phys. Chem. B* **105**, 7057 (2001); L. Bu and J. E. Straub, *ibid* **107**, 10634 and 12339 (2003); Y. Zhang, H. Fujisaki, and J. E. Straub, *ibid.* **111**, 3243 (2007).
7. I. Okazaki, Y. Hara, and M. Nagaoka, *Chem. Phys. Lett.* **337**, 151 (2001); M. Takayanagi, H. Okumura, and M. Nagaoka, *J. Phys. Chem. B*, **111**, 864 (2007).
8. T. Ishikura and T. Yamato, *Chem. Phys. Lett.* **432**, 533 (2006).
9. N. Ota and D. A. Agard, *J. Mol. Biol.* **351**, 345 (2005).
10. K. Sharp and J. J. Skinner, *Proteins* **65**, 347–361 (2006).
11. K. Moritsugu, O. Miyashita, and A. Kidera, *Phys. Rev. Lett.* **85**, 3970 (2000).
12. K. Moritsugu, O. Miyashita, and A. Kidera, *J. Phys. Chem. B* **107**, 3309 (2003).
13. K. Moritsugu and A. Kidera, *J. Phys. Chem. B* **108**, 3890 (2004).
14. D. M. Leitner, *Annu. Rev. Phys. Chem.* **59**, 233 (2008).
15. M. Toda, R. Kubo, and N. Saito, *Statistical Physics I: Equilibrium Statistical Mechanics*, 2nd edn., Springer, Heidelberg, 1992.
16. H. Fujisaki, K. Yagi, K. Hirao, and J. E. Straub, *Chem. Phys. Lett.* **443**, 6 (2007).
17. R. M. Hochstrasser, *Proc. Natl. Acad. Sci. U. S. A.* **104**, 14190 (2007).
18. W. Doster, S. Cusack, and W. Petry, *Nature* **337**, 754 (1989).
19. F. Tama and C. L. Brooks, *Annu. Rev. Biophys. Biomol. Struct.* **35**, 115 (2006).
20. R. M. Stratt, *Acc. Chem. Res.* **28**, 201 (1995).
21. P. H. Nguyen and G. Stock, *J. Chem. Phys.* **119**, 11350 (2003).
22. H. Fujisaki, Y. Zhang, and J. E. Straub, *J. Chem. Phys.* **124**, 144910 (2006); H. Fujisaki and J. E. Straub, *J. Phys. Chem. B* **111**, 12017 (2007).

23. T. W. Anderson, *An Introduction to Multivariate Statistical Analysis*, 3rd edn., Wiley-Interscience, New York, 2003.

24. I. M. Johnstone, *Proceedings of the International Congress of Mathematicians*, Madrid, Spain, 2006, http://arxiv.org/abs/math/0611589.

25. D. R. Brillinger, *Time Series: Data Analysis and Theory*, SIAM, Philadelphia, PA, 2001.

26. D. Thomson, *Proc. IEEE*, **70**, 1055 (1982).

27. D. Slepian and H. Pollak, *Bell Sys. Tech.* **40**, 43 (1961).

28. M. E. Mann and J. Park, *J. Geophys. Res.* **99**, 25819 (1994).

29. Y. Matsunaga, S. Fuchigami, and A. Kidera, *J. Chem. Phys.* **130**, 124104 (2009).

30. P. Hamm, M. H. Lim, and R. M. Hochstrasser, *J. Phys. Chem. B* **102**, 6123 (1998).

31. M. T. Zanni, M. C. Asplund, and R. M. Hochstrasser, *J. Chem. Phys.*, **114**, 4579 (2001).

32. L. P. DeFlores, Z. Ganim, S. F. Ackley, H. S. Chung, and A. Tokmakoff, *J. Phys. Chem. B* **110**, 18973 (2006).

# 6 Energy Flow Pathways in Photoreceptor Proteins

*Takahisa Yamato*

## CONTENTS

## 6.1 INTRODUCTION

When a drop of ink falls into a beaker of water, the ink diffuses in the water and the spot of ink disappears or becomes invisible to the naked eye after some time. This is a typical diffusion phenomenon that shows the spontaneous movement of particles/ substances from a region of higher concentration to a region of lower concentration. A close examination would confirm that the ink molecules are equally distributed in all directions, i.e., the diffusion of the ink spot is isotropic in the water medium.

In contrast to the water in the beaker, a protein molecule has a specific three-dimensional structure. A typical protein molecule is both anisotropic and inhomogeneous, with many atoms closely packed in the interior of the protein. A broad range of temporal and spatial dynamics occurs in the protein's interior, from small-scale rapid movements to large-scale slow movements. Native protein molecules also perform various biological functions, responding to a specific input stimulus such as light illumination, depending on their molecular types. Immediately after initial activation, the molecule is abruptly placed in a nonequilibrium condition, and the

**129**

free energy decreases until thermal equilibrium is achieved. During this relaxation, the protein molecule's shape changes, leading to its biological function.

The relaxation dynamics often proceeds in multiple steps. These include chemical reactions such as proton transfer and electron transfer associated with a series of different intermediate states. At the beginning of the relaxation, the potential energy of the external perturbation is localized to the active site of the protein. If the energy is then isotropically dissipated as in the case of ink diffusion, it simply heats the surrounding medium. However, in the native functional protein molecule, the excess energy localized at the active site is always converted to the conformational change or series of changes relevant to its biological function. The anisotropic nature of the energy relaxation is thus an important feature of biologically functional protein molecules.

Among the various biological functions of proteins, we are particularly interested in photochemistry, photophysics, photoenergy conversion, and photosignal transduction in living organisms. Using theoretical and computational techniques, we can investigate changes in these photosensory proteins after light illumination, and how these proteins convert light energy to cause conformational changes. In this chapter, we discuss these basic problems in terms of biophysics at the molecular level.

## 6.2   LIFE AND SUNLIGHT

Life can hardly exist without sunlight (Figure 6.1). In living organisms, electromagnetic light waves are detected by photoreceptor proteins, which typically consist of a polypeptide chain and one or more chromophore groups. The interaction between the electromagnetic field of the incident light and a photoreceptor protein is characterized by the transition dipole moment $\vec{m}_{ji} = \langle j| \vec{\mu} |i\rangle$, where $\vec{\mu}$ is the electric dipole moment. The nonzero value of $|\vec{m}_{ji}|$ indicates that the transition from state $|i\rangle$ to $|j\rangle$ is allowed, whereas the zero-value of $|\vec{m}_{ji}|$ indicates that the transition is forbidden. Fermi Golden Rule gives the transition rate between states as a result of the time-dependent perturbation theory of quantum mechanics. The dot product between $\vec{m}_{ji}$ and the polarization unit vector $\vec{\varepsilon}$, which is parallel to the electric field vector $\vec{E}(\vec{r}, t)$, determines the transition probability per unit time as $w = (\pi/h)^2 |\vec{m}_{ji} \cdot \vec{\varepsilon}|^2$, where $h$ is Planck's constant.

Some of the photoreceptor proteins experience a photoexcited electron transfer reaction at the chromophore site, while others change their molecular conformation as a consequence of the photoexcitation (Figure 6.2). These reactions can be described using the well-defined language of physics and chemistry.

## 6.3   COLOR TUNING

The Planck distribution formula describes the spectral intensity of the radiation field from a black body as

$$\rho(\nu, T)d\nu = \frac{8\pi h\nu^3}{c^3} \frac{1}{\exp(h\nu/k_{\mathrm{B}}T)} d\nu,$$

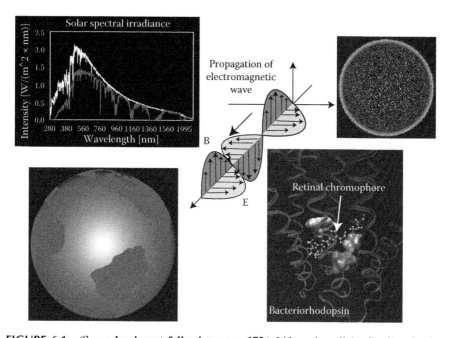

**FIGURE 6.1** **(See color insert following page 172.)** Life and sunlight. Profile of solar spectral irradiance is shown in the upper left panel. The white line is the extraterrestrial radiation that a location on earth would receive if there was no atmosphere or clouds. The magenta line is the radiation level after passing through 1.5 times the thickness of the atmosphere. Numerical data are taken from ASTM G-173 Reference Spectra (http://rredc. nrel.gov/solar/standards/am1.5). Light travels from the sun to the earth as electromagnetic wave propagation. Photoreceptor proteins in the cells of living organisms are able to interact with the electromagnetic waves. For instance, bacteriorhodopsin (lower right), one of the typical photoreceptors located in the purple membrane of *Halobacterium salinarum*, consists of the protonated Schiff base of the retinal chromophore and the surrounding polypeptide chain. It has a maximum optical absorption at 568 nm. Immediately after light illumination, the retinal chromophore undergoes ultrafast photoisomerization reaction from the all-*trans* form to the 13-*cis* form within 500 fs. The quantum yield of the reaction is as much as 0.64, indicating that two photons are sufficient to form the product state. The photoisomerization reaction triggers the photocycle consisting of K, L, M, N, and O intermediate states. As a consequence of the photocycle, one proton is transferred from the cytoplasmic side to the extracellular side, creating a proton concentration gradient. The bacterium synthesizes an ATP molecule using this proton concentration gradient across the membrane.

where

$\rho$ is the spectral energy density per unit volume per unit frequency
$c$ is the speed of light
$k_B$ is the Boltzmann's constant
$T$ is the absolute temperature

The extraterrestrial radiation fits the theoretical curve of the black body radiation of 5800 K (Figure 6.1). The UV region is absorbed by the ozone layer, and the IR region

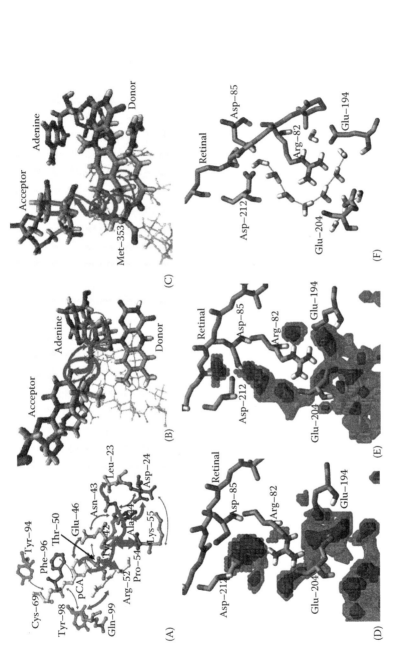

**FIGURE 6.2 (See color insert following page 172.)** Energy transfer, electron transfer, and proton transfer pathways of photoreceptor proteins. (A) Energy transfer pathway of PYP (From Ishikura, T. and Yamato, T. *Chem. Phys. Lett.*, 432(4–6), 533, 2006. With permission). Long-range intramolecular energy transfer pathways (red arrows) were investigated by the residue–residue energy conductivity analysis based on the Green–Kubo formula. The width of the arrow is proportional to the magnitude of the conductivity. (B,C) Electron transfer pathways in DNA photolyase (Miyazawa, Y. et al., *Biophys. J.* 94, 2194, 2008. With permission). Intramolecular photoinduced electron transfer pathways (red arrows) from the flavin cofactor to the UV-damaged DNA are analyzed by interatomic electron tunneling current. Two representative pathways are shown: (B) adenine and (C) methionine routes. The width of the arrow is proportional to the magnitude of the interatomic electron tunneling current. (D–F) Internal water distribution of bacteriorhodopsin and proton transfer pathway (Watanabe, H.C. et al., *Proteins* 75, 53, 2009. With permission). Density distributions of water molecules in the region from the Schiff base of retinal chromophore to the proton release site for light-adapted dark state (D) and the O-intermediate state model (E) of bacteriorhodopsin. The grid points with high population are shown in red, and the grid points with medium population are shown in blue. (F) Typical hydrogen bonding network of internal water.

is absorbed by carbon dioxide and other molecules. Thus, solar irradiance spectrum at the earth's surface has its maximum in the visible region (500 nm).

About three billion years ago, cyanobacteria started producing oxygen as a by-product of photosynthesis. Since the formation of the ozone layer—earth's astronaut suit—depends on oxygen molecules in the atmosphere, earth's surface was at that time exposed to UV light. UV light damages DNA molecules; for instance, cyclobutane pyrimidine dimer (CPD) is formed in DNA exposed to UV. To increase UV tolerance, cyanobacteria developed multiple strategies, among which is the DNA repair mechanism using DNA photolyase. This enzyme belongs to the cryptochrome/blue-light photoreceptor family. The DNA photolyase contains an FADH-cofactor, in which UV light activates an electron transfer reaction that is responsible for the covalent bond breaking in CPD. FADH is a reduced form of flavin adenine dinucleotide.

Photoreceptor proteins (Briggs and Spudich 2005) (Figure 6.3) are sophisticated molecular machines for detecting specific light frequencies with high sensitivity. As an example, bovine rhodopsin, a typical photoreceptor, has its absorption maximum at exactly 500 nm. Bovine rhodopsin is a visual pigment whose protein contains the protonated Schiff base of retinal as a chromophore for light sensing. The optical property of the rhodopsin is effectively adapted to the environment. Once the incident light energy is stored in the dark-state chromophore of the visual pigment, it undergoes an ultrafast photoisomerization reaction, triggering a series of conformational changes in the protein molecule.

In general, living organisms adapt to a diverse range of optical environments from deep sea to mountain tops by modulating the optical absorption spectra through mutations via the protein–chromophore interaction. The optical absorption maxima of different visual pigments cover a broad range of wavelengths from 360 to 560 nm. Several models have been proposed for the color tuning mechanism of rhodopsins.

## 6.4 PHOTOACTIVE YELLOW PROTEIN: PROTOTYPE OF SIGNAL TRANSDUCERS

Rhodopsins are not the only photoreactive proteins. Another example is the photoactive yellow protein (PYP) (Figure 6.4), a small water-soluble photoreceptor responsible for the negative phototaxis of *Halorhodospira halophila*. This protein contains *p*-coumaric acid (pCA) as a chromophore, which undergoes an ultrafast photoisomerization reaction immediately after light absorption. The photocycle involving different sequential intermediates, pG, pR, and pB, is then triggered by this photoreaction.

One of the key issues in studying photosensory receptors is to understand the molecular mechanism of the signal transducer function (conversion of light energy into specific structural changes required for biological functions). Since the atomistic three-dimensional structure was detailed (Borgstahl et al. 1995), PYP has attracted a number of researchers. It has served as the prototype of signal transfer proteins.

## 6.5 PRIMARY PHOTOCHEMISTRY OF PYP

Photoisomerization in photoreceptor proteins is significantly different from that in solution. For example, the all-*trans* retinal in bacteriorhodopsin (Figure 6.1) experiences the bond-specific photoisomerization reaction only at the C13=C14 bond.

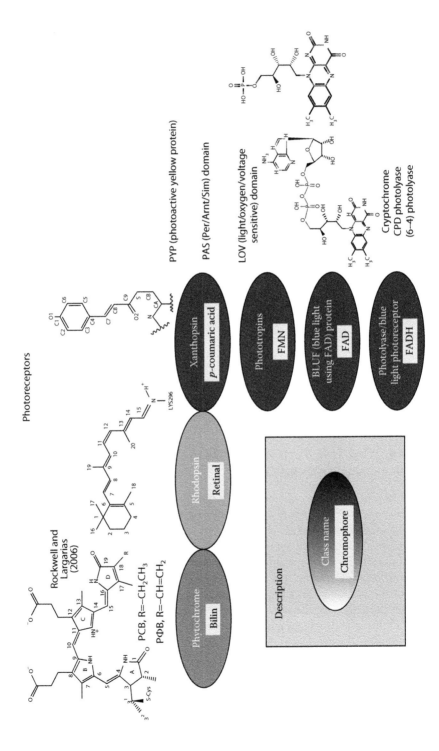

**FIGURE 6.3** Six classes of photoreceptor proteins. (From Rockwell, N.C. and Lagarias, J.C., *Plant cell* 18, 4, 2006. With permission.)

1. *Rhodopsins.* Rhodopsin family proteins contain a retinal chromophore surrounded by seven transmembrane α-helices. This family consists of visual rhodopsins, bacteriorhodopsins (Figure 6.1), and sensory rhodopsins. Rhodopsin family proteins provide photosignal transduction and light-activated ion pumping.

2. *Phytochromes.* A linear tetrapyrrole chromophore binds to them via a thioester linkage. Upon red light illumination, this chromophore experiences a *cis-to-trans* photoisomerization reaction. Phytochromes are responsible for red/far-red light-reversible plant responses. (From Rockwell, N.C. and Lagarias, J.C., *Plant cell* 18(1), 4, 2006. With permission.)

3. *Xanthopsins.* This family contains *p*-coumaric acid chromophores. PYP belongs to this family (Figure 6.4).

4. *Cryptochrome/photolyase blue light photoreceptor family.* The photoreceptors in this family are flavoproteins. They have a wide range of functions including circadian clock regulation, seed germination, and pigment accumulation.

5. *The light/oxygen/voltage/sensitive (LOV) domains of phototropins.* After light illumination of these domains, a flavin mononucleotide (FMN) chromophore forms a covalent bond with a cystein residue. This family of photoreceptors is responsible for the development of plants.

6. *Blue light using FAD (BLUF) domain-containing proteins.* This family of photoreceptors binds to flavin adenine dinucleotides.

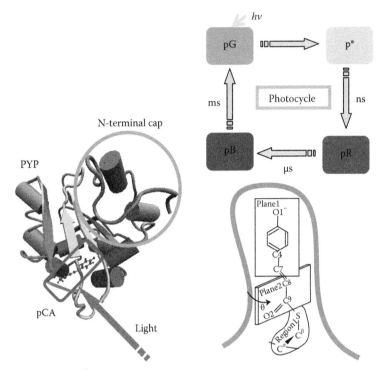

**FIGURE 6.4  (See color insert following page 172.)** Photoactive yellow protein. PYP derived from *H. halophila* is a water-soluble protein consisting of 125 amino acid residues (upper left). It is responsible for the negative phototaxis of *H. halophila* against blue light. PYP is a sensitive photon detector and uses the light energy to induce intramolecular chemical and conformational changes. The chromophore of PYP is *p*-coumaric acid (lower right). Although the chromophore is different from retinal, PYP demonstrates a photocycle (upper right) similar to those of the rhodopsin family proteins containing retinal. PYP is the first protein having a photocycle whose high-resolution structure is available. Analysis of the conformational change of PYP immediately after photoexcitation leads us to understand the initial stage of the photocycle of PYP in atomic detail. (From Yamada, A. et al., *Proteins* 55, 1070, 2004a. With permission.)

In contrast, the photoisomerization reaction of the all-*trans* retinal that is fully exposed to methanol solvent occurs at three different bonds. In addition, the isomerization reaction rate constant in the protein environment is much larger than that in the solution environment.

The primary photoreactions in the PYP have been experimentally studied both in the protein and in solution environments. The initial structural change of PYP was directly observed by time-resolved x-ray crystallography. Ultrafast fluorescence spectroscopy was performed on the initial process of the photoreaction of PYP. It was shown that the photoisomerization reaction of the PYP chromophore is completed within 1 ps. The initial photoreaction processes were analyzed by using time-resolved spectroscopic data. Although the reaction of the PYP chromophore in solution environments has been studied by several groups, the characterization of the dynamics in these environments is not fully understood.

Computational techniques are helpful in understanding the mechanism of the photoreaction in the protein environment. In 1998, Yamato et al. performed, for the first time, a molecular dynamics simulation to investigate the protein–chromophore interaction in PYP (Yamato et al. 1998). In general, however, the functionally important motion of a protein along the reaction coordinate is largely masked by the thermal fluctuation, making it difficult to analyze the effect of an external stimulus on the protein dynamics. We developed a new method to analyze the functionally important motion, free from thermal fluctuations, by comparing molecular dynamics simulation trajectories of the dark and excited states with the same initial conditions. This method was later applied to the response of a neuroreceptor upon ligand binding (Kubo et al. 2004), and extended to the ensemble perturbation method (Takayanagi et al. 2007) for the analysis of the anisotropic conformational relaxation of myoglobin upon flash photolysis.

Quantum mechanical study of the photoisomerization reaction of PYP began with calculations of the adiabatic potential energy surfaces of small model compounds of the chromophore (Yamada et al. 2001). However, to fully describe the protein–chromophore interaction, we had to include the entire protein molecule in the photoisomerization reaction. Recently, QM/MM (Hayashi and Ohmine 2000) and the ONIOM (Yamada et al. 2002; Vreven and Morokuma 2003), both of which are hybrid methods of high- and low-level calculations, have been used for large molecular systems. We performed the ONIOM (IMOMM) calculations on PYP to elucidate the role of the protein environment (Yamada et al. 2004a,b). We also investigated the origin of the force that drives photoisomerization.

## 6.6 MOLECULAR DESCRIPTION OF SIGNALING

Long-range intramolecular signaling is one of the characteristic features of the light energy conversion of photosensory receptors. For instance, pCA and the N-terminus region (N-terminal cap) are separated by the β-sheet in the molecule, and there are no direct interactions between them. However, after light illumination at pCA, partial unfolding occurs at the N-terminal region, which is important for the subsequent signal transduction processes. The molecular mechanism of this long-range intramolecular signaling has remained unsolved.

The solution may lie in molecular energy flow. The structure of a protein molecule is stabilized by various factors such as electrostatic interactions, hydrogen bonding, weak hydrogen bonding interactions, and nonspecific hydrophobic interactions. Partial unfolding of the local structure requires supply of sufficient amount of activation energy to the local site to overcome the energy barrier in the energy landscape from the folded state to the partially unfolded state. The spatial pattern of energy flow may be somehow related to the mechanism of the partial unfolding, which is necessary for the long-range intramolecular signaling of PYP.

Several groups have studied energy transfer in proteins theoretically and experimentally. To explore the energy transfer pathways in PYP, we developed a theoretical tool with which energy transport phenomena can be analyzed in atomistic detail. We proposed a new method to define microscopic energy conductivity in terms of interatomic energy flux (Ishikura and Yamato 2006).

## 6.7   INTERATOMIC ENERGY FLUX AND MICROSCOPIC ENERGY CONDUCTIVITY

In the molecular dynamics simulation, using molecular mechanics type force field functions, the potential energy, $V$, of a protein molecule is expressed as a function of coordinates, $q$, of the constituent atoms as

$$V = V(q_1, q_2, \ldots, q_{3N})$$

where $N$ is the total number of atoms. The total number of degrees of freedom becomes $(3N-6)$ after removing translational and rotational movements. Rabitz et al. described the local energy density $\varepsilon_i$ as the density of energy at each atom $i$ (Wang et al. 1998). Then, the derivative of the energy density is expressed in terms of the Liouville operator $\hat{L}$ as

$$\frac{d\varepsilon_i}{dt} = \hat{L}\varepsilon_i = \sum_j \left[ \dot{q}_j \frac{\partial}{\partial q_j} + \dot{p}_j \frac{\partial}{\partial p_j} \right] \varepsilon$$

Since the total energy should be conserved, the continuity equation holds for the derivative of $\varepsilon_i$ and the divergence of the energy flux $\vec{j}$ with the negative sign as

$$\frac{d\varepsilon_i}{dt} = -\nabla \cdot \vec{j} = -\sum_{k \neq i} J_{k \leftarrow i} = \sum_{k \neq i} J_{i \leftarrow k}$$

where $J_{i \leftarrow k}$ is the interatomic energy flux from atom $k$ to atom $i$. In terms of an operator

$$\hat{l}_j \equiv \left[ \dot{q}_j \frac{\partial}{\partial q_j} + \dot{p}_j \frac{\partial}{\partial p_j} \right],$$

we obtain the expression for the interatomic energy flux $J_{i \leftarrow j}$ as $J_{i \leftarrow j} = \hat{l}_j \, \varepsilon_i$. The energy flux between a pair of different groups $A$ and $B$ can be expressed as

$$J_{A \leftarrow B} = \sum_{i \in A}^{N_A} \sum_{j \in B}^{N_B} J_{i \leftarrow j},$$

where $N_A$ and $N_B$ are the numbers of atoms in $A$ and $B$, respectively. In general, when the energy density is low at site $A$ and high at site $B$ within a large molecule, a positive energy flux $J_{A \leftarrow B}$ should be observed to retain the equilibrium.

The derivation of the energy conductivity is based on the Green–Kubo formula. The energy exchange rate between these two sites is quantitatively measured by the energy conductivity $L_{AB}$, which is defined in terms of the time-correlation function of the energy flux as

$$L_{AB} = \frac{1}{k_B T} \int_0^\infty \left\langle J_{A \leftarrow B}(t) J_{A \leftarrow B}(0) \right\rangle dt$$

where
>   $k_B$ is the Boltzmann constant
>   $T$ is the absolute temperature

## 6.8  APPLICATIONS

### 6.8.1  INTRAMOLECULAR SIGNALING IN PYP

We reported a theoretical/computational analysis of the energy flow relevant to the long-range intramolecular cross talk between different regions in PYP, a photosensory receptor. To analyze the energy flow in atomic detail, we derived a theoretical expression for the interresidue energy conductivity in terms of the time-correlation function of the interatomic energy flux. The values of energy conductivities were numerically evaluated using a long molecular dynamics simulation trajectory of the PYP molecule in an aqueous solution environment. From these results, we detected several pathways for energy transfer relevant for PYP long-range intramolecular signaling (Ishikura and Yamato 2006), which agrees well with the model proposed by time-resolved x-ray crystallography (Ihee et al. 2005).

### 6.8.2  SAMPLING

The atomic coordinates of PYP were taken from the Protein Data Bank entry 2phy (Borgstahl et al. 1995). The polypeptide chain was immersed in a sphere of water molecules. The total number of atoms for this system became 10,779. We used the AMBER 99 force field (Wang et al. 2000) for the protein atoms and the TIP3P model (Jorgensen et al. 1983) for the water molecules. The force field for the chromophore was taken from a previous study (Yamato et al. 1998). The conjugate-gradient energy minimization was followed by a 500 ps molecular dynamics simulation with the temperature increasing from 0 to 300 K, and thermal equilibrium was reached. Then the simulation was continued for 5 ns at 300 K.

For energy calculations using the PRESTO program (Morikami et al. 1992), energy functions without truncation of long-range interactions were evaluated by the particle–particle, particle–cell method (Saito 1992). A part of the program code was imported to the original PRESTO program in our laboratory. The equations of motion were integrated using the VERLET algorithm with a time step of 1.0 fs. SHAKE constraints imposed on the covalent bonds for hydrogen atoms suppressed the rapid movements of these atoms. The stochastic boundary condition was applied to the spherical boundary surface. In the buffer region, within 8.0 Å from the spherical surface, all water molecules were treated as interacting Langevin particles, with a friction coefficient of $20\,ps^{-1}$ for nonhydrogen atoms.

### 6.8.3 Energy Transfer Pathways Based on Interresidue Energy Conductivity

At the beginning of the PYP photocycle, the light energy is stored in pCA. Since the carbonyl oxygen atom of pCA forms a hydrogen bond with the polypeptide backbone of the Cys69 residue, we defined the extended chromophore, pCA$^{ext}$, as consisting of pCA and Cys69. Figure 6.5 shows the thermal fluctuation of energy flux from pCA$^{ext}$ to the surrounding amino acid residues. We see that the amplitude of fluctuation depends on the strength of chromophore–residue interactions. If thermal equilibrium is assumed, for any residue pair, the theoretical value of the ensemble average of energy flux equals zero. To evaluate the energetic coupling between the residue pairs quantitatively, we analyzed interresidue energy conductivity based on the mathematical formula derived in the previous section. Figure 6.6 shows the numerical convergence of interresidue energy conductivities.

The values of these conductivities were derived from the 5 ns molecular dynamics simulation of PYP. The two residues, Thr70 and Pro68, have strong energetic couplings with pCA$^{ext}$ (Table 6.1), and Thr70 has the largest energy conductivity. This indicates the existence of a primary energy transfer pathway along the polypeptide backbone chain of PYP. Residues Tyr42, Thr50, and Glu46 also have large conductivities. These residues form a hydrogen bond network with pCA$^{ext}$, providing

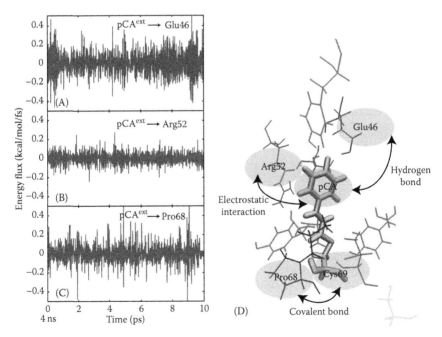

**FIGURE 6.5** Energy flux between chromophore and surrounding amino acid residues The values of energy flux from pCA$^{ext}$ to (A) Glu46, (B) Arg52, and (C) Pro68 are plotted as a function of time from $t = 4$ ns. (D) pCA$^{ext}$ and Glu46 are hydrogen bonded, while there is an attractive electrostatic interaction between the phenolate oxygen anion of pCA$^{ext}$ and the positively charged side chain of Arg52.

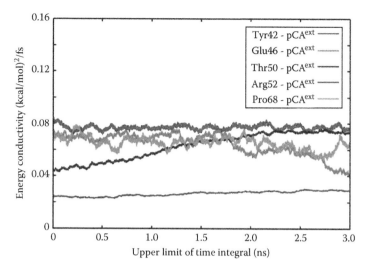

**FIGURE 6.6** **(See color insert following page 172.)** Convergence of energy conductivity. The dependence of energy conductivity, $L_{AB} = \dfrac{1}{k_B T} \displaystyle\int_0^\tau \left\langle J_{A \leftarrow B}(t) J_{A \leftarrow B}(0) \right\rangle dt$, on the upper limit of the time integral, $\tau$, is shown for A = Tyr42, Glu46, Thr50, Arg52, and Pro68, and B = pCA$^{ext}$.

important energy transfer pathways. The Arg52 residue, which forms a counterion to the phenolic oxygen anion of pCA, has slightly less conductivity than Tyr42, Thr50, and Glu46. The distance between the phenolic oxygen of pCA and the guanidinium group of Arg52 is approximately 6.5–7.0 Å, which led to the reduced conductivity of Arg52.

For the molecular mechanism of intramolecular signaling in PYP, Xie and coworkers (Xie et al. 1996) reported the importance of deprotonation of Glu46 during the PYP photocycle (the proteinquake model). Ihee et al. (2005) detailed the conformational changes using time-resolved x-ray crystallography. Furthermore, both theoretical (Groenhof et al. 2002; Itoh and Sasai 2004; Arai et al. 2005) and experimental (Kandori et al. 2000; Takeshita et al. 2002) studies characterized the structure and dynamics of PYP in the signaling state. In this chapter, we quantitatively analyze the anisotropic energy flow in PYP using microscopic energy conductivity. A two-dimensional map of energy conductivity shown in Figure 6.7A clearly shows that the values of inter-residue conductivity are large between adjacent residues. Figure 6.7B and C illustrates the anisotropic energy flow in PYP. Arrows represent the major pathways for energy transfer. Large values of interresidue energy conductivity for Asn43–Leu23 ($7.56 \times 10^{-2}$ (kcal/mol)$^2$/fs) and Ala44–Asp24 ($7.52 \times 10^{-2}$ (kcal/mol)$^2$/fs) indicate that the primary energy transfer pathway is (pCA → hydrogen bond network → helix α3 → N-terminal cap) (Ihee et al. 2005). Another pathway is via Lys55. Smaller values of interresidue energy conductivity—$3.1 \times 10^{-2}$ (kcal/mol)$^2$/fs (Lys55–Asp20) and $1.42 \times 10^{-2}$ (kcal/mol)$^2$/fs (Lys55–Asp24)—were observed for this pathway. Figure 6.7D illustrates a schematic view of the energy transfer pathways. For each path, the timescale of the energy transfer process was evaluated by exponential fitting of the time-correlation

**TABLE 6.1**

**Energy Conductivities between pCA^ext and the Surrounding Amino Acid Residues**

| Amino Acid | Energy Conductivity |
| --- | --- |
| Thr70 | 9.689 |
| Tyr42 | 7.810 |
| Thr50 | 7.289 |
| Pro68 | 6.404 |
| Glu46 | 4.263 |
| Ala67 | 2.964 |
| Arg52 | 2.887 |

*Note:*  Energy conductivities were calculated by the equation

$$k_B T \times L_{B,pCA_{ext}} = \int_0^{\tau} < J_{B \leftarrow pCA_{ext}}(0) J_{B \leftarrow pCA_{ext}}(t) >_T dt,$$

where

   $k_B$ is the Boltzmann's constant
   $T$ (=300 K) is the absolute temperature
   $J_{B \leftarrow pCA_{ext}}(t)$ is energy flux from the extended chromophore, pCA^ext, to residue $B$ at time $t$

Each value of $J_{B \leftarrow pCA_{ext}}(t)$ was recorded every 10 fs during the PYP molecular dynamics simulation. The ensemble average $\langle \ \rangle_T$ was taken over $2 \times 10^5$ samples and time integration was taken from $t = 0$ to $\tau$ (=3 ns). The unit of each value is $10^{-2} \times$ (kcal/mol)$^2$/fs.

function of energy flux. We observed binary behavior of the rapid (subpicosecond) and slow (several picoseconds) components for most of these pathways. It should be noted that the photolyzed heme group of carbonmonoxy myoglobin shows binary temperature relaxation of a rapid component of 3 ps and a slow component of 25 ps (Mizutani and Kitagawa 1997).

## 6.9   CONCLUSIONS

We studied the long-range intramolecular signaling mechanism in a photoreceptor protein, PYP, using molecular dynamics simulation. The anisotropic nature of the energy flow was demonstrated by the novel concept of interresidue energy conductivity. We found energy transfer pathways from the chromophore, which absorbs a photon at the beginning of the photoreaction, to the N-terminal cap region, where partial unfolding occurs to transduce the photosignal to an unknown second messenger.

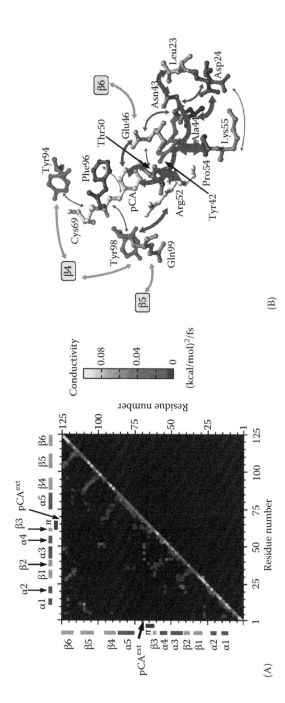

(continued)

**FIGURE 6.7** **(See color insert following page 172.)** Energy transfer pathways in PYP. (A) A two-dimensional map of interresidue energy conductivity. In the upper left triangle, the interresidue energy conductivities are shown in different colors indicating their magnitude. (B) Energy transfer pathways near pCA. pCA<sup>ext</sup> (bold yellow), consisting of pCA and Cys69, and the surrounding amino acid residues (thin). The red arrows indicate major energy transfer pathways. The line width of each arrow is proportional to the magnitude of energy conductivity. (Reprinted from Ishikura, T. and Yamato, T., *Chem. Phys. Lett.*, 432, 533, 2006. With permission.)

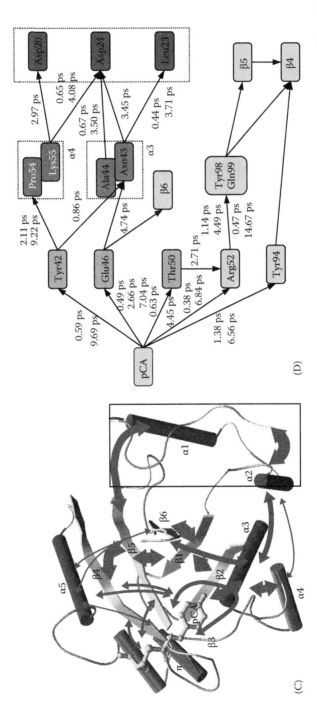

**FIGURE 6.7 (continued)** **(See color insert following page 172.)** (C) Energy flow pathways in PYP. (D) Energy transfer pathways from pCA$^{ext}$ (yellow) to the N-terminal cap region (red). The residues consisting of the hydrogen bond network (green) with pCA$^{ext}$, helix α3 (orange), and helix α4 (violet) are shown. For each path, the time-correlation function of energy flux was fitted to a single or double/triple exponential function/s. The time constants for these exponential functions are indicated.

Despite the availability of detailed information on the three-dimensional structure of PYP, it is difficult to understand the molecular mechanism of photosignal transduction. In this chapter, we attempted to examine the mechanism by theoretical biophysical computations. There are several problems to overcome before applying this method to other proteins. Among them is the convergence of interresidue energy conductivity. We observed poor convergence for some residue pairs. In the present study, we calculated interresidue energy conductivity only for the dark state (pG) of PYP. However, PYP experiences a photocycle that involves different intermediate states. The values of energy conductivity may be different between different intermediates. The application of energy conductivity analysis to the other intermediates would be interesting. Associated with the structural changes during the photocycle, the magnitude of energy conductivity may drastically change in the regulation region of PYP signaling. Another problem is the difference in timescales between the photocycle itself and the individual energy transfer processes of residue pairs. For instance, the last intermediate pB state is transformed into the initial pG state on a millisecond-to-second timescale, whereas the characteristic timescales of the elementary energy transfer processes between the residue pairs extend from subpicoseconds to several tens of picoseconds.

In general, it is very difficult to understand the relationships between the structures and functions of proteins. Many atoms are tightly packed in the protein interior, where a broad range of temporal and spatial dynamics occur, ranging from small-scale rapid movements to large-scale slow movements. In addition to these intrinsic features, proteins are characterized by additional features depending on their molecular functions, such as their response to physical stresses (e.g., heat or pressure), sensing of chemical stimuli, and interaction with external ligands. Various computational methods deal with protein structure prediction to understand the flow of genetic information from DNA to RNA to protein. However, the ultimate goal of life science research is not protein structure prediction, and the next important issue is to find a way to determine protein functions by exploiting their three-dimensional structures. How does a protein interact with an external ligand? Why does a protein sense stimulus and undergo conformational changes or chemical reactions? I hope that energy conductivity analysis could be a useful tool for studying such problems in terms of principles of biophysics.

## ACKNOWLEDGMENTS

The author would like to thank Professors John E. Straub and David M. Leitner for their kind invitation to make this contribution. The computations were performed at the Research Center for Computational Science, Okazaki Research Facilities, National Institute of Natural Sciences, and at Information Technology Center, Nagoya University. This work was supported by a Grant-in-Aid for the 21st Century Centers of Excellence (COE) program "Frontiers of Computational Science" at Nagoya University; Japan Science and Technology Agency (JST) for Core Research for Evolutional Science and Technology (CREST), "A Program System with Hierarchical Quantum Chemical Methods for Accurate Calculations of Biological Molecules" in the research area of "High Performance Computing for Multi-Scale Multi-Physics Phenomena"; and the Research Foundation for Opto-Science and Technology.

# REFERENCES

Arai, S., M. Togashi, et al. (2005). Molecular dynamics simulation of the M intermediate of photoactive yellow protein in the crystalline state. *Chemical Physics Letters* **414**(1–3): 230–233.

Borgstahl, G. E., D. R. Williams, et al. (1995). 1.4Å structure of photoactive yellow protein, a cytosolic photoreceptor: unusual fold, active site, and chromophore. *Biochemistry* **34**(19): 6278–6287.

Briggs, W. R., J. L. Spudich, Eds. (2005). *Handbook of Photosensory Receptors.* Weinheim: Wiley-VCH.

Groenhof, G., M. F. Lensink, et al. (2002). Signal transduction in the photoactive yellow protein. II. Proton transfer initiates conformational changes. *Proteins* **48**(2): 212–219.

Hayashi, S., I. Ohmine (2000). Proton transfer in bacteriorhodopsin: Structure, excitation, IR spectra, and potential energy surface analyses by an ab initio QM/MM method. *Journal of Physical Chemistry B* **104**(45): 10678–10691.

Ihee, H., S. Rajagopal, et al. (2005). Visualizing reaction pathways in photoactive yellow protein from nanoseconds to seconds. *Proceedings of the National Academy of Sciences USA* **102**(20): 7145–7150.

Ishikura, T., T. Yamato (2006). Energy transfer pathways relevant for long-range intramolecular signaling of photosensory protein revealed by microscopic energy conductivity analysis. *Chemical Physics Letters* **432**(4–6): 533–537.

Itoh, K., M. Sasai (2004). Dynamical transition and proteinquake in photoactive yellow protein. *Proceedings of the National Academy of Sciences USA* **101**(41): 14736–14741.

Jorgensen, W. L., J. Chandrasekhar, et al. (1983). Comparison of simple potential functions for simulating liquid water. *Journal of Chemical Physics* **79**(2): 926–935.

Kandori, H., T. Iwata, et al. (2000). Water structural changes involved in the activation process of photoactive yellow protein. *Biochemistry* **39**(27): 7902–7909.

Kubo, M., E. Shiomitsu, et al. (2004). Picosecond dynamics of the glutamate receptor in response to agonist-induced vibrational excitation. *Proteins* **54**(2): 231–236.

Miyazawa, Y., H. Nishioka, et al. (2008). Discrimination of class I cyclobutane pyrimidine dimer photolyase from blue light photoreceptors by single methionine residue. *Biophysics Journal* **94**(6): 2194–2203.

Mizutani, Y., T. Kitagawa (1997). Direct observation of cooling of heme upon photodissociation of carbonmonoxy myoglobin. *Science* **278**(5337): 443–446.

Morikami, K., T. Nakai, et al. (1992). Presto(protein engineering simulator)—A vectorized molecular mechanics program for biopolymers. *Computers & Chemistry* **16**(3): 243–248.

Rockwell, N.C., J.C. Lagarias (2006). The structure of phytochrom: A picture is worth a thousand spectra. *Plant cell* **18**(1): 4–14.

Saito, M. (1992). Molecular-dynamics simulations of proteins in water without the truncation of long-range Coulomb interactions. *Molecular Simulation* **8**(6): 321–333.

Takayanagi, M., H. Okumura, et al. (2007). Anisotropic structural relaxation and its correlation with the excess energy diffusion in the incipient process of photodissociated MbCO: High-resolution analysis via ensemble perturbation method. *Journal of Physical Chemistry B* **111**(4): 864–869.

Takeshita, K., Y. Imamoto, et al. (2002). Themodynamic and transport properties of intermediate states of the photocyclic reaction of photoactive yellow protein. *Biochemistry* **41**(9): 3037–3048.

van der Horst, M. A., K. J. Hellingwerf. (2004). Photoreceptor proteins, "star actors of modern times": A review of the functional dynamics in the structure of representative members of six different photoreceptor families. *Accounts of Chemical Research* **37**(1): 13–20.

Vreven, T., K. Morokuma. (2003). Investigation of the S-0 → S-1 excitation in bacteriorhodopsin with the ONIOM(MO: MM) hybrid method. *Theoretical Chemistry Accounts* **109**(3): 125–132.

Wang, J. M., P. Cieplak, et al. (2000). How well does a restrained electrostatic potential (RESP) model perform in calculating conformational energies of organic and biological molecules? *Journal of Computational Chemistry* **21**(12): 1049–1074.

Wang, Q., C. F. Wong, et al. (1998). Simulating energy flow in biomolecules: application to tuna cytochrome c. *Biophysics Journal* **75**(1): 60–69.

Watanabe, H. C., T. Ishikura, et al. (2009). Theoretical modeling of the O-intermediate structure of bacteriorhodopsin. *Proteins: Structure, Function, and Bioinformatics* **75**(1): 53–61.

Xie, A., W. D. Hoff, et al. (1996). Glu46 donates a proton to the 4-hydroxycinnamate anion chromophore during the photocycle of photoactive yellow protein. *Biochemistry* **35**(47): 14671–14678.

Yamada, A., T. Ishikura, et al. (2004a). Direct measure of functional importance visualized atom-by-atom for photoactive yellow protein: Application to photoisomerization reaction. *Proteins* **55**(4): 1070–1077.

Yamada, A., T. Ishikura, et al. (2004b). Role of protein in the primary step of the photoreaction of yellow protein. *Proteins* **55**(4): 1063–1069.

Yamada, A., T. Kakitani, et al. (2002). A computational study on the stability of the protonated Schiff base of retinal in rhodopsin. *Chemical Physics Letters* **366**(5–6): 670–675.

Yamada, A., S. Yamamoto, et al. (2001). Ab initio MO study on potential energy surfaces for twisting around C7 = C8 and C4-C7 bonds of coumaric acid. *Journal of Molecular Structure—Theochem* **536**(2–3): 195–201.

Yamato, T., N. Niimura, et al. (1998). Molecular dynamics study of femtosecond events in photoactive yellow protein after photoexcitation of the chromophore. *Proteins* **32**(3): 268–275.

# 7 Nonequilibrium Molecular Dynamics Simulation of Photoinduced Energy Flow in Peptides: Theory Meets Experiment

*Phuong H. Nguyen, Peter Hamm, and Gerhard Stock*

## CONTENTS

## 7.1 INTRODUCTION

Energy transport through molecular systems has received considerable interest, in particular, because to its importance for molecular electronics and the functioning of biological systems. For example, the energy transport through long-chain hydrocarbon molecules,[1] small molecules in solution,[2] and bridged azulene–anthracene compounds[3] has been studied. Also biological systems like reverse micelles[4] and proteins[5–8] have generated interest from both experimental and theoretical points of view. In spite of these efforts, the role of specific structural elements like α-helices and β-sheets in the energy transport is still not well understood.

In experiment, energy is usually induced into the biomolecule by photoexcitation of a chromophore. Prime examples include retinal in rhodopsin protein[9] or a synthetically attached azobenzene moiety.[10] These chromophores undergo (a) femtosecond internal conversion, which results in a highly vibrationally excited chromophore on a subpicosecond timescale.[11] Alternatively, one can directly pump a localized infrared (IR) vibration to induce energy into the system.[12] The photoexcitation is followed by (b) a picosecond vibrational energy redistribution from high-frequency, mostly localized, modes to low-frequency modes.[13,14] Delocalized harmonic modes with frequencies $\lesssim 200 \text{cm}^{-1}$ are believed to be the main reason of (c) transport of energy through a biomolecule, which also occurs on a picosecond timescale.[15–17] Furthermore, it also can take place through anharmonically coupled localized modes. Finally, due to heat transfer to the solvent, the (d) cooling of the molecules occurs.[5,18–20] These processes have been studied by transient IR spectroscopy by various groups. In particular, Hamm and coworkers have used isotope labeling of carbonyl oscillators at different positions along a peptide helix,[12,21] which act as local thermometers.[18]

The theoretical description of these experiments represents a challenging task. It includes the accurate modeling of the potential energy by an anharmonic force field as well as the appropriate dynamical description of the above described processes (a)–(d). Aiming at an atomistic description of biomolecules in aqueous solution, the first requirement leaves us with standard biophysical force field such as AMBER,[22] CHARMM,[23] GROMOS,[24] and OPLS,[25] which may only give an approximate model of the normal modes and the anharmonic couplings.[26] Furthermore, several aspects of the photoinduced dynamics are of quantum mechanical in nature, which is not straightforward to incorporate in a classical formulation.[27] There are various theories that describe the energy transport in a molecule, including formulations using time-dependent perturbation theory[28–30] or Golden Rule,[31] and harmonic transport theory.[15–17] However, it is not straightforward to apply these formulations to simulate the energy transfer observed, e.g., in a vis/IR experiment, because this also requires to consider all four processes (a)–(d) and not only of the energy transfer (c).

In a series of papers,[32–34] Stock and coworkers have combined the quasiclassical techniques used in the description of gas phase reactions[35] with biomolecular force fields used in molecular dynamics (MD) simulations.[36] This leads to nonequilibrium MD simulations, which mimic the laser excitation of the molecules by nonequilibrium phase-space initial condition for the solute and the solvent atoms. This approach is based on the following assumptions. Firstly, it is assumed that an empirical force field at least provides a qualitative modeling of the process. This is because the initial relaxation appears to be an ultrafast and generic process and because it can be expected that the strong interaction with the polar solvent smoothes out many details of the intramolecular force field. Secondly, quantum-mechanical effects are only included via the nonequilibrium initial conditions of the classical simulations. This means that the method represents a short-time approximation of quantum mechanics.

In this chapter, we wish to review this work. Being a critical aspect of the method, we discuss in Section 7.2 the generation of suitable nonequilibrium initial conditions representing IR or optical laser excitation of the system. As a simple first example, we study the vibrational energy relaxation of N-methylacetamide in $D_2O$ following the laser excitation of the amide I mode in its first excited state. We then turn to photoswitchable peptides,[10] which provide an exciting and relatively recent way to study heat transfer along the peptide backbone. In these experiments, the vibrational energy is locally deposited at the N-terminus of the helix by ultrafast internal conversion of a covalently attached, electronically excited, azobenzene moiety. We discuss the advantages and shortcomings of the nonequilibrium simulation method and attempt to relate to other approaches described in this book.

## 7.2 THEORY AND METHODS

We study the vibrational energy redistribution and heat transport processes in solvated biomolecules. With this in mind, we construct, for classical calculations, appropriate initial conditions that mimic infrared or optical laser excitation of the system (Sections 7.2.1 and 7.2.2), and perform nonequilibrium MD simulations of the various relaxation processes (Section 7.2.3).

### 7.2.1 SIMULATION OF INFRARED LASER EXCITATION

To construct appropriate initial conditions that mimic the infrared laser excitation of a specific normal mode into its first excited state, we consider a molecule containing $N$ atoms with Cartesian coordinates $\mathbf{r} = (\xi_1 \ldots \xi_{3N})^T$ and vibrational velocities $\mathbf{v} = (\dot{\xi}_1 \ldots \dot{\xi}_{3N})^T$. The total vibrational energy of the molecule is

$$E_{\text{vib}} = \sum_{i=1}^{3N} \frac{m_i}{2} \dot{\xi}_i^2 + V(\mathbf{r}). \tag{7.1}$$

To decompose the vibrational energy of a flexible molecule into single-mode contributions, it is useful to perform an instantaneous normal-mode analysis[37,38] of the vibrational dynamics. In this approach, we choose a structure at the instantaneous position $\bar{\mathbf{r}}(t)$ and consider the normal mode vibrations around this reference structure. We expand the molecular potential energy up to second order

$$V[\mathbf{r}(t)] = V[\bar{\mathbf{r}}(t)] + \sum_{i=1}^{3N} F_i(t)[\xi_i - \bar{\xi}_i(t)]$$

$$+ \frac{1}{2} \sum_{i,j=1}^{3N} H_{ij}(t)[\xi_i - \bar{\xi}_i(t)][\xi_j - \bar{\xi}_j(t)], \tag{7.2}$$

where

$F_i = \partial V / \partial \xi_i$ denotes the negative of the force in direction $\xi_i$

$H_{ij} = \partial^2 V / \partial \xi_i \partial \xi_j$ represents the elements of the Hessian matrix

Upon diagonalization of this Hessian matrix via $U^\dagger H U = \text{diag} (m_1 \omega_1^2 \ldots m_{3N} \omega_{3N}^2)$, we obtain $3N - 6$ instantaneous nonzero normal-mode frequencies $\omega_k$ and the dimensionless normal coordinate and normal momentum

$$q_k = \sqrt{m_k \omega_k / \hbar} \sum_{i=1}^{3N} U_{ik} (\xi_i - \bar{\xi}_i), \tag{7.3}$$

$$p_k = \sqrt{m_k / \hbar \omega_k} \sum_{i=1}^{3N} U_{ik} \dot{\xi}_i. \tag{7.4}$$

The total vibrational energy of the molecule defined in Equation 7.1 can then be decomposed into single-mode contributions as

$$E_{\text{vib}} = V[\bar{r}(t)] + \frac{1}{2} \sum_{k=1}^{3N-6} \hbar \omega_k \left( p_k^2 + q_k^2 + 2 a_k q_k \right), \tag{7.5}$$

where the coordinate shifts $a_k = (m_k \hbar \omega_k^3)^{-1/2} \sum_{i=1}^{3N} F_i U_{ik}$ account for the force term in Equation 7.2. Introducing shifted positions $q'_k = q_k + a_k$ and the energy $V_0 = V[\bar{r}(t)] - \frac{1}{2} \sum_k \hbar \omega_k a_k^2$, we finally obtain the instantaneous normal-mode representation of the vibrational energy

$$E_{\text{vib}} = V_0 + \frac{1}{2} \sum_{k=1}^{3N-6} \hbar \omega_k \left( p_k^2 + q_k^2 \right), \tag{7.6}$$

where $q'_k$ was redesignated as $q_k$. As the instantaneous normal-mode analysis is not necessarily performed at a minimum of the potential, several modes with imaginary frequency ($\omega_k^2 \leq 0$) may occur.[37,38] In the case of $N$-methylacetamide considered below, we observed several trajectories with up to three imaginary-frequency modes. All averaged normal-mode energies $\langle E_k \rangle$ of the nonequilibrium simulations (see Equation 7.8), though, were found to be positive.

In the second step, we employ the above introduced instantaneous normal modes to construct appropriate initial conditions for the nonequilibrium MD simulations. To this end, we represent the solute normal modes in terms of classical action-angle variables $\{n_k, \phi_k\}$[39]

$$q_k = \sqrt{(2n_k + \gamma)} \sin \phi_k,$$
$$\tag{7.7}$$
$$p_k = \sqrt{(2n_k + \gamma)} \cos \phi_k,$$

where the factor $\gamma = 1$ accounts for the zero-point energy of the mode. To obtain the initial positions and momenta of the initially excited mode (in our study, it is the amide I mode), we associate the action $n_k$ with the initial quantum state of the amide I mode, e.g., $n_k = 1$ for the first excited state. The initial actions of the remaining solute modes (which are at thermal equilibrium) may be sampled from the Boltzmann distribution $P(n_k) \propto e^{-n_k \hbar \omega_k / k_B T}$. In all cases, the vibrational phases $\phi_k$ are picked randomly from the interval $[0, 2\pi]$. This way an ensemble of normal-mode positions and momenta are calculated, which present a quasiclassical representation of the nonequilibrium quantum initial state of the solute molecule.[35,40]

### 7.2.2 SIMPLE MODEL OF AN AZOBENZENE PHOTOSWITCH

Moroder and coworkers have synthesized various peptides, in which an azobenzene unit was incorporated directly in the backbone.[41–44] This guarantees that the light-induced structural changes of the chromophore upon photoisomerization around the central N=N double bond are directly transferred into the peptide chain. Alternatively, the azobenzene photoswitch was covalently attached to the N-terminus of Aib-based $3_{10}$-helix.[21,45] By photoexciting the system by an ultrashort laser pulse, the subsequent conformational dynamics and the energy transfer in the peptide can be investigated by optical[46,47] or IR[21,48,49] spectroscopy. To facilitate an appropriate simulation of this process, we construct a simple model of the photoexcitation that mimics the experimental preparation.[32] The basic idea is explained in Figure 7.1, which shows a schematic view of the adiabatic potential-energy surfaces of the electronic ground-state $S_0$ and the first $n\pi^*$ electronic excited state $S_1$ of the azobenzene chromophore. Recent high-level ab initio calculations on azobenzene[50–52] have indicated that the torsion around the N=N double bond is the isomerization coordinate. The potential-energy curves shown in Figure 7.1 were obtained by assuming cosine functions for the corresponding diabatic potentials,[11] the energy gaps of which are designed to reproduce the maxima of the experimental *cis* and the *trans* absorption spectra and the ground-state *cis–trans* energy barrier.[41]

Also shown in Figure 7.1 is the N=N torsional potential-energy function as assumed in the GROMOS96 force field, $V_0(\varphi) = \frac{K_0}{2}(1 - \cos 2\varphi)$ with $K_0 = 84\,kJ/mol$. The force field model clearly underestimates the *cis–trans* energy barrier and also neglects energy difference of the *cis* and the *trans* configurations. At thermal energies ($k_B T \approx 2.5\,kJ/mol$ at $300\,K$) and short timescales, however, these differences are hardly relevant. As a minimal model of the *cis–trans* photoisomerization process, the dotted line in Figure 7.1 shows a potential-energy curve that is designed to connect $S_1$ *cis* state and the $S_0$ *trans* state by using a cosine function, $V_1(\varphi) = \frac{K_1}{2}(1 + \cos \varphi)$ with $K_1 = 320\,kJ/mol$. In the absence of an ab initio description of azobenzene chromophore and since we do not know the detailed phase-space distribution of the system after passing the conical intersection, the simple model by construction yields the correct experimental excess energy that is initially put in the N=N torsional degree of freedom.

We mimic the photoexcitation of the system by an ultrashort laser pulse by instantly switching from the ground-state N=N torsional potential $V_0(\varphi)$ to the excited-state potential $V_1(\varphi)$ (see Figure 7.1). Following this nonequilibrium preparation at time $t = 0$, the system isomerizes along excited-state N=N potential within $250\,fs$.

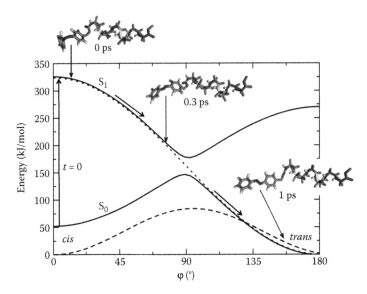

**FIGURE 7.1** **(See color insert following page 172.)** Scheme of the $S_0$ and the $S_1$ potential-energy curves of the azobenzene photoswitch as a function of the $N=N$ *cis–trans* isomerization coordinate φ. The solid lines represent the adiabatic potentials of a model which is designed to reproduce the experimental *cis* and *trans* absorption bands and the ground-state *cis–trans* energy barrier. The dashed line corresponds to the GROMOS force field potential of the $N=N$ torsion, the dotted line shows a model of the *cis–trans* photoisomerization potential, which simply connects the $S_1$ *cis* state and the $S_0$ *trans* state. Within this model, the photoexcitation of the system by an ultrashort laser pulse is simulated by instantly switching from the ground-state to the excited-state potential, which causes the isomerization of the system within ≈250 fs. Representative snapshots of the photoisomerization process reveal that azobenzene first undergoes a rotation along the $C-C-N=N$ and $N=N-C-C$ dihedral angles.

After the isomerization (i.e., for times ≥500 fs), the $N=N$ torsional potential is switched back to its ground state form, and a standard MD simulation is performed up to 1 ns. To obtain an impression of the photoisomerization process, selected snapshots at different times are shown in Figure 7.1. Although the $N=N$ torsional mode was chosen as reaction coordinate, interestingly, it is seen that the azobenzene first undergoes a rotation along the $C-C-N=N$ and $N=N-C-C$ dihedral angles, before the $C-N=N-C$ rotation is completed.

### 7.2.3 NONEQUILIBRIUM MD SIMULATIONS

Adopting the two models of laser-induced initial conditions developed above, a nonequilibrium MD description of the photoinduced relaxation processes can be rationalized as follows. As the system is in thermal equilibrium before the excitation, we first perform a standard canonical MD simulation using the GROMACS simulation program package.[53] From this trajectory, a number of (typically some hundred) statistically independent snapshots are chosen in which the positions and momenta of the solvent are stored as equilibrium initial conditions. Nonequilibrium initial conditions are then prepared by either applying Equation 7.7 (in the case of the IR laser excitation), or by switching the chromophore's potential energy surface (as shown in Figure 7.1). Following the preparation of the initial conditions, nonequilibrium simulations are performed. To correctly describe the initial cooling of the hot molecule in the solvent, no thermostat should be used. Hence, the trajectories were simulated at constant total energy (NVE ensemble), which requires a short time step of typically 0.2–0.5 fs.[32] The long-range electrostatic interactions were treated through the particle-mesh Ewald method,[54] whereby the nonbonded pair list was updated every 10 fs.

To study the vibrational energy relaxation of the amide I mode in $N$-methylacetamide, we employed the OPLS all-atom force field[25] to model the solute and the flexible simple-point-charge (SPC) water model[55] with doubled hydrogen masses to model the solvent $D_2O$. To investigate the photoinduced heat transfer in photoswitchable peptides, we used the GROMOS96 united atom force field 43a1.[24] Additional force field parameters for the azobenzene unit were derived from density functional theory as described in Ref. [32]. We employed a united-atom model[56] to describe the DMSO solvent, the SPC model[55] to describe water, and the rigid all-atom model of Ref. [57] to describe the chloroform solvent.

Following the simulations, the time-dependent observables of interest are obtained via an ensemble average over the initial distribution. For example, the average energy content of the $k$th normal mode is given by

$$\langle E_k(t) \rangle = \frac{1}{N_{\text{traj}}} \sum_{r=1}^{N_{\text{traj}}} E_k^{(r)}(t), \tag{7.8}$$

where

$E_k^{(r)}$ denotes the normal-mode energy pertaining to an individual trajectory

$N_{\text{traj}}$ is the number of trajectories

## 7.3 APPLICATIONS

### 7.3.1 VIBRATIONAL ENERGY RELAXATION OF THE AMIDE I MODE

As a first simple example, we apply the above explained methodology to study the vibrational energy redistribution of $N$-methylacetamide (NMA) in $D_2O$ following the laser excitation of the amide I mode in its first excited state. NMA has been used in numerous studies as a model system the peptide bond that links the various amino acids in a protein. The vibrational life time of the C=O vibration of NMA has been studied experimentally in Ref. [58], revealing a typical lifetime on the order of 1 ps (the decay is observed to occur in a biexponential, or nonexponential, manner). The relaxation rate does not change much whether the peptide bond is isolated (i.e., in NMA) or whether it is part of a larger peptide[58] or protein.[59] Furthermore, the decay is hardly affected by temperature,[59] and increases by less then a factor of two when decreasing the temperature below 100 K.

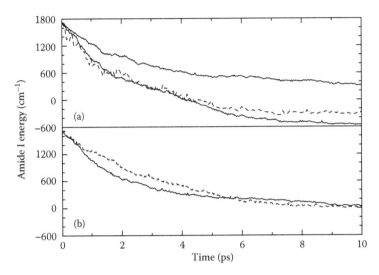

**FIGURE 7.2** Vibrational energy relaxation of the amide I mode of NMA in $D_2O$ following $v = 1 \rightarrow 0$ excitation as obtained for various quasiclassical initial conditions. For better comparison, the initially included zero-point energy is subtracted. (a) Comparison of action-angle initial conditions including zero-point energy in all solute modes (dashed line), in the amide I mode only (lower solid line), and no solute mode (upper solid line). Note that in the first two cases, the energy content of the amide I mode drops below zero for times $\gtrsim 4$ ps. (b) Including only 35% of the zero-point energy in all solute modes (dashed line) and in the amide I mode only (solid line), respectively, the amide I energy remains positive. (From Nguyen, P. H. and Stock, G., *J. Chem. Phys.* 119, 11350, 2003. With permission.)

Employing quasiclassical action-angle initial conditions, Equation 7.7, for the normal modes of NMA, we may assign quantum-mechanical zero-point energy (ZPE) to the solute vibrational modes. In the following, three cases are considered: (1) ZPE is included in all solute modes. (2) ZPE is included in the amide I mode only. (3) In the purely classical limit, no ZPE is included. In all cases, the classical thermal energy $k_BT$ is assigned to all solvent degrees of freedom. Figure 7.2(a) shows the time-dependent energy content of the amide I mode obtained for the three cases. For better comparison, the initially included ZPE is subtracted. In the purely classical case, the decay can be well fitted to a biexponential function with the decay times $\tau_1 = 1.9$ ps (80%) and $\tau_2 = 13.3$ ps. At longer times, the amide I energy converges to $k_BT \approx 200$ cm$^{-1}$ which is consistent with the equipartition theorem. Including ZPE in the amide I mode is seen to lead to a considerable enhancement of the initial relaxation process ($\tau_1 \approx 1.5$ ps). There is only little change, though, if the ZPE is also included in the remaining solute modes. Although we find a biexponential decay as observed in experiment, it is difficult to assess if it is for the same reasons, because the experimentally found decay times $\tau_1 = 0.45$ ps (80%) and $\tau_2 = 4$ ps are significantly shorter. Better agreement with experiment was recently achieved by a quantum-classical perturbative approach.[60]

It is noted, however, that for times $\gtrsim 4$ ps, the energy content of the amide I mode drops below its ZPE. This well-known artifact reflects the so-called ZPE problem of classical mechanics.[61] In quantum mechanics, each oscillator mode must hold an amount of energy that is larger or equal to the ZPE of this mode. In a classical trajectory calculation, on the other hand, energy can flow among the modes without this restriction. Numerous approaches have been proposed to fix the ZPE problem.[61] Here we adopt the method introduced in Ref. [62]. At the simplest level of the theory, only a fraction $\gamma$ ($0 \leq \gamma \leq 1$) of the full zero-point energy is included into the classical calculation. As an example, one may estimate the quantum correction $\gamma$ by simply requiring that the amide I energy remains larger than the ZPE for all times under consideration. From this procedure, we obtained $\gamma = 0.35$, which leads to the amide I relaxation shown in Figure 7.2(b) with time constants $\tau_1 = 1.8$ ps (75%) and $\tau_2 = 7.5$ ps (25%).

It should be stressed, though, that this procedure is only a remedy to fix the ZPE problem at short times. This is because in the long time limit, the energy of each normal mode decays to $k_B T$ in classical mechanics, regardless of the value of ZPE initially included. In other words, by incorporating ZPE into classical MD simulations, we find that the resulting trajectories describe the correct quantum-mechanical vibrational dynamics only up to a certain time. Starting at $t = 0$ with correct initial conditions ($\gamma = 1$), the classical approximation breaks down latest at $t \approx 4$ ps, when the energy content of the amide I mode drops below its ZPE. For $\gamma = 0.35$ this time is extended to $\approx 10$ ps, and for $\gamma = 0$, there is of course no problem with unrestricted flow of ZPE. However, if no ZPE is added, the phase-space the classical system can explore is too small compared to a quantum description, which results in an underestimation of the vibrational energy relaxation.[62] In that sense, nonequilibrium MD simulations can be regarded as short time approximation to quantum mechanics.

### 7.3.2 Photoinduced Heat Transfer in Bicyclic Peptides

As an experimentally particularly well-characterized molecular system,[41,46–49] we consider the octapeptide fragment H-Ala-Cys-Ala-Thr-Cys-Asp-Gly-Phe-OH (bcAMPB1), which was connected head-to-tail via (4-aminomethyl)-phenylazobenzoic acid as well as by a disulfide bridge, see Figure 7.3A. As bcAMPB1 can be dissolved in dimethyl sulfoxide (DMSO) but not in water, moreover a water-soluble photoswitchable peptide (bcAMPB2) was synthesized by replacing three aminoacids outside the active site motif by lysines[63] (see Figure 7.3B). According to nuclear magnetic resonance (NMR) experiments of bcAMPB1,[41] in equilibrium the *trans* azopeptide is predominantly in a single conformational state, while there are many conformations of similar energy in the equilibrium *cis* state of the peptide. These molecular systems have been studied extensively by time-resolved VIS-pump VIS-probe as well as VIS-pump IR-probe spectroscopy.[41,46–49] The original goal of these experiments, of course, was to follow the conformational dynamics of the peptide backbone responding to the photoisomerization of the azobenzene moiety (which is quite complex and occurs on various time scales). However, the huge amount of energy deposited by the photoisomerization resulted in a response in the IR spectrum at early delay times that was attributed to heating of the peptide, rather than a conformational change.[48] Thermal excitation of low-frequency modes in the vicinity of the peptide unit leads to a frequency shift of the C=O band, which is the result of anharmonic couplings between both types of modes.[18] These frequency shifts are a generic phenomenon,

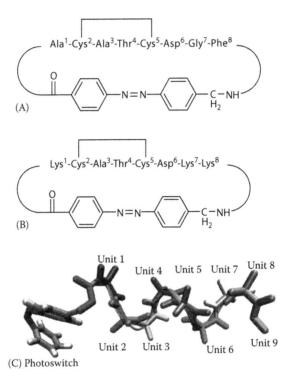

**FIGURE 7.3** Structures and amino acid labeling of the bicyclic azobenzene peptides bcAMPB1 (A) and bcAMPB2 (B), and the $3_{10}$-helix peptide PAZ-Aib$_8$-OMe (C).

which are very commonly observed in VIS-pump-IR-probe experiments. They are the basic mechanism behind the idea of using such high-frequency spectator modes as local thermometers.

Following photoexcitation, the azobenzene unit of the peptide undergoes nonadiabatic photoisomerization from the excited $S_1$ state of the *cis* isomer to the ground state $S_0$ of the *trans* isomer.[46,48] As shown in Ref. [32], the classical model of diabatically connecting initial and final states reproduces the 0.2 ps timescale of this process. Due to the isomerization the excess energy of $\approx$320 kJ/mol is rapidly redistributed to the vibrational modes of the peptide (by intramolecular vibrational relaxation) as well as to the surrounding solvent molecules (by vibrational cooling). Although the quantitative description of these processes in general requires a quantum-mechanical modeling,[28,31,64] a simple classical approach to study vibrational energy transport and cooling is to consider the kinetic energy of various parts of the molecular system.

The upper panels in Figure 7.4A show the time evolution of the kinetic energy of the azobenzene photoswitch and the octapeptide bcAMPB1 in DMSO, respectively. The same results but for bcAMPB2 peptide in D$_2$O are shown in Figure 7.4B. The general behavior of the results is similar for the two systems. Following photoexcitation, the kinetic energy of the azobenzene chromophore rises within only 100 fs. During the next 100 fs, the vibrational energy is transferred to the peptide.

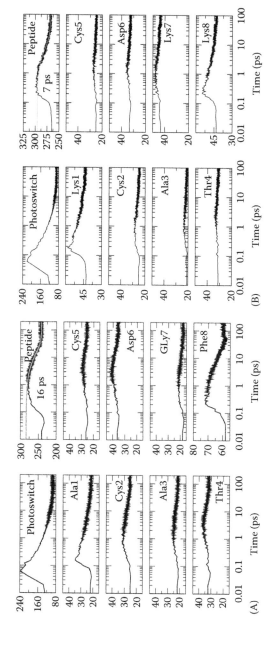

**FIGURE 7.4** Time evolution of the mean kinetic energy of various parts of bcAMPB1 in DMSO (A) and bcAMPB2 in $D_2O$ (B) following photoexcitation. In each graph, the upper panels show the kinetic energy of the azobenzene photoswitch (left) and the octapeptide (right). The decay of the latter can be well represented by a biexponential fit shown in gray. The lower panels show the time evolution of the kinetic energy pertaining to the individual peptide residues.

As a consequence, the peptide kinetic energy increases from 240 kJ/mol to about 290 kJ/mol, which is equivalent to the temperature jump of $\approx 100$ K found in experiment.[49] Subsequently, the vibrational energy of both azobenzene and peptide is dissipated into the solvent within 100 ps. A single exponential fit yields a decay time of 16 ps for the kinetic energy of the bcAMPB1 peptide in DMSO solvent. For bcAMPB2 peptide in $D_2O$, the cooling occurs significantly faster on a timescale of 7 ps. In experiment, cooling times of 7 and 4 ps are found for the solvents DMSO[49] and water,[44] respectively.

Since the cooling into the solvent is dominated by Coulomb interactions,[65] the faster relaxation of water may be attributed to its larger polarity as compared to DMSO. The overall underestimation of the efficiency of the cooling process by the MD simulations is, most likely, caused by two deficiencies of the description: The lack of polarizability of the classical force field and the neglect of internal degrees of freedom of the solvent. As a remedy, one may employ a polarizable and flexible all-atom model of the solvent, which adds more degrees of freedom, therefore increasing the heat capacity and the ability to cool the solute (albeit at the cost of other problems such as the classical description of high-frequency modes). In the light of these considerations, the agreement achieved by the present simple model appears satisfactory.

To study the energy transport along the peptide, the lower panels in Figure 7.4 show the time evolution of the kinetic energy pertaining to the individual peptide residues. The two end residues Ala1 and Phe8 of bcAMPB1 or Lys1 and Lys8 of bcAMPB2 are directly connected to the photoswitch, and are seen to receive most of the transferred energy from the photoswitch. While this process is ultrafast ($\approx 0.2$ ps) and carries substantial vibrational energy, the subsequent transfer to the remaining peptide residues is significantly slower (1–2 ps) and less prominent.

### 7.3.3 HEAT TRANSPORT ALONG A PEPTIDE HELIX

Motivated by the early delay-time results of Ref. [48] (see previous paragraph), we designed a molecular system aimed to study the energy transport through a peptide with well-defined structure, i.e., without the complication of an additional conformational change.[12,21,66] Figure 7.3C shows the essential idea: Vibrational energy is locally deposited at the N-terminus of an Aib-rich $3_{10}$-helix by ultrafast internal conversion of a covalently attached, electronically excited, azobenzene moiety. The subsequent heat flow through the helix is detected with subpicosecond time resolution by employing vibrational probes as local thermometers at various distances from the heat source. Certain amide I vibrations are singled out from the main amide I band by site-selective isotope labeling to provide a local thermometer.

In a first paper,[67] we showed that the concept does indeed work, and established the heat conduction properties of the model system. In contrast to our expectation, however, the helix is not a particularly good heat conductor; the heat conduction along the helix (heat diffusivity $2 \text{ Å}^2\text{ps}^{-1}$) only slightly exceeds that into the solvent. In a second piece of work,[12] the energy transport efficiency was compared after excitation of the azobenzene chromophore with a 3 eV (UV) photon to that after excitation of a peptide C=O oscillator with a 0.2 eV (IR) photon. The experiments showed that heat transport through the peptide after excitation with low-energy photons is at least four times more efficient than after UV excitation.

To explain this somewhat surprising finding, nonequilibrium MD simulations of the heat transport were performed for both high- and low-energy excitation conditions.[12,21] The main results of these simulations are comprised in Figure 7.5B, which shows the time evolution of the kinetic energy per atom of the peptide units along

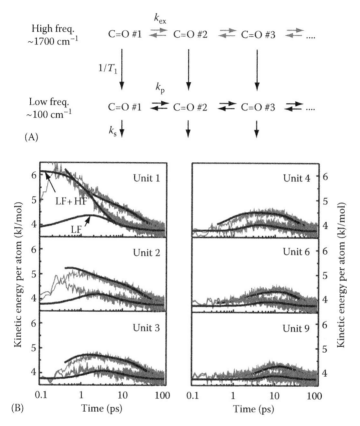

**FIGURE 7.5** (A) Schematic representation of the energy transport along a peptide helix. The gray arrows describe direct (excitonic) population transfer with a rate $k_{ex}$, while the black arrows illustrate energy transport through low-frequency modes with a rate $k_p$. Cooling to the solvent occurs with rate $k_s$. The lifetime of the high-frequency modes is $T_1$. (B) Time evolution of the dynamics of PAZ-Aib8-OMe from the MD simulations. The mean kinetic energy of selected residues after high-energy (gray) and low-energy (red) excitation is plotted as a function of time. The black lines show global fits according to a simple rate equation model (panel A). After low-energy excitation, the C=O stretch vibration of unit 1 is excited, a high-frequency (HF) mode. Energy dissipates out of this mode with a time constant of 1.5 ps into low-frequency (LF) modes around unit 1. After high-energy excitation, in contrast, it is assumed that energy is deposited into the low-frequency (LF) modes directly in a at least randomized, if not thermalized, manner. In both cases, energy transport through the peptide units take place with a time constant of 0.4 ps. Cooling to the solvent occurs on a 18 ps timescale. (From Backus, E. et al., *J. Phys. Chem. B,* 112, 9091, 2008. With permission.)

the helix. In the case of high-energy excitation, following the photoexcitation at time $t = 0$, the kinetic energy is deposited into the azobenzene photoswitch within less than 0.1 ps. This excess energy is then transferred to the Aib peptide (30%) as well as directly to the solvent (70%).[21] The peak of the photoinduced energy (gray lines) reaches the first unit at about 0.3 ps. The time-delayed increase of the kinetic energies of the subsequent peptide units nicely illustrates the propagation of energy (or heat) along the peptide backbone. Using a simple rate equation model with an exchange rate between two peptide units $k_p$ and a dissipation rate to the solvent $k_s$ (see Figure 7.5A) the time traces can be well fitted resulting in $k_p = (0.5\,\text{ps})^{-1}$ and $k_s = (18\,\text{ps})^{-1}$. Compared to the experimental findings $k_p = (2\,\text{ps})^{-1}$ and $k_s = (7\,\text{ps})^{-1,\,12}$ the energy transport along the peptide in simulation is thus about five times faster and the cooling rate is about two times slower.

In the case of low-energy excitation (red lines), the kinetic energy of the C=O oscillators is lower by a factor 3 than in the case of high-energy excitation. Due to the signal-to-noise ratio obtained from the average over 400 trajectories, the energy transport is hard to extract for distances larger than four peptide units from the excitation. Nevertheless, by using the same rate equation model (see Figure 7.5A), the data can again be modeled with a propagation rate $k_p \approx (0.2–0.6\,\text{ps})^{-1}$ and a cooling time to the solvent of $k_s = (18\,\text{ps})^{-1}$ (black lines in Figure 7.5B).

The above results certainly appear surprising on a first account. On one hand, our experiments have shown that the energy transport is significantly more efficient for low-energy excitation (although there is much less energy available) than for high-energy excitation. On the other hand, the MD calculations have revealed quite similar energy propagation rates in both cases. These findings may be caused by three effects. First, in contrast to UV pumping, in the low-energy pump experiment we start with a direct excitation of C=O vibrations, which may delocalize significantly and form excitonic states. A careful consideration of this possibility suggests, however, that this effect should not considerably contribute to the overall transport.[12] Second, given the large energy gradients $\Delta T/T$ caused by high-energy excitation, the thermal diffusivity of the peptide may itself depend on the temperature. Adopting a simple harmonic model of one-dimensional diffusive transport,[15,16] it is found that this effect is mostly caused by the heat capacity of the normal modes and may account for at most a factor 2 in the reduction of the heat transport for UV excitation. Because the classical heat capacity of a harmonic system does not depend on temperature, classical MD simulations cannot account for this effect.

Finally, given a hopping rate between adjacent sites as fast as $k_p = (0.5\,\text{ps})^{-1}$, we must conclude that the usual assumption, that thermalization within one individual peptide unit is significantly faster than thermalization between units, breaks down. In fact, we know, for example, that the depopulation rate of the initially excited C=O vibration is $1/T_1 = (1.2\,\text{ps})^{-1}$, and similar timescales are expected for subsequent relaxation steps as energy cascades down[13,14] from higher to lower energy states within one peptide unit. Following high-energy excitation, higher-frequency states, which tend to be localized, get thermally excited at a temperature that locally exceeds that of the surrounding by a large amount. As these localized modes hardly contribute to energy transfer, the subsequent relaxation cascade to delocalized low-frequency modes may represent the rate-limiting step rather than hopping from site to site.[13,14]

At the low excitation levels of the IR experiment, on the other hand, repopulation between vibrational modes does not really occur and cooling within the individual peptide units will not be rate-limiting. This difference between high- and low-energy excitation is again largely underestimated by the MD simulations. Most likely, this is a consequence of our simple model employed to describe the initial *cis–trans* photoisomerization of azobenzene. Disregarding virtually all aspects of multidimensional nonadiabatic photodynamics,[27] our simplistic ansatz to initially deposit the entire photon energy in the central N=N torsion of azobenzene naturally effects that a large fraction of the initial energy is transferred directly into the solvent. To obtain a more realistic modelling of the initial step, true nonadiabatic ab inito MD simulations are required.[69–71]

## 7.4 DISCUSSION

In this chapter, we have shown that nonequilibrium MD simulations may provide a versatile tool to describe photoinduced vibrational energy redistribution and heat transport in solvated biomolecules such as peptides. While there is a reasonable qualitative agreement between MD simulations and experiment, with respect to the overall process, both may differ in the timescales of the various relaxation processes. For example, we have found that the amide I vibrational energy redistribution and the cooling of the photoexcited peptide is modeled somewhat (by a factor of 2) too slow, while the heat propagation through a peptide helix was predicted too fast in the case of UV excitation. The first thought that comes into one's mind is the accuracy of the force fields. Varying force field parameters, one may be able to figure out the mechanism of energy transport. Preliminary results suggest that transport through the helix is essentially mediated by covalent bonds, but cooling into the solvent is dominated by Coulomb interactions.[65] The latter are weaker and play essentially no role in the intramolecular energy propagation process. Since cooling into the solvent proceeds in all three dimensions, however, both processes may still compete. One may expect that the too slow cooling rate would increase when the polarizability of the solvent molecules were included.

Clearly, the parameters of empirical force fields have *not* been designed to model energy relaxation processes. Vibrational energy relaxation depends to a large extent on specific resonances which were not taken care of in the design of biomolecular force field such as AMBER,[22] CHARMM,[23] GROMOS,[24] and OPLS[25] (for the same reason, these force fields cannot describe vibrational spectroscopy). Furthermore, although empirical force fields are intrinsically anharmonic (e.g., through nonbonding interactions), the anharmonicity of high-frequency vibrations is poorly accounted for—another important factor in energy relaxation process. Given these deficiencies, the agreement we obtain between MD simulation and experiment may be not so bad, and one might hope that at least some of the problems could be fixed by tuning force fields parameters.

A more severe problem might be the classical mechanics nature of MD simulations. It is nowadays established that quantum mechanics does not play a major role in the nuclear dynamics of condensed phase systems at room temperature (apart from, maybe, in proton transfer reactions), but the same is not necessarily true for thermal properties. In fact, the deviation of a classical from a quantum-mechanical heat capacity is a historic result that triggered the development of quantum mechanics in the first place.[72] This deviation is not a small effect, and it is still very evident at room temperature. That is, the heat capacity of all modes that are larger than $\hbar\omega \gtrsim k_B T \approx 200\,\text{cm}^{-1}$ is not described correctly in a classical MD simulation. All stretching, bending, and torsional degrees of freedom of individual groups fall into this category; only collective conformational modes of, for example, the peptide backbone are typically below the critical cut-off.

On the other hand, it has been argued that it is mostly the modes $\hbar\omega \lesssim 200\,\text{cm}^{-1}$, which are delocalized, that contribute to heat transport.[15–17,64] To the extent this is correct, the nonclassical modes may not play any role in the heat transport properties, and the wrongly described heat capacity partially cancels out when calculating the heat diffusivity. Nevertheless, a classical simulation will distribute thermal energy very differently over the various degrees of freedoms of a protein than a quantum description would do. This will in particular be relevant when the transient temperature jumps are large, $\Delta T \gtrsim T$. In fact, we have found experimentally a strong dependence of the energy propagation rate on the amount of initial energy and the way we deposit it.[12] This dependence could not be reproduced by the MD simulation, and the classical nature of the latter might be one reason for this failure.

## 7.5  CONCLUDING REMARKS

In many photobiological processes such as the photoisomerization of retinal in rhodopsin during the process of vision[9] or the quenching of excitation energy in antenna complexes by carotenoids in the photosynthetic apparatus,[73] an energy equivalent of a visible photon is dissipated on a subpicosecond timescale. A similar situation is expected for newly developed molecular electronic devices or nanostructures. Such processes result in enormous temperature gradients as large as many 100 K over length scales of a few chemical bonds only. Under these extreme conditions, heat transport is no longer diffusive in the same way as known from macroscopic physics. This is since thermalization *within* an individual building block of a macromolecule is not necessarily significantly faster than thermalization *between* building blocks. In this chapter, we have laid the foundation for a deeper understanding of these processes, based both on theory and experiment.

Heat transport properties have also been brought into relation to allosteric regulation, i.e., the regulation of the properties of an active site of a protein after binding of a substrate at a distant second (allosteric) site. Allosteric regulation is a fundamental process responsible for signaling in biological systems. For example, Ranganathan and coworkers have identified an evolutionarily conserved pathway of (free) energetic connectivity in certain protein families, using sequence analysis.[74–76] Later, Agard and coworkers[77] as well as Sharp and Skinner[78] identified the very same sequence of amino acids as a preferred energy transport channel, based on nonequilibrium MD simulations. However, a simplistic direct comparison between both pieces of work contains a caveat: While large-scale structural changes of proteins during allosteric signaling are expected to occur on a microsecond or even much slower timescale, the energy transport discussed in Refs. [77,78] is a picosecond phenomenon. Very generally speaking, vibrational energy in disordered condensed phase systems dissipates on a few picosecond timescale. Hence, signaling in allosteric regulation certainly is not directly related to transport of vibrational energy.

Nevertheless, often the existence of fast fluctuations is a prerequisite of a slow process. The most prominent example is transition state theory, where the preexponential factor can be interpreted as a fast "search rate," whereas the actual reaction rate is significantly slower, dictated by a barrier that has to be surmounted (certainly, this picture cannot applied in a one-to-one manner to the problem of allostery). Hence, the chain of reasoning might go the other way around. That is, the connectivity of sites responsible for allostery, the nature of which is essentially unknown at this point, might manifest itself also as a preferred pathway of vibrational energy. In other words, while transport of vibrational energy may not be the reason for allostery, it might still indicate possible allosteric connectivity in proteins. Theoretical simulations have revealed that the transport of vibrational energy in proteins can be highly specific and very anisotropic.[15–17,20,64,77–80]. A high specificity would appear to be an important prerequisite for allosteric regulation. As there is virtually no experimental evidence for these speculations, a future goal is to design a versatile experimental approach to deposit and to detect vibrational energy at essentially any pair of positions in a protein. Together with a detailed structure–function analysis, these investigations will form the foundation for a better understanding of allostery on an atomistic level.

## ACKNOWLEDGMENTS

We gratefully thank our coworkers Ellen Backus, Virgiliu Botan, Yuguang Mu, Sang-Min Park, and Claudio Toniolo for a fruitful and pleasant collaboration. Furthermore, we thank David Leitner and Josef Wachtveitl for inspiring and helpful discussions. This work has been supported by the Frankfurt Center for Scientific Computing, the Fonds der Chemischen Industrie, the Deutsche Forschungsgemeinschaft, and the Swiss National Science Foundation.

## REFERENCES

1. Z. Wang, J. A. Carter, A. Lagutchev, Y. K. Koh, N.-H. Seong, D. G. Cahill, and D. D. Dlott, *Science* **317**, 787 (2007).
2. G. M. Wang, E. M. Sevick, E. Mittag, D. J. Searles, and D. J. Evans, *Phys. Rev. Lett.* **89**, 050601 (2002).
3. D. Schwarzer, P. Kutne, C. Schroder, and J. Troe, *J. Chem. Phys.* **121**, 1754 (2004).
4. J. C. Deàk, Y. S. Pang, T. D. Sechler, Z. H. Wang, and D. D. Dlott, *Science* **306**, 473 (2004).
5. M. Tesch and K. Schulten, *Chem. Phys. Lett.* **169**, 97 (1990).
6. T. Lian, B. Locke, Y. Kholodenko, and R. M. Hochstrasser, *J. Phys. Chem.* **98**, 11648 (1994).
7. P. Li and P. M. Champion, *Biophys. J.* **66**, 430 (1994).
8. Y. Mizutani and T. Kitagawa, *Science* **278**, 443 (1997).
9. P. Kukura, D. W. McCamant, S. Yoon, D. B. Wandschneider, and R. A. Mathies, *Science* **310**, 1006 (2005).
10. C. Renner and L. Moroder, *Chem. Bio. Chem.* **7**, 869 (2006).
11. W. Domcke and G. Stock, *Adv. Chem. Phys.* **100**, 1 (1997).

12. E. Backus, P. H. Nguyen, V. Botan, R. Pfister, A. Moretto, M. Crisma, C. Toniolo, G. Stock, and P. Hamm, *J. Phys. Chem. B* **112**, 9091 (2008).
13. J. R. Hill and D. D. Dlott, *J. Chem. Phys.* **89**, 830 (1988).
14. R. Rey, K. B. Moller, and J. T. Hynes, *Chem. Rev.* **104**, 1915 (2004).
15. X. Yu and D. M. Leitner, *J. Phys. Chem. B* **107**, 1698 (2003).
16. X. Yu and D. M. Leitner, *J. Chem. Phys.* **122**, 054902 (2005).
17. D. M. Leitner, *Annu. Rev. Phys. Chem.* **59**, 233 (2008).
18. P. Hamm, S. M. Ohline, and W. Zinth, *J. Chem. Phys.* **106**, 519 (1997).
19. E. R. Henry, W. A. Eaton, and R. M. Hochstrasser, *Proc. Natl. Acad. Sci. U. S. A.* **83**, 8982 (1986).
20. D. E. Sagnella and J. E. Straub, *J. Phys. Chem. B* **105**, 7057 (2001).
21. V. Botan, E. Backus, R. Pfister, A. Moretto, M. Crisma, C. Toniolo, P. H. Nguyen, G. Stock, and P. Hamm, *Proc. Natl. Acad. Sci. U. S. A.* **104**, 12749 (2007).
22. W. D. Cornell, P. Cieplak, C. I. Bayly, I. R. Gould, K. M. Merz, Jr., D. M. Ferguson, D. C. Spellmeyer, T. Fox, J. W. Caldwell, and P. A. Kollman, *J. Am. Chem. Soc.* **117**, 5179 (1995).
23. A. D. MacKerell, Jr., D. Bashford, M. Bellott, R. L. Dunbrack, J. D. Evanseck, M. J. Field, S. Fischer, J. Gao, H. Guo, S. Ha, et al., *J. Phys. Chem. B* **102**, 3586 (1998).
24. W. F. van Gunsteren, S. R. Billeter, A. A. Eising, P. H. Hünenberger, P. Krüger, A. E. Mark, W. R. P. Scott, and I. G. Tironi, *Biomolecular Simulation: The GROMOS96 Manual and User Guide* (Vdf Hochschulverlag AG an der ETH Zürich, Zürich, Switzerland, 1996).
25. W. L. Jorgensen, D. S. Maxwell, and J. Tirado-Rives, *J. Am. Chem. Soc.* **118**, 11225 (1996).
26. S. K. Gregurick, G. M. Chaban, and R. B. Gerber, *J. Phys. Chem. A* **106**, 8696 (2002).
27. G. Stock and M. Thoss, *Adv. Chem. Phys.* **134**, 243 (2005).
28. H. Fujisaki and J. E. Straub, *Proc. Natl. Acad. Sci. U. S. A.* **102**, 6726 (2005).
29. H. Fujisaki, Y. Zhang, and J. Straub, *J. Chem. Phys.* **124**, 144910 (2006).
30. H. Fujisaki and J. Straub, *J. Phys. Chem. B* **111**, 12017 (2007).
31. S. A. Egorov, K. F. Everitt, and J. L. Skinner, *J. Phys. Chem. A* **103**, 9494 (1999).
32. P. H. Nguyen and G. Stock, *Chem. Phys.* **323**, 36 (2006).
33. P. H. Nguyen and G. Stock, *J. Chem. Phys.* **119**, 11350 (2003).
34. P. H. Nguyen, R. D. Gorbunov, and G. Stock, *Biophys. J.* **91**, 1224 (2006).
35. R. Schinke, *Photodissociation Dynamics* (Cambridge University Press, Cambridge, U.K., 1993).
36. D. Frenkel and B. Smit, *Understanding Molecular Simulations* (Academic Press, San Diego, CA, 2002).
37. M. Buchner, B. M. Ladanyi, and R. M. Stratt, *J. Chem. Phys.* **97**, 8522 (1992).
38. T. Keyes, *J. Phys. Chem. A* **101**, 2921 (1997).
39. H. Goldstein, *Classical Mechanics* (Addison-Wesley, Reading, MA, 1980).
40. W. L. Hase, *Advances in Classical Trajectory Methods*, Vol. 1 (Jai Press, London, 1992).
41. C. Renner, J. Cramer, R. Behrendt, and L. Moroder, *Biopolymers* **54**, 501 (2000).
42. A. Cattani-Scholz, C. Renner, C. Cabrele, R. Behrendt, D. Oesterheld, and L. Moroder, *Angew. Chem. Int. Ed. Engl.* **41**, 289 (2002).
43. C. Renner, U. Kusebauch, M. Löweneck, A. G. Milbradt, and L. Moroder, *J. Peptide Res.* **65**, 4 (2004).
44. H. Satzger, C. Root, C. Renner, R. Behrendt, L. Moroder, J. Wachtveitl, and W. Zinth, *Chem. Phys. Lett.* **396**, 191 (2004).
45. C. Toniolo, M. Crisma, F. Formaggio, and C. Peggion, *Biopolymers (Pept. Sci.)* **60**, 396 (2001).
46. J. Wachtveitl, S. Spörlein, H. Saltger, B. Fonrobert, C. Renner, R. Behrendt, D. Oesterhelt, L. Moroder, and W. Zinth, *Biophys. J.* **86**, 2350 (2004).

47. S. Spörlein, H. Carstens, C. Renner, R. Behrendt, L. Moroder, P. Tavan, W. Zinth, and J. Wachtveitl, *Proc. Natl. Acad. Sci. U. S. A.* **99**, 7998 (2002).
48. J. Bredenbeck, J. Helbing, A. Sieg, T. Schrader, W. Zinth, C. Renner, R. Behrendt, L. Moroder, J. Wachtveitl, and P. Hamm, *Proc. Natl. Acad. Sci. U. S. A.* **100**, 6452 (2003).
49. J. Bredenbeck, J. Helbing, C. Renner, L. Moroder, J. Wachtveitl, and P. Hamm, *J. Phys. Chem. B* **107**, 8654 (2003b).
50. T. Ishikawa, T. Noro, and T. Shoda, *J. Chem. Phys.* **115**, 7503 (2001).
51. A. Cembran, F. Bernardi, M. Garavelli, L. Gagliardi, and G. Orlandi, *J. Am. Chem. Soc.* **126**, 3234 (2004).
52. C. Ciminelli, G. Granucci, and M. Persico, *Chem. Eur. J.* **9**, 2327 (2004).
53. E. Lindahl, B. Hess, and D. van der Spoel, *J. Mol. Mod.* **7**, 306 (2001).
54. T. Darden, D. York, and L. Petersen, *J. Chem. Phys.* **98**, 10089 (1993).
55. H. J. C. Berendsen, J. P. M. Postma, W. F. van Gunsteren, and J. Hermans, in *Intermolecular Forces*, ed. B. Pullman (D. Reidel Publishing Company, Dordrecht, the Netherlands 1981), pp. 331–342.
56. H. Liu, F. Müller-Plathe, and W. F. van Gunsteren, *J. Am. Chem. Soc.* **117**, 4363 (1995).
57. I. G. Tironi and W. F. van Gunsteren, *Mol. Phys.* **83**, 381 (1994).
58. P. Hamm, M. Lim, and R. M. Hochstrasser, *J. Phys. Chem. B* **102**, 6123 (1998).
59. K. A. Peterson, C. W. Rella, J. R. Engholm, and H. A. Schwettman, *J. Phys. Chem. B* **103**, 557 (1999).
60. H. Fujisaki and G. Stock, *J. Phys. Chem.* **129**, 134110 (2008).
61. Y. Guo, D. L. Thompson, and T. D. Sewell, *J. Chem. Phys.* **104**, 576 (1996).
62. G. Stock and U. Müller, *J. Chem. Phys.* **111**, 65 (1999).
63. C. Renner, J. Cramer, R. Behrendt, and L. Moroder, *Biopolymers* **63**, 382 (2002).
64. D. M. Leitner, *Adv. Chem. Phys.* **130B**, 205 (2005).
65. S. M. Park, P. H. Nguyen, and G. Stock, submitted (2009).
66. E. Backus, P. H. Nguyen, V. Botan, R. Pfister, A. Moretto, M. Crisma, C. Toniolo, G. Stock, and P. Hamm, *J. Phys. Chem. B* **112**, 15487 (2008).
67. V. Botan, E. H. G. Backus, R. Pfister, A. Moretto, M. Crisma, C. Toniolo, P. H. Nguyen, G. Stock, and P. Hamm, *Proc. Natl. Acad. Sci. U. S. A.* **104**, 12749 (2007).
68. That is, in addtion to the transport along the peptide states describing low frequency excitation, we explicitly included a "high-frequency state" for group # 1 and its $1/T_1$ decay into the low frequency modes of group # 1 [see Figure 2.5(a)].
69. N. L. Doltsinis and D. Marx, *Phys. Rev. Lett.* **88**, 166402 (2002).
70. A. Toniolo, C. Ciminelli, M. Persico, and T. J. Martínez, *J. Chem. Phys.* **123**, 234308 (2005).
71. C. Nonnenberg, H. Gaub, and I. Frank, *Comp. Phys. Comm.* **7**, 1455 (2006).
72. A. Einstein, *Ann. Phys.* **22**, 180 (1907).
73. G. Cerullo, D. Polli, G. Lanzani, S. D. Silvestri, H. Hashimoto, and R. J. Cogdell, *Science* **298**, 2395 (2002).
74. S. W. Lockless and R. Ranganathan, *Science* **286**, 295 (1999).
75. A. I. Shulman, C. Larson, D. J. Mangelsdorf, and R. Ranganathan, *Cell* **116**, 417 (2004).
76. G. M. Suel, S. W. Lockless, M. A. Wall, and R. Ranganathan, *Nat. Struct. Biol.* **10**, 59 (2003).
77. N. Ota and D. A. Agard, *J. Mol. Biol.* **351**, 345 (2005).
78. K. Sharp and J. J. Skinner, *Proteins* **65**, 347 (2006).
79. T. Ishikura and T. Yamato, *Chem. Phys. Lett.* **432**, 533 (2006).
80. Y. Gao, M. Koyama, S. F. El-Mashtoly, T. Hayashi, K. Harada, Y. Mizutani, and T. Kitagawa, *Chem. Phys. Lett.* **429**, 239 (2006).

# 8 Energy Flow Analysis in Proteins via Ensemble Molecular Dynamics Simulations: Time-Resolved Vibrational Analysis and Surficial Kirkwood–Buff Theory

*Masataka Nagaoka, Isseki Yu,*
*and Masayoshi Takayanagi*

## CONTENTS

## 8.1 INTRODUCTION

In 1965, Jacques Monod stated in his Nobel lecture that "the ambition of molecular biology is to interpret the essential properties of organisms in terms of molecular structures" [1]. Similarly, it would be no exaggeration to say that most computational biologists naturally believe that "the ambition of *computational* molecular biology is to interpret the essential *functions* of organisms in terms of molecular *dynamics.*"

A national project called "The Next-Generation Supercomputer Project" was recently started in Japan to design and build the fastest supercomputer in the world, one that reaches 10 petaflops, by 2012 [2]. In relation to this, a number of leading research projects are being undertaken to develop full-electronic calculation programs, especially those using density functional theory. This includes software development with ProteinDF [3] and fragment molecular orbital (FMO) [4] methods. Worldwide, intense competition has brought the simulation timescale for protein dynamics up to the millisecond scale, with the ultimate aim at clarifying the physiological functions of supergiant biomolecules [5].

Even if the equilibrium structure of a biomolecule is precisely understood, however, further investigation of molecular dynamics is often needed to clarify the molecular mechanisms of protein functions at the atomic level. An example of this is allostery in the oxygen binding of hemoglobin (Hb), which occurs in the low-affinity deoxy state (T state) or in the high-affinity ligated state (R state) [6]. As the first step to understanding Hb allostery, a number of experimental [7–14] and theoretical [15–23] studies have been carried out not only for Hb itself but also for a ligand photolysis process involving myoglobin (Mb), which has a structure very similar to an Hb subunit. While it is true that Mb itself does not show allostery, it should be possible to clarify the characteristics of allostery by comparing Mb with subunits of allosteric Hb.

In relation to protein functions, it is also true that solvents, generally cosolvents, play important roles in keeping the protein structures stable. However, protein dynamics should also be assumed to be significantly influenced by the molecular motions around or with proteins themselves. For example, cosolvents such as zwitterionic compatible solutes (CSs) are known to protect protein function against environmental stresses, such as salinity or high temperature [24–29]. A number of experiments have indicated that addition of CS restricts the structural fluctuation of proteins, and enhances their kinetic stability [25,26]. It has been also indicated experimentally that some CSs, e.g., ectoine, amino acids, and sugars, increase the melting temperatures $T_m$ of proteins, thereby enhancing the thermodynamic stabilities of their folded (native) structures [24c,27,28]. These effects are explained by the preferential exclusion model where CSs are preferentially excluded from the protein surfaces. Considering the above, it is not so difficult to accept the observation that even a single protein exhibits multiple functions by changing its conformation through dynamic interactions with its environment. One example is the dynamic polymorphism of Ras that regulates signal transduction pathway function by dynamically interacting with various effectors [30].

In this chapter, we would like to introduce three strategic methodologies related to energy flow in proteins that are based on the ensemble molecular dynamics (EMD) method (Section 8.2.1), together with their individual applications (Section 8.3.1). First, to gain some microscopic insights into the intramolecular vibrational energy flow of heme cooling, we describe a time-resolved vibrational analysis (TRVA) (Section 8.2.2) of a partial moiety of a protein, conducted with time-resolved power spectra (TRPS) obtained from the Fourier transform (FT) of time-segmental velocity autocorrelation functions (TSVAFs) (Section 8.2.2.2).

In fact, protein studies via normal mode analysis (NMA) were performed independently as early as 20 years ago by groups led by Gō [31] and Karplus [32,33]. Such vibrational analyses for rather large proteins, including in some cases the entire ribosome, are now possible thanks to ever faster computer systems. Today we also have ElNémo, a web interface to the elastic network model that provides a fast and simple tool to compute, visualize, and analyze low-frequency normal modes of large macromolecules (http://www.igs-server.cnrs-mrs.fr/elnemo/index.html) [34]. Earlier, instantaneous normal mode analysis (INMA) was introduced in the 1980s and applied to the study of the microscopic state of liquids, in which there is active movement on the microscopic level even though the liquid might be in equilibrium [35]. However, if one would like to know the energy flow in proteins, certain quantities, energy, frequencies, and other aspects need to be determined time-dependently.

Second, we look at the structural change and energy flow in proteins immediately after some perturbation, e.g., photolysis (less than 100 ps time regime), which can be precisely elucidated by the perturbation EMD (PEMD) method (Section 8.2.1.2). This method lets us make a high-accuracy analysis and detect subtle changes that are almost completely obscured by the large fluctuation in a single MD trajectory. Focusing such attention on the statistical reliability of their MD calculations is actually a recent tendency among computational biochemists [22].

Third, to specify the spatial difference in the degree of preferential exclusion, we define and analyze the preferential exclusion parameters of ectoine for two solutes with the aid of the Kirkwood–Buff (KB) theory [36,37] (Section 8.2.3). To obtain the spatial profiles of the preferential exclusion parameter, we modified the calculation procedure of the KB integral from that in the original KB theory so as to enable systematic discussion of the dependence of the KB integral on the distance from the molecular surface of a solute molecule, i.e., the surficial KB integral [38–41].

For applications of the strategic methodologies, we consider, first of all, the photolysis of carbonmonoxymyoglobin (MbCO), one of the most thoroughly studied Mb-related phenomena, because of its low number of side reactions and other advantageous properties. Then, after introducing a newly prepared repulsive potential function for the (d, d) state, we performed a number of MD simulations of the CO photodissociation process both in complete vacuum and in aqueous solution. The subsequent vibrational cooling of the heme was then analyzed in the heme–globin system, while evaluating the cooling rates of the heme. In addition, we applied the TRVA treatment to the active center (the heme and an imidazole ring of the proximal histidine) to examine the energy flow mechanism in Mb after the photolysis of MbCO in aqueous solution, with the help of INMA (Section 8.3.1.2).

Such structural changes and vibrational energy relaxation of the heme group and the CO ligand escape process after photolysis have been investigated by picosecond time-resolved resonance Raman spectroscopy [7–9], transient grating (TG) technique [11,12], MD simulations [15–23], and other methods. The most remarkable is the direct observation of the structural changes driven by photolysis with time-resolved x-ray crystallography [14]. It enables us to literally "watch" the CO escape and concomitant structural changes mainly occurring in the vicinity of the heme group from 100 ps to several microseconds after photolysis. In Section 8.3.1.3, through the direct application of the PEMD method, it is shown clearly that incipient globin dynamics appears after photolysis that occurs faster than the time resolution of the time-resolved x-ray crystallography [14]. This result should fill some of the gaps in experimental findings.

Next, in Section 8.3.2, we describe long-time full atomic MD simulations (longer than 150 ns) for a small peptide, met-enkephalin (M-Enk), in addition to a larger chymotrypsin inhibitor 2 (CI2), both in ectoine aqueous solutions with the same concentration and in pure water at room temperature. To determine the spatial distribution of each solvent component, the atom number densities of water molecules and ectoine molecules around both solutes were analyzed. We found that one dominant structure of M-Enk does not preferentially exclude the ectoine molecules from its surface as CI2 does. To understand the reason for this difference in ectoine exclusion, in addition to the effect of direct interaction between M-Enk and ectoine, the influence of hydration (i.e., property alteration of the hydration layer near the solute surface) on the development of ectoine preferential exclusion around each solute was examined at the molecular level.

## 8.2  THEORETICAL METHODS

### 8.2.1  Ensemble Molecular Dynamics Method

To obtain sufficient data through which we can deduce some statistically significant conclusions, we have proposed and developed several methods, including the ensemble MD (EMD) method and the PEMD method [17].

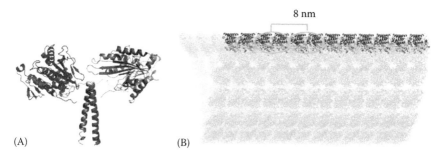

**FIGURE 1.1**  Kinesin and microtubule. (A) Conventional kinesins are homodimer, each of the monomer is made of head, neck-linker, and neck-helix domain. The neck-linkers of two heads are colored in green and yellow, respectively. The neck-helices from the two monomers associate the two subunits. (B) A microtubule with 13 protofilaments, each of which is made of an 8 nm periodic head-to-tail alignment of the tubulin dimer subunits. A single protofilament that is used as a track for kinesin is colored in red. Kinesins take steps hand-over-hand along the protofilament.

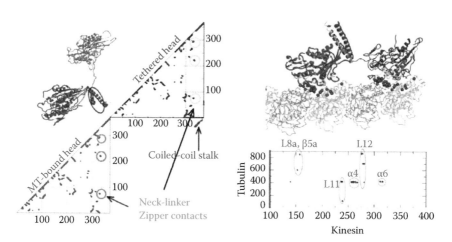

**FIGURE 1.4**  Native contact maps for kinesin dimer (left) and the interface between the kinesin and tubulin binding site (right). The neck-linker zipper contacts that play several important roles for the kinesin function are enclosed in the circles (the green circles for the MT-bound head, and the yellow circles for the tethered head).

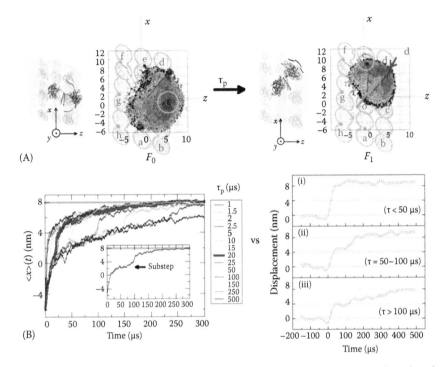

**FIGURE 1.5** Stepping dynamics of kinesin motors. (A) Free energy surfaces, projected on the MT surface, are obtained by sampling the position of tethered head while the neck-linker zipper contacts in the MT-bound head are switched off ($F_0(x,y,z)$) or switched on ($F_1(x,y,z)$). The variation from $F_0(x,y,z)$ to $F_1(x,y,z)$ can be interpreted as a power stroke (neck-linker zippering) dynamics of the kinesin motor, which leads to the stepping dynamics when combined with the diffusional motion of the tethered head. (B) The average time traces from the BD simulation of a quasiparticle with varying $\tau_p$ (the panel on the left), and the average time traces measured using optical tweezers by Higuchi and coworkers (the three panels on the right) [50]. Higuchi and coworkers divided the individual SM time traces into three groups depending on the stepping time ((i) $t < 50\,\mu s$, (ii) $t = 50-100\,\mu s$, (iii) $t > 100\,\mu s$) and averaged over each ensemble. (Adapted from Nishiyama, M. et al., *Nat. Cell Biol.*, 3, 425, 2001.).

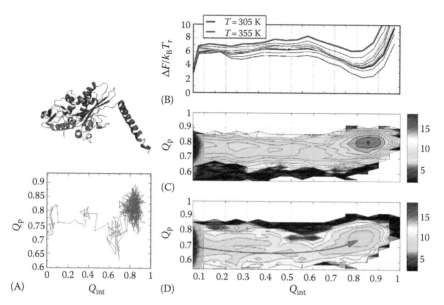

**FIGURE 1.6** PMF between the tethered kinesin head and the MT binding site "e" as a function of $Q_p$ and $Q_{int}$. $Q_p$ is the fraction of native contacts of the MT binding motifs whose structures are colored in orange in A. $Q_{int}$ is the fraction of native contacts for the binding interface between the kinesin and MT (the contact map for the interface is shown in Figure 1.4). (A) The MT binding motifs of the kinesin head domain are colored in orange (top). An exemplary binding trajectory as a function of $Q_p$ and $Q_{int}$ (panel on the bottom). Note that $Q_p$ value reaches the minimum value prior to the binding. (B) One-dimensional projection of free energy profile as a function of $Q_{int}$ at varying temperatures. (C, D) Two-dimensional free energy surface for the binding event as a function of $Q_p$ and $Q_{int}$. The partial unfolding along the binding process is more clearly manifested at a higher temperature. An average binding route suggested from the free energy is drawn in (D).

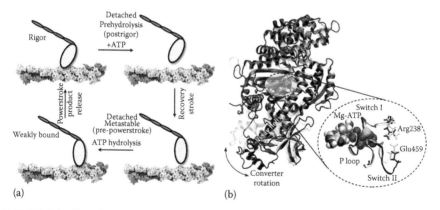

**FIGURE 2.1** Function and structure of conventional myosin. (a) The Lymn–Taylor kinetic scheme for the functional cycle[11,16,57]; only a myosin monomer is shown for simplicity. The process of interest here is the recovery stroke. (b) Structural differences between two x-ray conformations (postrigor,[14] in blue, and pre-powerstroke,[13] in green) of the *Dictyostelium discoideum* myosin motor domain. The most visible transition is the converter rotation, although there are also notable changes in the nucleotide binding site and the relay helix that connects the converter and the nucleotide binding site. With ADP·VO$_4^-$ bound, the nucleotide binding site of the pre-powerstroke state has a closed configuration (in yellow); with ATP bound, the nucleotide binding site in the postrigor state is open (in blue) due to displacement of the Switch II loop.

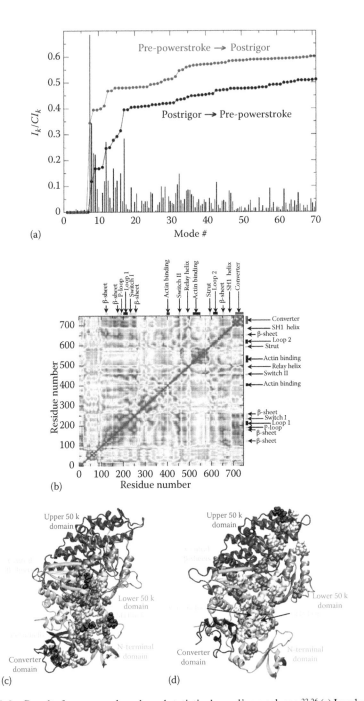

**FIGURE 2.2** Results from normal mode and statistical coupling analyses.[22,26] (a) Involvement and cumulative involvement coefficients (Equations 2.1 and 2.2) for the recovery stroke using normal modes from the postrigor (black) and pre-powerstroke (red) structures. (b) Covariance matrix the Cα atoms calculated using normal modes for the postrigor structure. Red (blue) indicates positive (negative) correlation. (c) Hinge and bending residues in the postrigor structure identified by DynDom[34] based on using normal modes with involvement coefficients larger than 0.10. (d) Mapping of the 52 strongly coupled core residues (in van der Waals form) identified by SCA to the pre-powerstroke structure. In (c) and (d), residues are colored based on residue type; blue: basic, orange: acidic, ice-blue: polar, white: nonpolar.

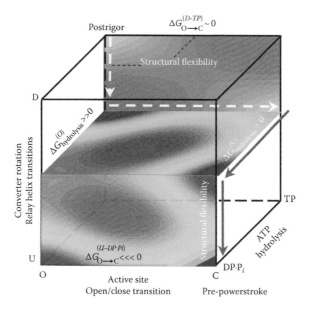

**FIGURE 2.5** A schematic view for the coupling between three major events (converter rotation/relay helix kinking, active site closure and ATP hydrolysis) in the recovery stroke based on results from simulation studies. The scheme highlights the intrinsic structural flexibility of the converter,[22,26] the coupling between open/close transition of the active site to both ATP hydrolysis and converter rotation,[22–24] and that turning on ATP hydrolysis requires extensive structural changes beyond active site closure (SwII displacement).[25]

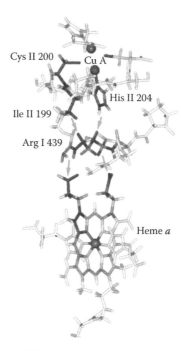

**FIGURE 4.9** $Cu_A$ to heme $a$ ET transition in CcO [51]. Electron transfer occurs via two interfering paths running from $Cu_A$, via Arg439, and to heme $a$ group, and includes a jump between subunits I and II. Water at the interface between the subunits facilitates electron tunneling between the subunits. The color coding is used to show atomic probabilities for tunneling electron to pass a given atom.

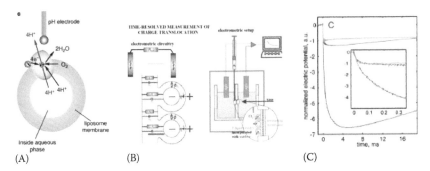

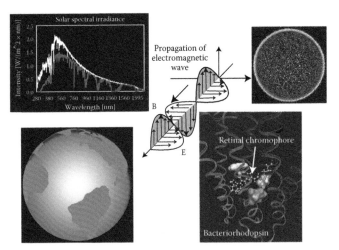

**FIGURE 4.10** (A) Schematics of artificial liposome and reconstituted CcO; after injection of an electron, several proton transfer reactions occur that result in a change of membrane potential. (B) The measuring electrochemical cell (Wikström and Konstantinov, private communication). The liposomes are adsorbed on one side of the measuring membrane. The charge transfer across the liposome membrane (± separation) results in potential difference across the measuring membrane; (C) Typical signal observed in such measurements: initial potential drop, green curve, is due to electron injection—ET Cu$_A$ to heme $a$ (normal to membrane)—followed by two or more slow kinetic phases due to motion of protons across the membrane. Experimental data are the rates and amplitudes of the signal.

**FIGURE 6.1** Life and sunlight. Profile of solar spectral irradiance is shown in the upper left panel. The white line is the extraterrestrial radiation that a location on earth would receive if there was no atmosphere or clouds. The magenta line is the radiation level after passing through 1.5 times the thickness of the atmosphere. Numerical data are taken from ASTM G-173 Reference Spectra (http://rredc.nrel.gov/solar/standards/am1.5). Light travels from the sun to the earth as electromagnetic wave propagation. Photoreceptor proteins in the cells of living organisms are able to interact with the electromagnetic waves. For instance, bacteriorhodopsin (lower right), one of the typical photoreceptors located in the purple membrane of *Halobacterium salinarum*, consists of the protonated Schiff base of the retinal chromophore and the surrounding polypeptide chain. It has a maximum optical absorption at 568 nm. Immediately after light illumination, the retinal chromophore undergoes ultrafast photoisomerization reaction from the all-*trans* form to the 13-*cis* form within 500 fs. The quantum yield of the reaction is as much as 0.64, indicating that two photons are sufficient to form the product state. The photoisomerization reaction triggers the photocycle consisting of K, L, M, N, and O intermediate states. As a consequence of the photocycle, one proton is transferred from the cytoplasmic side to the extracellular side, creating a proton concentration gradient. The bacterium synthesizes an ATP molecule using this proton concentration gradient across the membrane.

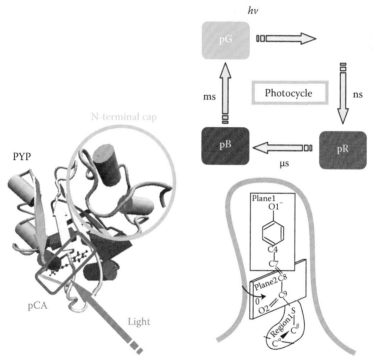

**FIGURE 6.4** Photoactive yellow protein. PYP derived from *H. halophila* is a water-soluble protein consisting of 125 amino acid residues (upper left). It is responsible for the negative phototaxis of *H. halophila* against blue light. PYP is a sensitive photon detector and uses the light energy to induce intramolecular chemical and conformational changes. The chromophore of PYP is *p*-coumaric acid (lower right). Although the chromophore is different from retinal, PYP demonstrates a photocycle (upper right) similar to those of the rhodopsin family proteins containing retinal. PYP is the first protein having a photocycle whose high-resolution structure is available. Analysis of the conformational change of PYP immediately after photoexcitation leads us to understand the initial stage of the photocycle of PYP in atomic detail. From Yamada, A. et al., *Proteins* 55(4):1070, 2004a. With permission.

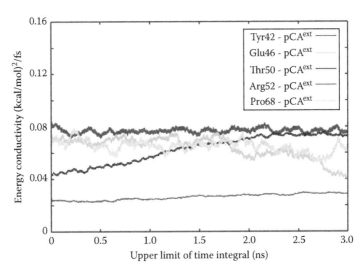

**FIGURE 6.6** Convergence of energy conductivity. The dependence of energy conductivity, $L_{AB} = \dfrac{1}{k_{B}T}\displaystyle\int_{0}^{\tau}\big\langle J_{A\leftarrow B}(t)J_{A\leftarrow B}(0)\big\rangle\,dt$, on the upper limit of the time integral, $\tau$, is shown for A =Tyr42, Glu46, Thr50, Arg52, and Pro68, and B = pCA$^{ext}$.

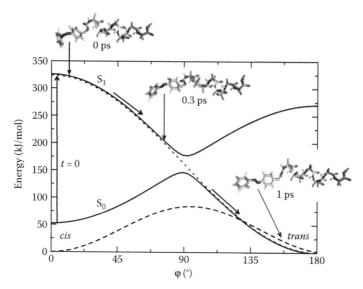

**FIGURE 7.1** Scheme of the $S_0$ and the $S_1$ potential-energy curves of the azobenzene photo-switch as a function of the N=N *cis–trans* isomerization coordinate φ. The solid lines represent the adiabatic potentials of a model which is designed to reproduce the experimental *cis* and *trans* absorption bands and the ground-state *cis–trans* energy barrier. The dashed line corresponds to the GROMOS force field potential of the N=N torsion, the dotted line shows a model of the *cis–trans* photoisomerization potential, which simply connects the $S_1$ *cis* state and the $S_0$ *trans* state. Within this model, the photoexcitation of the system by an ultrashort laser pulse is simulated by instantly switching from the ground-state to the excited-state potential, which causes the isomerization of the system within ≈250 fs. Representative snapshots of the photoi-somerization process reveal that azobenzene first undergoes a rotation along the C–C–N=N and N=N–C–C dihedral angles.

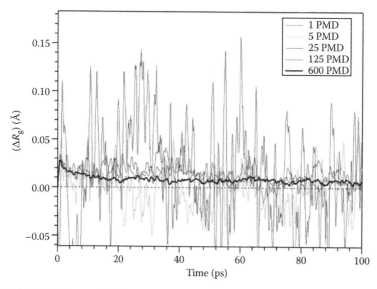

**FIGURE 8.5** $R_g$ changes of a single PMD trajectory (1 PMD), ensemble-averaged over 5, 25, 125, and 600 PMD trajectories (5, 25, 125, and 600 PMD). From Takayanagi, M. et al., *J. Phys. Chem. B,* 111, 864, 2007. With permission.

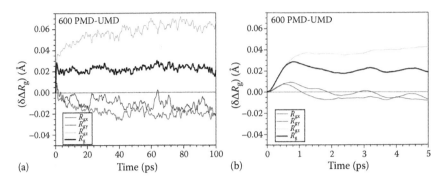

**FIGURE 8.6** $R_g$ changes calculated by the EPMD method using the 600 pairs of PMD and UMD trajectories; (a) 0–100 ps and (b) 0–5 ps. From Takayanagi, M. et al., *J. Phys. Chem. B*, 111, 864, 2007. With permission.

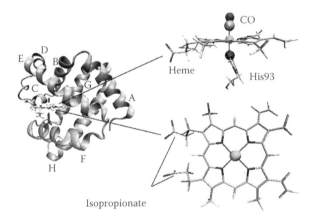

**FIGURE 9.1** Structure of carbonmonoxy myoglobin (MbCO). The heme group is detailed in the side and top views. The eight helices, the ligand CO, and the His93 residue covalently bonded to heme are labeled.

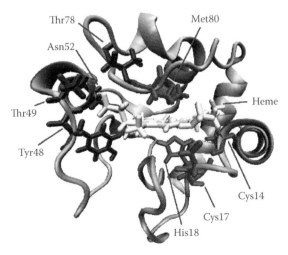

**FIGURE 9.2** Structure of cytochrome $c$ (Cyt $c$) with heme, residues covalently bonded to heme (Cys14, Cys17, His18, and Met80), and residues hydrogen bonded to heme (Tyr48, Thr49, Asn52, and Thr78).

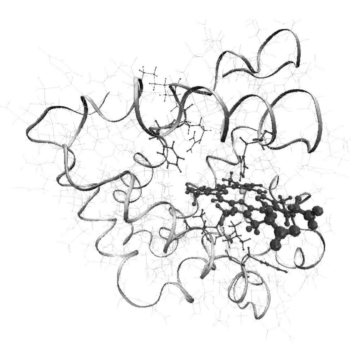

**FIGURE 9.3** The demonstration of the heme cooling pathways following the ligand photo-dissociation in MbCO. The protein structure is shown in ribbon, the heme moiety and protein residues involved in heme cooling pathways are shown in CPK model. The direct interaction between the heme isopropionate side chains and the solvent molecules was found to be the dominant energy transfer pathway. The energy transfer from the heme to the protein was found to follow three channels: (1) "through space" energy transfer mediated by collisions of hot heme atoms with surrounding protein atoms, (2) "through bond" energy transfer through the Fe-His93 connection, and (3) "through projectile" energy transfer mediated by collisions with the CO ligand as it dissociates from the heme and collides with the residue atoms forming the heme distal pocket.

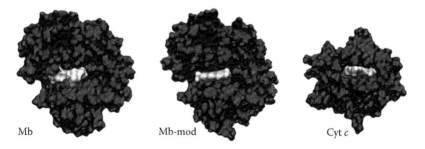

Mb                    Mb-mod                    Cyt $c$

**FIGURE 9.4** Depiction of the heme exposure to solvent in native myoglobin (Mb), modified-heme myoglobin (Mb-mod), and cytochrome $c$ (Cyt $c$) with surface filling model. The protein residues are shown in dark colors and the heme moiety in light colors.

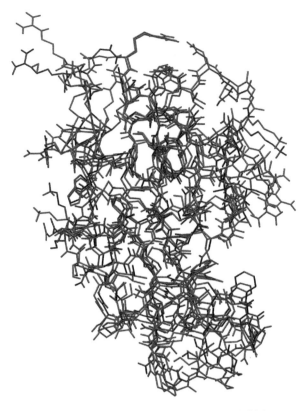

**FIGURE 10.2** An example of coordinate shift caused by the initial energy minimization (illustrated by lysozyme crystal structure, PBD code: 3lzt). The final minimized structure (in red) has 1.34 Å RMSD from the native structure (in blue).

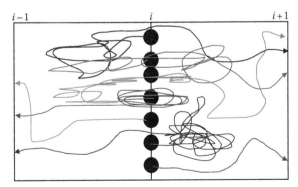

**FIGURE 13.3** A schematic representation of sampling terminating trajectories between milestones $i$ and $i \pm 1$. The milestones are planes. Each trajectory is drawn with a different color and initiated from a lack circle at milestone $i$. The probability of termination at milestone $i+1$ is 4/7 and to terminate at $i$ is 3/7.

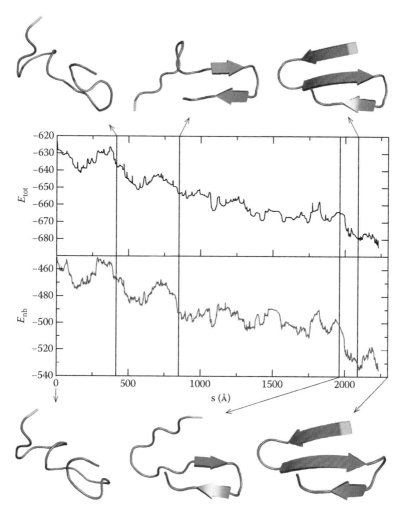

**FIGURE 14.1** Total potential energy (upper) and nonbonding component (lower) as a function of the integrated path length, $s$, for a selected folding pathway in Beta3s.[134] The energy is in units of kcal mol⁻¹. Vertical lines indicate roughly where formation of the the two hairpins starts and finishes. These four conformations are illustrated, together with the two endpoints. The C-termini are colored red and the N-termini are green. The C-terminal hairpin forms first and then the N-terminal strand docks against it to complete the β-sheet.

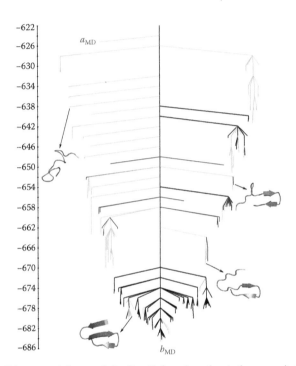

**FIGURE 14.2** Disconnectivity graph for Beta3s based on the stationary points that appear in the 250 "fastest" discrete paths (potential energy in kcal mol$^{-1}$). The denatured and folded end points are labelled $a_{MD}$ and $b_{MD}$, respectively. The branches are colored according to whether the corresponding local minima appear in the fastest or slowest of these paths, emphasising that only minor differences in the folding pathway occur within this set. Green branches lead to minima present on both the fastest and slowest paths in the set, red branches correspond to minima on the fastest path, but not the slowest, and blue branches correspond to minima on the slowest path, but not the fastest. The intervening configurations from the fastest path illustrated in Figure 14.1 are also included here.

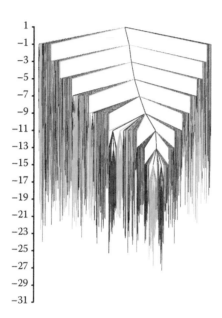

**FIGURE 14.5**  Free energy disconnectivity graph for NtrC at 298 K for regrouping threshold $\Delta F_{\text{barrier}} = 5\,\text{kcal mol}^{-1}$. The energy is in kcal mol$^{-1}$. The branches corresponding to groups of structures classified as inactive (open) and active (closed) are colored green and blue, respectively.[167]

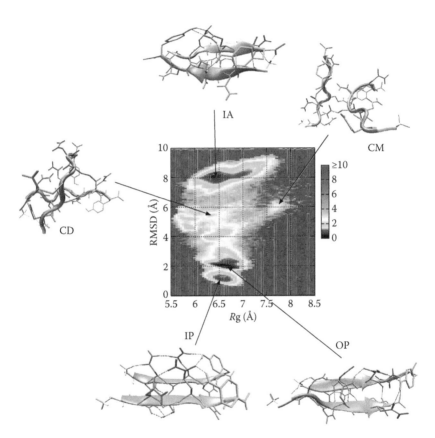

**FIGURE 14.7** Free energy surface calculated at 298 K for the GNNQQNY dimer in terms of the radius of gyration, $R_g$, and $C_\alpha$-RMSD from the amyloid β-sheet.[166] The free energy (kcal mol$^{-1}$) is indicated by shading according to the scale on the right. Representative structures are indicated for five of the minima.

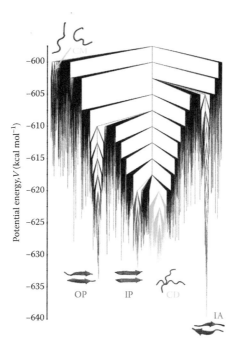

**FIGURE 14.8** Potential energy disconnectivity graph for GNNQQNY.[166] Branches are colored for minima with $C_\alpha$-RMSD < 1 Å from the lowest minima in the IP, OP, IA, and CD sets and $V < -620$ kcal mol$^{-1}$. CD structures correspond to conformations where the two peptides are roughly orthogonal and the CM structures correspond to loosely associated dimers.

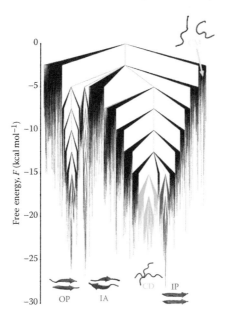

**FIGURE 14.9** Free energy disconnectivity graph for the dimeric phase of GNNQQNY at 298 K.[166] Branches are colored for free energy groups containing minima of the PES with $C_\alpha$-RMSD < 1 Å from the lowest minima in the IP, OP, IA, and CD sets and $F < -15$ kcal mol$^{-1}$.

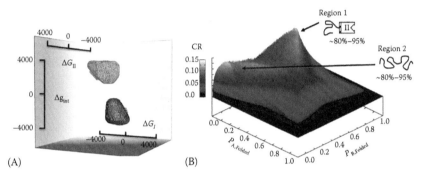

**FIGURE 15.3** Allosteric coupling is optimized in proteins that contain ID. (A) The parameter combinations of $\Delta G_I$, $\Delta G_{II}$, and $\Delta g_{int}$ that generate a CR ≥ 0.10. A wide range of parameter combinations was sufficient to elicit a high CR values. The absence of points in the middle region of the 3D plot indicates that although a wide range of parameter values can combine to produce a high CR, significant interaction energy (i.e., for this case $|\Delta g_{int}| \geq 1\,kcal/mol$) was a prerequisite to coupling. All energies are presented in calories per mole. (B) Plot of the summed probability of those states of the model that are competent to bind ligand A ($P_{A,Folded} = P_N + P_1$) vs. the summed probability of those states of the model that are competent to bind ligand B ($P_{B,Folded} = P_N + P_2$), shown as a function of CR (color-coded as indicated). The maximum CRs were observed in two regions. In region 1, domain I is unfolded and domain II is folded. Binding of ligand A folds domain I, but because of unfavorable domain coupling, domain II unfolds. In region 2, both domains are unfolded. Binding of ligand A folds domain 1, and as a result of favorable domain coupling, domain II folds.

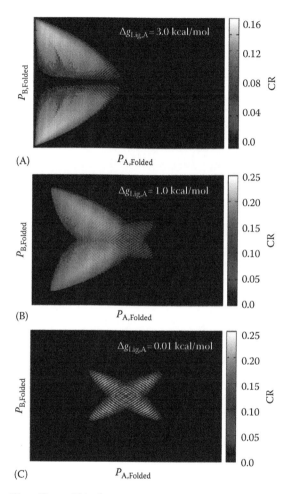

**FIGURE 15.4** The effect of binding energy on the optimum distribution for allosteric coupling. Each plot shows the summed probability of those states of the model that are competent to bind ligand A ($P_{A,Folded} = P_N + P_1$) versus the summed probability of those states of the model that are competent to bind ligand B ($P_{B,Folded} = P_N + P_2$), color-coded as a function of CR (as indicated in the plots). Shown is the dependence of the calculated CR values on the binding free energies ($\Delta g_{Lig,A}$) modeled as (A) 3.0 kcal/mol, (B) 1.0 kcal/mol, and (C) 0.01 kcal/mol. For clarity in the plots, CR values less than 0.10, 0.15, and 0.20 were zeroed to highlight the maxima for $\Delta g_{Lig,A} = 3.0$, 1.0, and 0.01, respectively.

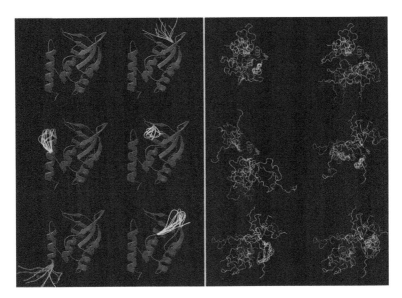

**FIGURE 15.6** The conformational ensemble as calculated by the COREX/BEST algorithm. Shown is a sample of 12 (out of more than $10^6$) of the more probable conformational states calculated by COREX/BEST for SNase. Ensemble states on the left were chosen among those that retained most of the native structure, whereas those on the right were chosen among the mostly unfolded states. Segments of the protein colored red indicate regions that would be folded; regions colored yellow would be unfolded. To illustrate the concept that the unfolded segments freely sample accessible conformational space, these regions were modeled in the figure as multiple flexible loops.

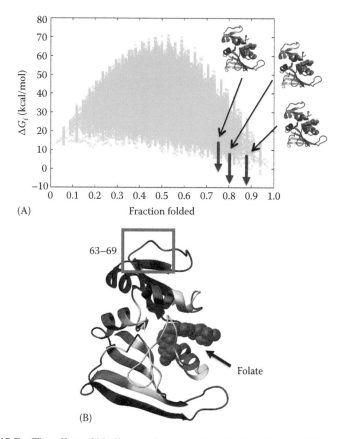

(A)

ΔG_i (kcal/mol)

Fraction folded

63–69

Folate

(B)

**FIGURE 15.7** The effect of binding on the ensemble. (A) The theory of linked functions explains that the effect of increasing concentration of ligand is to stabilize those states of the ensemble that have greater affinity for that ligand [63,64]. To simulate folate binding, those ensemble states of DHFR that were competent to bind folate, which was approximated as those states in which the folate binding site remained folded, were preferentially stabilized relative to all other states of the ensemble [35]. This is represented graphically in the figure by the blue arrows. Each green point in the plot represents one state in the DHFR ensemble. Here, the fraction of residues folded in each state is shown as a function of the calculated state stability, $\Delta G_i$. (B) Redistribution of the ensemble states in response to folate binding had the effect of stabilizing some regions of DHFR while destabilizing (promoting unfolding) other regions. The cartoon structure of DHFR was color-coded such that regions colored red were stabilized the most due to folate binding, green represents modest stability increases, whereas blue is the least. Regions colored purple indicate a negative effect; that folate binding decreased the stability of the protein in these regions, and is highlighted in the figure for the loop containing residues 63–69.

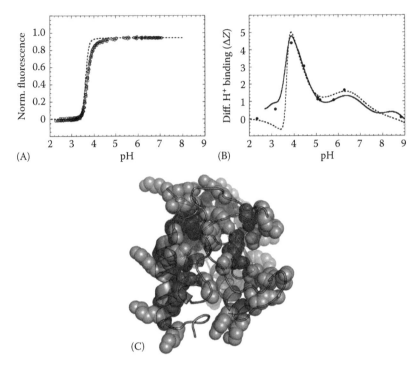

(A)

(B)

(C)

**FIGURE 15.8**   The effect of proton (H⁺) binding on the ensemble. (A) The experimental unfolding of SNase (open circles) as monitored by pH titration of the intrinsic fluorescence of Trp-140 [46]. The stippled line overlaid on the experimental data represents the calculated unfolding of the SNase ensemble as determined by $\langle \text{Fraction\_Native} \rangle = \sum_i \text{Fraction\_Native}_i \cdot P(\text{pH})_i$, where $P(\text{pH})_i$ is given by Equation 15.7. The *Fraction Native* value of each state $i$ was calculated as the number of residues folded divided by the total number of residues of SNase. The change in fluorescence due to the acid unfolding of SNase was normalized to a value of 1 for direct comparison to the ensemble simulation. (B) The experimental difference in H⁺ binding between GdnHCl-unfolded and native SNase was determined by the continuous (solid line) and batch (solid circles) potentiometric techniques [46]. The calculated H⁺ binding difference (stippled line) was determined by $\langle Z \rangle = \sum_i Z_i \cdot P(\text{pH})_i$, where $Z_i$ was calculated as the number of H⁺ bound to each ensemble state i as a function of pH. (C) The positions of the residues that can titrate H⁺ in SNase are shown by the semitransparent space-fill representation of the side chain atoms. Of the 61 titratable groups (59 residues + the N- and C-termini) in SNase, only 18 (colored red) were calculated to contribute to its pH-dependent stability in the ensemble simulations [46,49]. Interestingly, those 18 titratable groups were not clustered in any obvious manner in the native structure of SNase and are intermixed among the other titratable groups that apparently do not play a dominant role in its pH-dependent stability.

Human α-lactalbumin

Native (N) state                    Molten globule (MG) state

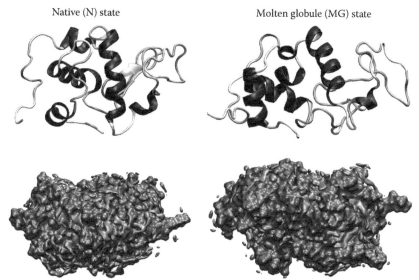

**FIGURE 16.2** Molecular graphics images of the native conformation and a selected molten globule conformer from MD simulations of HαLA in aqueous solution [5]. The top two images are cartoon representations of the protein backbone, with α-helices represented by purple ribbons, β-sheets by yellow ribbons, and reverse turns and random coil by cyan and white tubes. The MG conformer is less compact and contains less secondary structure than the native structure. The bottom two images depict the protein molecules in magenta colored surface representation, and water isodensity surfaces, contoured at ~1.5 times the bulk density, in gray. The solvation of the MG conformer is more diffuse, with less accumulation of water at high density in the nooks and crannies, than that of the native structure.

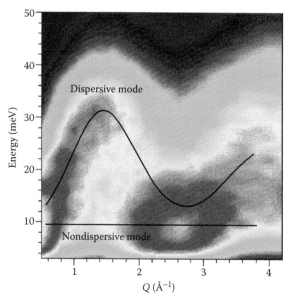

**FIGURE 16.3** Contour plots of the longitudinal current spectra, $C_L(Q,E) = (E^2/Q^2)S(Q,E)$, computed from coherent dynamic structure factors for protein hydration water in the RNase crystal at 300 K [60]. The black lines trace the maxima of the two Brillouin side peaks in the spectra, one of which is dispersive (i.e., the excitation energy is wavevector transfer/length scale dependent), and the other nondispersive.

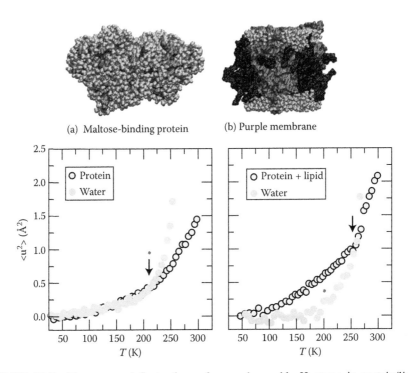

(a) Maltose-binding protein    (b) Purple membrane

**FIGURE 16.5** Mean-squared fluctuations of nonexchangeable H atoms in protein/lipid (filled circles) and water molecules (open circles) in: (a) the soluble protein, maltose-binding protein (MBP) [64], and (b) the protein BR and its surrounding lipids in purple membranes (PM) [9]. The MSFs were measured by neutron scattering experiments that probe motions on up to nanosecond time scales. In each system, protein/lipid motion and water motion were measured separately by employing selective deuteration. Approximate locations of dynamical transitions in the protein/lipid data are indicated by arrows, and in the water data by asterisks. The accompanying molecular graphics images are from MD simulations, carried out in concert with the scattering experiments, of a hydrated powder model of MBP and a unit cell of PM [9,10]. The images depict water molecules in cyan, protein molecules in gray, and lipid molecules in black.

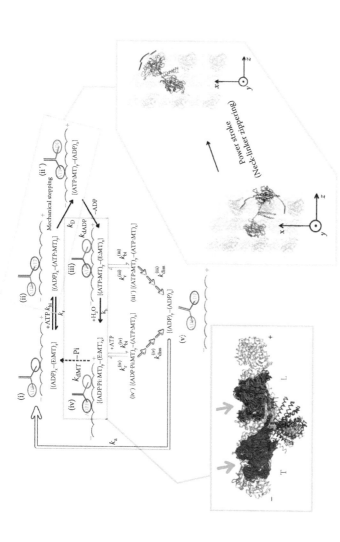

**FIGURE 1.3** Mechanochemical cycle of conventional kinesin (kinesin-1) and the summary of our theoretical study. (A) During the kinetic step within the blue box, ATP binding to the leading head is inhibited, which leads to the high level of processivity of the kinesin motor. This is explained by the mechanochemistry due to the asymmetric strain-induced regulation mechanism between the two motor domains on the MT. The thermal ensemble of structures from the simulations shows that the nucleotide binding pocket of the leading head is more disordered than that of the trailing head, both of which are indicated by the green arrows. The perturbed configuration of the catalytic site of the leading head remains bound to the MT. The tension built on the neck-linker of the leading head disrupts the ATP binding pocket from its native-like pocket topology. (B) The kinetic step from (ii) to (ii') enclosed in the green box denotes the stepping dynamics of the kinesin motor, which is explained by the combined processes of power stroke and diffusional search of the next binding site. Because of the multiplicity of MT binding sites, the pattern of time traces involving stepping dynamics is affected by the rate of power stroke (see Figure 1.5 and text for more details).

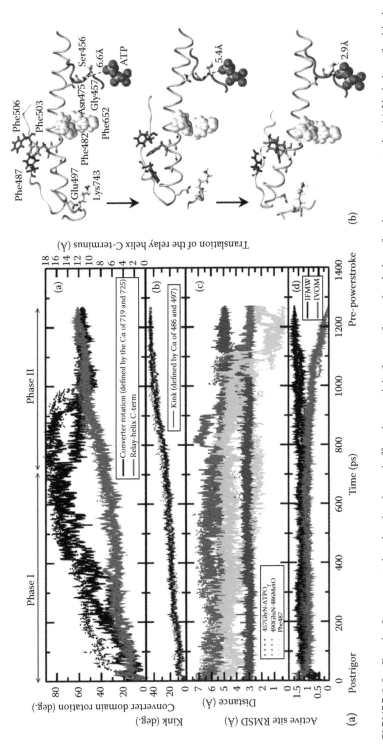

**FIGURE 2.3** Results from targeted molecular dynamics[22] and potential of mean force simulations for the recovery stroke. (a) Variation of critical structural parameters along the three TMD trajectories for illustrating the sequence of events along the approximation transition paths. (b) Three snapshots (at 0.0 s, 630.0 and 1270.0 ps) from one of the three TMD simulations, in the same format as Figure 2 in Ref.[39] for comparison, to illustrate the proposed coupling between the small motion of SwII with the large translation of the relay helix C-terminus.

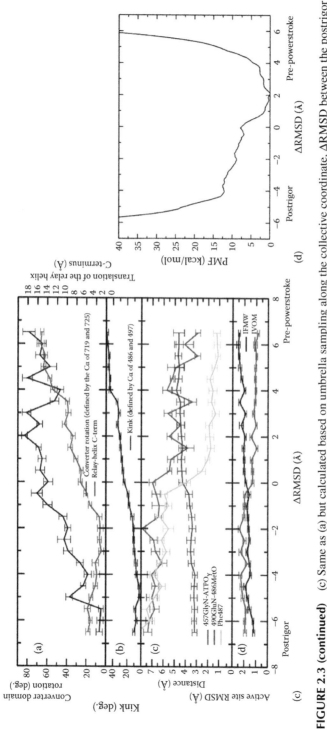

**FIGURE 2.3 (continued)** (c) Same as (a) but calculated based on umbrella sampling along the collective coordinate, ΔRMSD between the postrigor and pre-powerstroke states; (d) Calculated PMF for the recovery stroke along ΔRMSD.

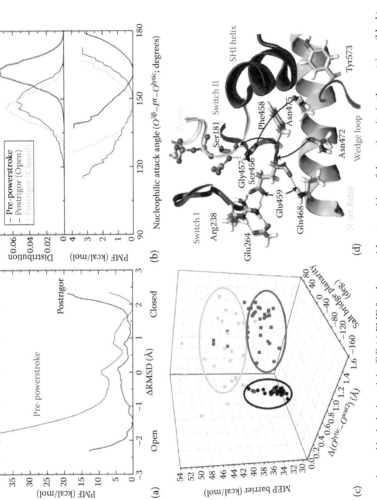

**FIGURE 2.4** Active site properties and hydrolysis activity.[23,25] (a) PMF for the open/close transition of the active site in the postrigor (black) and pre-powerstroke (red) structures; (b) Comparison of the distribution and corresponding PMF for the nucleophilic attack angle based on equilibrium simulation for the pre-powerstroke state and two postrigor structures. (c) Minimum energy path (MEP) barriers for the first step of ATP hydrolysis calculated starting from snapshots collected from equilibrium simulations of the pre-powerstroke state and a closed postrigor structure (see text). The barriers are plotted against the Arg238–Glu459 salt bridge planarity and the differential distance between the lytic water and Wat2 in the reactant and transition state. The black dots indicate data for the pre-powerstroke state; the blue, green, and red set indicate data from the closed postrigor simulations with different behaviors of Wat2.[25] (d) Key hydrogen-bonding interactions in the active site region of the closed postrigor structure. The arrows indicate interactions that are broken when SwII is displaced to close the active site in the postrigor state; i.e., rearrangements in the N-terminus of the relay helix and wedge loop are required to form the stable active site as in the pre-powerstroke state.

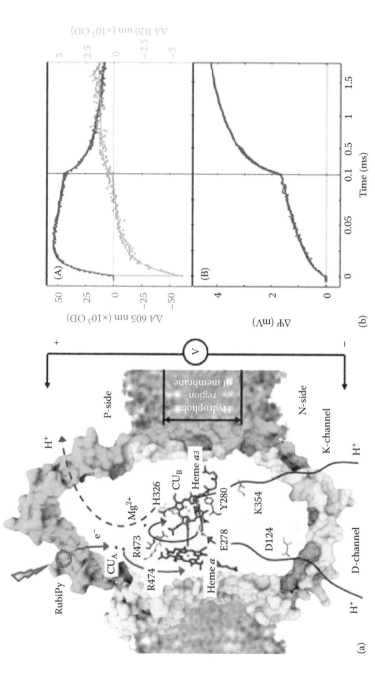

**FIGURE 4.11** (a) Schematics of CcO incorporated in the membrane, and electron and proton transfer reactions in the enzyme in the experiment of Belevich et al. [17]. Two key subunits are shown together with the four redox centers Cu$_A$, heme $a$, heme $a3$, and Cu$_B$. Upon laser flash, an electron from Ru–bpy complex is injected into the enzyme, which drives the enzyme from O to E state generating membrane potential. The electron transfer path is indicated in red. Proton transfer is in blue. Protons are delivered via two channels D and K. The pumped proton is released on the P-side of the membrane from an yet unknown PLS group above heme $a3$. (b) Upper panel is kinetics of absorption changes of heme $a$ and Cu A centers; lower panel—kinetics of the membrane potential. The kinetic curves reveal four kinetic phases with rate constants: 10 μs, 150 μs, 800 μs, and 2.4 ms, and corresponding amplitudes: A1 = 12%, A2 = 42%, A3 = 30%, A4 = 16%.

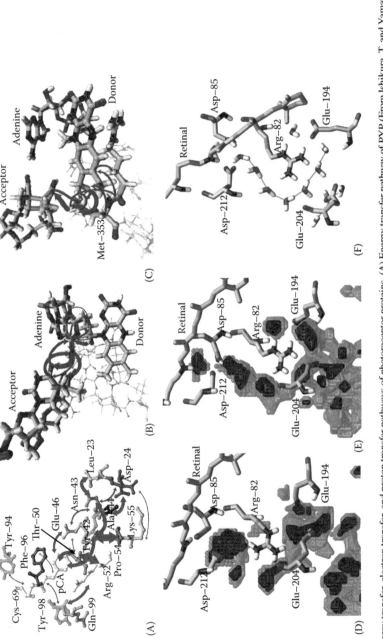

**FIGURE 6.2** Energy transfer, electron transfer, and proton transfer pathways of photoreceptor proteins. (A) Energy transfer pathway of PYP (From Ishikura, T. and Yamato, T., *Chem. Phys. Lett.*, 432(4–6), 533, 2006. With permission). Long-range intramolecular energy transfer pathways (red arrows) were investigated by the residue–residue energy conductivity analysis based on the Green-Kubo formula. The width of the arrow is proportional to the magnitude of the conductivity. (B,C) Electron transfer pathways in DNA photolyase (Miyazawa, Y. et al., *Biophys. J.* 94, 2194, 2008. With permission). Intramolecular photoinduced electron transfer pathways (red arrows) from the flavin cofactor to the UV-damaged DNA are analyzed by interatomic electron tunneling current. Two representative pathways are shown: (B) adenine and (C) methionine routes. The width of the arrow is proportional to the magnitude of the interatomic electron tunneling current. (D–F) Internal water distribution of bacteriorhodopsin and proton transfer pathway (Watanabe, H.C. et al., *Proteins* 75, 53, 2009. With permission). Density distributions of water molecules in the region from the Schiff base of retinal chromophore to the proton release site for light-adapted dark state (D) and the O-intermediate state model (E) of bacteriorhodopsin. The grid points with high population are shown in red, and the grid points with medium population are shown in blue. (F) Typical hydrogen bonding network of internal water.

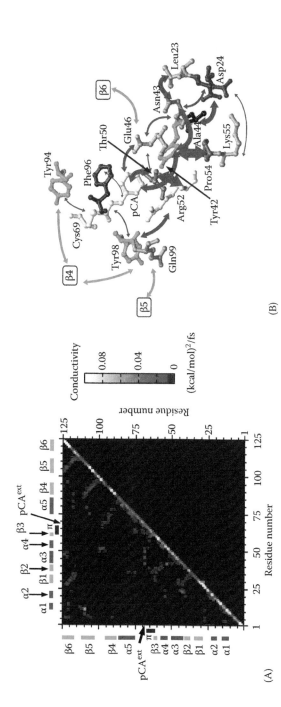

**FIGURE 6.7** Energy transfer pathways in PYP. (A) A two-dimensional map of interresidue energy conductivity. In the upper left triangle, the interresidue energy conductivities are shown in different colors indicating their magnitude. (B) Energy transfer pathways near pCA. pCA^ext (bold yellow), consisting of pCA and Cys69, and the surrounding amino acid residues (thin). The red arrows indicate major energy transfer pathways. The line width of each arrow is proportional to the magnitude of energy conductivity. Reprinted from Ishikura, T. and Yamato, T., *Chem. Phys. Lett.*, 432(4–6): 533, 2006. With permission.

(continued)

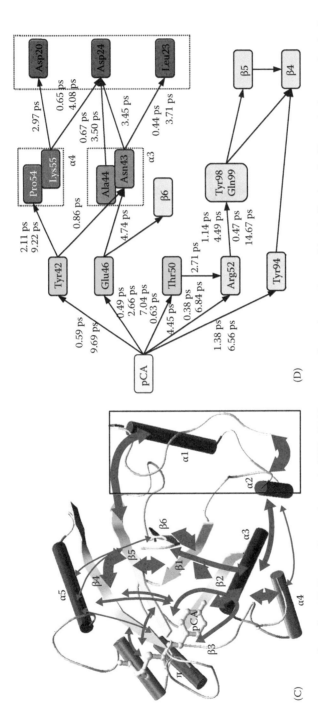

**FIGURE 6.7 (continued)** (C) Energy flow pathways in PYP. (D) Energy transfer pathways from pCA$^{ext}$ (yellow) to the N-terminal cap region (red). The residues consisting of the hydrogen bond network (green) with pCA$^{ext}$, helix α3 (orange), and helix α4 (violet) are shown. For each path, the time-correlation function of energy flux was fitted to a single or double/triple exponential function/s. The time constants for these exponential functions are indicated.

### 8.2.1.1  Ensemble MD Method: A Fair Attack Rarely Launched

The mean of a sample ensemble with number of trajectories $N_{\text{traj}}$ is

$$\langle A \rangle_{N_{\text{traj}}} = \frac{1}{N_{\text{traj}}} \sum_{I=1}^{N_{\text{traj}}} A^{I} \tag{8.1}$$

and the standard error of this mean is

$$\sigma_{\text{M}} = \frac{\sigma}{\sqrt{N_{\text{traj}}}} \tag{8.2}$$

where $\sigma$ is the standard deviation defined by

$$\sigma^{2} = \frac{1}{N_{\text{traj}}} \sum_{I=1}^{N_{\text{traj}}} \left( A^{I} - \langle A \rangle_{N_{\text{traj}}} \right)^{2}. \tag{8.3}$$

According to a standard statistics textbook [42], if one assumes the confidence level $c$ (%), the confidence interval (CI) for the real mean $\bar{A}$ is, when $\sigma$ is estimated,

$$\langle A \rangle_{N_{\text{traj}}} - t(c) \cdot s_{\text{M}} \leq \bar{A} \leq \langle A \rangle_{N_{\text{traj}}} + t(c) \cdot s_{\text{M}} \tag{8.4}$$

where $s_{\text{M}}$ is an estimate of $\sigma_{\text{M}}$ and $t(c)$ is a number depending on both the degree of freedom $N_{\text{traj}}-1$ and the confidence level $c$. For example, if a confidence level of 95% were assumed, $t(c)$ would become ~1.96 and the inequality (Equation 8.4) becomes

$$\langle A \rangle_{N_{\text{traj}}} - 1.96 \cdot s_{\text{M}} \leq \bar{A} \leq \langle A \rangle_{N_{\text{traj}}} + 1.96 \cdot s_{\text{M}}. \tag{8.5}$$

Now, let us assume that $\langle A \rangle_{N_{\text{traj}}}$ could be expressed by the following numeric form with $r$ digit-length,

$$\langle A \rangle_{N_{\text{traj}}} \cong \overbrace{a.b \cdots c}^{r\,\text{digits}} \times 10^{p} \tag{8.6}$$

where $|a|$ is a positive single-figure integer, $b$, ... and $c$ are all nonnegative single-figure integers, and $p$ is an integer. Then, for a positive integer $q$ ($<r$), if one could find an $N_{\text{traj}}$ that fulfills the relation

$$t(c) \cdot s_{\text{M}} \leq 10^{-(q-p)} \tag{8.7}$$

the sample mean $\langle A \rangle_{N_{\text{traj}}}$ would become a good estimate of $\bar{A}$ with a precision of at least $q$ digit lengths, and should be regarded as close enough to $\bar{A}$, e.g., the experimentally observed value. Thus, we hereafter call this method, which estimates the ensemble average of an observable $A$ by using sufficiently large $N_{traj}$ trajectories so that the inequality (Equation 8.4) might be fulfilled, the $c$-CI EMD method, or simply the EMD method.

### 8.2.1.2 Perturbation Ensemble MD Method: Ensemble MD Simulations with and without Perturbation

An application of the EMD method, the perturbation ensemble MD (PEMD) method, which is a perturbative treatment with EMD method, provides us with a very accurate means to estimate physical quantity changes induced by a perturbation (e.g., the photoexcitation of heme [17b]). To start with, many pairs of MD simulations are executed with and without the perturbation, i.e., perturbed MD (PMD) and unperturbed MD (UMD), respectively. Note that all the initial atomic positions and velocities for a couple of PMD and UMD calculations are identical except for those in the perturbed region. Next, $N_{traj}$ pairs of numerical values of the physical quantity $A$ are calculated from the PMD and UMD simulations,

$$\left( A^{PMD,I}, A^{UMD,I} \right), \quad I = 1, 2, \ldots, N_{traj} \tag{8.8}$$

where $N_{traj}$ is the number of each set of trajectories, and $A^{PMD,I}$ and $A^{UMD,I}$ are those values of the physical quantity $A$ calculated through the $I$th PMD and UMD trajectory, i.e., $\{r^{PMD,I}(t)\}$ and $\{r^{UMD,I}(t)\}$ $(I = 1, 2, \ldots, N_{traj})$, respectively. Then, the variation of $A$ between the pair of PMD and UMD simulations for the trajectory number $I$ is calculated:

$$\delta A^I = A^{PMD,I} - A^{UMD,I}. \tag{8.9}$$

Finally, the variation is ensemble-averaged over the $N_{traj}$ trajectories:

$$\langle \delta A \rangle = \frac{1}{N_{traj}} \sum_{I=1}^{N_{traj}} \delta A^I = \frac{1}{N_{traj}} \sum_{I=1}^{N_{traj}} \left( A^{PMD,I} - A^{UMD,I} \right). \tag{8.10}$$

In addition, the time change of $A$ at time $t$ with respect to that at the initial time is defined as

$$\Delta A^I(t) = A^I(t) - A^I(0) \tag{8.11}$$

and the ensemble-averaged time change and its perturbative variation are thus calculated as

$$\langle \Delta A(t) \rangle = \frac{1}{N_{traj}} \sum_{I=1}^{N_{traj}} \Delta A^I(t) = \frac{1}{N_{traj}} \sum_{I=1}^{N_{traj}} \left( A^I(t) - A^I(0) \right) \tag{8.12}$$

and

$$\langle \delta\Delta A(t) \rangle = \frac{1}{N_{\text{traj}}} \sum_{I=1}^{N_{\text{traj}}} \delta\Delta A^I(t)$$

$$= \frac{1}{N_{\text{traj}}} \sum_{I=1}^{N_{\text{traj}}} \left\{ \left( A^{\text{PMD},I}(t) - A^{\text{PMD},I}(0) \right) - \left( A^{\text{UMD},I}(t) - A^{\text{UMD},I}(0) \right) \right\}. \quad (8.13)$$

### 8.2.2 TIME-RESOLVED VIBRATIONAL ANALYSIS

#### 8.2.2.1 Instantaneous Normal Modes in Nonequilibrium Processes

In order to investigate the vibrational energy flow in protein microscopically, based on the EMD method in the previous section, it is possible to pursue a TRVA straightforwardly by repeatedly diagonalizing the instantaneous Hessian matrix (IHM) $H^I(t)$ with elements

$$H^I_{i\alpha, j\beta}(t) = \frac{\partial^2 V(r^I(t))}{\partial r^I_{i\alpha}(t) \partial r^I_{j\beta}(t)}, \quad \alpha, \beta = x, y, z \quad (8.14)$$

along the $I$th MD trajectory $\{r^I(t) | I = 1, 2, ..., N_{\text{traj}}\}$. In this way, $N_{\text{traj}}$ sets of instantaneous normal modes with vibrational frequencies $\{\omega^I_r(t) | r = 1, 2, ..., 3N\}$ are obtained at an elapsed time $t$ since a given nonequilibrium process started. In Equation 8.14, $V(r^I(t)) = V(r^I_1(t), r^I_2(t), ..., r^I_i(t), ..., r^I_N(t))$ is the potential energy where $r^I_i(t)$ is the $i$th atom position vector of a specific $N$ atomic portion in a protein in the $I$th MD trajectory and $r^I_{i\alpha}$ and $r^I_{j\beta}$ are the $\alpha$- and $\beta$-components of the $i$th and $j$th atom position vector, respectively. Thus, $\{\omega^I_r(t)\}$ as a whole yields the time-resolved density of state $D(\omega, t)$ at $t$:

$$D(\omega, t) = \left\langle \frac{1}{3N} \sum_{r=1}^{3N} \delta(\omega - \omega^I_r(t)) \right\rangle_{N_{\text{traj}}} \quad (8.15)$$

where

$$\int d\omega D(\omega, t) = 1. \quad (8.16)$$

One might notice that the above treatment basically follows the INM approach by Stratt and coworkers [35], which was originally developed to show the instantaneous harmonic collective characteristics of molecular liquids in equilibrium. However, its extension to nonequilibrium processes is straightforward on the condition that one can, using the EMD method, evaluate the instantaneous Hessian matrix over a plausible nonequilibrium ensemble, rather than evaluating the Hessian matrix over an equilibrium ensemble of molecular liquid configurations [35]. The variables appearing in the original INM treatment should then become essentially time-dependent, e.g., $\{\omega^I_r(t)\}$.

### 8.2.2.2 Time-Segmental Velocity Autocorrelation Function and Its Fourier Transform

Instead of using the direct INM approach, a set of INMs can be obtained, in a coarse-grained manner, through the Fourier transform (FT) of the following TSVAF $V_{i\alpha}^L(t)$ in each MD trajectory:

$$V_{i\alpha}^L(t) \equiv \frac{1}{V_0^L} \int_L^{L+\Delta\tau} \upsilon_{i\alpha}(t'+t)\upsilon_{i\alpha}(t')dt' \tag{8.17}$$

where

$$V_0^L \equiv \left\langle \int_L^{L+\Delta\tau} \upsilon_{i\alpha}^0(t')\upsilon_{i\alpha}^0(t')dt' \right\rangle_{N_{\text{traj}}} \tag{8.18}$$

and

$$L = 0, \Delta\tau, 2\Delta\tau, \ldots, (M-1) \cdot \Delta\tau \ (=T) \tag{8.19}$$

Thus, $V_{i\alpha}^L(t)$ s are calculated every $\Delta\tau$ interval (or time-segment) with a time resolution determined according to the interval in which the trajectory data are stored [16]. In Equation 8.17, $\upsilon_{i\alpha}$ is the $\alpha$-component of the $i$th atom velocity vector in the case of PMD simulations while $\upsilon_{i\alpha}^0$ is the $\alpha$-component in the case of reference equilibrium MD simulations without perturbations, i.e., a set of UMD simulations. Therefore, $V_{i\alpha}^L(t)$ shows the relative correlation strength in comparison to the reference correlation $V_0^L$ (Equation 8.18), which is the average within the $L$th $\Delta\tau$ time-segment over the latter UMD simulations ($N_{\text{traj}}$ numbers of trajectories). The present way to calculate the velocity correlation is to find the fundamental frequencies without combination tones in each $\Delta\tau$ time-segment.

After executing FT of $V_{i\alpha}^L(t)$:

$$\tilde{V}_{i\alpha}^L(\nu) \equiv \int_L^{L+\Delta\tau} V_{i\alpha}^L(t)e^{2\pi i c\nu t}dt \tag{8.20}$$

and averaging over both the number of atoms ($i = 1, 2, \ldots, N$) and their Cartesian components ($\alpha = x, y, z$), TRPS

$$I^L(\nu) \equiv \frac{1}{3N} \sum_{i,\alpha} \left| \tilde{V}_{i\alpha}^L(\nu) \right|^2 \tag{8.21}$$

are obtained in the range from 0 to $1/(2c\,\Delta t)$ cm$^{-1}$ with $1/(c\,\Delta\tau)$ cm$^{-1}$ resolution for the time step of MD simulation $\Delta t$ and the velocity of light $c$. Finally, arithmetic-averaging is done again over $N_{\text{traj}}$ sets of PMD simulations, i.e., $I_{\text{ave}}^L(\nu)$. If $\Delta\tau$ can be regarded as being small enough, $I_{\text{ave}}^L(\nu)$ should essentially become $D(\omega, t)$ (Equation 8.15), assuming the equality $2\pi c\nu = \omega$.

Although $I^L(\nu)$ itself does not show the vibrational mode vector of a specific portion of a protein, we can supplement the power spectrum with a NMA of the specific portion. The information of $\left|\tilde{V}_{i\alpha}^L(\nu)\right|^2$ is also helpful in discussing the microscopic mechanism of energy flow in protein.

### 8.2.3 SURFICIAL KIRKWOOD–BUFF THEORY

#### 8.2.3.1 Solvent-Atom Number Density

In this section, assuming mixed solvent systems, the subscript $s$ of a quantity stands for a certain solvent component, while $\alpha$ stands a protein as the solute molecule. Then, the time-resolved solvent atom number density (TRSAND) $\rho_{\alpha s}(r, t)$ of a certain solvent component $s$ is defined as a function of the distance $r$ from the solute molecule $\alpha$ (see Figure 8.1) by

$$\rho_{\alpha s}(r, t) = \left\langle \frac{\partial N_{\alpha s}(r, t)}{\partial V_\alpha(r, t)} \right\rangle_{N_{\text{traj}}} = \left\langle \frac{\partial N_{\alpha s}(r, t)}{\partial r} \Big/ \frac{\partial V_\alpha(r, t)}{\partial r} \right\rangle_{N_{\text{traj}}} \quad (8.22)$$

where $\langle\cdots\rangle_{N_{\text{traj}}}$ denotes the ensemble average over the PMD trajectories at a time $t$, $V_\alpha(r, t)$ is the instantaneous volume (see Figure 8.1), and $N_{\alpha s}(r, t)$ is the instantaneous integrated coordination number defined as [39]

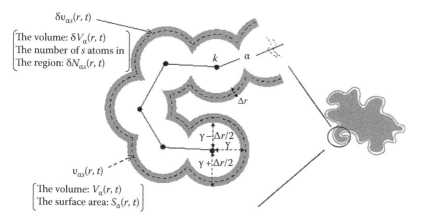

**FIGURE 8.1**  Schematic representation of the region $\upsilon_\alpha(r, t)$, and key quantities. The distance $r$ from (or between) certain atom(s) is that from (or between) the center of the atom(s). The (center of) constituent atoms $k$'s and their bonds in the part of the solute molecule $\alpha$ are represented as black points and black lines. The whole shape of the solute molecule $\alpha$ is also represented by the right ameba-like shape colored in gray. The boundary of $\upsilon_\alpha(r, t)$ is defined as a surface whose minimum distance to any solute atom $k$ of the solute molecule $\alpha$ is $r$, and is represented as the dashed curve. The volume and surface area of $\upsilon_\alpha(r, t)$ are denoted by $V_\alpha(r, t)$ and $S_\alpha(r, t)$, respectively. The region wedged between the two regions $\upsilon_\alpha(r + \Delta r/2, t)$ and $\upsilon_\alpha(r - \Delta r/2, t)$, i.e., $\delta\upsilon_\alpha(r, t)$, is represented as a gray band. The volume of $\delta\upsilon_\alpha(r, t)$ and the number of those atoms of the solvent component $s$ contained in $\delta\upsilon_\alpha(r, t)$ are denoted by $\delta V_\alpha(r, t)$ and $\delta N_{\alpha s}(r, t)$, respectively. (From Yu, I. et al., *J. Phys. Chem. B*, 111, 10231, 2007. With permission.)

$$N_{\alpha s}(r,t) = \sum_{i=1}^{M^s} h(r - \sigma_{\alpha s}(r_i(t))) \tag{8.23}$$

$$\sigma_{\alpha s}(\mathbf{r}_i) = \min_{k \in \alpha} \left| \mathbf{r}_i - \mathbf{r}_k^\alpha \right| \tag{8.24}$$

Here, $h(x)$ is a step function:

$$h(x) = \begin{cases} 1 & (x \geq 0), \\ 0 & (x < 0). \end{cases} \tag{8.25}$$

Then, $h(r - \sigma_{\alpha s}(r_i(t)))$ takes 1 if the constituent atom $i$ of the solvent component $s$ is included within the region $\upsilon_\alpha(r,t)$ (the interior of the dashed closed curve in Figure 8.1), i.e., if $r - \sigma_{\alpha s}(r_i) > 0$, where $\sigma_{\alpha s}(r_i)$ is defined as the minimum distance between the atom $i$ of the solvent component $s$ and any solute atom $k$ of the solute molecule $\alpha$. Thus, $N_{\alpha s}(r,t)$ is an integral number that counts all the constituent atoms of the solvent component $s$ that are included within the region $\upsilon_\alpha(r,t)$, whose volume and surface area are denoted by $V_\alpha(r,t)$ and $S_\alpha(r,t)$, respectively. $M^s$ denotes the total number of constituent atoms in the solvent component $s$. $\delta\upsilon_\alpha(r,t)$ indicates the region wedged between two regions: $\upsilon_\alpha(r + \Delta r/2, t)$ and $\upsilon_\alpha(r - \Delta r/2, t)$ (gray band in Figure 8.1). The volume and the number of atoms of the solvent component $s$ that is contained in $\delta\upsilon_\alpha(r,t)$ are denoted by $\delta V_\alpha(r,t)$ and $\delta N_{\alpha s}(r,t)$, respectively.

If the whole system is in equilibrium, then $\rho_{\alpha s}(r,t)$ becomes time-independent and trajectory number independent. It then becomes simply the solvent atom number density (SAND) $\rho_{\alpha s}(r)$ (see Figure 8.1) estimated numerically by

$$\rho_{\alpha s}(r) = \left\langle \frac{\partial N_{\alpha s}(r,t)}{\partial V_\alpha(r,t)} \right\rangle_T = \left\langle \frac{N_{\alpha s}(r + \Delta r/2, t) - N_{\alpha s}(r - \Delta r/2, t)}{V_\alpha(r + \Delta r/2, t) - V_\alpha(r - \Delta r/2, t)} \right\rangle_T \tag{8.26}$$

where $\langle \cdots \rangle_T$ denotes the time-average defined as $1/T \int_0^T (\cdots) dt$ [40].

### 8.2.3.2 Excess Coordination Number and Preferential Exclusion Parameter

The excess coordination number of atoms in solvent component $s$ around a certain atom site $k$ in the solute molecule $\alpha$, denoted as $N_{\alpha s}^{\text{ex}}$, is defined as [36,37]:

$$N_{\alpha s}^{\text{ex}} = \rho_s(\infty)G_{\alpha s} = \rho_s(\infty)\int [g_{ks}(\mathbf{r}) - 1] d\mathbf{r} \tag{8.27}$$

where $\rho_s(\infty)$ is the atom number density of solvent component $s$ in the bulk solvent phase and $G_{\alpha s}$ is the KB integral [36,37]. $g_{ks}(\mathbf{r})$ in RHS of Equation 8.27, is the pair correlation function between the site $k$ and the atoms in $s$ when they are separated by $\mathbf{r}$,

$$g_{ks}(\boldsymbol{r}) = g_{ks}(\boldsymbol{r}_s - \boldsymbol{r}_k) = \frac{\rho_{ks}(\boldsymbol{r}_k, \boldsymbol{r}_s)}{\rho_k(\infty)\rho_s(\infty)} = \frac{\zeta_{ks}(\boldsymbol{r}_k, \boldsymbol{r}_s)}{\rho_s(\infty)} \tag{8.28}$$

where $\zeta_{ks}(\boldsymbol{r}_k, \boldsymbol{r}_s)$ is the average atomic pair number density of an atom in solvent component $s$ at $\boldsymbol{r}_s$ assuming that the atom site $k$ in the solute molecule $\alpha$ is at $\boldsymbol{r}_k$.

Originally, $G_{\alpha s}$ was defined by $G_{ks}^R(R)$ in the infinity limit of the radius $R$ that denotes a large distance including the whole inhomogeneous phase around the solute molecule $\alpha$:

$$G_{\alpha s} = \lim_{R \to \infty} G_{ks}^R(R) \tag{8.29}$$

where the $R$-dependent KB integral is defined as

$$
\begin{aligned}
G_{ks}^R(R) &= \int_0^R [g_{ks}(r) - 1] 4\pi r^2 dr \\
&= \int_0^R g_{ks}(r) 4\pi r^2 dr - \frac{4\pi R^3}{3}
\end{aligned}
\tag{8.30}
$$

with the use of $g_{ks}(r)$ the radial distribution function (RDF) between the atom $k$ and the atoms in the solvent component $s$, which is related to $g_{ks}(r)$ as

$$\int_V g_{ks}(r) dr = \int_0^\infty g_{ks}(r) \cdot 4\pi r^2 dr. \tag{8.31}$$

The common KB integral $G_{ks}^R(R)$ (i.e., Equation 8.30) is hereafter called the radial KB integral. Unlike the $r$ used in Section 8.2.3.1, $r$ in Equation 8.30 indicates the distance from the center of a certain atom $k$ in the solute molecule $\alpha$. $G_{ks}^R(R)$ multiplied by $\rho_s(\infty)$, i.e., $\rho_s(\infty)G_{ks}^R(R)$, should be related to $N_{\alpha s}^{ex}$, i.e., the excess coordination number of atoms in the solvent component $s$ around the solute molecule $\alpha$:

$$\lim_{R \to \infty} \rho_s(\infty) G_{ks}^R(R) = \rho_s(\infty) G_{\alpha s} = N_{\alpha s}^{ex} \tag{8.32}$$

where $G_{\alpha s}$ ($= G_{ks}^R(\infty)$) indicates the KB parameter of the solvent component $s$ around the solute molecule $\alpha$.

However, in the numerical estimation using MD simulations there would be a practical difficulty, in trying to include the whole inhomogeneous phase, in that $R$ might exceed half of the simulation box length used in our model systems. Moreover, if the profile of the radial KB integral (Equation 8.30) depended sensitively on the location of the atom site $k$, selected arbitrarily in the solute molecule $\alpha$, a direct estimation of the profile might not be suitable for discussion of the spatial difference between the solvent distributions around two kinds of solute molecules with significant differences of size, conformation, and surface topography. Hence, we have defined here another type of $R$-dependent KB integral, $G_{\alpha s}^S(R)$, i.e., the surficial KB integral, which enables us to distinguish the profiles of $N_{\alpha s}^{ex}(R)$ as a function of the distance $R$. This is the distance not from a certain atom site in the solute molecule $\alpha$, but from the surface of the solute molecule $\alpha$, which can be defined by a collection of the solute $\alpha$ atoms closest to each solvent atom:

$$G_{\alpha s}^S(R) = \int_{r=0}^{r=R} \left\langle \left( \frac{1}{\rho_s(\infty)} \frac{\partial N_{\alpha s}(r,t)}{\partial V_\alpha(r,t)} - 1 \right) S_\alpha(r,t) \right\rangle_T dr \tag{8.33}$$

Here $r$ denotes the distance from the solute molecule $\alpha$ in common with Figure 8.1. Thus, the $N_{\alpha s}^{ex}(R)$ profile can be expressed by the coordination number $N_{\alpha s}(R,t)$ in the region $\upsilon_\alpha(R,t)$, and its volume $V_\alpha(R,t)$, defined in Section 8.2.3.1:

$$N_{\alpha s}^{ex}(R) = \rho_s(\infty)G_{\alpha s}^S(R) = \left\langle N_{\alpha s}(R,t) - \rho_s(\infty)V_\alpha(R,t) \right\rangle_T \tag{8.34}$$

Using this definition, one can clearly see the inhomogeneous profile of the $R$-dependent KB integral and the excess coordination number of the solvent species, which reflect the spatial/structural characteristics around/of each individual solute molecule.

According to the KB theory [36,37,41], the preferential exclusion parameter of a cosolvent $s = e$ to water $s = w$ for a solute molecule $\alpha$ can be defined as

$$\nu_{\alpha e}(R) \equiv \rho_e(\infty)\left\{ G_{\alpha e}^S(R) - G_{\alpha w}^S(R) \right\}$$

$$= N_{\alpha e}^{ex}(R) - \frac{\rho_e(\infty)}{\rho_w(\infty)} N_{\alpha w}^{ex}(R) \tag{8.35}$$

where $N_{\alpha e}^{ex}(R)$ and $N_{\alpha w}^{ex}(R)$ are the excess coordination numbers of atoms in the cosolvent molecules e and water molecules w associated with the solute molecule $\alpha$, respectively. A negative value in $\nu_{\alpha e}(R)$ means that the preferential exclusion of ectoine occurs up to the distance $R$ from the surface of the solute molecule $\alpha$, while a positive value in $\nu_{\alpha e}(R)$ means that the preferential binding of ectoine occurs. Although $\nu_{\alpha e}(R)$ is referred to as a preferential binding parameter [41] in some cases, in the present work it is referred to entirely as the preferential exclusion parameter.

## 8.3  APPLICATIONS

### 8.3.1  MbCO → Mb* + CO Photodissociation Process

We focus here on the MbCO → Mb* + CO process, that is, the photodissociation of MbCO and the subsequent energy flow (cooling) of the vibrationally excited heme, provided the electronic state of heme stays in the (d, d) state after the photodissociation (see Figure 8.2). The formation of the (d, d) state occurs in a very short time (<350 fs) [13]. On the other hand, its lifetime is assumed to be long due to the fact that the (d, d) state of NiOEP (OEP: octaethylporphyrin) has a time constant of about 330 ps [8].

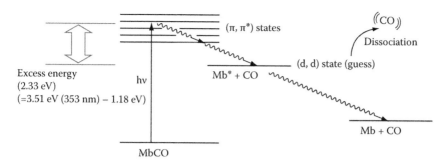

**FIGURE 8.2**   Schematic energy diagram of the photodissociation process of MbCO. Each level indicates an electronic state of the heme in myoglobin. For CO binding myoglobin (MbCO), the heme changes its electronic state by photon ($hv$) absorption. Heme of the $(\pi, \pi^*)$ states reaches the (d, d) state in a very short time via radiationless processes. In this state, the heme has excess vibrational energy and CO is dissociated (Mb* + CO). The electronic state of the heme is finally changed to the ground state (Mb + CO).

### 8.3.1.1   MD Simulations: Computational Detail

To simulate the photolysis of a CO ligand molecule by the absorption of one photon of wavelength 353 nm (3.51 eV (=81 kcal/mol)), the force field parameters of the heme group were changed from those of the ground state ("Liganded" in Ref. 15a) to those of the excited state ("Unliganded hard" in Ref. 15a). In the ground state, the heme is six-coordinated, planar, and CO ligated. In the excited state, it is five-coordinated, domed, and in the deoxy form. In addition, we introduced a repulsive potential function between the iron atom and the ligand molecule and deposited the excess kinetic energy 2.33 eV (=54 kcal/mol) into 24 atoms of the porphin ring. This large amount of energy is desirable for the informative analysis of PMD simulations. The remaining energy of the photon, 1.18 eV (=3.51–2.33 eV (=27 kcal/mol)), was assumed to be used to change the electronic ground state to the (d, d) excited state of the heme, corresponding to the energy difference between the $^1A_1$ and the $^1B_2$ states at the equilibrium bond distance of $^1A_1$. It was obtained by the Super-CI CASSCF calculation [43].

#### 8.3.1.1.1   For the Time-Resolved Vibrational Analysis

In the MD calculations for the TRVA of the MbCO $\rightarrow$ Mb* + CO process in vacuo, the Dreiding potential function [44] was used to describe interatomic interactions in the protein. For amino acids, the force field parameters of the Dreiding 2.21 and partial charges estimated by Gasteiger [45] were utilized. For both CO-liganded and unliganded heme, we adopted the force field parameters determined by Henry et al. [15]. For the unliganded heme in the (d, d) state during the PMD simulation, the "unliganded hard set" for the ground state [15a] was used, with the assumption that the parameters of the (d, d) state are similar to those of the ground state. The partial charges used for the heme were those of AMBER5 [46], while the partial charges for CO were those reproducing its dipole moment. In all the MD calculations, the smoothing cutoff of 25.0–25.5 Å was applied with respect to nonbonding potentials.

Characteristically, the repulsive potential function introduced into the Dreiding potential function was newly prepared as a function of the Fe–CO bond length, using the energy curve obtained in a previous molecular orbital (MO) study by Chiba et al. [43] for the $^1B_2$ ($3d_\pi \rightarrow 3d_{z^2}$) excited state of the $C_{2v}$ symmetrical FeP(py)CO (FeP: porphinato-iron; py: pyridine). In this super-CI CASSCF calculation, 14 electrons were distributed in all possible ways among 12 active MOs. The main components of the active orbitals are five 3d orbitals of Fe, six 2p orbitals of CO, and a 2s orbital of C of a CO molecule.

The geometrical structure of MbCO (2543 atoms) was taken from a crystal structure for the sperm whale MbCO [47] (PDB ID #: 2MB5 [48]). After replacing the deuterium atoms of MbCO in the structural data by hydrogen atoms, the PDB structure was subjected to 500 cycles of the conjugate gradient energy minimization. MD simulations were then carried out for MbCO by integrating the equations of motion of all the atoms using the leapfrog Verlet algorithm with a time step of 2 fs. The initial velocities of atoms were assigned according to the Maxwell–Boltzmann distribution at 300 K. The temperature was maintained at 300 K by velocity scaling at each interval of 250 steps (0.5 ps). After a 1000 ps-long equilibration run, the MbCO system was allowed to evolve freely without the velocity scaling. After we obtained a 50 ps long trajectory of MbCO, an ensemble consisting of eight sets of coordinates and velocities was prepared for both EUMD and EPMD simulations with photodissociation. The time separation between these different initial sampling sets was 5 ps.

Using each initial set, the Mb* + CO system was evolved for the following 50 ps. For the PMD simulations on the photon absorption, we instantaneously changed the force field parameters of the CO-liganded heme into those of the unliganded heme, adding the repulsive potential between CO and the heme. At the same time, by adding 24 randomly generated excess velocity vectors to the original velocity ones, the total kinetic energy of 20 carbon and 4 nitrogen atoms of the porphin ring was increased to represent the injection of additional vibrational energy, i.e., the excess energy, into the heme as was done by Henry et al. [15b]. The method to partition the remaining energy to heme atoms as kinetic energies will be explained separately in Section 8.3.1.1.2. For comparison, a set of UMD simulations for MbCO initiated by using $\upsilon_i^{initial}$ were also done. The Cerius$^2$ OFF [49] program was used for all the MD calculations. Trajectory data of the simulations were stored at every 2 fs interval.

In the INM analysis (see Section 8.2.2), assuming that the specific portion of MbCO might be the active center, i.e., a heme and the imidazole ring of His93 (as a supermolecule), the atomic coordinates are chosen for the instant that the photoexcitation (or perturbation) should be applied in an MD simulation, and we attach one hydrogen atom to the $N^\gamma$ part of imidazole of the active center because of a broken bond. The NMA was then carried out with the same potential function, force field parameters, and atomic charges as were used in the PMD simulations described in the previous subsection. Using the above, we will now discuss the vibrational energy relaxation, flow, and cooling of the excited heme in terms of the microscopic mechanisms.

### 8.3.1.1.2    Initial Velocity Setting in the Excited State: Total Angular Momentum Conservation

Apart from dealing with the electronically excited state of heme for the first time, we adopted more reasonable, newly developed energy partitioning by applying the following procedure: just after photon absorption, the velocities of the 24 atoms in vibrationally excited heme in the (d, d) state were set to

$$v_i^{excited} = v_i^{initial} + e\boldsymbol{u}_i, \quad i = 1, 2, \dots, 24 \tag{8.36}$$

where $v_i^{initial}$ is the velocity vector of the $i$th atom in the porphin ring just before the absorption. Assuming that $\boldsymbol{u}_i$ s are unit vectors whose directions were obtained by a uniform pseudorandom number generator, a scalar value $e$ was adjusted so as to reproduce an increase of 2.33 eV in the kinetic energy of the 24 atoms as a whole. Here, this $\boldsymbol{u}_i$ s and $e$ were generated to conserve not only the total linear momentum but also the total angular momentum of the 24 atoms before and after the photon absorption [50].

### 8.3.1.1.3    For the Anisotropic Structural Relaxation Process

For the purpose of investigating the anisotropic structural relaxation of the Mb* + CO system in aqueous solution, another set of EUMD and EPMD simulations were separately carried out with the AMBER7 program using the parm99 force field [51]. However, for the heme and the heme-bound CO ligand parameters, we used, as explained in the Section 8.3.1.1.1, the same bond, angle, and dihedral parameters as developed by Henry et al. [15a], while for the electrostatic and van der Waals parameters, we used those determined by Giammona [52]. The atomic charges of the dissociated CO molecule were calibrated to mimic its dipole moment 0.11 D [53], i.e., $-0.021e$ and $+0.021e$ for the carbon and oxygen atom, respectively. The SHAKE method was used to constrain the hydrogen-heavy atom bond distances, and the particle mesh Ewald (PME) procedure was used to handle long-range electrostatic interactions. The integration time-step was 2 fs. All the MD simulations were performed using periodic boundary conditions: The MbCO structure (PDB ID#: 2MB5 [48]) was solvated with 2986 TIP3P water molecules [54] and 9 chloride counterions yielding a periodic box size of ~51 × 47 × 47 Å$^3$.

Next, after a 400 ps equilibration NPT run at ambient condition (300 K, 1 atm), a 600 ps NPT run was performed. We obtained 600 snapshots by storing them at every 1 ps interval. Then, from each of the 600 snapshots a couple of PMD and UMD simulations were performed in the EMD sense, for 100 ps in the NVE condition, with and without the preparation of the photolyzed state described in the previous section.

### 8.3.1.2    Energy Flow in Photoexcited MbCO

From the total kinetic energy of the 24 porphin ring atoms, we calculated the instantaneous kinetic temperature

$$\langle T(t) \rangle_{N_{\text{traj}}} \equiv \frac{2}{3Nk_B} \left\langle \frac{1}{2} \sum_{i=1}^{24} m_i \upsilon_i(t)^2 \right\rangle_{N_{\text{traj}}}$$  (8.37)

where $m_i$ and $\upsilon_i$ are the mass and the velocity vector of the $i$th atom, respectively [55–57]. When the excess energy was injected into the heme, the temperature exhibited a very abrupt increase up to 1058 K. From the temperature curve after the photoexcitation, ensemble-averaged over the PMD simulations for the time range of 0–50 ps, a double-exponential was fitted, resulting in two relaxation time constants of 0.84 ps (66%) and 18.7 ps (34%) for the fast and slow contributions, respectively.

Figure 8.3 shows the averaged TRPS (Equation 8.21), with a clear time-dependent change in the spectral intensity of the active center. In the figure, a spectrum (–5 to 0 ps) of MbCO in equilibrium is also included for comparison to the spectra of the post-photodissociation (Mb* + CO). Because these TRPS were classically calculated, the shapes of spectra might be inadequate, especially for the region of high frequencies. However, the peak positions should indicate reasonably enough the peaks of the vibrational bands.

In the experimental results [9a] of Mizutani and Kitagawa, $\nu_4$ (1350 cm$^{-1}$), $\nu_3$ (1461 cm$^{-1}$), and $\nu_5$ (1115 cm$^{-1}$) bands of heme exist. Although the $\nu_4$ peak in their resonance Raman spectroscopic experiment is typically observed to show immediate generation and double-exponential decay, it is not so clear in the present simulations. This is partly because the excess vibrational energy was equally distributed among the heme atoms without considering any quantum mechanical selection rules, and partly because the estimated spectrum did not directly correspond to the resonance Raman spectrum.

In contrast, in the present simulations, the spectral intensities of vibrational modes were estimated from TRVAFs to investigate the whole modes. In doing this, as shown in Figure 8.3, we found a number of strong peaks, for example those at ~630, ~1850, and ~3200 cm$^{-1}$. Moreover, the TRPS showed both generation and disappearance of those bands. For instance, the peak at ~1850 cm$^{-1}$ disappears in 10 < $t$ < 15 ps. On the contrary, in 20 < $t$ < 25 ps, a peak appears at ~1950 cm$^{-1}$. Further, one can find that the ~1500 cm$^{-1}$ peak has a comparative level of intensity in the range of 5–20 ps. Thus, the spectral intensities of these vibrational peaks show fast mutual exchanges of magnitudes. This means that there should exist a number of nonlinear vibrational couplings, which should couple multiple vibrational modes of the heme with those of the globin or solvent molecules, if they exist.

Peak shifts are also observed. The peaks (627 and 1501 cm$^{-1}$ in 0 < $t$ < 5 ps) are red-shifted by 13 cm$^{-1}$. Blue-shifts of 7 cm$^{-1}$ are also found in the peaks (1301 cm$^{-1}$ in 0 < $t$ < 5 ps, 934 cm$^{-1}$ in 5 < $t$ < 10 ps). These shifts probably resulted from switching vibrational states from the CO-liganded heme to the unliganded heme.

To clarify the correspondence between spectral peaks and vibrational modes obtained by FTs of TSVAFs, we also performed an INM analysis for the active center (see Section 8.3.1.1.1). It was found that the frequencies of $\nu_7$, $\nu_{32}$, $\nu_6$, and $\nu_5$ are 616, 772, 800, and 1070 cm$^{-1}$, respectively. These are quite similar to experimental values of MbCO [9a] and/or those of NiOEP [58]. The relatively small differences, therefore, would come almost certainly from whether or not the active center is buried in the globin.

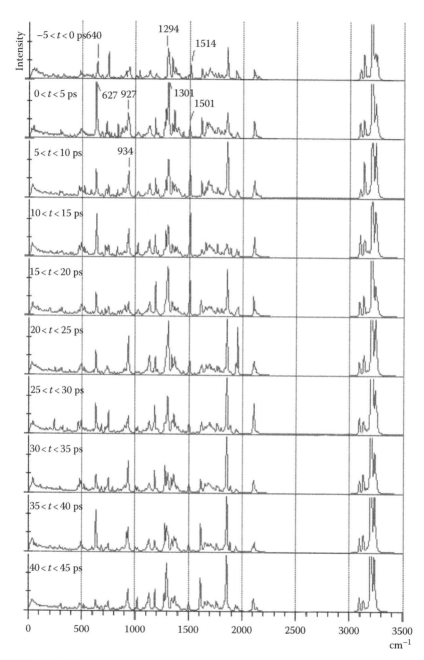

**FIGURE 8.3** Time-resolved power spectra (TRPS) estimated by Fourier transforms (FTs) of the time-segmental velocity autocorrelation functions (TSVAFs) of the active center. The spectra indicate the intensities of the vibrational modes of the active center in the globin. The intensities over 3500 cm$^{-1}$ are zero (not shown). The photodissociation is achieved at $t = 0$ ps, after the previous equilibrium MD stimulation is denoted by negative time duration. The intensities of power spectra are depicted as functions of the vibrational frequency. The ordinate axes indicate the absolute FT intensity in arbitrary units. The spectrum ($-5 < t < 0$ ps) is that of MbCO, while the other spectra are those of Mb* + CO. (From Okazaki, I. et al., *Chem. Phys. Lett.*, 337, 151, 2001. With permission.)

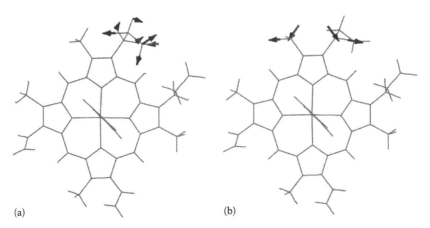

(a)                                                          (b)

**FIGURE 8.4** Two vibrational normal mode vectors of the propionate groups of the heme. The arrows illustrate the vibrational vector; (a) mode of frequency ~1850 cm$^{-1}$ and (b) high-frequency mode of ~3200 cm$^{-1}$. (From Okazaki, I. et al., *Chem. Phys. Lett.*, 337, 151, 2001. With permission.)

Based on the INM analyses, we observed the following, which led to an interesting conjecture. Two propionate groups in the heme (Figure 8.4), with vibrational modes at ~1850 cm$^{-1}$ ($v_{1850}$) and ~3200 cm$^{-1}$ ($v_{3200}$), have strong intensities and are localized as shown in Figures 8.3 and 8.4. On the other hand, it is well known that the vibrational frequencies of a water molecule (H$_2$O) are 3657 cm$^{-1}$ ($v_1^W$; symmetric stretching), 1595 cm$^{-1}$ ($v_2^W$; symmetric bending), and 3756 cm$^{-1}$ ($v_3^W$; asymmetric stretching). Therefore, the vibrational modes of the propionate groups should couple with those of water because not only should $v_{1850}$ in the propionate groups couple with $v_1^W$ and/or $v_3^W$, but also $v_{3200}$ should couple with $v_2^W$ by overtone. Furthermore, since parts of the propionate groups, especially carbonyls (–COO), are exposed to the water solvent, the excess vibrational energy could diffuse through those groups to the water solvent from the heme.

Recently, the above conjecture has been clearly proved experimentally by Mizutani and Kitagawa's research group [10] and they have concluded that "The amputation of the heme's propionates results in an increase in the time constant of the $v_7$ band by a factor of 3.9 ± 1.6 … [and] the heme vibrational relaxation is significantly affected by amputation of the propionates and strongly support the theoretical proposal that a possible doorway for energy release from the vibrationally excited heme involves the interaction of its propionate groups with neighboring solvent molecules" [10b].

### 8.3.1.3   Anisotropic Expansion of Photoexcited MbCO

Figure 8.5 is a plot of the change in $R_g$ (the radius of gyration of MbCO) of a single PMD trajectory, with those changes ensemble-averaged over 5, 25, 125, and 600 PMD trajectories in the PEMD method. It can be understood that the fluctuation was significantly reduced by increasing the number of trajectories for the average. By averaging $R_g$ over the ensemble of 600 values, the $R_g$ change was naturally obtained with small fluctuation and we can detect subtle changes to an accuracy of 0.03 Å. Such changes were almost completely obscured in a single PMD value due to fluctuations larger than 0.1 Å.

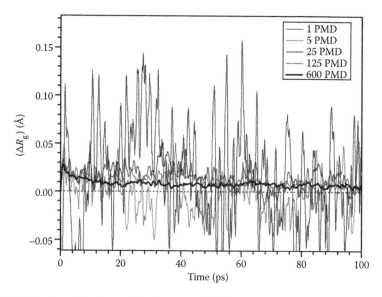

**FIGURE 8.5** **(See color insert following page 172.)** $R_g$ changes of a single PMD trajectory (1 PMD), ensemble-averaged over 5, 25, 125, and 600 PMD trajectories (5, 25, 125, and 600 PMD). (From Takayanagi, M. et al., *J. Phys. Chem. B*, 111, 864, 2007. With permission.)

Accordingly, it can be said that the $R_g$ changes were obtained under the influence of the photolysis only and that they show sufficiently small fluctuations (maximum 95% CI is ±0.02 Å) (Figure 8.6). Comparing the three component values, therefore, it was found that the behaviors of in-plane ($R_{gx}$ and $R_{gy}$) and out-of-plane ($R_{gz}$) components are specifically different. While those in the former show contractions after the short-time small expansions within 1 ps (peak at about 0.7 ps), the latter component clearly rises during the simulation time with a rapid increase up to 0.04 Å immediately after the photolysis. This anisotropic structural change, which mainly occurs in the $z$-direction, was induced by two mechanisms: one is the displacement

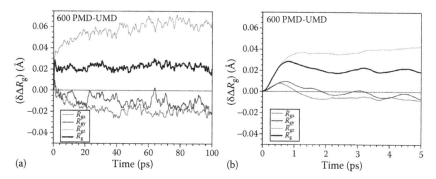

**FIGURE 8.6** **(See color insert following page 172.)** $R_g$ changes calculated by the EPMD method using the 600 pairs of PMD and UMD trajectories; (a) 0–100 ps and (b) 0–5 ps. (From Takayanagi, M. et al., *J. Phys. Chem. B*, 111, 864, 2007. With permission.)

of the heme iron atom induced by the heme conformational change from a planar form to a domed one and the other is the photolyzed CO ligand collision with the distal pocket residues. It should be noted that, among and during 100 ps PMD simulations, most of the dissociated COs appeared in the primary docking site except for a few trajectories in which COs migrated to the Xe4 site [20]. These push the proximal histidine or distal residues outward along the $z$-direction and result in $z$-direction expansion. In fact, the structure ensemble-averaged over the PMD trajectories provides displacements of residues in the heme vicinity that are as clear as those measured by the time-resolved x-ray crystallography [14].

It was found that most of the obtained anisotropic changes occur within 1 ps, followed by relatively small changes. This agrees with the results of TG spectroscopy by Goodno et al. [11b]. They also demonstrated asymmetric conformational relaxation in a protein within 500 fs, whose out-of-plane ($z$-direction) expansion is larger than the in-plane ($x$- and $y$-directions) changes. However, their analysis could not definitively determine whether the in-plane changes were expansion or contraction because the observed signals were the sum of the contributions from the globin strain and the solvent heating. Since our calculations, meanwhile, include only the contribution from the structural change of Mb and reveal contraction in the $x$- and $y$-directions, the present observation may be reasonably considered affirmative evidence that Mb partially shows the physical properties of an elastic medium. That is to say, if Mb expands along one direction, it contracts at the same time along perpendicular directions to keep its atomic number density constant.

### 8.3.2 CHYMOTRYPSIN INHIBITOR 2 AND MET-ENKEPHALIN IN ECTOINE AQUEOUS SOLUTION

It is widely known that the cosolvents influence the structural stability of proteins and alter dynamic properties of water molecules [26,39]. It is, therefore, natural that the energy flow in protein should also be strongly influenced in the mixed solvent environment. When the energy flow from the protein to the ambient solvent is investigated from the view point of the solvent structural dynamics, it is, of course, necessary to develop some theoretical frameworks, which are able to evaluate the spatial distributions of each solvent component around the large and flexible macromolecules. For this purpose, we developed the surficial KB theory [40], and applied it to examine the role of ectoine [24], a zwitterionic CS that protects protein function against environmental stresses.

It has been also indicated experimentally that ectoine enhances the thermodynamic stabilities of their folded (native) structures [24c]. This observation has been explained by the preferential exclusion model, which states that CS molecules are expelled from the protein surface [28,29] and the growth of the preferential exclusion corresponds with the increase of excess chemical potential of the protein [28,29]. In fact, our previous MD simulation also indicated numerically that ectoine molecules are preferentially excluded near the CI2 surface [39]. Thus, to understand how CS molecules interact microscopically with proteins, and whether the addition of CS might indirectly stabilize them irrespective of their molecular properties, the hydration structures have been studied not only for CI2 but also for a smaller peptide, met-enkephalin (M-Enk), in ectoine aqueous solutions of the same concentration and in pure water at room temperature [40].

### 8.3.2.1  MD Simulations: Computational Detail

In the present systems, all the MD calculations were performed using the AMBER7 program [51] and the force field parameter set, parm99 [59], was used for all the molecules. The partial atomic charges of ectoine molecules were taken from our previous study [38]. The starting conformation of the zwitterionic form of M-Enk that we adopted was an x-ray structure in which the peptide backbone is extended [60].

MD simulation of M-Enk in 1.5 M ectoine aqueous solution was performed as follows. First, the M-Enk was set in a periodic boundary box ($37.7 \times 34.2 \times 31.3\,\text{Å}^3$) filled with TIP3P water molecules [54] and ectoine molecules (this system was named MEm$^L$). Second, other solvation models of M-Enk in 2.4 M ectoine aqueous solution and pure water (i.e., MEm$^H$ and MEp) were independently constructed. Ectoine concentrations in both MEm$^L$ and MEm$^H$ were within the concentration generally used for experimental works of ectoine-type CSs [24c]. After the starting structures were minimized molecular mechanically for 2000 cycles to reduce any bad contact pairs, each system was initially equilibrated for the first 200 ps under the NPT condition at 300 K and 1 atm. With these equilibration procedures, pressure, temperature, total energy, and the density in each system were saturated at constant values. Then, further simulation of 50 ns was performed for each system at 300 K under the NVT condition. Trajectories were numerically integrated by the Verlet method with a time step of 2 fs. The electrostatic interactions were treated by the PME method [61]. All the bonds involving hydrogen atoms were constrained by the SHAKE method [62]. For equilibration, both temperature and pressure were regulated using the Berendsen algorithm [63]. To investigate the difference in solvent distributions, MD simulation for CI2 in pure water (CI2p) and in the same solvent condition as MEm$^L$ (CI2m) were performed for 2 ns at 300 K following the same procedure as in our previous simulation of CI2 [39].

We adopted 1.5 M for the present ectoine concentration with reference to experimental work with hydroxyectoine in aqueous solution [36,37]. The weight molarity of ectoine in CI2m was 1.8 mol/kg (the solution density: $1.05\,\text{g/cm}^3$), while that in the experiment with 1.5 M hydroxyectoine was 1.4 mol/kg (the solution density: $1.3\,\text{g/cm}^3$).

### 8.3.2.2  Hydration Structure around Chymotrypsin Inhibitor 2 and Met-Enkephalin

We discuss the preferential exclusion of ectoine using both the surficial KB integrals $G^S_{\alpha s}(R)$ (Equation 8.32) and the preferential exclusion parameter $\nu_{\alpha c}(R)$ (Equation 8.34) given by the KB theory [19–20]. Because CI2 did not show any significant structural change during the CI2m simulation, the entire time duration of the CI2m simulation was utilized for the analysis. In contrast, because the structure of M-Enk changed clearly, only a part of the bent period (11–20 ns) in the MEm$^L$ simulation was utilized. The values of several parameters in the KB theory defined originally at infinite distance from the solute molecule $\alpha$ were approximately substituted at $R = 10.0\,\text{Å}$, e.g., $G^S_{\alpha s}(R = 10.0\,\text{Å})$ was identified with the original KB integral $G_{\alpha s}$ (Equation 8.28).

Let us remember again that KB integral $G_{\alpha w}$ or $G_{\alpha e}$ itself is related to the extent to which the individual solvent component $s$ (water ($s = w$) or ectoine ($s = e$)) is excluded or bound as an overall tendency around a solute molecule $\alpha$ (M-Enk ($\alpha = m$) or CI2 ($\alpha = c$)). It is understood that both negative values of $G_{mw}$ and $G_{cw}$, i.e., $-728.45$ and $-5827.35$ cm$^3$/mol, indicate that the exclusion of water molecules mainly comes from the van der Waals volume. The present value of $G_{cw}$ ($=-5827.35$ cm$^3$/mol) could be reasonably compared to that of a similar size of protein in "pure" water obtained in a previous theoretical work [64]: Imai et al. developed a theoretical treatment of the KB theory combined with 3D-RISM theory and calculated the partial molar volume for several proteins in pure water [64]. According to their results, the KB integral of water for ubiquitin, which has almost the same solvent-accessible surface area (SASA) and slightly larger amino acid chain length than CI2, was evaluated as $-6081$ cm$^3$/mol. Although it is meaningless to directly compare the values of the KB integral of two solutes with different molecular sizes, we would like to emphasize that the opposite sign of KB integrals indicates a clear opposite tendency in the ectoine distributions around two solutes, i.e., the binding for M-Enk[b] (positive value in $G_{me}$) and the exclusion for CI2 (negative value in $G_{ce}$).

Next, we discuss further the preferential exclusion of ectoine using the preferential exclusion parameter $\nu_{\alpha e}$ (Equation 8.34). Since the positive value of $\nu_{\alpha e}$ indicates the preferential binding of ectoine as $\nu_{me}$ around M-Enk[b], in the case of CI2, despite the local exclusion of ectoine around CI2 shown by $G_{ce}$, the present positive $\nu_{ce}$ value indicates that ectoine molecules show not the preferential exclusion but the preferential binding as an overall tendency in the CI2m simulation. However, it could be said that a considerable difference exists between two solutes in the characteristics of ectoine exclusion. For instance, while $\nu_{\alpha e}$s take on very similar positive values for both solutes, their value per unit SASA for CI2, i.e., $\nu_{ce}/\overline{A}_c$, is about six times smaller than that for M-Enk[b] [40]. This is clear evidence that the surface of CI2 excludes ectoine molecules somewhat more strongly than does the surface of M-Enk[b]. Furthermore, another significant difference was found in the ectoine exclusion style between those two $\nu_{\alpha e}(R)$ profiles. In Figure 8.7, $\delta\nu_{\alpha e}(r)$, the difference of $\nu_{\alpha e}(R)$ defined as

$$\delta\nu_{\alpha e}(r) = \nu_{\alpha e}(r + \Delta r/2) - \nu_{\alpha e}(r - \Delta r/2) \tag{8.38}$$

is shown, where a negative and a positive value in $\delta\nu_{\alpha e}(r)$ mean that the preferential exclusion (black) and the preferential binding (gray) of ectoine occur at distance $r$ from the surface of solute molecule $\alpha$, respectively. The large black area in Figure 8.7b shows that the preferential exclusion of ectoine is developed near the surface of CI2. On the other hand, $\delta\nu_{\alpha e}(r)$ shows the preferential binding of ectoine at almost any distance from the surface of M-Enk[b] to bulk solvent phase (Figure 8.7a).

From these analyses, it seems probable that a stronger preferential exclusion of ectoine is developed around CI2 than around M-Enk[b]. It was, therefore, unexpected that the computational result, i.e., $\nu_{\alpha e}$ values, showed that not only M-Enk but also CI2 exhibited preferential binding of ectoine, considering the generally believed functions of the CSs on proteins, i.e., preferential exclusion. We therefore discuss the factors that weaken the preferential exclusion of ectoine, focusing on the molecular force field used in the present study.

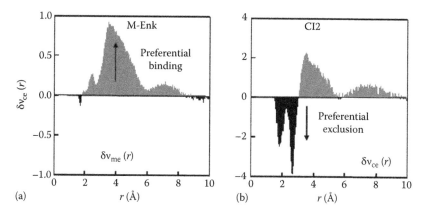

**FIGURE 8.7** Differences of preferential exclusion parameter of ectoine: (a) around M-Enkb and (b) around CI2. (From Yu, I. et al., *J. Phys. Chem. B*, 111, 10231, 2007. With permission.)

Smith and coworkers previously compared the KB integrals of several other solutes (urea, acetone, amides, and NaCl) in water [65] with experimental data, and it is well-known that when some generally used force fields are applied to these solutes they tend to show too much self-association in TIP3P [54] and SPC [66] water. At the same time, in most of these cases, the KB integrals of water molecules around the cosolvent molecules are considerably smaller than the experimental values. The excessive self-association of cosolvent molecules is attributable to their lack of affinity for water molecules. Smith et al. reported that many solute models using ab initio gas-phase-derived charge distributions might not accurately represent polarized solution charge distributions [65].

These facts may be one reason that excessive association between the ectoine molecules and the solute seemed to occur in our simulation, in which the partial atomic charges of ectoine were optimized by the ab initio method in the gas phase [38]. In other words, these findings suggest that in reality ectoine molecules may accumulate water molecules more strongly, making it more difficult for ectoine molecules to access the protein surface. One may expect, therefore, that ectoine molecules with more appropriate charge distributions would increase the water density a little more than is estimated by present MD simulations, and reduce the $v_{ce}$ value reasonably so as to reproduce the preferential exclusion of ectoine for CI2 (cf. Equation 8.34).

Thus, it is expected that the preferential exclusion of ectoine should significantly alter the dynamic property of water molecules around the protein. In fact, our previous simulation showed that the diffusion constant of the water molecules in the first hydration layer of CI2 significantly slows down by the addition of ectoine [39]. This implies that, in the ectoine aqueous solution, the energy flow via the translational motion of solvent molecules should be weakened and be probably different in the characteristics in comparison with that in the pure water. For further studies, the surficial KB approach must be helpful in analyzing the energy flow mechanism in the mixed solvent environment.

## 8.4 CONCLUSIONS

The task of theoretically or computationally describing the structures and dynamics of proteins remains challenging even today, when the most advanced computer technology allows multiscale simulation for multichemistry of small-to-medium size molecules. However, to investigate typical phenomena in modern biochemistry regarding proteins (for example, ligand binding or dissociation in heme proteins, such as Mb and Hb), and preferential exclusion of CS molecules around protein surfaces, a combinatorial computational chemical approach using electronic state (ES) calculation and MD simulation has become a very powerful and realistic strategy.

Several strategic methodologies were introduced with those applications in this chapter. First, the PEMD method was defined and used in the photodissociation process, Mb*CO → Mb + CO, and then the biphasic relaxation constants were evaluated and found to be in good agreement with the previous results. Second, by using the Fourier transform of the TSVAFs and the INMA complementarily, it was shown that the vibrational excitation of the propionate groups in the heme probably play a decisive role in the fast channel of energy transfer from heme proteins to the solvent water. Recently, this theoretical observation has been strongly supported experimentally [10].

Further, in order to detect the subtle anisotropic structural change of MbCO triggered by the ligand photolysis, time changes of $R_g$ and its three Cartesian components, $R_{gx}$, $R_{gy}$, and $R_{gz}$, were calculated for 100 ps within the PEMD methodology, for the first time [17b]. This method works very well and provides us with accurate enough $R_g$ changes free from large fluctuations. The results drawn from this method were found to be in very good agreement with the previous observation in a TG experiment, indicating the same anisotropic expansion within 500 fs, i.e., Mb largely expands in the perpendicular direction of the heme plane and slightly contracts in the other directions. This agreement demonstrates the high accuracy of the PEMD method in detecting subtle changes that are completely obscured under their thermal fluctuations. Moreover, in several previous studies devoted to the excess energy relaxation from the heme to the surrounding solvent after photolysis, it was shown that the propionate side chains serve an important role in the fast energy transfer process [16,19], while it has also been long supposed that the obtained anisotropic structural changes, or collective motions of the globin matrix, contribute to the fast process [11,12]. Considering these studies, it was quantitatively shown here that the contributions to the fast energy relaxation are important and should be nonnegligible [17b].

Third, to obtain the spatial profiles for the preferential solvation around a solute molecule, we proposed another definition of the KB integral around the solute molecule with finite dimensions, i.e., the surficial KB integral, to systematically examine the solvation on the distance from the molecular surface of the solute molecule. With the aid of surficial KB treatment, the present MD simulations of chymotrypsin inhibitor 2 (CI2) showed the preferential exclusion of ectoine (i.e., a CS) near the protein surface, while it is generally considered that CSs are preferentially excluded from the surface of proteins [28,29]. On the other hand, the preferential exclusion of ectoine was found to be significantly weakened near the surface of much smaller solute met-enkephalin (M-Enk) in its bent structure [39,40].

In the present microscopic investigation, it is true that some results seem to be contradictory. However, they may also be viewed as evidence that there is still room for improvement in the molecular force field in the estimation of preferential exclusion parameters $\nu_{\alpha c}$ with the use of MD simulations. For example, it may be conjectured that improvement of the partial atomic charge distribution of ectoine and alteration of the water force field, e.g., from TIP3P to some other water model, would bring about an increase in the water accumulation around the ectoine molecules. This would improve the preferential exclusion parameters $\nu_{\alpha c}$ so that they are in much better agreement with the experimental ones.

Finally, we would like to note that now is the most appropriate and perhaps the best time for computational biologists to work to realize, on their own initiative, "the ambition of molecular biology to interpret the essential functions of organisms in terms of molecular *dynamics*" [1] since the computational equipment and knowledge have become sufficiently developed, based on a long history of application to molecular sciences [2–5]. It will become, therefore, very important in the near future to study the protein functions theoretically together with their dynamics. Such computational biological studies on energy flow and signaling in proteins, e.g., allostery and cell function including communication among binding sites, will surely lead to new and innovative scientific findings in relation to physics, chemistry, biology, and other fields.

## ACKNOWLEDGMENTS

The work discussed in this chapter was carried out partly with Prof. Isao Okazaki (Hirosaki University), and with help from Prof. Yusuke Hara (Waseda University). Profs. Yasutaka Mizutani (Osaka University) and Teizo Kitagawa (Toyota Physical & Chemical Research Institute) are acknowledged for their informative and supportive discussions from the experimental side. M.N. is also grateful to Prof. John E. Straub for fruitful discussions, especially in the early stage of Mb-related research, and to Prof. David M. Leitner for his kind invitation to us to contribute to this volume. This work was supported by a Grant-in-Aid for Scientific Research (C) (#20550011) from the Ministry of Education, Culture, Sports, Science and Technology (MEXT) in Japan and partially by a Grant-in-Aid for the 21st Century COE (Centers of Excellence) program "Frontiers of Computational Science" at Nagoya University. Support was also received from the Japan Science and Technology Agency (JST) for the Core Research for Evolutional Science and Technology (CREST) "Multi-Scale Simulation of Condensed-Phase Reacting Systems" in the research area "High Performance Computing for Multi-scale and Multi-Physics Phenomena."

## REFERENCES

1. J. Monod, *Science*, **154**, 475 (1966).
2. http://www.nsc.riken.jp/index-eng.html
3. (a) F. Sato, T. Yoshihiro, I. Okazaki, and H. Kashiwagi, *Chem. Phys. Lett.*, **310**, 523 (1999); (b) T. Yoshihiro, F. Sato, and H. Kashiwagi, *Chem. Phys. Lett.*, **346**, 313 (2001).
4. (a) K. Kitaura, E. Ikeo, T. Asada, T. Nakano, and M. Uebayasi, *Chem. Phys. Lett.*, **313**, 701 (1999); (b) http://staff.aist.go.jp/d.g.fedorov/fmo/main.html

5. B. Borrell, *Nature*, **451**, 240 (2008).

6. (a) M.F. Perutz, A.J. Wilkinson, M. Paoli, and G.G. Dodson, *Annu. Rev. Biophys. Biomol. Struct.*, **27**, 1 (1998); (b) W.A. Eaton, E.R. Henry, J. Hofrichter, and A. Mozzarelli, *Nat. Struct. Biol.*, **6**, 351 (1999).

7. S. Franzen, B. Bohn, C. Poyart, and J.L. Martin, *Biochemistry*, **34**, 1224 (1995).

8. Y. Mizutani, Y. Uesugi, and T. Kitagawa, *J. Chem. Phys.*, **111**, 8950 (1999).

9. (a) Y. Mizutani and T. Kitagawa, *Science*, **278**, 443 (1997). (b) Y. Mizutani and T. Kitagawa, *J. Phys. Chem. B*, **105**, 10992 (2001). (c) Y. Mizutani and T. Kitagawa, *Bull. Chem. Soc. Jpn.*, **75**, 623 (2002); (d) A. Sato and Y. Mizutani, *Biochemistry*, **44**, 14709 (2005).

10. (a) Y. Gao, M. Koyama, S.F. El-Mashtoly, T. Hayashi, K. Harada, Y. Mizutani, and T. Kitagawa, *Chem. Phys. Lett.*, **429**, 239 (2006); (b) M. Koyama, S. Neya, and Y. Mizutani, *Chem. Phys. Lett.*, **430**, 404 (2006).

11. (a) R.J.D. Miller, *Ann. Rev. Phys. Chem.*, **42**, 587 (1991); (b) G.D. Goodno, V. Astinov, and R.J.D. Miller, *J. Phys. Chem. A*, **103**, 10630 (1999).

12. (a) M. Sakakura, S. Yamaguchi, N. Hirota, and M. Terazima, *J. Am. Chem. Soc.*, **123**, 4286 (2001). (b) Y. Nishihara, M. Sakakura, Y. Kimura, and M. Terazima, *J. Am. Chem. Soc.*, **126**, 11877 (2004).

13. J. Rodriguez and D. Holten, *J. Chem. Phys.*, **91**, 3525 (1989).

14. (a) F. Schotte, M. Lim, T.A. Jackson, A.V. Smirnov, J. Soman, J.S. Olson, G.N. Phillips Jr., M. Wulff, and P.A. Anfinrud, *Science*, **300**, 1944 (2003); (b) F. Schotte, J. Soman, J.S. Olson, M. Wulff, and P.A. Anfinrud, *J. Struct. Biol.*, **147**, 235 (2004); (c) G. Hummer, F. Schotte, and P.A. Anfinrud, *Proc. Natl. Acad. Sci. U. S. A.*, **101**, 15330 (2004).

15. (a) E.R. Henry, M. Levitt, and W.A. Eaton, *Proc. Natl. Acad. Sci. U. S. A.*, **82**, 2034 (1985); (b) E.R. Henry, W.A. Eaton, and R.M. Hochstrasser, *Proc. Natl. Acad. Sci. U. S. A.*, **83**, 8982 (1986).

16. I. Okazaki, Y. Hara, and M. Nagaoka, *Chem. Phys. Lett.*, **337** (2001) 151.

17. (a) M. Takayanagi, I. Yu, and M. Nagaoka, *Chem. Phys. Lett.*, **421**, 300 (2006); (b) M. Takayanagi, H. Okumura, and M. Nagaoka, *J. Phys. Chem. B*, **111**, 864 (2007).

18. J.E. Straub and M. Karplus, *Chem. Phys.*, **158**, 221 (1991).

19. (a) D.S. Sagnella and J.E. Straub, *J. Phys. Chem. B*, **105**, 7057 (2001); (b) L. Bu and J.E. Straub, *J. Phys. Chem. B*, **107**, 10634 (2003).

20. (a) C. Bossa, M. Anselmi, D. Roccatano, A. Amadei, B. Vallone, M. Brunori, and A. Di Nora, *Biophys. J.*, **86**, 3855 (2004); (b) C. Bossa, A. Amadei, I. Daidone, M. Anselmi, B. Vallone, M. Brunori, and A. Di Nora, *Biophys. J.*, **89**, 465 (2005); (c) M. Anselmi, M. Aschi, A. Di Nora, and A. Amadei, *Biophys. J.*, **92**, 3442 (2007)

21. (a) J. Cohen, A. Arkhipov, R. Braun, and K. Schulten, *Biophys. J.*, **91**, 1844 (2006); (b) J. Cohen and K. Schulten, *Biophys. J.*, **93**, 3591 (2007).

22. R. Elber and Q.H. Gibson, *J. Phys. Chem. B*, **112**, 6147 (2008).

23. D.M. Leitner, *Ann. Rev. Phys. Chem.*, **59**, 233 (2008).

24. (a) E.A. Galinski, *Experientia*, **49**, 487 (1993); (b) E.A. Galinski, H.P. Pfeiffer, and H.G. Truper, *Eur. J. Biochem.*, **149**, 135 (1985); (c) S. Knapp, R. Ladenstein, and E.A. Galinski, *Extremophiles*, **3**, 191 (1999).

25. P. Lamosa, D.L. Turner, R. Ventura, C. Maycock, and H. Santos, *Eur. J. Biochem.*, **270**, 4604 (2003).

26. R.L. Foord and R.J. Leatherbarrow, *Biochemistry*, **37**, 2969 (1998).

27. P.H. Yancey, M.E. Clark, S.C. Hand, R.D. Bowlus, and G.N. Somero, *Science*, **217**, 1214 (1982).

28. J.C. Lee and S.N. Timasheff, *J. Biol. Chem.*, **256**, 7193 (1981).

29. (a) T. Arakawa and S.N. Timasheff, *Arch. Biochem. Biophys.*, **224**, 169 (1983); (b) T. Arakawa and S.N. Timasheff, *Biophys. J.*, **47**, 411 (1985).

30. M. Sugawa, Y. Arai, A. Hikikoshi, Iwane, Y. Ishii, and T. Yanagida, *Biosystems*, **88**, 243 (2007).

31. (a) N. Go, T. Noguti, and T. Nishikawa, *Proc. Natl Acad. Sci. U. S. A.*, **80**, 3696 (1983); (b) A. Kidera and N. Go, *Proc. Natl Acad. Sci. U.S.A.*, **87**, 3718 (1990).
32. B. Brooks, and K. Karplus, *Proc. Natl Acad. Sci. U.S.A.*, **80**, 6571 (1983).
33. (a) Y. Seno and N. Go, *J. Mol. Biol.*, **216**, 95 (1990); (b) *J. Mol. Biol.*, **216**, 111 (1990)
34. K. Suhre and Y.-H. Sanejpuand, *Nuc. Acids Res.*, **32**, 610 (2004).
35. M. Buchuner, B.M. Landanyi, and R.M. Stratt, *J. Chem. Phys.*, **97**, 8522 (1992).
36. J.G. Kirkwood and F.P. Buff, *J. Chem. Phys.*, **19**, 774 (1951).
37. (a) A. Ben-Naim, *Statistical Thermodynamics for Chemist and Biochemist* (Plenum, New York, 1992); (b) A. Ben-Naim, *Molecular Theory of Solutions* (Oxford, New York, 2006).
38. K. Suenobu, M. Nagaoka, T. Yamabe, and S. Nagata, *J. Phys. Chem. A*, **102**, 7505 (1998).
39. I. Yu and M. Nagaoka, *Chem. Phys. Lett.*, **388**, 316 (2004).
40. (a) I. Yu, Y. Jindo, and M. Nagaoka, *J. Phys. Chem. B*, **111**, 10231 (2007); (b) I. Yu, M. Takayanagi, and M. Nagaoka, ibid., **113**, 3543 (2009).
41. M. Aburi and P.E. Smith, *J. Phys. Chem. B*, **108**, 7382 (2004).
42. H. Cramér, *The Elements of Probability Theory and Some of Its Applications* (John Wiley & Sons, New York, 1955)
43. (a) K. Chiba, F. Sato, and H. Kashiwagi, *private communication* (1993); (b) K. Chiba, F. Sato, and H. Kashiwagi, *Proc. of Int. Symp. Adv. Comput. Life-Sci.*, **1**, 73 (1992).
44. S.L. Mayo, B.D. Olafson, and W.A. Goddard III, *J. Phys. Chem.*, **94**, 8897 (1990).
45. J. Gasteiger and M. Marsili, *Tetrahedron*, **36**, 3219 (1980) (the procedure actually used in Cerius$^2$ is that in polygraf 1.0).
46. D.A. Case, D.A. Pearlman, J.W. Caldwell, T.E. Cheatham III, W.S. Ross, C.L. Simmerling, T.A. Darden, K.M. Merz, R.V. Stanton, A.L. Cheng, J.J. Vincent, M. Crowley, D.M. Ferguson, R.J. Radmer, G.L. Seibel, U.C. Singh, P.K. Weiner, and P.A. Kollman, AMBER 5 (University of California, San Francisco, CA, 1997).
47. X. Cheng and B.P. Schoenborn, *Acta Crystallogr.*, **B46**, 195 (1990).
48. Protein Data Bank, The Research Collaboratory for Structural Bioinformatics. http://www.rcsb.org/pdb/home/home.do.
49. Molecular Simulations Inc., San Diego, CA, (1998).
50. http://act.jst.go.jp/index_e.html
51. D.A. Case, D.A. Pearlman, J.W. Caldwell, T.E. Cheatham III, J. Wang, W.S. Ross, C.L. Simmerling, T.A. Darden, K.M. Merz, R.V. Stanton, A.L. Chen, J.J. Vincent, M. Crowley, V. Tsui, H. Gohlke, R.J. Radmer, Y. Duan, J. Pitera, I. Massova, G.L. Seibel, U.C. Singh, P.K. Weiner, and P.A. Kollman, AMBER 7 (University of California, San Francisco, CA, 2002).
52. D.A. Giammona, PhD thesis, University of California, Davis, CA, 1984.
53. The Chemical Society of Japan, Ed., *Kagaku-Binran* (*Handbook of Chemistry*) *Basic Volume*, 4th ed., Vol. II (Maruzen, Tokyo, 1993), p. 27 (in Japanese).
54. W.L. Jorgensen, J. Chandrasekhar, J.D. Madura, R.W. Impey, and M.L. Klein, *J. Chem. Phys.*, **79**, 926 (1983).
55. M.P. Allen and D.J. Tildesley, *Computer Simulation of Liquids* (Oxford, New York, 1987).
56. D.J. Evans and G.P. Morriss, *Statistical Mechanics of Nonequilibrium Liquids* (Academic Press, New York, 1990).
57. T. Okamoto and M. Nagaoka, *Chem. Phys. Lett.*, **407**, 444 (2005).
58. M. Abe, T. Kitagawa, and Y. Kyogoku, *J. Chem. Phys.*, **69**, 4526 (1978).
59. J. Wang, P. Cieplak, and P.A. Kollman, *J. Comp. Chem.*, **21**, 1049 (2000).
60. J. Griffin, D. Langs, G. Smith, T. Blundell, I. Tickle, and S. Bedarkar, *Proc. Natl. Acad. Sci. U.S.A.*, **83**, 3272 (1986).
61. T. Darden, D. York, and L. Pedersen, *J. Chem. Phys.*, **98**, 10089 (1993).
62. (a) J.P. Ryckaert, G. Ciccotti, and H.J.C. Berendsen, *J. Computat. Phys.*, **23**, 327 (1977); (b) H.J.C. Berendsen, J.P.M. Postma, W.F. van Gusteren, A. Dinola, and J.R. Haak, *J. Chem. Phys.*, **81**, 3684 (1984).

63. D.V. Spoel and H.J.C. Berendsen, *Biophys. J.*, **72**, 2032 (1997).

64. T. Imai, A. Kovalenko, and F. Hirata, *J. Phys. Chem. B*, **109**, 6658 (2005).

65. (a) S. Weerasinghe and P.E. Smith, *J. Phys. Chem., B*, **107**, 3891 (2003); (b) *J. Phys. Chem.*, **118**, 10663 (2003); (c) *J. Phys. Chem.*, **119**, 11342 (2003); (d) K. Myungshim and P.E. Smith, *J. Comput. Chem.*, **27**, 1477 (2006).

66. H.J.C. Berendsen, J.P.M. Postma, W.F. Van Gunsteren, and J. Hermans, In *Intermolecular Forces*; Pullman, B., Ed. (Reidel, Dordrecht, the Netherlands, 1981).

# Part III

Vibrational Energy Flow in
Proteins and Nanostructures:
Normal Mode-Based Methods

# 9 Directed Energy Funneling in Proteins: From Structure to Function

*Yong Zhang and John E. Straub*

## CONTENTS

## 9.1   INTRODUCTION

Proteins are designed to serve function. The functioning process is generally accompanied by protein dynamics, structural change, and energy flow. The cooperative quaternary structural motions in hemoglobin (Hb), induced by the tertiary structure dynamics following oxygen binding/release, is one classic example of such a process [1]. Myoglobin (Mb) functions as an oxygen storage protein with a well-defined structure similar to that of a Hb subunit. In sperm whale Mb, the heme group, the active site embedded in Mb, is covalently bonded to residue His93 on proximal side and a small ligand on the distal side (see Figure 9.1). Ligand dissociation can occur when the ligand–heme complex absorbs a visible or UV photon, which can cause vibrational excitation of the ligand, heme, and the surrounding residues [2,3] as well as local and global protein structural changes [4,5], making Mb and its mutants an excellent model system for studies of protein structure, dynamics, and function [6–8]. Essential insight into the intricate connections between protein structure and function has been obtained by monitoring the timescales and mechanisms of vibrational energy relaxation (VER) associated with these functional dynamic processes [9–23].

### 9.1.1   VER TIMESCALE AND PATHWAYS

The timescale of VER in Mb can span several orders of magnitude with the exact timescale dependent on the functional dynamic process being probed. For example, energy decay from the excited C–D stretching motion or protein backbone C=O

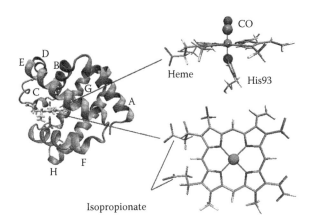

**FIGURE 9.1**   (See color insert following page 172.) Structure of carbonmonoxy myoglobin (MbCO). The heme group is detailed in the side and top views. The eight helices, the ligand CO, and the His93 residue covalently bonded to heme are labeled.

stretching motion (amide I mode) was found to occur on a 1 picosecond or subpico-second timescale [24–28], whereas vibrational relaxation of photolyzed carbon mon-oxide in the heme pocket can last from tens to hundreds of picoseconds in different Mb mutants [7,29,30]. Starting from a basis of harmonic normal modes and includ-ing the contribution of cubic anharmonicity in the potential energy, providing non-linear coupling, Leitner and coworkers estimated the thermal diffusivity of Mb at 300 K to be 14 $\text{Å}^2$ $\text{ps}^{-1}$, approximately the value for water. The thermal conductivity at 300 K was found to be 2.0 mW $\text{cm}^{-1}$ $\text{K}^{-1}$, about one-third the value for water [31]. These calculations demonstrated the promise of normal mode-based models includ-ing nonlinear mode coupling in calculating rates of energy transfer in proteins.

Much attention has been paid to the process of VER from the excited heme moiety. Time-resolved Raman spectra of Mb were measured by Martin and coworkers follow-ing a 500 nm pulse excitation [32]. It was found that excess energy initially deposited in the heme is transferred to protein modes in roughly 5 ps. Within 15 ps, the excess energy of the heme was found to have fully dissipated. Anfinrud and coworkers studied heme cooling following photoexcitation of heme in Mb using femtosecond time-resolved near-IR absorbance spectroscopy and observed a single exponential decay with time constant of 6.2 ± 0.5 ps [33]. Kitagawa and coworkers monitored the mode specific behavior of heme vibrational relaxation using time-resolved resonance Raman ($\text{TR}^3$) spectroscopy [34–36]. From the temporal changes of the anti-Stokes Raman intensity, the decay time constants were found to be 1.1 ± 0.6 ps for the $v_4$ band and 1.9 ± 0.6 ps for the $v_7$ band, implying a thermal decay of the heme within 2 ps [36,37].

Hochstrasser and coworkers measured the $D_2O$ stretching mode shift following photoexcitation of heme in Mb solution using femtosecond IR spectroscopy [38]. It was estimated that 60% of the deposited energy was transferred on a timescale of 7.5 ± 1.5 ps, while 40% of the energy relaxed on a longer timescale of 20 ps. The fast component was suggested to proceed through unidentified "collective motions" of the protein or direct energy transfer from the heme to solvent through the heme isopropionate side chains and the slow relaxation attributed to energy transfer from heme to water through the protein via a classical diffusion process.

Miller and coworkers studied the energy flow between the protein and solvent using transient phase grating spectroscopy, which can monitor the temperature of the solvent matrix [39]. They demonstrated that energy transfer to the surrounding sol-vent occurred within 20 ps mediated by the protein backbone. A nanosecond time-scale was also observed, which was attributed to the slow conformational relaxation of the protein globin [40].

This brief survey conveys the broad range of timescales and variety of functional motions that have been revealed through studies of VER process in heme proteins.

### 9.1.2 PROTEIN DYNAMICS AND FUNCTION

In carbonmonoxy Mb, the hexa-coordinated heme iron atom lies near the porphy-rin plane defined by the four pyrrole nitrogen atoms tethered above and below by the proximal His93 and the distal ligand. Following photolysis of CO, the heme becomes penta-coordinate and the iron atom moves out of the porphyrin plane by roughly 0.3 Å [4,5]. By comparing the crystal structures of carbonmonoxy Mb and deoxymyoglobin, it was found that, in addition to the iron atom in the heme, atoms in the protein globin were displaced following photolysis, mainly in the helices E and F which hold the heme [4]. His93, the only protein residue bonded to the heme, belongs to helix F. The heme tilting motion relative to the His93 imidazole side chain was suggested to play an essential role in this globin's functionally important motions. This reorientational motion of the heme and E and F helices serves to define the conformational change in the heme pocket that translates into the slight but essential diminishment in the preference to binding carbon monoxide over oxygen.

The dynamics of the protein globin motion following ligand photolysis was studied by Kitagawa, Mizutani, and their coworkers using time-resolved UV resonance Raman spectroscopy [41,42]. It was found that the observed motion in the H helix is transmitted from the F helix and FG corner, which in turn is induced by the heme iron out-of-plane doming motion. Whereas the structural changes of the A helix is not compatible with the structural change in the proximal side of the heme. A pathway involving displacement of the distal Val68 was found to be plausible.

Based on femtosecond coherence spectroscopic studies [43], Miller and coworkers suggested a mechanism of cascading energy transfer following ligand photodissociation in Mb where Fe–CO bond breaking following photoexcitation leads to excitation of the heme ring in-plane breathing-like $v_4$ and $v_7$ modes, which in turn couple with the iron out-of-plane doming motion. The doming motion is then coupled to low-frequency protein backbone modes that induce the mesoscale motions in the helices E and F. This detailed scenario suggests how protein structure encodes and is translated to function.

### 9.1.3 ROLE OF HEME LOCAL ENVIRONMENT

Cytochrome $c$ (Cyt $c$) is a widely studied protein in which the heme group is connected to protein residues by four covalent bonds as well as hydrogen bonds (see Figure 9.2). In the pioneering in vacuo molecular dynamics simulations of Henry et al. [44], the simulated heme cooling was fitted as (1) a single exponential decay, with a time constant in the range of 8–13 ps for both Mb and Cyt $c$ or (2) biexponential decay with a fast time constant of 1–4 ps for both proteins and slow decay of 20 ps for Cyt $c$ and 40 ps for Mb. The authors appreciated that the hemes in the two proteins have distinctly different exposure to the solvent in solution, which they anticipated may play an essential role in defining the overall energy transfer pathways.

Champion and coworkers studied the electronic and vibrational relaxation dynamics of heme in Fe(II) Cyt $c$ using resonance Raman saturation spectroscopy [45]. It was found that the heme cooling follows a single exponential decay with a time constant of roughly 4 ps. In a later study by Champion and coworkers [46], photodissociation of the distal ligand of the heme, Met80, was observed in Fe(II) Cyt $c$. A multistep heme cooling process characterized by time constants of 0.1, 0.8, and 2.8 ps was reported. Recent work by Kruglik and coworkers using 550 nm photon excitation confirmed the photodissociation of Met80 in reduced Cyt $c$ [47]. The oxidized form of Cyt $c$ was also studied with no evidence of ligand photodissociation.

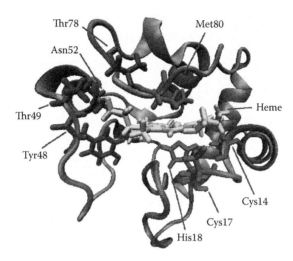

**FIGURE 9.2    (See color insert following page 172.)** Structure of cytochrome *c* (Cyt *c*) with heme, residues covalently bonded to heme (Cys14, Cys17, His18, and Met80), and residues hydrogen bonded to heme (Tyr48, Thr49, Asn52, and Thr78).

The heme cooling was found to be single exponential for both Fe(II) Cyt *c* and Fe(III) Cyt *c* with a time constant of roughly 7 ps, similar to the observation of heme cooling in Mb.

Metalloporphyrins have been widely studied as a model of the heme moiety in proteins. [48–50] Intriguing mode-specific energy transfer has been reported by Kitagawa and coworkers in which the $v_4$ and $v_7$ bands of nickel octaethylporphyrin (NiOEP) were reported to relax according to a double exponential decay with fast time constant $2.6 \pm 0.5$ ps in several nonpolar solvents [51]. A solvent-dependent relaxation rate was observed for iron(III) meso-tetra(4-sulfonatephenyl)porphyrin (FeTSPP), which is soluble in polar solvents [52]. In water solution, the anti-Stokes $v_4$ band was observed to display a single exponential decay with a time constant of $1.9 \pm 0.4$ ps. In methanol–benzene mixtures, an additional slow decay was observed, with a time constant dependent on the methanol:benzene ratio, indicating the sensitivity of VER to the local solvation environment.

In the background of these pioneering studies, fundamental questions related to heme cooling mechanisms, protein dynamics following ligand photolysis, and the relation to functionally important protein dynamics may be addressed. In this chapter, we summarize the results of our efforts to identify the underlying principles governing energy flow in these systems using theoretical and computational tools developed to address these questions. Our use of classical molecular dynamics to study heme cooling as a function of protein topology and solvent environment is described in the next section. That discussion is followed by an overview of our use of time-dependent perturbation theory applied at the density functional theory (DFT) level to study state-to-state intramolecular energy transfer in several heme models. We conclude with our perspective on this state of the field and its most prominent open questions.

## 9.2 LONGER TIME HEME COOLING DYNAMICS AS A PROBE OF ENERGY FLOW AND SIGNALING

### 9.2.1 MOLECULAR DYNAMICS SIMULATIONS

Molecular dynamics (MD) simulations were carried out using the CHARMM simulation program [53] with the all-hydrogen parameter set of the CHARMM force field (version 22) [54]. The initial structure of the protein was taken from the Protein Data Bank (PDB) and the simulation system was prepared by introducing the protein into a previously equilibrated solvent box [55]. The solvent molecules that overlap with the protein were removed and the system was carefully equilibrated at 300 K. The equilibrated system was then allowed to evolve for 200 ps during which coordinates were saved every 20 ps for a total of 10 configurations. Each of these configurations was re-equilibrated and followed by a microcanonical ensemble (constant volume and energy) production run from the nonequilibrium (photoexcited) or equilibrated (nonphotoexcited) state. To simulate the absorption of a photon, a quantity of excess kinetic energy equivalent to the corresponding photon was deposited in the heme uniformly among all atoms. The molecular dynamics was simulated using the Verlet algorithm [56] with periodic boundary conditions and a time step of 1.0 fs. The nonbonded potential was truncated using a group switching function extending from 9.5 to 11.5 Å.

### 9.2.2 HEME COOLING RATE AND PATHWAYS IN NATIVE MB

For the native sperm whale Mb [5], with approximately 88 kcal mol$^{-1}$ of excess kinetic energy deposited in the 73 heme atoms, the increase in temperature of the heme is over 400 K. The average time dependence of the nonequilibrium relaxation of the excess kinetic energy within the heme computed over simulated trajectories is well modeled by a single exponential with a relaxation time of 5.9 ps [57]. This observation agrees with the experimental results of Anfinrud and coworkers in both the single-exponential nature of the process and the exact value of the time constant [33].

In Mb, the heme is buried in the protein matrix with roughly 90 van der Waals interactions with heme pocket residues. The only covalent link between the heme and the protein is the bond between the heme Fe and the proximal His93. The spatial dependence of the kinetic energy relaxation from the excited heme was identified by visualizing the protein residues that have a significant amount of excess kinetic energy. The energy transfer from the heme to the protein was found to follow three channels: (1) "through space" energy transfer mediated by collisions of hot heme atoms with surrounding protein atoms, (2) "through bond" energy transfer through the Fe–His93 connection, and (3) "through projectile" energy transfer mediated by collisions with the CO ligand as it dissociates from the heme and collides with the residue atoms, forming the heme distal pocket (see Figure 9.3).

In the gas phase molecular dynamics study of heme cooling [44], the heme cooling was found to follow a double exponential decay with time constants of 1–4 ps and

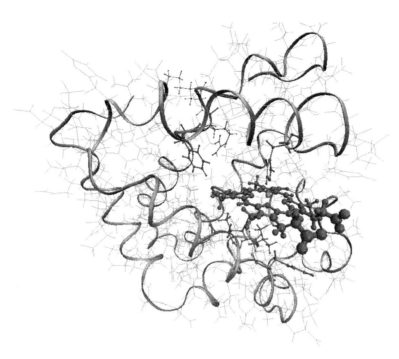

**FIGURE 9.3** **(See color insert following page 172.)** The demonstration of the heme cooling pathways following the ligand photodissociation in MbCO. The protein structure is shown in ribbon, the heme moiety and protein residues involved in heme cooling pathways are shown in CPK model. The direct interaction between the heme isopropionate side chains and the solvent molecules was found to be the dominant energy transfer pathway. The energy transfer from the heme to the protein was found to follow three channels: (1) "through space" energy transfer mediated by collisions of hot heme atoms with surrounding protein atoms, (2) "through bond" energy transfer through the Fe-His93 connection, and (3) "through projectile" energy transfer mediated by collisions with the CO ligand as it dissociates from the heme and collides with the residue atoms forming the heme distal pocket.

40 ps for the fast and slow processes, with roughly half of the excess energy was dissipated in each process. Similar conclusions were drawn by Nagaoka and coworkers in their simulation analysis of heme cooling in Mb employing the inventive time-resolved vibrational analysis technique [13]. Their work is reviewed and extended in a separate chapter of this volume. When solvent molecules are explicitly included in the simulation [57], the heme cooling follows a single exponential decay with a faster rate, indicating the essential role played by solvent in the relaxation process.

The ergodic measure is a useful means of determining the timescale for the self-averaging of a given property in a many-body system [58–61]. The computation of the kinetic energy metric, a particular form of the ergodic measure, is an effective means of assigning timescales for atomic kinetic energy relaxation in molecular systems and numerically more accurate than integrating over the corresponding velocity autocorrelation function [58]. From the nonphotolysis trajectories, the kinetic energy diffusion constants were found to be $1.87\,ps^{-1}$ for the protein, $6.76\,ps^{-1}$ for the solvent, and 7.49 ps$^{-1}$ for the heme [58]. Clearly, the kinetic energy redistribution within heme or solvent are far faster than in the protein. While this does not demonstrate energy transfer from one moiety to another (for example, from heme to solvent), it does indicate that solvent can play an essential role in energy flow out of the heme.

### 9.2.3 ROLE OF HIS93 AND ISOPROPIONATE GROUPS EXPLORED BY HEME COOLING IN MUTATED AND MODIFIED MB

Residue mutation or modification is a popular technique used to probe the dynamic and thermodynamic properties of proteins. Applying the same protocol as the native Mb, the process of heme cooling was simulated for (1) the Mb His93Gly mutant, in which the only covalent bond between the heme moiety and the protein is disrupted and (2) Mb with a modified heme, in which the two isopropionate groups are replaced by hydrogen atoms and the direct interaction between the heme and solvent are removed.

For the H93G Mb mutant, the kinetic energy relaxation from the excited heme was found to follow a single exponential decay with the time constant of 5.9 ps [62], essentially identical to the timescale for heme cooling of photolyzed wild-type Mb [57], suggesting that the covalent bond to the heme does not play a primary role in the process of heme cooling for Mb in room temperature aqueous solution. In contrast, the heme cooling time constant for the modified-heme Mb was found to be 8.8 ps, indicating that the rate is 50% slower relative to the native wild-type Mb due to the amputation of the heme's two isopropionate side chains. This is strong evidence that the two isopropionate side chains play an essential role in funneling excess kinetic energy from the heme of wild-type Mb to the surrounding environment (see Figure 9.3). In addition to direct energy transfer to the solvent, a second heme cooling pathway was visualized in the modified Mb system, which can be recognized as a specific through space energy transfer channel where energy flows through residues Phe138, Lys139, Arg140, eventually reaches solvent molecules [62] (see Figure 9.3).

The H93G mutant and modified Mb proteins serve as excellent model systems and have been used in theory and experiment to probe the fundamental principles of energy transfer and heme cooling. Similar Mb models were studied in subsequent experiments. Champion and coworkers studied the spectral dynamics of the photoexcited horse heart Mb, both the native and H93G mutant [63]. It was found that the transient relaxation of each compound is completed within 10 ps; the optical transient of H93G was observed to be quite similar to that of native deoxyMb. When the protoheme in Mb was replaced by an iron porphine, in which all heme side chains were replaced by hydrogens, the relaxation is also completed within 10 ps but is obviously slower than the native Mb, consistent with simulation results [62].

Using time-resolved Raman spectroscopy, Kitagawa and coworkers investigated the vibrational relaxation of the $v_4$ and $v_7$ modes of the heme following photoexcitation in wild-type sperm whale Mb and its modifications, in which one or the other isopropionate side chain is selectively replaced by a methyl group [64]. Both modified hemes have similar relaxation rate ($1.5 \pm 0.1$ ps for the $v_4$ mode and $4.9 \pm 0.3$ ps for the

$v_7$ mode, respectively), but slower than that in the native protein (1.3 ± 0.1 ps for the $v_4$ mode and 3.4 ± 0.3 ps for the $v_7$ mode). Further, Mizutani and coworkers studied the etioheme-substituted Mb, in which the protoheme side chains were replaced by methyl or ethyl groups [65]. Etioheme is similar to protoheme except that etioheme lacks hydrogen bonding side chains. For the $v_4$ and $v_7$ modes, the decay rates of the anti-Stokes intensities of etioheme Mb (with time constants of 1.7 ± 0.2 and 7.0 ± 1.9 ps, respectively) are slower than in native Mb (1.2 ± 0.4 s and 1.8 ± 0.4 ps, respectively). These observations support the conjecture that the coupling between the heme isopropionate groups and solvent plays an essential role in the heme cooling in photoexcited Mb, supporting the simulation-inspired predictions [57,62].

## 9.2.4 THE INTERACTION BETWEEN ISOPROPIONATE AND SOLVENTS FROM HEME COOLING OF MB IN VARIOUS SOLVENTS

To further our understanding of interaction between the heme isopropionate side chains and the solvent, the kinetic energy relaxation of the excited heme in Mb was simulated in various solvents, including normal water ($H_2O$), heavy water ($D_2O$), normal glycerol (Gly-$h_8$), deuterated glycerol (Gly-$d_8$), and a nonpolar solvent. Two forms of the heme, one native and one lacking acidic side chains, were employed in the simulations which, together with different solvents, provided various interaction feature between the heme and the environment [55].

As expected, slower heme cooling was observed for the modified heme relative to native heme (see Table 9.1). For Mb in pure aqueous solution, the kinetic energy relaxation time constants were observed to be 5.1 ± 0.2 and 8.3 ± 0.3 ps for the native heme and the modified heme, respectively. A small change was observed from our previous simulation results of 5.9 ± 0.2 ps for native heme [57] and 8.8 ± 0.3 ps for modified heme, [62] due to the use of the refined penta-coordinate heme parameters for the heme following photolysis [66]. The solvent-dependent heme cooling rates were observed for both native and modified Mb (see Table 9.1).

### TABLE 9.1
### Excess Kinetic Energy Relaxation Time Constants in "Heme Cooling" Following Ligand Photolysis of CO in Myoglobin Simulated Using Classical Molecular Dynamics at 300 K

| Solvent | Time Constant (ps) | |
|---|---|---|
| | Native Heme Mb | Modified Heme Mb |
| $H_2O$ | 5.1 ± 0.2 | 8.3 ± 0.3 |
| $D_2O$ | 6.5 ± 0.3 | 8.5 ± 0.4 |
| Gly-$h_8$ | 4.7 ± 0.1 | 7.3 ± 0.2 |
| Gly-$d_8$ | 4.3 ± 0.2 | 6.7 ± 0.3 |
| Nonpolar | 6.7 ± 0.3 | 7.7 ± 0.3 |

Based on the two-step VER mechanism, invoked previously to explain the linear dependence of vibrational cooling rate on the thermal diffusivity of the solvent [67,68], the kinetic energy diffusion in water is expected to be faster than in glycerol as water has a higher kinetic energy diffusion coefficient than glycerol. The calculated kinetic energy diffusion coefficient of nonpolar solvent is smaller than that of water but larger than glycerol [55]. Obviously, the predicted heme cooling rates do not follow the two-step mechanism. The contradiction between heme cooling rates and macroscopic solvent properties suggests that the solvent-dependent heme cooling rate results from the detailed interaction between the heme and solvent.

To understand the microscopic mechanism of heme cooling following ligand photolysis, three intermolecular energy transfer mechanisms [50,69] were considered: (1) energy transfer via hydrogen bonding (HB) interaction, (2) direct vibration–vibration (V–V) energy transfer via resonant interaction; and (3) energy transfer via vibration–translation (V–T) or vibration–rotation (V–R) interaction, in other words, via thermal collisions.

The two isopropionate side chains of the native heme are negatively charged in the simulation model. In water and glycerol solutions, hydrogen bonds form with solvent molecules. In the nonpolar solution, the two side chains form hydrogen bonds with protein residues.

In aqueous solution, relaxation time constants for native Mb show a slowdown of 1.4 ps upon solvent deuteration. This indicates that the V–V pathway is involved in energy transfer between the heme and normal water. The only observed difference between $H_2O$ and $D_2O$ in the classical approximation is the distribution of vibrational frequencies or the density of states of accepting modes. The modes observed between 700 and 1100 cm$^{-1}$ for normal water are absent for deuterated water, suggesting that these medium frequency modes are important to the mechanism of heme cooling. For glycerol solutions, a dependence of the heme cooling rate on deuteration was observed for both native and modified heme. This suggests that the V–V energy transfer mechanism makes important contributions in both cases, a conjecture supported by the abundance of medium frequency modes in both solvents.

V–T and V–R energy transfer, which can be thought of as thermal collisions, contribute to the energy transfer process in all cases. Due to the smaller molecular size, the nonpolar solvent has a higher density of contacts with the heme than other solvents for both native and modified Mb. Higher collision frequency with the heme atoms resulting from a larger heme–solvent interface in the native heme relative to the modified form leads to more efficient energy transfer.

The relative effectiveness of the three channels for each simulated system is summarized in Table 9.2. The observed solvent-dependent heme cooling rates and mechanisms are well explained by these microscopic interactions [55].

It should be noted that when we say three intermolecular energy transfer mechanisms, it does not mean that they are isolated from each other. These "channels" work collaboratively. For example, it has been suggested that the V–T/V–R (collision) energy transfer mechanism is more effective when the solute and solvent molecules are "held close" by electrostatic interaction or hydrogen bonding [70,71]. In addition, it has been reported that librational motions, defined as the hydrogen bond–hindered water molecule rotational motions with frequency 600–950 cm$^{-1}$, play a

**TABLE 9.2**

**Summary of the Availability of Energy Transfer Channels for the Relaxation of Excited Heme in Myoglobin**

| Solvent | Native Heme Mb | | | Modified Heme Mb | | |
|---------|------|------|--------|------|------|--------|
| | HB | V–V | V–T/V–R | HB | V–V | V–T/V–R |
| $H_2O$ | Yes | Yes | Yes | No | No | Yes |
| $D_2O$ | Yes | No | Yes | No | No | Yes |
| Gly-$h_8$ | Yes | Strong | Yes | No | Yes | Yes |
| Gly-$d_8$ | Yes | Strong | Yes | No | Yes | Yes |
| Nonpolar | Yes | No | Strong | No | No | Strong |

key role in the ultrafast energy dissipation from the excited water molecule in liquid water [72]. The modes with frequencies 700–1100 cm$^{-1}$ observed in our simulations of normal water are similar in frequency to the experimentally observed librational motions of liquid water, and probably play a key role in the heme cooling process. Similar modes in heavy water have lower frequencies and by our estimate play a less essential role in the heme cooling process.

### 9.2.5 ROLE OF PROTEIN TOPOLOGY IN HEME COOLING

In Mb, the heme is connected to the protein globin through a single covalent bond between Fe and the proximal His93. The heme interacts with the surrounding protein residues mainly through van der Waals interactions. In Cyt $c$, the heme group has a distinctly different local environment. The heme is covalently bonded to four protein residues, Cys14, Cys17, His18, and Met80; in addition, the two isopropionate side chains, which can interact with the solvent directly and were observed to play an essential role in the heme cooling of Mb, are embedded in the protein and form hydrogen bonds with residues Tyr48, Thr49, Asn52, and Thr79 (see Figure 9.2).

Our effective protocol for the use of MD simulation of heme cooling in Mb was applied to Cyt $c$ to probe the timescales and pathways of heme cooling in both reduced cytochrome $c$ and oxidized cytochrome $c$ following heme photoexcitation. Five different solvent environments, including normal water, heavy water, normal glycerol, deuterated glycerol, and a nonpolar solvent, were employed [73]. The simulated kinetic energy relaxation time constants for each system are summarized in Table 9.3. When the statistical uncertainty is considered, the heme cooling in Cyt $c$ has the same energy relaxation time constants in all solvent environments, a result distinct from the dramatic solvent dependence of the heme cooling rate of Mb [55].

This diversity of solvent-dependent heme cooling rate and mechanism in Mb and Cyt $c$ is a direct result of the difference in heme local environment in the two proteins. In Cyt $c$, all three direct energy transfer pathways between the heme and solvent, important in Mb, are missing or substantially less effective. The two isopropionate side chains of heme in Cyt $c$ are embedded in the protein and energy

**TABLE 9.3**
**Excess Heme Kinetic Energy Relaxation Time Constants Following Photoexcitation of Cytochrome $c$ Simulated Using Classical Molecular Dynamics at 300K**

|  | Time Constant (ps) | |
| --- | --- | --- |
| Solvent | Fe(II) Cyt $c$ | Fe(III) Cyt $c$ |
| $H_2O$ | $6.4 \pm 0.4$ | $7.0 \pm 0.4$ |
| $D_2O$ | $6.2 \pm 0.3$ | $6.8 \pm 0.3$ |
| Gly-$h_8$ | $6.8 \pm 0.4$ | $7.3 \pm 0.4$ |
| Gly-$d_8$ | $7.2 \pm 0.4$ | $7.4 \pm 0.4$ |
| Nonpolar | $7.0 \pm 0.5$ | $7.0 \pm 0.7$ |

In each solvent, $38\,kcal\,mol^{-1}$ of excess kinetic energy was deposited in the heme of Fe(II) Cyt $c$ after Met80 photodissociation, while $52\,kcal\,mol^{-1}$ was deposited in Fe(III) Cyt $c$ in the absence of ligand photodissociation. The simulation data were well fitted by a single-exponential function in all cases.

transfer through hydrogen bonds to solvent was not possible as no hydrogen bonds with solvent were formed. The lack of deuteration dependence in the heme cooling rate suggests the absence of an effective vibration–vibration energy transfer mechanism between heme and solvent in the Cyt $c$ systems [50]. The only direct energy transfer pathway between Cyt $c$ heme and solvent is thermal collision, which is also less effective due to the poor solvent contact when compared to Mb (see Figure 9.4), as indicated by the solvent radial probability function [55,73]. This is consistent with the observation that the librational modes play a key role in the vibration–vibration energy transfer mechanism [72].

The deeper reason for the diversity of the solvent-dependent heme cooling between Mb and Cyt $c$ is that the two proteins are designed to support varying

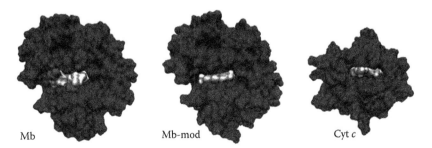

FIGURE 9.4 **(See color insert following page 172.)** Depiction of the heme exposure to solvent in native myoglobin (Mb), modified-heme myoglobin (Mb-mod), and cytochrome $c$ (Cyt $c$) with surface filling model. The protein residues are shown in dark colors and the heme moiety in light colors.

protein function. Mb supports the reversible binding and release of dioxygen ligand to the heme Fe atom. That function is served by a somewhat open conformation of the heme moiety to solvent. In fact, it has been suggested that the hydrogen bonding of the heme isopropionate side chains plays an essential role in controlling ligand escape from the heme pocket [74].

In contrast, cytochrome $c$ acts as an essential component of the electron transfer chain between the two large enzymes, cytochrome reductase and cytochrome oxidase. Consider how the heme environment has been designed to support the electron transfer function in cytochrome $c$ and cytochrome $c$ peroxidase (CcP) complex [75–81]. In this complex, one electron transfers from the Cyt $c$ heme to the CcP heme. It was found in the yeast Cyt $c$/CcP crystal structure that two hydrophobic resides from CcP, Ala193, and Ala194 have close contact with the methyl group of the Cyt $c$ heme side chain [76]. This methyl group, the docking site of Cyt $c$, is centered on the pyrrole ring C of the heme and is exposed to the solvent. It was suggested that the electron is transferred through this hydrophobic interaction, following a pathway consisting of the Gly192 and Trp191 residues of CcP, to reach the CcP heme. The weak interaction between the heme side chains and solvent molecules in Cyt $c$, as discussed above, preserved the methyl group for the docking process essential for the protein function. This mechanism is similar to that identified by Friesner et al. as "hydrophobic enclosure" [82].

## 9.3    SHORTER TIME INTRAMOLECULAR STATE-TO-STATE ENERGY TRANSFER IN THE INITIAL PHASES OF HEME COOLING

Classical molecular dynamics provides an effective tool for probing longer time, "coarse-grained" vibrational energy flow in proteins. A detailed understanding of shorter-term energy flow demands a mode-specific and possibly quantum dynamical approach.

### 9.3.1    QUANTUM TIME-DEPENDENT PERTURBATION THEORY ANALYSIS OF STATE-TO-STATE ENERGY TRANSFER

In order to study the short time vibrational energy transfer behavior of a vibrationally excited system, we employ a non-Markovian time-dependent perturbation theory [83]. Our approach builds on the successful application of Markovian time-dependent perturbation theory by Leitner and coworkers to explore heat flow in proteins and glasses, and Tokmakoff, Fayer, and others, in modeling vibrational population relaxation of selected modes in larger molecules. In a separate chapter in this volume, Leitner provides an overview of the development of normal mode-based methods, such as the one employed here, for the study of energy flow in solids and larger molecular systems.

This approach allows for a fully quantum mechanical treatment of the dynamics, avoiding the use of "quantum correction factors" used to denote classical dynamical approaches, with the concession that the potential energy surface must be expanded, ignoring higher order nonlinearity in the mode coupling. The potential energy surface is expanded with respect to the normal coordinates of the system, $q_S$, and bath, $q_\alpha$, and their frequencies up to third and fourth order nonlinear coupling:

$$H = H_S + H_B - q_S \delta F + q_S^2 \delta G \tag{9.1}$$

$$H_S = \frac{p_S^2}{2} + V(q_S) \tag{9.2}$$

$$H_B = \sum_\alpha \left( \frac{p_\alpha^2}{2} + \frac{\omega_\alpha^2 q_\alpha^2}{2} \right) \tag{9.3}$$

$$\delta F = \sum_{\alpha,\beta} C_{S\alpha\beta} \left( q_\alpha q_\beta - \langle q_\alpha q_\beta \rangle \right) \tag{9.4}$$

$$\delta G = \sum_{\alpha,\beta} C_{SS\alpha\beta} \left( q_\alpha q_\beta - \langle q_\alpha q_\beta \rangle \right) + \sum_\alpha C_{SS\alpha} q_\alpha \tag{9.5}$$

where

$H_S$ ($H_B$) is the system (bath) Hamiltonian
$C_{S\alpha\beta}$ ($C_{SS\alpha\beta}$) are the third (fourth) order coupling terms

Contributions from terms like $C_{S\alpha\beta\gamma}$ are small and ignored in the summation. Starting from the von Neumann–Liouville equation, a reduced density matrix for the system mode is derived using the time-dependent perturbation theory after tracing over the bath degrees of freedom. The commonly employed Markov approximation is avoided and no assumption of a separation in timescales between system and bath mode relaxation is invoked in this theory [84]. The ultimate result for the evolution of the ground state vibrational population is

$$(\rho_S)_{00}(t) \cong \frac{2}{\hbar^2} \sum_{\alpha,\beta} \left[ C_{--}^{\alpha\beta} u_t (\tilde{\omega}_S - \omega_\alpha - \omega_\beta) + C_{++}^{\alpha\beta} u_t (\tilde{\omega}_S + \omega_\alpha + \omega_\beta) + C_{+-}^{\alpha\beta} u_t (\tilde{\omega}_S - \omega_\alpha + \omega_\beta) \right]$$

$$+ \frac{2}{\hbar^2} \sum_\alpha \left[ C_-^\alpha u_t (\tilde{\omega}_S - \omega_\alpha) + C_+^\alpha u_t (\tilde{\omega}_S + \omega_\alpha) \right] \tag{9.6}$$

where $u_t(\Omega)$ is defined as

$$u_t(\Omega) = \int_0^t dt' \int_0^{t'} dt'' \cos\Omega(t'-t'') = \frac{1 - \cos\Omega t}{\Omega^2} \tag{9.7}$$

The coefficients, $C_{--}^{\alpha\beta}$, $C_{++}^{\alpha\beta}$, $C_{+-}^{\alpha\beta}$, $C_-^\alpha$, and $C_+^\alpha$, can be derived from the nonlinear coupling constants $C_{S\alpha\beta}$ and $C_{SS\alpha\beta}$ as

$$\vec{C}^{\alpha\beta} = \begin{pmatrix} C_{--}^{\alpha\beta} & C_{+-}^{\alpha\beta} \\ C_{+-}^{\alpha\beta} & C_{++}^{\alpha\beta} \end{pmatrix} = \left\{ (q_S)_{10} C_{S\alpha\beta} - (q_S^2)_{10} C_{SS\alpha\beta} \right\}^2 \vec{S}^{\alpha\beta} \tag{9.8}$$

$$\vec{S}^{\alpha\beta} = \frac{\hbar^2}{2\omega_\alpha\omega_\beta} \begin{pmatrix} (1+n_\alpha)(1+n_\beta) & 2(1+n_\alpha)n_\beta \\ 2(1+n_\alpha)n_\beta & n_\alpha n_\beta \end{pmatrix} \tag{9.9}$$

$$\vec{C}^\alpha = \begin{pmatrix} C_-^\alpha \\ C_+^\alpha \end{pmatrix} = (q_S^2)_{10}^2 C_{SS\alpha}^2 \vec{R}^\alpha \tag{9.10}$$

$$\vec{R}^\alpha = \frac{\hbar}{2\omega_\alpha} \begin{pmatrix} 1+n_\alpha \\ n_\alpha \end{pmatrix} \tag{9.11}$$

where $n_\alpha = 1/(e^{\beta\hbar\omega_\alpha} - 1)$ is the thermal phonon number. When the system mode is excited to the $v = 1$ state, VER is described by the decay of the reduced density matrix element $\rho_{11}(t) = 1 - \rho_{00}(t) \approx \exp[-\rho_{00}(t)]$ with only the normal mode frequencies of the model system and the nonlinear coupling constants $C_{S\alpha\beta}$ and $C_{SS\alpha\beta}$ as input. The dominant terms that define the rate of energy transfer for specific channels will result from resonant coupling between the system mode and a combination of bath modes. The relative importance of a specific channel will depend on the frequency resonance as well as the strength of the nonlinear coupling constants. The role of these Fermi resonances in energy flow in proteins has been explored using low-temperature classical molecular dynamics simulations by Kidera and coworkers [85–87] and by Leitner using quantum mechanical perturbation theory [22,23]. The relative importance of resonant energy transfer, as well as the applicability of normal mode based models, in describing energy flow in proteins is explored in depth by Kidera and coworkers in a separate chapter in this volume. In an additional chapter, Segal explores the importance of resonant energy transfer in describing heat flow in nanostructures using a variety of related theoretical approaches.

### 9.3.2 PROBING MODE-SPECIFIC VER IN METALLOPORPHYRINS

The timescales and mechanisms of mode-specific VER in several metalloporphyrin models, including imidazole ligated ferrous iron porphine (FeP-Im), iron protoporphyrin IX (Fe-heme), iron porphine (Fe-P), nickel protoporphyrin IX (Ni-heme), and nickel octaethylporphyrin (NiOEP) (see Figure 9.5), were studied using the non-Markovian time-dependent perturbation theory at the B3LYP/6–31G(d) level [88–90]. All calculations were carried out using the Gaussian03 package [91]. The system mode vibrational energy transfer rate constant was derived by fitting the reduced density matrix element $\rho_{11}(t)$ time profile to a single exponential decay function. The results are summarized in Table 9.4. The energy transfer pathways were identified by calculating the third order Fermi resonance parameters defined as [25,92]

$$r_{S\alpha\beta} = \frac{|C_{S\alpha\beta}|}{\hbar|\tilde{\omega}_S - \omega_\alpha - \omega_\beta|} \sqrt{\frac{\hbar}{2\tilde{\omega}_S}} \sqrt{\frac{\hbar}{2\omega_\alpha}} \sqrt{\frac{\hbar}{2\omega_\beta}} \tag{9.12}$$

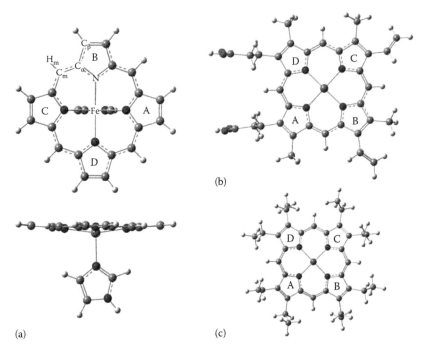

**FIGURE 9.5** Depiction of the optimized structure of (a) FeP-Im (top view and side view), (b) Fe-heme, and (c) NiOEP at the B3LYP/6-31G(d) level. The four pyrrole rings, A, B, C, and D, are labeled. Ni-heme has a structure similar to that of Fe-heme. FeP and NiP have structure similar to FeP-Im without the axial ligand (not shown here).

where $\tilde{\omega}_S$ and $\omega_\alpha$ are the frequencies of the system mode and bath modes, respectively. Large values of $r_{S\alpha\beta}$ indicate relaxation channels with strong nonlinear coupling, energetic resonance, or a balance between the two. The higher order parameters were found to be much smaller and were neglected in the analysis.

### 9.3.2.1 VER of FeP-Im Modes

The optimized structure of imidazole ligated iron porphine was found to be similar to the five-coordinate heme in Mb [89]. It has been suggested that the iron out-of-plane (Fe-oop) motion is the first event to follow ligand dissociation in Mb or Hb [93] and these Fe-oop motions, which are believed to be excited during this process, have received considerable attention due to their role in conformational transitions associated with protein function [43,94–97]. In the imidazole ligated ferrous iron porphine, five Fe-oop modes were identified. The relaxation of these modes, as well as the in-plane $v_4$ and $v_7$ modes, which were also found to be excited following heme photoexcitation [34,35,43], were studied to gain insight into the timescale and pathways of vibrational population relaxation.

Applying the time-dependent perturbation theory described above, the vibrational energy transfer time constant was found to be $1.7 \pm 0.2\,\mathrm{ps}$ for the excited $v_4$ mode and roughly $2.9\,\mathrm{ps}$ for the $v_7$ mode, in good agreement with the experimentally

**TABLE 9.4**

**Summary of System Mode Frequencies (Harmonic and Corrected Anharmonic), VER Time Constants, and the Assignments Derived from Calculations at B3LYP/6-31G(d) Level for the Metalloporphyrin Models for Use in the Time-Dependent Perturbation Theory**

| Mode Number | Frequency (cm⁻¹) | | $T_1$ (ps) | | Assignment |
| --- | --- | --- | --- | --- | --- |
| | Harm | Anharm | Simulation | Experiment | |
| FeP-Im | | | | | |
| 95 | 1380.7 | 1372.6 | $1.7 \pm 0.2$ | | $\nu_4$ |
| 44 | 733.1 | 729.9 | $2.9 \pm 0.0$ | | $\nu_7$ |
| 23 | 348.5 | 349.2 | $7.9 \pm 0.7$ | | $\gamma_7$ |
| 18 | 242.2 | 238.2 | $2.5 \pm 0.1$ | | $\gamma_{242}$ |
| 16 | 230.7 | 233.5 | $4.9 \pm 0.2$ | | $\gamma_{230}$ |
| 11 | 163.0 | 170.2 | $7.9 \pm 0.9$ | | $\nu$(Fe-Im) |
| 5 | 64.5 | 75.8 | $66.5 \pm 81.1$ | | Doming |
| FeP | | | | | |
| 78 | 1403.3 | 1398.2 | $2.1 \pm 0.2$ | | $\nu_4$ |
| 34 | 739.6 | 735.5 | $3.6 \pm 0.4$ | | $\nu_7$ |
| 15 | 360.8 | 362.1 | $28.6 \pm 14.2$ | | $\gamma_7$ |
| Fe-heme | | | | | |
| 147 | 1414.1 | 1414.0 | $1.2 \pm 0.1$ | $1.1 \pm 0.6^a$ | $\nu_4$ |
| 73 | 697.6 | 699.8 | $2.1 \pm 0.1$ | $1.9 \pm 0.6^a$ | $\nu_7$ |
| 47 | 351.1 | 353.2 | $17.0 \pm 4.5$ | | $\gamma_7$ |
| Ni-heme | | | | | |
| 148 | 1421.1 | 1418.7 | $1.1 \pm 0.0$ | | $\nu_4$ |
| 147 | 1419.9 | 1418.8 | $1.2 \pm 0.1$ | | $\nu_4$ |
| 74 | 699.7 | 703.1 | $2.5 \pm 0.1$ | | $\nu_7$ |
| 47 | 348.4 | 357.6 | $5.4 \pm 0.2$ | | $\gamma_7$ |
| NiOEP | | | | | |
| 157 | 1422.5 | 1421.2 | $1.1 \pm 0.0$ | $11 \pm 2^b$ | $\nu_4$ |
| 74 | 715.5 | 718.0 | $7.0 \pm 0.4$ | $10 \pm 2^b$ | $\nu_7$ |
| 51 | 348.6 | 359.6 | $13.0 \pm 2.4$ | | $\gamma_7$ |

[a] Time-resolved resonance Raman spectroscopy data for carboxy Mb [35].
[b] Time-resolved resonance Raman spectroscopy data for NiOEP in benzene [51].

determined values for the heme in MbCO following photodissociation ($1.1 \pm 0.6$ ps for $\nu_4$ mode and $1.9 \pm 0.6$ ps for $\nu_7$ mode) [35]. Considering the use of a model system and approximations inherent to the theory, no better agreement could be expected.

For the excited $\nu_4$ and $\nu_7$ modes, both of which are porphine in-plane breathing-like motions, as shown in Figure 9.6, most bath modes involved in the important

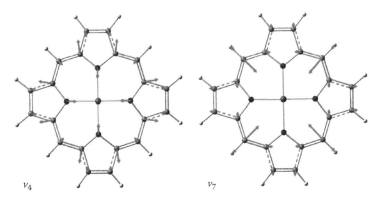

**FIGURE 9.6** Depiction of the computed porphine $v_4$ and $v_7$ modes. The ligand imidazole and porphyrin side chains are not shown.

energy transfer pathways (defined as $r_{S\alpha\beta} \geq 0.05$) are associated with porphine out-of-plane motions (not to be confused with Fe out-of-plane motions). A few modes associated with ligand imidazole motions were found to be involved in the energy transfer pathways from the excited $v_4$ mode. No such mode was found for the VER from the excited $v_7$ mode. Furthermore, no direct energy exchange between the $v_4$ and $v_7$ modes was observed (see Table 9.5), although the frequency of the $v_4$ mode is nearly in 2:1 resonance with the $v_7$ mode (see Table 9.4).

The overtone of the $\gamma_7$ mode is found to be the most important VER pathway from the excited $v_7$ mode. $\gamma_7$ is an iron out-of-plane mode associated with a porphine methine wagging motion, excited by the iron displacement out of the porphine plane following diatomic ligand photolysis in Mb. The coupling between the $\gamma_7$ mode and coupled lower frequency bath modes are relatively weak and associated with small Fermi resonance parameters. The $\gamma_7 + \gamma_7$ to $v_7$ energy transfer channel was found to be the single most important pathway for energy transfer from the excited $\gamma_7$ mode.

**TABLE 9.5**

**Summary of the Energy Transfer (ET) between the $v_4$ and $v_7$ and the $v_7$ and $\gamma_7$ Modes in the Porphyrin Models Studied**

|         | ET between $v_4$ and $v_7$ | ET between $v_7$ and $\gamma_7$ |
|---------|:---------:|:---------:|
| FeP-Im  | No  | Yes |
| FeP     | No  | Yes |
| Fe-heme | No  | Yes |
| Ni-heme | Yes | No  |
| NiOEP   | No  | No  |

Only those pathways with third-order Fermi resonance parameters larger than 0.05 are considered.

This resonant coupling between the $\gamma_7$ and $v_7$ modes provides a possible population excitation mechanism for the $v_7$ mode (see Table 9.5). When electronic state changes are involved, however, such as those associated with ligand photodissociation in heme proteins, the energy transfer mechanism captured by our theory, consisting of nonlinear coupling between vibrational modes on the ground electronic surface, can be overwhelmed by the electron–nuclear coupling. The electron–nuclear coupling leads to forces orders of magnitude larger than the vibrational mode–mode coupling, which has been the focus of this study, and acts on shorter timescales [98,99].

The $v_7$ mode has been observed to be highly excited following the ligand dissociation in the MbCO coherence spectrum. No similar signal was observed for deoxyMb [43]. Based on femtosecond coherence spectroscopy, Champion and coworkers have suggested that low frequency vibrational modes associated with electronic rearrangements in the heme iron following ligand photodissociation can be excited simultaneously [98,99]. We have assumed the separation of the population decay and dephasing processes in our time-dependent perturbation theory. As such, it is beyond the reach of our theory to address the questions concerning mode–mode coherent coupling in these system.

Four Fe out-of-plane modes, $\gamma_{242}$, $\gamma_{230}$, and the well-known $v(Fe\text{-}Im)$ and doming motions, are found to be weakly coupled to the low-frequency bath modes, as indicated by the third order coupling constants. Energy flow from these modes is less effective than from the excited $v_4$ and $v_7$ modes. For these Fe-oop modes, the imidazole ligand motion appears to be involved in the energy transfer pathways. These low-frequency bath modes are presumed to be coupled to the delocalized protein backbone motions in Mb following photodissociation, triggering large-scale protein conformational change, especially in the F helix, toward the deoxyMb equilibrium structure. The weak coupling and the low frequency characteristic of these modes is the origin of the relatively long timescale related to the protein structure relaxation [6,37,100,101]. On the other hand, the $v_4$, $v_7$, and $\gamma_7$ modes are decoupled from other Fe-oop and low-frequency modes, including those involving imidazole motions. This decoupling can block energy transfer between the high-frequency modes (such as $v_7$) and low-frequency protein backbone motions through the heme–imidazole connection. This observation provides a possible mode-specific explanation for the similar heme cooling rates observed in native Mb and the H93G mutant [57,62,63].

To clarify the role of the imidazole ligand in the VER process, the VER properties of the $v_4$ and $v_7$ modes in iron porphine (FeP), a porphyrin model without side chains or axial ligands, were examined [88]. Relaxation of the excited $v_4$ and $v_7$ modes in FeP is predicted to be slower than relaxation in FeP-Im, consistent with the fact that there are fewer bath modes in FeP relative to FeP-Im. However, we find the identified energy transfer pathways for FeP to be similar to those for FeP-Im for the $v_4$ and $v_7$ modes.

### 9.3.2.2  Isopropionate Side Chains Essential to $v_4$ and $v_7$ Modes VER

We have further investigated the VER of the $v_4$ and $v_7$ modes in Fe-heme with a particular focus on the role of side chains in the mechanism of VER [88]. No axial ligand was included to save computational expense. This simplification of our model is justified by our observations that the imidazole makes a minor contribution to VER from the $v_4$ and $v_7$ modes. The VER time constants were found to be $1.2 \pm 0.1\,\mathrm{ps}$

for the excited $v_4$ mode and $2.1 \pm 0.1$ ps for the $v_7$ mode. These values agree well with the experimentally determined values for heme in MbCO following photodissociation of CO observed by Mizutani and Kitagawa ($1.1 \pm 0.6$ ps for $v_4$ mode and $1.9 \pm 0.6$ ps for $v_7$ mode) [35] and are consistent with the fact that the heme is a relatively isolated moiety in the Mb protein. This also indicates that the heme side chains are more important than the axial ligands for VER of the $v_4$ and $v_7$ modes. The energy transfer pathways from each excited system mode were identified through the computed third order Fermi resonance parameters. Similar features to FeP-Im or FeP are observed for these bath modes, including the absence of direct energy exchange between the $v_4$ and $v_7$ modes and effective energy flow between the $v_7$ and $\gamma_7$ mode (see Table 9.5). In addition, the heme side chains provide additional degrees of freedom that may act as energy accepting bath modes. The bath modes associated with these side chain motions have been observed to be involved in the energy transfer pathways for the excited $v_4$ and $v_7$ modes.

For each bath mode, the contribution to the norm from each atom can be calculated from the normalized eigenvectors. Modes with a total contribution larger than 0.5 from the side chain atoms were identified as "side chain dominant modes"; those with a contribution larger than 0.8 from the side chain atoms were identified as "side chain localized modes." Similar definitions can be made for porphine core atoms or the two isopropionate groups. Table 9.6 summarizes the results for the dominant modes or the localized modes for those modes involved in the important energy transfer pathways in $v_4$ and $v_7$ modes relaxation.

---

**TABLE 9.6**

**Summary of the Number of Modes Involved in the Energy Transfer Pathways from the Excited $v_4$ and $v_7$ Modes of Fe-Heme as Derived from B3LYP/6-31G(d) Calculations**

| Moiety | # of Bath Modes | |
|---|---|---|
| | $v_4$ Mode | $v_7$ Mode |
| Total | 40 | 18 |
| Porphine core $\geq 0.5$ | 19 (47.5%) | 5 (27.8%) |
| All side chains $\geq 0.5$ | 22 (55.0%) | 13 (72.2%) |
| Isopropionates $\geq 0.5$ | 9 (22.5%) | 4 (22.2%) |
| Other side chains $\geq 0.5$ | 8 (20.0%) | 2 (11.1%) |
| Porphine core $\geq 0.8$ | 6 (15.0%) | 1 (5.6%) |
| All side chains $\geq 0.8$ | 10 (25.0%) | 4 (22.2%) |
| Isopropionates $\geq 0.8$ | 5 (12.5%) | 2 (11.1%) |
| other side chains $\geq 0.8$ | 0 (0.0%) | 0 (0.0%) |

For each moiety, the contribution to the norm is noted.

As this analysis indicates, most bath modes involved in the important energy transfer pathways from the excited $v_4$ and $v_7$ modes are delocalized motions (60.0% for $v_4$ and 72.2% for $v_7$). This would allow the energy, initially deposited in the localized $v_4$ and $v_7$ modes, to be redistributed efficiently throughout the heme or porphyrin. The calculations also indicate that the two isopropionate groups in the heme play essential, direct roles in VER of the excited $v_4$ and $v_7$ modes. This is consistent with the conclusion of previous classical MD simulations [55,57,62] and experimental studies [63–65] that the two isopropionate groups form the dominant pathways for directed "energy funneling" in the mechanism of rapid heme cooling in Mb.

Both $v_4$ and $v_7$ are porphine core in-plane modes. The $v_4$ mode is primarily an inner-ring breathing motion while the $v_7$ mode is an outer-ring breathing motion. Our calculations indicate that the two modes share many similar VER features. However, unique features are also observed for each mode.

For the excited $v_4$ mode, roughly half (19 out of 40, 47.5%) of the energy accepting modes are dominated by porphine core motions, six of which (15% of total) are localized in the porphine core. In contrast, for the excited $v_7$ mode, less porphine core motion is involved (5 out of 18, 27.8%, are porphine core dominant modes and one porphine core localized mode). Seven modes with ≥0.5 contribution from the porphine core in-plane motions are involved in the important VER pathways from the excited $v_4$ mode, two of which are localized porphine core in-plane motions. In contrast, for the excited $v_7$ mode, no porphine in-plane mode was found to be essential to the VER pathway. These differences are consistent with the fact that the $v_4$ mode is associated with displacement of the porphine inner-ring atoms, mainly Fe–N and N–C stretching motions, whereas the $v_7$ mode is associated with the outer-ring methine group motion. Our earlier discussion of VER in five-coordinate FeP-Im led to the conjecture that the $v_7$ mode can be excited through its coupling to the overtone of the $\gamma_7$ mode. No similar coupling was found for the $v_4$ mode, leading to the complementary conjecture that the $v_4$ mode is excited together with the iron out-of-plane motions.

### 9.3.2.3 Nonplanar Porphyrin Structure Tunes the $v_4$ and $v_7$ Modes VER Pathways

The iron porphyrin models discussed above all share a nearly planar porphine core, similar to the heme in Mb, whereas the heme in Cyt $c$ is distorted from the planar structure by the covalent bonds and hydrogen bonds with surrounding protein residues. To address the effect of environmentally induced heme distortion on the mechanism of VER, rather than including the surrounding protein residues explicitly (too expensive for any reasonable high level of theory), we replaced the heme's central iron atom by nickel. Due to the smaller size of nickel relative to iron, the porphine structure is distorted in the Ni-substituted Ni-heme in a manner similar to the heme distortion observed in cytochrome $c$. Another nickel porphyrin model, nickel octaethylporphyrin (NiOEP), was also optimized to a structure with similar nonplanar porphine core geometry [90].

In Ni-heme, the $v_4$ mode was found to be split by 1.2 cm$^{-1}$. The fitted VER time constants for both modes are roughly the same, ~1.1 ps. This result is comparable to that observed in Fe-heme, apparently unaffected by the replacement of the iron by nickel. The VER time constant of the excited $v_7$ mode was found to be $2.5 \pm 0.1$ ps, also similar to that observed for Fe-heme ($2.1 \pm 0.1$ ps).

It is intriguing and significant that the energy relaxation pathways show distinctly different features. In contrast to our results for Fe-heme, ~60% of the modes defining the pathway of VER from the excited $v_4$ modes of Ni-heme are dominated by porphine core motion. For the excited $v_7$ mode of Ni-heme, more side chain–dominated motions (60.9% of all) are involved. For the $v_4$ and $v_7$ modes, the two isopropionate groups of the Ni-heme are found to be involved in the VER pathways, as was observed in Fe-heme, but their role is similar to those of the other six side chains in the VER processes in Ni-heme. Distinct from the mechanism of $v_4$ mode relaxation in Fe-heme and other Fe porphyrin models, the $v_7$ mode in the nonplanar Ni-heme was found to be involved in the important VER pathways from the excited $v_4$ modes and forms pathways with large third order Fermi resonance parameters, defining a plausible mechanism of effective energy exchange between the $v_4$ and $v_7$ modes in Ni-heme. In addition, the effective energy exchange pathway between the $v_7$ mode and the overtone of the $\gamma_7$ mode, observed in planar iron porphyrin models, was predicted to be insignificant in the nonplanar Ni-heme (see Table 9.5).

The VER time constants for the excited $v_4$ and $v_7$ modes of NiOEP were predicted to be $1.1 + 0.0$ and $7.0 \pm 0.4$ ps, respectively, somewhat faster than the experimentally derived timescale for NiOEP of $11 \pm 2$ ps for the $v_4$ mode and $10 \pm 2$ ps for the $v_7$ mode [51]. This predicted difference is in line with the experimentally determined relaxation times for NiOEP in benzene solution, an environment in which the NiOEP is expected to have a slightly different structure and mode couplings than in vacuo.

For the excited $v_4$ mode, both porphine core motions and side chain motions are predicted to be involved in the VER pathways. We draw similar conclusions for the excited $v_7$ mode. In spite of the distortion of the porphine core, comparable to that observed in Ni-heme, the $v_7$ mode is not predicted to be involved in the energy transfer pathways from the excited $v_4$ mode in NiOEP. The $\gamma_7$ overtone is strongly coupled with the $v_7$ mode in NiOEP, but the third order Fermi resonance parameter is smaller than observed for other pathways. This result is similar to the predictions for the nonplanar Ni-heme (see Table 9.5).

Through these studies, we have identified the important VER pathways using the third order Fermi resonance parameters, which are mainly determined by the product of the third order coupling constant $C_{S\alpha\beta}$ and the frequency resonance parameter $1/|\tilde{\omega}_S - \omega_\alpha - \omega_\beta|$. Comparing the corresponding $C_{S\alpha\beta}$ and $1/|\tilde{\omega}_S - \omega_\alpha - \omega_\beta|$ values of planar iron porphyrins and nonplanar nickel porphyrins, it was observed that the absence of the $v_7$ mode in the important energy transfer pathways from the excited $v_4$ mode in planar iron porphyrins is due to the weak coupling or/and bad frequency match, whereas the absence of the $v_7$ to $\gamma_7$ overtone energy transfer channel results from weak coupling between these modes in nonplanar nickel porphyrins. Our analysis provides strong support for the conjecture that changes in porphine structure, induced by the protein structure and topology, lead to modulation of the coupling constants and "tuning" of the important VER pathways for energy relaxation from the excited $v_4$ and $v_7$ modes.

### 9.3.3 Short Time State-to-State VER and Longer Time Heme Cooling

In past studies, we applied classical molecular dynamics simulation to explore "macroscopic" heme cooling and predicted a timescale of ~6 ps [55,57,62], in good agreement with experimental measurements [33]. On the other hand, the mode-specific quantum dynamical studies indicate a faster timescale (~2 ps for the $v_4$ and $v_7$ modes) [35,88–90]. One possible reason for this difference in timescales is that the initially excited $v_4$ and $v_7$ modes do not dissipate their excess energy directly to the environment.

Using a master equation to model the VER process as a multistep reaction, the excess energy flow kinetics in FeP was examined, where the third order Fermi resonance parameters served as approximate reaction rate constants [88]. It was found that the subsequent relaxation is slow relative to relaxation of the initially excited system mode, providing an explanation for the observed difference in relaxation timescales.

This conclusion raises an additional question. In our classical molecular dynamics simulations, the excess kinetic energy was uniformly added to all heme atoms. Does the observed simulated heme cooling rate depend upon the mode of excitation? No dependence on the mode of excitation was observed in the classical simulations. In addition to the $v_4$ and $v_7$ modes, other modes, including iron out-of-plane motions, are initially excited, leading to a broad excitation of heme atoms in the initially excited state. It is reasonable to assume that, ignoring coherence, that state of excitation is similar to our uniform heating protocol.

## 9.4 CONCLUSIONS AND FUTURE CHALLENGES

Using classical molecular dynamics, we have studied the relatively long timescale heme cooling in Mb following photoexcitation. The derived timescales are in good agreement with experimental results. The essential role of the heme isopropionate side chains in the decay process has been firmly established through our careful analysis of the simulation results for the wild-type and designed protein mutants and modifications. The mechanism of interaction between the heme side chains and solvent molecules was explored using dynamical simulations in various solvents. These simulation results provide support for the conjecture that protein structure dictates the energy transfer mechanism and timescale in support of the corresponding protein functions.

Our application of time-dependent perturbation theory with density functional theory has led to a detailed picture of mode specific energy transfer for a number of model heme compounds in vacuo. Our calculations allow for the interpretation of previous theoretical and experimental observations at a mode-specific level. Our results support the conjecture that the protein structure encodes a "tuning" of the energy transfer timescales and mechanisms in a way that facilitates protein function and signaling.

It is important to recognize that our analysis of quantum mechanical energy transfer is one that is essentially carried out at zero temperature. At higher temperatures, the relative applicability of single reference normal mode models is debatable and there is no doubt that "off-resonance" energy transfer may play an increasingly important role. These essential concerns are discussed in detail by Kidera and coworkers in a separate chapter in this volume. An additional concern that is carefully explored by Leitner, in his chapter in this volume, is the validity of ignoring higher order nonlinear coupling. His results are cautionary and suggest that high order nonlinear interactions will be required to accurately address energy transfer at temperatures approaching those of relevance to cellular function.

In spite of these insights, our understanding of the intimate way in which protein structure is translated into protein dynamics and function is still in its nascent stages. For example, our classical simulations have focused on the intermolecular energy transfer (IET) between the heme and solvent; our DFT calculations have focused mainly on the intramolecular vibrational transfer (IVR) within the heme moiety. However, an understanding of the competition between IET and IVR is an essential component of the mechanism of VER for molecules with various sizes [102]. Identification of the IVR threshold is another essential component of our understanding of the reaction kinetics [103,104]. At this time, relatively little is known regarding the IVR threshold appropriate to heme cooling in proteins.

Our studies raise the question of how we can identify the limitations of classical models of energy flow in proteins and the point at which a quantum mechanical treatment is required. Answers to these questions await improvement in existing theories, the development of new theories, and the introduction of new model systems including other heme proteins.

## ACKNOWLEDGMENT

We are grateful for the generous support of the research by the National Science Foundation (Grant No. CHE-0316551 and CHE-0750309) and Boston University's Center for Computer Science.

## REFERENCES

1. Dickerson, R. E.; Geis, I., *Hemoglobin: Structure, Function, Evolution, and Pathology*. The Benjamin/Cummings Publishing Company, Inc., Menlo Park, CA, 1983.
2. Asplund, M. C.; Zanni, M. T.; Hochstrasser, R. M., Two-dimensional infrared spectroscopy of peptides by phase-controlled femtosecond vibrational photon echoes. *Proc. Natl. Acad. Sci. U.S.A.* **2000**, 97, 8219–8224.
3. Kholodenko, Y.; Volk, M.; Gooding, E.; Hochstrasser, R. M., Energy dissipation and relaxation processes in deoxy myoglobin after photoexcitation in the Soret region. *Chem. Phys.* **2000**, 259, 71–87.
4. Kachalova, G. S.; Popov, A. N.; Bartunik, H. D., A steric mechanism for inhibition of CO binding to heme proteins. *Science* **1999**, 284, 473–476.
5. Schlichting, I.; Berendzen, J.; Phillips, G. N., Jr.; Sweet, R. M., Crystal structure of photolyzed carbonmonoxy-myoglobin. *Nature* **1994**, 371, 808–812.
6. Li, H.; Elber, R.; Straub, J. E., Molecular dynamics simulation of nitric oxide recombination to myoglobin mutants. *J. Biol. Chem.* **1993**, 268, 17908–17916.
7. Hill, J. R.; Dlott, D. D.; Rello, C. W.; Peterson, K. A.; Decatur, S. M.; Boxer, S. G.; Fayer, M. D., Vibrational dynamics of carbon monoxide at the active sites of mutant heme proteins. *J. Phys. Chem.* **1996**, 100, 12100–12107.

8. Rector, K. D.; Rella, C. W.; Hill, J. R.; Kwok, A. S.; Sligar, S. G.; Chien, E. Y. P.; Dlott, D. D.; Fayer, M. D., Mutant and wild-type myoglobin-CO protein dynamics: Vibrational echo experiments. *J. Phys. Chem. B* **1997**, 101, 1468–1475.

9. Münck, E.; Champion, P. M., Heme proteins and model compounds: Mossbauer absorption and emission spectroscopy. *Ann. N Y Acad. Sci.* **1975**, 244, 142–162.

10. Sage, J. T.; Paxson, C.; Wyllie, G. R. A.; Sturhahn, W.; Durbin, S. M.; Champion, P. M.; Alp, E. E.; Scheidt, W. R., Nuclear resonance vibrational spectroscopy of a protein active-site mimic. *J. Phys.: Condens. Matter* **2001**, 13, 7707–7722.

11. Elber, R.; Karplus, M., Multiple conformational states of proteins: A molecular dynamics analysis of myoglobin. *Science* **1987**, 235, 318–321.

12. Elber, R.; Karplus, M., Enhanced sampling in molecular dynamics: Use of the time-dependent Hartree approximation for a simulation of carbon monoxide diffusion through myoglobin. *J. Am. Chem. Soc.* **1990**, 112, 9161–9175.

13. Okazaki, I.; Hara, Y.; Nagaoka, M., On vibrational cooling upon photodissociation of carbonmonoxymyoglobin and its microscopic mechanism from the viewpoint of vibrational modes of heme. *Chem. Phys. Lett.* **2001**, 337, 151–157.

14. Nagy, A. M.; Raicu, V.; Miller, R. J. D., Nonlinear optical studies of heme protein dynamics: implications for proteins as hybrid states of matter. *Biochim. Biophys. Acta* **2005**, 1749, 148–172.

15. Hill, J. R.; Tokmakoff, A.; Peterson, K. A.; Sauter, B.; Zimdars, D.; Dlott, D. D.; Fayer, M. D., Vibrational dynamics of carbon monoxide at the active site of myoglobin: Picosecond infrared free-electron laser pump-probe experiments. *J. Phys. Chem.* **1994**, 98, 11213–11219.

16. Rector, K. D.; Jiang, J.; Berg, M. A.; Fayer, M. D., Effects of solvent viscosity on protein dynamics: Infrared vibrational echo experiments and theory. *J. Phys. Chem. B* **2001**, 105, 1081–1092.

17. Finkelstein, I. J.; Goj, A.; McClain, B.; Massari, A. M.; Merchant, K. A.; Loring, R. F.; Fayer, M. D., Ultrafast dynamics of myoglobin without the distal histidine: Stimulated vibrational echo experiments and molecular dynamics simulations. *J. Phys. Chem. B* **2005**, 109, 16959–16966.

18. Straub, J. E.; Karplus, M., Molecular dynamics study of the photodissociation of carbon monoxide from myoglobin: Ligand dynamics in the first 10 ps. *Chem. Phys.* **1991**, 158, 221–248.

19. Bu, L.; Straub, J. E., Simulating vibrational energy flow in proteins: Relaxation rate and mechanism for heme cooling in cytochrome *c*. *J. Phys. Chem. B* **2003**, 107, 12339–12345.

20. Bu, L.; Straub, J. E., Vibrational frequency shifts and relaxation rates for a selected vibrational mode in cytochrome *c*. *Biophys. J.* **2003**, 85, 1429–1439.

21. Ota, N.; Agard, D. A., Intramolecular signaling pathways revealed by modeling anisotropic thermal diffusion. *J. Mol. Biol.* **2005**, 351, 345–354.

22. Leitner, D. M., Vibrational energy transfer in helices. *Phys. Rev. Lett.* **2001**, 87, 188102.

23. Leitner, D. M., Energy flow in proteins. *Annu. Rev. Phys. Chem.* **2008**, 59, 233–259.

24. Chin, J. K.; Jimenez, R.; Romesberg, F. E., Direct observation of protein vibrations by selective incorporation of spectroscopically observable carbon-deuterium in cytochrome *c*. *J. Am. Chem. Soc.* **2001**, 123, 2426–2427.

25. Cremeens, M. E.; Fujisaki, H.; Zhang, Y.; Zimmermann, J.; Sagle, L. B.; Matsuda, S.; Dawson, P. E.; Straub, J. E.; Romesberg, F. E., Effects toward developing direct probes of protein dynamics. *J. Am. Chem. Soc.* **2006**, 128, 6028–6029.

26. Hamm, P.; Lim, M.; Hochstrasser, R. M., Structure of the amide I band of peptide measured by femtosecond nonlinear-infrared spectroscopy. *J. Phys. Chem. B* **1998**, 102, 6123–6138.

27. Peterson, K. A.; Rella, C. W.; Engholm, J. R.; Schwettman, H. A., Ultrafast vibrational dynamics of the myoglobin amide I band. *J. Phys. Chem. B* **1999**, 103, 557–561.

28. Fujisaki, H.; Straub, J. E., Vibrational energy relaxation of isotopically labeled amide I modes in cytochrome *c*: Theoretical investigation of vibrational energy relaxation rates and pathways. *J. Phys. Chem. B* **2007**, 111, 12017–12023.

29. Sagnella, D. E.; Straub, J. E.; Jackson, T. A.; Lim, M.; Anfinrud, P. A., Vibrational population relaxation of carbon monoxide in the heme pocket of photolyzed carbonmonoxy myoglobin: Comparison of time-resolved mid-IR absorbance experiments and molecular dynamics simulations. *Proc. Natl. Acad. Sci. U. S. A.* **1999**, 96, 14324–14329.

30. Sagnella, D. E.; Straub, J. E., A study of vibrational relaxation of B-state carbon monoxide in the heme pocket of photolyzed carboxymyoglobin. *Biophys. J.* **1999**, 77, 70–84.

31. Yu, X.; Leitner, D. M., Vibrational energy transfer and heat conduction in a protein. *J. Phys. Chem. B* **2003**, 107, 1698–1707.

32. Petrich, J. W.; Poyart, C.; Martin, J. L., Photophysics and reactivity of heme proteins: A femtosecond absorption study of hemoglobin, myoglobin, and protoheme. *Biochemistry* **1988**, 27, 4049–4060.

33. Lim, M.; Jackson, T. A.; Anfinrud, P. A., Femtosecond near-IR absorbance study of photoexcited myoglobin: Dynamics of electronic and thermal relaxation. *J. Phys. Chem.* **1996**, 100, 12043–12051.

34. Mizutani, Y.; Kitagawa, T., Direct observation of cooling of heme upon photodissociation of carbonmonoxy myoglobin. *Science* **1997**, 278, 443–446.

35. Mizutani, Y.; Kitagawa, T., Ultrafast dynamics of myoglobin probed by time-resolved resonance Raman spectroscopy. *Chem. Rec.* **2001**, 1, 258–275.

36. Kitagawa, T.; Haruta, N.; Mizutani, Y., Time-resolved resonance Raman study on ultrafast structural relaxation and vibrational cooling of photodissociated carbonmonoxy myoglobin. *Biopolymers* **2002**, 61, 207–213.

37. Kruglik, S. G.; Mojzes, P.; Mizutani, Y.; Kitagawa, T.; Turpin, P.-Y., Time-Resolved resonance Raman study of the exciplex formed between excited Cu-Porphyrin and DNA. *J. Phys. Chem. B* **2001**, 105, 5018–5031.

38. Lian, T.; Locke, B.; Kholodenko, Y.; Hochstrasser, R. M., Energy flow from solute to solvent probed by femtosecond IR spectroscopy: Malachite green and heme protein solutions. *J. Phys. Chem.* **1994**, 98, 11648–11656.

39. Miller, R. J. D., Vibrational energy relaxation and structural dynamics of heme proteins. *Annu. Rev. Phys. Chem.* **1991**, 42, 581–614.

40. Genberg, L.; Heisel, F.; McLendon, G.; Miller, R. J. D., Vibrational energy relaxation processes in heme proteins: Model systems of vibrational energy dispersion in disordered systems. *J. Phys. Chem.* **1987**, 91, 5521–5524.

41. Gao, Y.; El-Mashtoly, S. F.; Pal, B.; Hayashi, T.; Harada, K.; Kitagawa, T., Pathway of information transmission from heme to protein upon ligand binding/dissociation in myoglobin revealed by UV resonance Raman spectroscopy. *J. Biol. Chem.* **2006**, 281, 24637–24646.

42. Sato, A.; Gao, Y.; Kitagawa, T.; Mizutani, Y., Primary protein response after ligand photodissociation in carbonmonoxy myoglobin. *Proc. Natl. Acad. Sci. U. S. A.* **2007**, 104, 9627–9632.

43. Armstrong, M. R.; Ogilvie, J. P.; Cowan, M. L.; Nagy, A. M.; Miller, R. J. D., Observation of the cascaded atomic-to-global length scales driving protein motion. *Proc. Natl. Acad. Sci. U. S. A.* **2003**, 100, 4990–4994.

44. Henry, E. R.; Eaton, W. A.; Hochstrasser, R. M., Molecular dynamics simulations of cooling in laser-excited heme proteins. *Proc. Natl. Acad. Sci. U. S. A.* **1986**, 83, 8982–8986.

45. Li, P.; Sage, J. T.; Champion, P. M., Probing picosecond processes with nanosecond lasers: Electronic and vibrational relaxation dynamics of heme proteins. *J. Chem. Phys.* **1992**, 97, 3214–3227.

46. Wang, W.; Ye, X.; Demidov, A. A.; Rosca, F.; Sjodin, T.; Cao, W.; Sheeran, M.; Champion, P. M., Femtosecond multicolor pump-probe spectroscopy of ferrous cytochrome *c*. *J. Phys. Chem. B* **2000**, 104, 10789–10801.

47. Negrerie, M.; Cianetti, S.; Vos, M. H.; Martin, J.-L.; Kruglik, S. G., Ultrafast heme dynamics in ferrous versus ferric cytochrome *c* studied by time-resolved resonance Raman and transient absorption spectroscopy. *J. Phys. Chem. B* **2006**, 110, 12766–12780.

48. Rodriguez, J.; Kirmaier, C.; Holten, D., Time-resolved and static optical properties of vibrationally excited porphyrins. *J. Chem. Phys.* **1991**, 94, 6020–6029.

49. Rodriguez, J.; Holten, D., Ultrafast vibrational dynamics of a photoexcited metalloporphyrin. *J. Chem. Phys.* **1989**, 91, 3525–3531.

50. Mizutani, Y.; Kitagawa, T., Vibrational energy relaxation of metalloporphyrins in a condensed phase probed by time-resolved resonance Raman spectroscopy. *Bull. Chem. Soc. Jpn.* **2002**, 75, 623–639.

51. Mizutani, Y.; Uesugi, Y.; Kitagawa, T., Intramolecular vibrational energy redistribution and intermolecular energy transfer in the (d,d) excited state of nickel octaethylporphyrin. *J. Chem. Phys.* **1999**, 111, 8950–8962.

52. Mizutani, Y.; Kitagawa, T., A role of solvent in vibrational energy relaxation of metalloporphyrins. *J. Mol. Liq.* **2001**, 90, 233–242.

53. Brooks, B. R.; Bruccoleri, R. E.; Olafson, B. D.; States, D. J.; Swaminathan, S.; Karplus, M., CHARMM: A program for macromolecular energy, minimization, and dynamics calculations. *J. Comput. Chem.* **1983**, 4, 187–217.

54. MacKerell, A. D., Jr.; Bashford, D.; Bellott, M.; Dunbrack, R. L.; Evanseck, J. D.; Field, M. J.; Fischer, S.; Gao, J.; Guo, H.; Ha, S.; Joseph-McCarthy, D.; Kuchnir, L.; Kuczera, K.; Lau, F. T. K.; Mattos, C.; Michnick, S.; Ngo, T.; Nguyen, D. T.; Prodhom, B.; Reiher, W. E., III; Roux, B.; Schlenkrich, M.; Smith, J. C.; Stote, R.; Straub, J.; Watanabe, M.; Wiorkiewicz-Kuczera, J.; Yin, D.; Karplus, M., All-atom empirical potential for molecular modeling and dynamics studies of proteins. *J. Phys. Chem. B* **1998**, 102, 3586–3616.

55. Zhang, Y.; Fujisaki, H.; Straub, J. E., Molecular dynamics study on the solvent dependent heme cooling following ligand photolysis in carbonmonoxy myoglobin. *J. Phys. Chem. B* **2007**, 111, 3243–3250.

56. Verlet, L., Computer "experiments" on classical fluids. I. Thermodynamical properties of Lennard-Jones molecules. *Phys. Rev.* **1967**, 1, 98–103.

57. Sagnella, D. E.; Straub, J. E., Directed energy "funneling" mechanism for heme cooling following ligand photolysis or direct excitation in solvated carbonmonoxy myoglobin. *J. Phys. Chem. B* **2001**, 105, 7057–7063.

58. Sagnella, D. E.; Straub, J. E.; Thirumalai, D., Time scales and pathways for kinetic energy relaxation in solvated proteins: Application to carbonmonoxy myoglobin. *J. Chem. Phys.* **2000**, 113, 7702–7711.

59. Thirumalai, D.; Mountain, R. D., Ergodic convergence properties of supercooled liquids and glasses. *Phys. Rev. A* **1990**, 42, 4574–4587.

60. Straub, J. E.; Rashkin, A. B.; Thirumalai, D., Dynamics in rugged energy landscapes with applications to the S-peptide and ribonuclease A. *J. Am. Chem. Soc.* **1994**, 116, 2049–2063.

61. Straub, J. E.; Thirumalai, D., Exploring the energy landscape in proteins. *Proc. Natl. Acad. Sci. U.S.A.* **1993**, 90, 809–813.

62. Bu, L.; Straub, J. E., Vibrational energy relaxation of "tailored" hemes in myoglobin following ligand photolysis supports energy funneling mechanism of heme "cooling". *J. Phys. Chem. B* **2003**, 107, 10634–10639.

63. Ye, X.; Demidov, A.; Rosca, F.; Wang, W.; Kumar, A.; Ionascu, D.; Zhu, L.; Barrick, D.; Wharton, D.; Champion, P. M., Investigations of heme protein absorption line shapes, vibrational relaxation, and resonance Raman scattering on ultrafast time scales. *J. Phys. Chem. A* **2003**, 107, 8156–8165.

64. Gao, Y.; Koyama, M.; El-Mashtoly, S. F.; Hayashi, T.; Harada, K.; Mizutani, Y.; Kitagawa, T., Time-resolved Raman evidence for energy 'funneling' through propionate side chains in heme 'cooling' upon photolysis of carbonmonoxy myoglobin. *Chem. Phys. Lett.* **2006**, 429, 239–243.

65. Koyama, M.; Neya, S.; Mizutani, Y., Role of heme propionates of myoglobin in vibrational energy relaxation. *Chem. Phys. Lett.* **2006**, 430, 404–408.

66. Meuwly, M.; Becker, O. M.; Stote, R.; Karplus, M., NO rebinding to myoglobin: A reactive molecular dynamics study. *Biophys. Chem.* **2002**, 98, 183–207.

67. Iwata, K.; Hamaguchi, H.-O., Microscopic mechanism of solute-solvent energy dissipation probed by picosecond time-resolved Raman spectroscopy. *J. Phys. Chem. A* **1997**, 101, 632–637.

68. Okazaki, T.; Hirota, N.; Terazima, M., Energy conversion process from the photoexcited electronic states studied by the temperature lens and acoustic peak delay methods in solution. *J. Mol. Liq.* **2001**, 90, 243–249.

69. Scherer, P. O. J.; Seilmeier, A.; Kaiser, W., Ultrafast intra- and intermolecular energy transfer in solutions after selective infrared excitation. *J. Chem. Phys.* **1985**, 83, 3948–3957.

70. Gnanakaran, S.; Hochstrasser, R. M., Vibrational relaxation of HgI in ethanol: Equilibrium molecular dynamics simulations. *J. Chem. Phys.* **1996**, 105, 3486–3496.

71. Ladanyi, B. M.; Stratt, R. M., On the role of dielectric friction in vibrational energy relaxation. *J. Chem. Phys.* **1999**, 111 (5), 2008–2018.

72. Ashihara, S.; Huse, N.; Espagne, A.; Nibbering, E. T. J.; Elsaesser, T., Ultrafast structural dynamics of water induced by dissipation of vibrational energy. *J. Phys. Chem. A* **2007**, 111, 743–746.

73. Zhang, Y.; Straub, J. E., Diversity of solvent dependent energy transfer pathways. *J. Phys. Chem. B* **2009**, 113, 825–830.

74. Belogortseva, N.; Rubio, M.; Terrell, W.; Miksovska, J., The contribution of heme propionate groups to the conformational dynamics associated with CO photodissociation from horse heart myoglobin. *J. Inorg. Biol.* **2007**, 101, 977–986.

75. Northrup, S. H.; Boles, J. O.; Reynolds, J. L., Brownian dynamics of cytochrome *c* and cytochrome *c* peroxidase association. *Science* **1988**, 241, 67–70.

76. Pelletier, H.; Kraut, J., Crystal structure of a complex between electron transfer partners, cytochrome *c* peroxidase and cytochrome *c*. *Science* **1992**, 258, 1748–1755.

77. Pappa, H. S.; Tajbaksh, S.; Saunders, A. J.; Pielak, G. J.; Poulos, T. L., Probing the cytochrome *c* peroxidase-cytochrome *c* electron transfer reaction using site specific cross-linking. *Biochemistry* **1996**, 35, 4837–4845.

78. Nocek, J. M.; Zhou, J. S.; Forest, S. D.; Priyadarshy, S.; Beratan, D. N.; Onuchic, J. N.; Hoffman, B. M., Theory and practice of electron transfer within protein-protein complexes: Application to the multidomain binding of cytochrome *c* by cytochrome *c* peroxidase. *Chem. Rev.* **1996**, 96, 2459–2489.

79. Leesch, V. W.; Bujons, J.; Mauk, A. G.; Hoffman, B. M., Cytochrome *c* peroxidase-cytochrome *c* complex: Locating the second binding domain on cytochrome *c* peroxidase with site-directed mutagenesis. *Biochemistry* **2000**, 39, 10132–10139.

80. Rosenfeld, R. J.; Hays, A.-M. A.; Musah, R. A.; Goodin, D. B., Excision of a proposed electron transfer pathway in cytochrome *c* peroxidase and its replacement by a ligand-binding channel. *Protein Sci.* **2002**, 11, 1151–1159.

81. Gray, H. B.; Winkler, J. R., Electron tunneling through proteins. *Quart. Rev. Biophys.* **2003**, 36, 341–372.

82. Friesner, R. A.; Murphy, R. B.; Repasky, M. P.; Frye, L. L. G. J. R.; Halgren, T. A.; Sanschagrin, P. C.; Mainz, D. T., Extra precision Glide: Docking and scoring incorporating a model of hydrophobic enclosure for protein-ligand complexes. *J. Med. Chem.* **2006**, 49, 6177.

83. Fujisaki, H.; Zhang, Y.; Straub, J. E., Time-dependent perturbation theory for vibrational energy relaxation and dephasing in peptides and proteins. *J. Chem. Phys.* **2006**, 124, 144910.

84. Mikami, T.; Okazaki, S., Path integral influence functional theory of dynamics of coherence between vibrational states of solute in condensed phase. *J. Chem. Phys.* **2004**, 121 (20), 10052.

85. Moritsugu, K.; Miyashita, O.; Kidera, A., Vibrational energy transfer in a protein molecule. *Phys. Rev. Lett.* **2000**, 85 (18), 3970–3973.

86. Moritsugu, K.; Miyashita, O.; Kidera, A., Temperature dependence of vibrational energy transfer in a protein molecule. *J. Phys. Chem. B* **2003**, 107 (14), 3309–3317.

87. Moritsugu, K.; Kidera, A., Protein motions represented in moving normal mode coordinates. *J. Phys. Chem. B* **2004**, 108 (12), 3890–3898.

88. Zhang, Y.; Straub, J. E., Direct evidence for mode-specific vibrational energy relaxation from quantum time-dependent perturbation theory. II. The $\nu_4$ and $\nu_7$ modes of iron protoporphyrin~IX and iron porphine. *J. Chem. Phys.* **2009**, 130, 095102.

89. Zhang, Y.; Fujisaki, H.; Straub, J. E., Direct evidence for mode-specific vibrational energy relaxation from quantum time-dependent perturbation theory. I. Five-coordinate ferrous iron porphyrin model. *J. Chem. Phy.* **2009**, 130, 025102.

90. Zhang, Y.; Straub, J. E., Direct evidence for mode-specific vibrational energy relaxation from quantum time-dependent perturbation theory. III. The $\nu_4$ and $\nu_7$ modes of non-planar nickel porphyrin models. *J. Chem. Phys.* **2009**, 130, 215101.

91. Frisch, M. J.; Trucks, G. W.; Schlegel, H. B.; M. A. Robb, G. E. S.; Cheeseman, J. R.; Montgomery, J. J. A.; K. N. Kudin, T. V.; Burant, J. C.; Millam, J. M.; Iyengar, S. S.; V. Barone, J. T.; Mennucci, B.; Cossi, M.; Scalmani, G.; G. A. Petersson, N. R.; Nakatsuji, H.; Hada, M.; Ehara, M.; R. Fukuda, K. T.; Hasegawa, J.; Ishida, M.; Nakajima, T.; Honda, Y.; H. Nakai, O. K.; Klene, M.; Li, X.; Knox, J. E.; Hratchian, H. P.; V. Bakken, J. B. C.; Adamo, C.; Jaramillo, J.; Gomperts, R.; O. Yazyev, R. E. S.; Austin, A. J.; Cammi, R.; Pomelli, C.; P. Y. Ayala, J. W. O.; Morokuma, K.; Voth, G. A.; Salvador, P.; V. G. Zakrzewski, J. J. D.; Dapprich, S.; Daniels, A. D.; O. Farkas, M. C. S.; Malick, D. K.; Rabuck, A. D.; J. B. Foresman, K. R.; Ortiz, J. V.; Cui, Q.; Baboul, A. G.; J. Cioslowski, S. C.; Stefanov, B. B.; Liu, G.; Liashenko, A.; I. Komaromi, P. P.; Martin, R. L.; Fox, D. J.; Keith, T.; C. Y. Peng, M. A. A.-L.; Nanayakkara, A.; Challacombe, M.; B. Johnson, P. M. W. G.; Chen, W.; Wong, M. W.; Gonzalez, C.; Pople, J. A., Gaussian 03, Revision C.02. In.

92. Fujisaki, H.; Yagi, K.; Hirao, K.; Straub, J. E., Quantum dynamics of *N*-methylacetamide studied by the vibrational configuration interaction method. *Chem. Phys. Lett.* **2007**, 443, 6–11.

93. Franzen, S.; Poyart, C.; Martin, J. L., Evidence for sub-picosecond heme doming in hemoglobin and myoglobin: A time-resolved resonance Raman comparison of carbon-monoxy and deoxy species. *Biochemistry* **1995**, 34, 1224–1237.

94. Deak, J.; Chiu, H.-L.; Lewis, C. M.; Miller, R. J. D., Ultrafast phase grating studies of heme proteins: Observation of the low-frequency modes directing functionally important protein motions. *J. Phys. Chem. B* **1998**, 102, 6621–6634.

95. Srajer, V.; Reinisch, R.; Champion, P. M., Protein fluctuations, distributed coupling, and the binding of ligands to heme proteins. *J. Am. Chem. Soc.* **1988**, 110, 6656–6670.

96. Ye, X.; Lonascu, D.; Gruia, F.; Yu, A.; Benabbas, A.; Champion, P. M., Temperature-dependent heme kinetics with nonexponential binding and barrier relaxation in the absence of protein conformational substates. *Proc. Natl. Acad. Sci. U. S. A.* **2007**, 104, 14682–14687.

97. Klug, D. D.; Zgierski, M. Z.; Tse, J. S.; Liu, Z.; Kincaid, J. R.; Czarneck, K., Doming modes and dynamics of model heme compounds. *Proc. Natl. Acad. Sci. U. S. A.* **2002**, 99, 12526–12530.

98. Rosca, F.; Kumar, A. T. N.; Ionascu, D.; Ye, X.; Demidov, A. A.; Sjodin, T.; Wharton, D.; Barrick, D.; Sligar, S. G.; Yonetani, T.; Champion, P. M., Investigations of anharmonic low-frequency oscillations in heme proteins. *J. Phys. Chem. A* **2002**, 106, 3540–3552.

99. Zhu, L.; Sage, T.; Champion, P. M., Observation of coherent reaction dynamics in heme proteins. *Science* **1994**, 266, 629–632.

100. Findsen, E. W.; Scott, T. W.; Chance, M. R.; Friedman, J. M.; Ondrias, M. R., Picosecond time-resolved Raman studies of photodissociated carboxymyoglobin. *J. Am. Chem. Soc.* **1985**, 107, 3355–3357.

101. Xie, X.; Simon, J. D., Protein conformational relaxation following photodissociation of CO from carbonmonoxymyoglobin: Picosecond circular dichroism and absorption studies. *Biochemistry* **1991**, 30, 3682–3692.

102. Elsaesser, T.; Kaiser, W., Vibrational and vibronic relaxation of large polyatomic molecules in liquids. *Annu. Rev. Phys. Chem.* **1991**, 42, 83–107.

103. Leitner, D. M., Heat transport in molecules and reaction kinetics: The role of quantum energy flow and localization. *Adv. Phys. Chem. B* **2005**, 130, 205–256.

104. Leitner, D. M.; Gruebele, M., A quantum model of restricted vibrational energy flow on the way to the transition state in unimolecular reactions. *Mol. Phys.* **2008**, 106, 433–442.

# 10 A Minimalist Network Model for Studying Biomolecular Vibration

*Mingyang Lu and Jianpeng Ma*

## CONTENTS

## 10.1 INTRODUCTION

In recent years, a series of coarse-grained methods for normal mode analysis, named elastic network models (ENMs), have become highly popular for describing vibrational motions of biomolecules [1–3]. An ENM represents a biomolecular structure as a network of nodes connected by simple harmonic springs. In the simplest ENM, pairs of $\alpha$ carbons within a certain cutoff distance, e.g., 13Å, are connected with simple harmonic springs of equal stiffness [4,5]. Other variations of ENMs involve different choices of nodes [6–10] or different forms of spring stiffness [11–13]. Although seemingly oversimplified, ENMs can accurately produce the lowest frequency normal modes that contribute most to the large-amplitude, collective and functional motions, owing to the dominant effects of molecular shape on the lowest frequency modes [14–16].

ENMs have a wide range of applications in computational biophysics. They have been applied to explain large conformational changes of biomolecular complexes [2,3], to help model anisotropic thermal parameters for x-ray crystallographic refinement [17,18], and to enhance sampling efficiency [19,20]. A major application of network models is that they are advantageous in analyzing vibrational energy transfers between residues in proteins [21]. For instance, additional nonlinear terms have been included in the Hamiltonian of ENMs to study discrete breathers [22], energy transfer among normal modes [23], and slow energy relaxation [24]. In another study, ENM has been directly applied for nonlinear relaxation dynamics in the overdamped limit [25]. Gaussian network model (GNM) has been used to study the effect of protein size on spectral and fractal dimension of proteins [26]. In addition, a specialized network model has been developed to simulate information transfer with Markov process [27].

From the studies of network models, it is clear that the distribution of stiffness of the structure plays an important role in vibrational energy transfer [22,27]. In general, however, ENMs lack details of molecular interactions [12], which may limit the application of ENMs in studying vibrational energy transfer.

Some other normal mode methods [28–31] retain molecular interaction information by using conventional molecular mechanics force fields such as CHARMM [32–34] or AMBER [35–38]. In a recent study, a force-field-based normal mode method, rotations–translations of blocks (RTB) [28], has been found to perform better than ENMs in predicting atomic anisotropic displacement parameters (ADPs) from crystal structures [12]. However, force-field-based methods usually lack a network structure and need a tedious initial energy minimization. The latter often results in large structural deviation. Figure 10.1 shows the typical distribution of the root mean square deviations (RMSDs) of protein heavy atoms from their original positions after minimization. The average magnitude of structure shift is around 1.5 Å, but it can be as large as 4–5 Å for some ultraflexible systems. Figure 10.2 shows an example of structure deviation

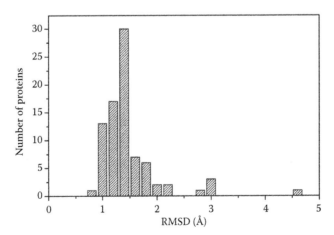

**FIGURE 10.1**   Histogram of heavy atom RMSD between protein structures before and after initial energy minimization. The plot is generated from 83 ultrahigh resolution (<1 Å) protein crystal structures (>50 residues, <50% sequence identity). The protein test set and minimization protocol are described in greater detail in reference [39]. (From Figure 1 in Lu, M. and Ma, J., *Proc. Natl. Acad. Sci. U. S. A.*, 105, 15358, 2008.)

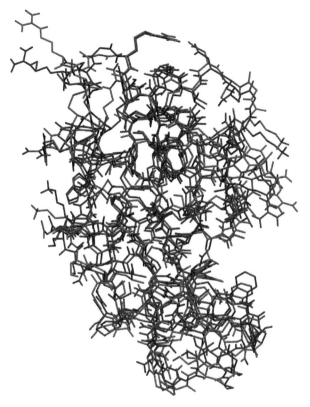

**FIGURE 10.2**    **(See color insert following page 172.)** An example of coordinate shift caused by the initial energy minimization (illustrated by lysozyme crystal structure, PBD code: 3lzt). The final minimized structure (in red) has 1.34 Å RMSD from the native structure (in blue).

before and after minimization. The structural shift caused by initial energy minimization may also result in errors in studying vibrational energy transfer.

To overcome the aforementioned problems, we recently developed a normal mode scheme [39] that features a network topology while retaining detailed molecular interaction information. The method also does not require initial energy minimization, which is achieved by slightly modifying the force field so that the current structure is at the energy minimum. Since the global molecular motions are not sensitive to small changes in the local stiffness [14], this new scheme can avoid initial energy minimization without compromising the quality of the lowest frequency vibrational modes. The method is called the minimalist network model (MNM) for normal mode analysis, in which the second derivative matrix, the Hessian matrix, is slightly modified to be positive semidefinite. It is named "minimalist" since it utilizes a minimal representation of molecular interactions. MNM differs from ENMs in that it maintains important molecular interaction details directly derived from a molecular force field. It also differs from conventional normal mode analysis in that it avoids initial energy minimization. Moreover, MNM has no additional parameters upon coarse-graining, and the Hessian modification step is much faster than energy minimization. It is worth noting that MNM also produces a network structure whose local interactions between node pairs contain rich information about the strength, direction, and stability of the structures.

In this chapter, we first introduce a general pairwise decomposition scheme starting from the RTB method. Next, we show how to modify the Hessian matrix to bypass initial energy minimization. To test MNM, we compared it to various other normal mode analysis methods in terms of similarity of modes and ability to predict the ADPs from x-ray crystallographic experiments. According to our results, MNM not only produces reliable eigenvectors for low-frequency modes, but it also models the experimental ADPs consistently better than any other available method. Since MNM directly calculates modes from the unminimized structures without sacrificing important details of molecular interaction, we expect this method to be a powerful tool for applications such as structural refinement, vibrational energy transfer, and conformational sampling. The contents of this chapter are mainly modified from the original publication of the MNM method [39].

## 10.2   THEORY AND METHODS

MNM utilizes the RTB method as a starting point, followed by a novel pairwise decomposition (PD) scheme. In this section, we first review conventional normal mode analysis and the RTB method, and then we introduce the PD scheme. The MNM Hessian is finally derived from the PD method.

### 10.2.1   CONVENTIONAL NORMAL MODE ANALYSIS AND THE RTB METHOD

For a molecule of $N$ atoms with its structure at a local energy minimum, the normal modes can be calculated from a $3N \times 3N$ mass-weighted second derivative matrix $\mathbf{H}$, the Hessian matrix, defined in a molecular force field such as CHARMM [32–34] or AMBER [35–38]. For each mode, the eigenvalue $\lambda$ and the $3N \times 1$ eigenvector $\mathbf{r}$ satisfy the eigenvalue equation, $\mathbf{Hr} = \lambda\mathbf{r}$.

In this study, conventional all-atom normal mode analysis was performed with the CHARMM package. Like other studies in the literature [29], the united atom CHARMM19 force field [33] with the EEF1 solvation model [40] was used. Before normal mode calculation, the structures underwent many cycles of energy minimization by the adopted-basis Newton–Raphson method with decreasing harmonic constraints.

To reduce the computational cost of normal mode calculations on supramolecular complexes, a coarse-grained normal mode method called the RTB method [28] was developed. In RTB, the whole molecule is divided into $n$ rigid-body blocks whose motion is modeled by six translational and rotational external degrees of freedom for each block. Thus, the $6n \times 6n$ RTB Hessian matrix, $\mathbf{H}_{RTB}$, is related to the all-atom Hessian by $\mathbf{H}_{RTB} = \mathbf{P}^T\mathbf{HP}$, where $\mathbf{P}$ is a $3N \times 6n$ orthogonal projection matrix. For any $6n \times 1$ rigid body displacement vector $\mathbf{x}$, the corresponding atomic displacement vector $\mathbf{r}$ is given by

$$\mathbf{r} = \mathbf{P}\,\mathbf{x} \qquad\qquad (10.1)$$

In this study, each block was selected to contain all the atoms in a residue.

## 10.2.2 THE PAIRWISE DECOMPOSITION (PD) SCHEME

In the PD scheme, the interactions of the whole molecule are decomposed into pairwise interactions of small subsystems (blocks). For an isolated molecule of $n$ blocks whose structure is at a local energy minimum, such that any external motions of the whole molecule produce no net forces, the RTB Hessian $\mathbf{H}_{RTB}$ satisfies

$$\mathbf{H}_{RTB}\Omega = \mathbf{0} \tag{10.2}$$

where $\Omega$ is the $6n \times 6$ eigenvector matrix for the six zero-frequency translational–rotational modes. Equation 10.2 is extremely useful in this study, because the matrix $\Omega$ can be calculated without actually diagonalizing $\mathbf{H}_{RTB}$. According to Equation 10.1, the matrix $\Omega$ obeys $\Omega = \mathbf{P}^T\mathbf{P}_1$, where $\mathbf{P}_1$ is the $3N \times 6$ projection matrix between atomic displacement vectors and external translation–rotational vectors (i.e., $\mathbf{P}$ for the case of $n = 1$).

By definition, the PD Hessian $\mathbf{H}_{PD}$ satisfies

$$\frac{1}{2}\mathbf{x}^T\mathbf{H}_{PD}\mathbf{x} = \frac{1}{2}\sum_{i<j}\begin{pmatrix}\mathbf{x}_i\\\mathbf{x}_j\end{pmatrix}^T\mathbf{H}_{ij}\begin{pmatrix}\mathbf{x}_i\\\mathbf{x}_j\end{pmatrix} \tag{10.3}$$

where

$\mathbf{x}_i$ is the $6 \times 1$ rigid body displacement subvector of $\mathbf{x}$ for block $i$
$\mathbf{H}_{ij}$ is the $12 \times 12$ decomposed Hessian matrix for an isolated system of blocks $i$ and $j$

$\mathbf{H}_{ij}$ is derived from $\mathbf{H}_{RTB}$ according to

$$\mathbf{H}_{ij} = \begin{pmatrix} -\frac{1}{2}(\mathbf{K}_{ij}\Gamma_{ij}^{-1} + (\Gamma_{ij}^{-1})^T\mathbf{K}_{ij}^T) & \frac{1}{2}(\mathbf{K}_{ij} + (\Gamma_{ij}^{-1})^T\mathbf{K}_{ij}^T\Gamma_{ij}) \\ \frac{1}{2}(\mathbf{K}_{ij}^T + \Gamma_{ij}^T\mathbf{K}_{ij}\Gamma_{ij}^{-1}) & -\frac{1}{2}(\mathbf{K}_{ij}^T\Gamma_{ij} + \Gamma_{ij}^T\mathbf{K}_{ij}) \end{pmatrix} \tag{10.4}$$

where

$\Gamma_{ij} = \Omega_i\,\Omega_j^{-1}$
$\Omega_i$ is the $6 \times 6$ nonsingular submatrix of $\Omega$ for block $i$
$\mathbf{K}_{ij} = \partial^2 E/\partial\mathbf{x}_i\,\partial\mathbf{x}_j$ is the $6 \times 6$ submatrix of $\mathbf{H}_{RTB}$
$E$ is the total energy

It is easy to confirm that each $\mathbf{H}_{ij}$ satisfies Equation 10.2 for the case of $n = 2$, i.e.,

$$\mathbf{H}_{ij}\begin{pmatrix}\Omega_i\\\Omega_j\end{pmatrix} = \mathbf{0} \tag{10.5}$$

thus proving that $\mathbf{H}_{ij}$ represents a typical Hessian matrix of an isolated system of two blocks. Hence, $\mathbf{H}_{PD}$, by including all components of $\mathbf{H}_{ij}$, represents the Hessian of the whole molecule. This treatment automatically produces six external translational–rotational modes. While the PD scheme is intended for minimized structures, Equation 10.4 can still be used to calculate $\mathbf{H}_{ij}$ for unminimized structures, but with one major caveat: $\mathbf{H}_{PD}$ derived on unminimized structures is not necessarily positive semidefinite.

### 10.2.3 PERTURBATION ANALYSIS ON THE PD SCHEME

The PD scheme is designed to reproduce the molecular interactions of RTB. Thus, we apply perturbation theory to further assess the difference between the PD and RTB schemes. According to perturbation theory, for the normal modes of a molecule with eigenvalues $\lambda_{(k)}$ and eigenvectors $\mathbf{x}_{(k)}$ ($k$ represents the mode indices), and a small change $\Delta\mathbf{H} = \mathbf{H}_{PD} - \mathbf{H}_{RTB}$,

$$\Delta\lambda_{(k)} = \mathbf{x}_{(k)}^{T}\, \Delta\mathbf{H}\mathbf{x}_{(k)} \tag{10.6a}$$

$$\Delta\mathbf{x}_{(k)} = \sum_{k \neq l} c_{(kl)}\mathbf{x}_{(l)} \tag{10.6b}$$

$$c_{(kl)} = \frac{\mathbf{x}_{(k)}^{T}\, \Delta\mathbf{H}\mathbf{x}_{(l)}}{\lambda_{(k)} - \lambda_{(l)}} \tag{10.6c}$$

It can be shown that

$$\left\langle \frac{\Delta\lambda_{(k)}}{\lambda_{(k)}} \right\rangle \approx 0 \tag{10.7a}$$

$$\sigma\!\left(\frac{\Delta\lambda_{(k)}}{\lambda_{(k)}}\right) \propto \frac{1}{\sqrt{n}} \tag{10.7b}$$

$$\left|\left\langle c_{(kl)}\right\rangle\right| \leq \gamma \frac{\lambda_{(k)} + \lambda_{(l)}}{\left|\lambda_{(k)} - \lambda_{(l)}\right|} \tag{10.7c}$$

where
  $n$ in Equation 10.7b is the number of blocks when the blocks are roughly uniform in size
  $\gamma$ is a small scaling factor

Proofs for Equation 10.7 are described in the following. The RTB Hessian $\mathbf{H}_{RTB}$ can have the same form as Equation 10.3:

$$\frac{1}{2}\mathbf{x}^{T}\mathbf{H}_{RTB}\mathbf{x} = \frac{1}{2}\sum_{i<j}\begin{pmatrix}\mathbf{x}_i \\ \mathbf{x}_j\end{pmatrix}^{T}\mathbf{H}_{ij}^{0}\begin{pmatrix}\mathbf{x}_i \\ \mathbf{x}_j\end{pmatrix} \tag{10.8}$$

where

$$\mathbf{H}_{ij}^0 = \begin{pmatrix} -\mathbf{K}_{ij}\Gamma_{ij}^{-1} & \mathbf{K}_{ij} \\ \mathbf{K}_{ij}^T & -\mathbf{K}_{ij}^T\Gamma_{ij} \end{pmatrix}$$

Compared to $\mathbf{H}_{ij}$, $\mathbf{H}_{ij}^0$ is no longer symmetric, but it also satisfies Equation 10.5. For an RTB mode whose eigenvalue is $\lambda$ and eigenvector is $\mathbf{x}$, Equations 10.4 and 10.6 give

$$\Delta\lambda = \mathbf{x}^T\Delta\mathbf{H}\mathbf{x}$$

$$= \sum_{i<j} \begin{pmatrix} \mathbf{x}_i \\ \mathbf{x}_j \end{pmatrix}^T (\mathbf{H}_{ij} - \mathbf{H}_{ij}^0) \begin{pmatrix} \mathbf{x}_i \\ \mathbf{x}_j \end{pmatrix}$$

$$= \frac{1}{2}\sum_{i<j} \begin{pmatrix} \mathbf{x}_i \\ \mathbf{x}_j \end{pmatrix}^T \begin{pmatrix} \mathbf{K}_{ij}\Gamma_{ij}^{-1} - (\Gamma_{ij}^{-1})^T\mathbf{K}_{ij}^T & (\Gamma_{ij}^{-1})^T\mathbf{K}_{ij}^T\Gamma_{ij} - \mathbf{K}_{ij} \\ \Gamma_{ij}^T\mathbf{K}_{ij}\Gamma_{ij}^{-1} - \mathbf{K}_{ij}^T & \mathbf{K}_{ij}^T\Gamma_{ij} - \Gamma_{ij}^T\mathbf{K}_{ij} \end{pmatrix} \begin{pmatrix} \mathbf{x}_i \\ \mathbf{x}_j \end{pmatrix}$$

$$= \sum_{i<j} \mathbf{x}_i^T \left[ (\Gamma_{ij}^{-1})^T\mathbf{K}_{ij}^T\Gamma_{ij} - \mathbf{K}_{ij} \right]\mathbf{x}_j$$

$$= 2\sum_{i<j} \mathbf{x}_i^T\Delta\mathbf{K}_{ij}\mathbf{x}_j \qquad (10.9)$$

where the summation $\sum_{i<j}$ is over all block pairs $i, j$ in the range of the interaction cutoff, and $\Delta\mathbf{K}_{ij}$ is the difference between $\frac{1}{2}(\mathbf{K}_{ij} + (\Gamma_{ij}^{-1})^T\mathbf{K}_{ij}^T\Gamma_{ij})$ and the RTB $\mathbf{K}_{ij}$. Now let

$$\varepsilon_{ij} = \frac{\mathbf{x}_i^T\Delta\mathbf{K}_{ij}\mathbf{x}_j}{\mathbf{x}_i^T\mathbf{K}_{ij}\mathbf{x}_j}$$

and assume that the Frobenius norm of $\Delta\mathbf{K}_{ij}$ is much smaller than that of $\mathbf{K}_{ij}$ (note that the Frobenius norm of $\mathbf{K}_{ij}$ is invariant upon any translational and rotational transformation). We also assume that $\varepsilon_{ij}$ satisfies a random distribution whose mean and standard deviation are $\bar{\varepsilon}$ and $\delta$, respectively. For different modes, both $\bar{\varepsilon}$ and $\delta$ are functions of $\lambda$ and they are also very small in magnitude. Hence,

$$\Delta\lambda = 2\sum_{i<j} \mathbf{x}_i^T\Delta\mathbf{K}_{ij}\mathbf{x}_j$$

$$= 2\sum_{i<j} \varepsilon_{ij}\mathbf{x}_i^T\mathbf{K}_{ij}\mathbf{x}_j \qquad (10.10)$$

$$\langle\Delta\lambda\rangle \approx n_{ij}\bar{\varepsilon}\langle\mathbf{x}_i^T\mathbf{K}_{ij}\mathbf{x}_j\rangle$$

where $n_{ij}$ is twice the number of block pairs calculated in the summation, which is approximately proportional to $n$. Since the eigenvectors $\mathbf{x}$ are normalized,

$$\lambda = \mathbf{x}^T \mathbf{H}_{RTB} \mathbf{x}$$

$$= \sum_{i,j} \mathbf{x}_i^T \mathbf{K}_{ij} \mathbf{x}_j$$

$$= \sum_{i<j} \mathbf{x}_i^T (\mathbf{K}_{ij} + \mathbf{K}_{ij}^T) \mathbf{x}_j + \sum_i \mathbf{x}_i^T \mathbf{K}_{ii} \mathbf{x}_i \qquad (10.11)$$

Thus, $\left| \left\langle \mathbf{x}_i^T \mathbf{K}_{ij} \mathbf{x}_j \right\rangle \right| \approx \alpha \dfrac{\lambda}{n_{ij}}$, where is a scaling factor introduced here because the summation $\sum_{i<j}$ includes no contributions from the diagonal blocks. Note that the value of $\alpha$ is also slightly dependent on frequency. After all, $\langle \Delta\lambda/\lambda \rangle \approx \alpha \bar{\varepsilon} \approx 0$, which proves Equation 10.7a. As for the standard deviation (Equation 10.7b)

$$\sigma(\Delta\lambda) \approx \sqrt{n_{ij}\delta^2 \left( \alpha \frac{\lambda}{n_{ij}} \right)^2} = \frac{\alpha\delta\lambda}{\sqrt{n_{ij}}},$$

which proves

$$\sigma\left( \frac{\Delta\lambda}{\lambda} \right) \propto \frac{1}{\sqrt{n}}.$$

For Equation 10.7c, we consider two different RTB modes $k$ and $l$ ($k \neq l$):

$$c_{(kl)} = \frac{\mathbf{x}_{(k)}^T \Delta\mathbf{H}\mathbf{x}_{(l)}}{\lambda_{(k)} - \lambda_{(l)}}$$

$$= \frac{(\mathbf{x}_{(k)} + \mathbf{x}_{(l)})^T \Delta\mathbf{H}(\mathbf{x}_{(k)} + \mathbf{x}_{(l)}) - \mathbf{x}_{(k)}^T \Delta\mathbf{H}\mathbf{x}_{(k)} - \mathbf{x}_{(l)}^T \Delta\mathbf{H}\mathbf{x}_{(l)}}{2(\lambda_{(k)} - \lambda_{(l)})}$$

$$= \frac{1}{\lambda_{(k)} - \lambda_{(l)}} \sum_{i<j} [(\mathbf{x}_{i,(k)} + \mathbf{x}_{i,(l)})^T \Delta\mathbf{K}_{ij}(\mathbf{x}_{j,(k)} + \mathbf{x}_{j,(l)}) - \mathbf{x}_{i,(k)}^T \Delta\mathbf{K}_{ij}\mathbf{x}_{j,(k)} - \mathbf{x}_{i,(l)}^T \Delta\mathbf{K}_{ij}\mathbf{x}_{j,(l)}]$$

$$= \frac{1}{\lambda_{(k)} - \lambda_{(l)}} \sum_{i<j} \mathbf{x}_{i,(k)}^T (\Delta\mathbf{K}_{ij} + \Delta\mathbf{K}_{ij}^T) \mathbf{x}_{j,(l)} \qquad (10.12)$$

where $\mathbf{x}_{i,(k)}$ denotes the component of $\mathbf{x}_{(k)}$ for block $i$. Similar to the proof for Equation 10.7a, the distribution of $\varepsilon_{ij}$ is also applied, and according to the orthogonality condition of different modes $\mathbf{x}_{(k)}^T \mathbf{H}_{RTB} \mathbf{x}_{(l)} = 0$:

$$\left|\langle c_{ab}\rangle\right| \approx \frac{\overline{\varepsilon}}{\left|\lambda_{(k)} - \lambda_{(l)}\right|} \left|\sum_{i<j} \mathbf{x}_{i,(k)}^{T} (\mathbf{K}_{ij} + \mathbf{K}_{ij}^{T}) \mathbf{x}_{j,(l)}\right|$$

$$= \frac{\overline{\varepsilon}}{\left|\lambda_{(k)} - \lambda_{(l)}\right|} \left|\sum_{i} \mathbf{x}_{i,(k)}^{T} \mathbf{K}_{ii} \mathbf{x}_{i,(l)}\right| \qquad (10.13)$$

Meanwhile, since $\mathbf{H}_{RTB}$ is a symmetric positive semidefinite (SPSD) matrix, $\mathbf{x}^{T}\mathbf{H}_{RTB}$ $\mathbf{x} \geq 0$ for any vector $\mathbf{x} = (0, 0\ldots, \mathbf{x}_{i}, \ldots 0)^{T}$. Thus, $\mathbf{x}_{i}^{T}\mathbf{K}_{ii}\,\mathbf{x}_{i} \geq 0$ for any $\mathbf{x}_{i}$, which indicates that $\mathbf{K}_{ii}$ is also SPSD for any $i$. Thus, $(\mathbf{x}_{i,(k)} - \mathbf{x}_{i,(l)})^{T}\mathbf{K}_{ii}(\mathbf{x}_{i,(k)} - \mathbf{x}_{i,(l)}) \geq 0$, and following the above derivation,

$$\sum_{i} \mathbf{x}_{i,(k)}^{T} \mathbf{K}_{ii} \mathbf{x}_{i,(l)} = \sum_{i} [\mathbf{x}_{i,(k)}^{T} \mathbf{K}_{ii} \mathbf{x}_{i,(k)} + \mathbf{x}_{i,(l)}^{T} \mathbf{K}_{ii} \mathbf{x}_{i,(l)} - (\mathbf{x}_{i,(k)} - \mathbf{x}_{i,(l)})^{T} \mathbf{K}_{ii} (\mathbf{x}_{i,(k)} - \mathbf{x}_{i,(l)})]/2$$

$$\leq \left(\sum_{i} \mathbf{x}_{i,(k)}^{T} \mathbf{K}_{ii} \mathbf{x}_{i,(k)} + \sum_{i} \mathbf{x}_{i,(l)}^{T} \mathbf{K}_{ii} \mathbf{x}_{i,(l)}\right)/2$$

$$\leq \kappa(\lambda_{(k)} + \lambda_{(l)})/2 \qquad (10.14)$$

where
  $\kappa$ is the maximum value of $|1 - \alpha_{(k)}|$ and $|1 - \alpha_{(l)}|$
  $\alpha_{(k)}$ and $\alpha_{(l)}$ are the scaling factors (introduced above) for modes $k$ and $l$, respectively

In the same way, the lower bound can be estimated as $\sum_{i} \mathbf{x}_{i,(k)}^{T} \mathbf{K}_{ii} \mathbf{x}_{i,(l)} \geq -\kappa(\lambda_{(k)} + \lambda_{(l)})/2$. Finally, we obtain Equation 10.7c

$$\left|\langle c_{kl}\rangle\right| \leq \overline{\varepsilon}\kappa/2 \frac{\lambda_{(k)} + \lambda_{(l)}}{\left|\lambda_{(k)} - \lambda_{(l)}\right|} \qquad (10.15)$$

where $\gamma = \overline{\varepsilon}\kappa/2$.

Equation 10.7 indicates that the PD scheme can produce normal modes with almost the same eigenvalues as RTB. Moreover, Equation 10.7c suggests that the low-frequency eigenvector subspace is conserved, because each of the low-frequency eigenvectors from PD can be approximately expressed as a linear combination of the RTB eigenvectors with similar eigenvalues.

## 10.2.4 THE MINIMALIST NETWORK MODEL METHOD

The MNM method further modifies the Hessian matrix based on the PD scheme, and it guarantees the matrix to be positive semidefinite. This process is fundamentally equivalent to modifying the molecular interactions defined in the original force field. In MNM, all PD $\mathbf{H}_{ij}$ s are changed to their nearest (in terms of the Frobenius norm) SPSD matrices $\mathbf{H}_{ij}^{+}$.

A symmetric real matrix $\mathbf{M}$ may be decomposed into two parts $\mathbf{M}^+ + \mathbf{M}^-$, where $\mathbf{M}^+ = \mathbf{U}\Lambda^+ \mathbf{U}^T$ is the nearest SPSD matrix [41]. Here, $\mathbf{U}$ is the eigenvector matrix of $\mathbf{M}$, and $\Lambda^+$ is the positive part of the diagonal matrix of $\mathbf{M}$ after diagonalization (i.e., the elements for the negative eigenvalues are forced to be zero).

$\mathbf{H}_{ij}^+$ ensures that the interaction energy of each pair of blocks is at a local minimum, and so the total interaction energy of the whole system, as the sum of all pairwise interactions, is also at the local minimum. Note that this feature of default local minimum energy of the system is also satisfied by ENMs [5,6]. Thus, unlike the PD scheme, MNM can be applied to an un-minimized structure by modifying the $\mathbf{H}_{ij}$ s from the matrix $\mathbf{H}_{\mathrm{RTB}}$ of the original structure. Even so, a short relaxation process was performed to eliminate possible bad contacts that may affect the true molecular interactions. The relaxation consists of 100 steps of steepest-descent energy minimization: the first 50 steps are with weak harmonic constraints and the other 50 steps are without these constraints. Such a structure relaxation causes 0.13 Å RMSD from the initial structure on average, which is much smaller than that of deep energy minimization in conventional normal mode calculations.

Similar to Equation 10.3, the MNM Hessian matrix may be obtained from

$$\frac{1}{2}\mathbf{x}^T\mathbf{H}_{\mathrm{MNM}}\mathbf{x} = \frac{1}{2}\sum_{i<j}\begin{pmatrix}\mathbf{x}_i\\\mathbf{x}_j\end{pmatrix}^T \mathbf{H}_{ij}^+ \begin{pmatrix}\mathbf{x}_i\\\mathbf{x}_j\end{pmatrix} \tag{10.16}$$

Note that any $\mathbf{H}_{ij}^+$ satisfies Equation 10.5 and $\mathbf{H}_{\mathrm{MNM}}$ satisfies Equation 10.2. Finally, the eigenvectors of MNM (in the same form as those of RTB) can be transformed to the all-atom representation by Equation 10.1.

## 10.3 RESULTS

Since no standard criterion exists for measuring the accuracy of normal modes calculated from native structures without energy minimization, we assessed the quality of the MNM normal modes in modeling the protein atomic displacement parameters (ADPs) by fitting the modes against the experimental data.

There are two ways to model ADPs by normal modes in the literature. One is via direct calculation, whereby ADPs are directly derived from the atomic thermal fluctuations predicted by normal modes [12,42]. The other way, adopted in this study, is by a fitting procedure, in which ADPs are obtained by optimization against the experimental data, for example, diffraction data or experimental ADPs. The latter approach is chosen because it can produce more realistic ADPs according to several normal-mode-based x-ray crystallographic refinements [17,18,43,44] and other theoretical studies [45].

ADPs of each atom consist of three diagonal parameters $U_{xx}$, $U_{yy}$, and $U_{zz}$, and three off-diagonal parameters $U_{xy}$, $U_{xz}$, and $U_{yz}$. In the fitting procedure, the $3 \times 3$ ADP matrix for each atom can be modeled by normal modes as

$$\mathbf{U}^{\mathrm{mode}} = \sum_k \sum_l \sigma_{kl}\mathbf{v}_{(k)}\mathbf{v}_{(l)}^T \tag{10.17}$$

where

$\sigma_{kl}$ is the $kl$th element of $m \times m$ matrix $\Sigma = \mathbf{SS}^{\mathrm{T}}$

$\mathbf{S}$ is an $m \times m$ lower triangular matrix introduced to make $\Sigma$ positive semidefinite

$m$ is the number of chosen modes

$\mathbf{v}_{(k)}$ represents the three components of the atom in the $k$th normal mode eigenvector (the atom index is omitted for simplicity)

The summations are both over the $m$ low-frequency normal modes.

In the optimization process, the target function is the average Kullack–Leibler (KL) distance for all heavy atoms, which characterizes the difference between two Gaussian probability distributions defined by the theoretical and experimental ADPs [30,42,46]. Given the eigenvalues ($\omega_{(p)}^{\mathrm{mode}}$, $\omega_{(p)}^{\mathrm{data}}$, $p \in 1,2,3$) and eigenvectors ($\mathbf{a}_{(p)}^{\mathrm{mode}}$, $\mathbf{a}_{(p)}^{\mathrm{data}}$, $p \in 1,2,3$) of the theoretical ("mode") and experimental ("data") ADP matrices, the KL distance can be calculated from

$$D_{\mathrm{KL}} = -\frac{3}{2} + \frac{1}{2}\sum_{p=1}^{3}\ln\frac{\omega_{(p)}^{\mathrm{mode}}}{\omega_{(p)}^{\mathrm{data}}} + \frac{1}{2}\sum_{p=1}^{3}\sum_{q=1}^{3}\frac{\omega_{(p)}^{\mathrm{data}}}{\omega_{(q)}^{\mathrm{mode}}}\left|(\mathbf{a}_{(p)}^{\mathrm{data}})^{\mathrm{T}}\mathbf{a}_{(q)}^{\mathrm{mode}}\right|^{2} \tag{10.18}$$

the derivation of which is in Ref. [42].

The details of the optimization procedure can be found in Ref. [39]. Compared to recent studies, where only Cα data were used [12,42,45], the current study includes all valid data for the heavy atoms, which makes the experimental dataset much larger than that of any of the recent studies. The resulting data-to-parameter ratio is high enough for proper fitting of the normal mode parameters to the ADPs; for example, there were about 6000 ADP data points for a protein of 100 residues, and there were 126 parameters for normal-mode-based fitting with 20 normal modes (including six zero-frequency modes representing the external motions).

To compare with MNM, we also evaluate the performance of RTB and an all-atom-based ENM (elNémo [47], with slight modification to include hydrogen atoms). The protein test set, the same as that used in Ref. [12], consists of 83 ultrahigh resolution (1 Å resolution or higher) crystal structures of various sizes (the number of residues ranges from 55 to 1300). The fitting results are shown in Table 10.1. The left-most column shows the average optimized KL distances between the ADPs of various normal mode methods and experimental data. The smallest KL distance (best fitting) is found to be by MNM. The method elNémo, when applied to native structures, has comparable average KL distance to RTB, but the KL distance is worse when elNémo is applied to structures after initial energy minimization.

Meanwhile, to cross-validate the fitting quality, we also evaluated the Pearson's linear correlation coefficient between normal-mode-optimized and experimental ADPs. Here, four types of correlation coefficients are assessed and the results are listed in Table 10.1; all four correlation coefficients for MNM are consistently better than those for RTB. The coefficients for two elNémo applications are similar to each other, yet they are worse than those of MNM and RTB.

**TABLE 10.1**

**Comparison of Performance of Various Normal Mode Analyses**

| NMA Method | KL Distance | Pearson's Correlation Coefficient with Experimental Data | | | |
|---|---|---|---|---|---|
| | | All ADPs | Diagonal ADPs | Off-Diagonal ADPs | Cα Isotropic B |
| MNM | 0.117 | 0.850 | 0.687 | 0.496 | 0.874 |
| RTB | 0.126 | 0.836 | 0.647 | 0.456 | 0.845 |
| elNémo | 0.127 | 0.792 | 0.608 | 0.415 | 0.835 |
| elNémo (mini) | 0.132 | 0.801 | 0.604 | 0.415 | 0.824 |
| MNM (zero) | 0.117 | 0.850 | 0.687 | 0.497 | 0.874 |
| RTB (zero) | 0.125 | 0.838 | 0.653 | 0.461 | 0.848 |

*Note:* The results are averaged for 83 ultrahigh-resolution protein crystal structures (see Section 10.3). The first column shows the minimum KL distance between calculated and experimental ADPs. The four columns on the right-most end represent the Pearson's correlation coefficients of various groups of ADPs for the minimum KL distance. elNémo was applied to unminimized ("elNémo") and minimized structures ("elNémo (mini)"). The last two rows (with "(zero)") show the results of modified versions of MNM and RTB, in which the zero modes calculated from the minimized structures were replaced by the zero modes of the unminimized native structures.

Moreover, to check the quality of fitting, we also compared the calculated ADP correlation coefficients with the correlations of experimental ADPs for the same protein but in different crystal environments. According to the results in the first row of Table 10.1 and in Tables 1 and 2 in Ref. [42], MNM is found to give worse ADP correlations than those between the same protein in identical crystal form (upper limit in the experiment), but better than those between the same protein in different crystal forms. Since the fitting procedure is sensitive enough to detect ADP differences of the same magnitude as the differences in experimental values due to crystal packing, the modes of MNM are reasonable for modeling experimental ADPs.

Our fitting procedure includes six zero-frequency modes to model the overall rigid-body motions. However, the RTB zero modes of minimized structures do not necessarily characterize the rigid-body motions of crystal structures. Thus, for both MNM and RTB, we changed the zero modes to those obtained from crystal structures, and we performed the fitting procedure again. Note that the new zero modes are slightly nonorthogonal to the original low-frequency normal modes calculated on the minimized structures, but the fitting process may be insensitive to the orthogonality. As shown in the last two rows in Table 10.1, the modification of the zero modes yields no apparent improvement for MNM because relaxation causes almost no change to the structures, while the RTB fit improved only slightly.

We also applied conventional all-atom normal modes (via the CHARMM package) for fitting tests on several proteins after initial energy minimization. Oddly, the fitting results are often worse than those for RTB (data not shown), presumably because RTB prevents surface side-chains from undergoing large deviations during initial minimization.

Finally, we also assessed the atomic improvements in KL distances from RTB to MNM as a function of atomic structural deviation upon minimization. As shown in

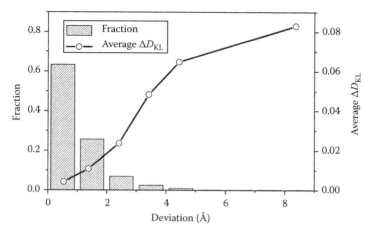

**FIGURE 10.3** Average KL distance improvement from RTB to MNM upon energy minimization. The changes in KL distance from RTB to MNM ($\Delta D_{KL} = D_{KL}^{RTB} - D_{KL}^{MNM}$) were calculated and averaged over atoms with similar atomic positional deviations. This figure presents both the histogram of atomic positional deviations ($Y$-axis to the left) and the average improvement in KL distance ($Y$-axis to the right) as a function of atomic positional deviation. Minimized structures are used for RTB and un-minimized structures for MNM. (From Figure 5 in Lu, M. and Ma, J., *Proc. Natl. Acad. Sci. U.S.A.*, 105, 15358, 2008.)

Figure 10.3, the average KL distance improves for all RMSD regions, whereas larger improvements are found for the atoms with larger structural deviations. Besides, the magnitude of $\Delta D_{KL}$ is substantial compared with the magnitude of KL distances in the second column of Table 10.1. Once more, this finding underscores the impact of structure shifts caused by energy minimization. Preventing this shift becomes particularly important for functionally relevant regions, e.g., protein active site loops, which are often found in the very flexible regions of the protein.

## 10.4 CONCLUDING DISCUSSION

In this chapter, we introduced a novel coarse-grained normal mode method, the MNM [39]. Starting from the RTB blocking scheme and any molecular force field, MNM uses the PD scheme to approximate the pairwise interactions between blocks. Finally, the decomposed Hessian is modified to include only the quadratic portion of the local interaction energy. MNM not only avoids the lengthy and sometime erroneous initial energy minimization, but it also retains detailed molecular interaction information during coarse-graining.

According to the direct comparison of normal mode vectors, MNM is found to preserve the lowest frequency normal modes. More importantly, MNM outperforms all other normal mode methods in predicting experimental ADPs measured by x-ray crystallography. This is in part due to the elimination of structural shifts during initial energy minimization.

Just like RTB, MNM can be applied to proteins at various coarse-graining levels. Although the results only show the application of MNM at the residue level, the PD and MNM schemes can also be applied using multiresidue blocks, e.g., secondary structure elements, as in the case of supramolecular complexes. In this way, multiresidue MNM was found to produce similar results as the single-residue version for an all-alpha protein SLT70 (PDB code: 1qsa), in which each helix or any other residue is chosen as a block.

MNM can also be applied to other biomolecules such as nucleic acids, sugars, and lipids. MNM should be particularly powerful for mixed systems, e.g., protein–nucleic acid complexes or protein–lipid complexes, for which the blocks for nonprotein components can be chosen according to chemical intuition, e.g., each nucleotide in DNA is a block.

Though the pairwise interactions of blocks in MNM, like two nodes in ENMs [5], are modeled as quadratic functions, the physical nature of block–block interaction energies in MNM is distinct from that between two nodes in ENM. Unlike ENM, the two blocks in MNM are not modeled as two mass points, and thus they can have much more complex relative motions between them, e.g., torsional motions, thereby reflecting more detailed molecular interactions. Hence, MNM is fundamentally distinct from ENMs that introduce a nonuniform distribution of force constants between nodes [30,48–50].

More detailed molecular interactions presumably permit MNM to be more applicable to the analysis of vibrational energy transfer. MNM captures more detailed interaction strengths due to the use of a realistic molecular force field. Furthermore, because the interactions between blocks are not solely along the directions that connect them, MNM can provide rich information on the relative translation and rotation of the block pairs during energy transfers. Moreover, by comparing the Hessian matrix of PD and MNM, we can also deduce the stability of each pairwise interaction along the network. Thus, by utilizing strength, direction and stability information, MNM is a promising model for studying vibrational energy transfer [21]. It will also be interesting to use MNM to characterize energy landscape and compare with previous studies [51,52].

## ACKNOWLEDGMENTS

J.M. acknowledges support from the National Institutes of Health (R01-GM067801), the National Science Foundation (MCB-0818353), and the Welch Foundation (Q-1512).

## ABBREVIATIONS

| | |
|---|---|
| ADP | anisotropic displacement parameter |
| AMBER | assisted model building with energy refinement |
| CHARMM | Chemistry at HARvard Molecular Mechanics |
| KL | Kullack–Leibler |
| MNM | minimalist network model |
| RMSD | root mean square deviation |
| RTB | rotations–translations of blocks |

## REFERENCES

1. Ma, J., New advances in normal mode analysis of supermolecular complexes and applications to structural refinement, *Curr. Protein Pept. Sci.*, *5*, 119 (2004).
2. Ma, J., Usefulness and limitations of normal mode analysis in modeling dynamics of biomolecular complexes, *Structure*, *13*, 373 (2005).
3. Bahar, I. and Rader, A. J., Coarse-grained normal mode analysis in structural biology, *Curr. Opin. Struct. Biol.*, *15*, 586 (2005).
4. Haliloglu, T., Bahar, I., and Erman, B., Gaussian dynamics of folded proteins, *Phys. Rev. Lett.*, *79*, 3090 (1997).
5. Atilgan, A. R., Durell, S. R., Jernigan, R. L., Demirel, M. C., Keskin, O., and Bahar, I., Anisotropy of fluctuation dynamics of proteins with an elastic network model, *Biophys. J.*, *80*, 505 (2001).
6. Tirion, M. M., Large amplitude elastic motions in proteins from a single-parameter, atomic analysis, *Phys. Rev. Lett.*, *77*, 1905 (1996).
7. Doruker, P., Jernigan, R. L., and Bahar, I., Dynamic of large proteins through hierarchical levels of coarse-grained structures, *J. Comput. Chem.*, *23*, 119 (2002).
8. Kurkcuoglu, O., Jernigan, R. L., and Doruker, P., Collective dynamics of large proteins from mixed coarse-grained elastic network model, *QSAR Combinatorial Sci.*, *24*, 443 (2005).
9. Ming, D., Kong, Y., Lambert, M., Huang, Z., and Ma, J., How to describe protein motion without amino-acid sequence and atomic coordinates, *Proc. Natl. Acad. Sci. U. S. A.*, *99*, 8620 (2002).
10. Tama, F., Wriggers, W., and Brooks, C. L., Exploring global distortions of biological macromolecules and assemblies from low-resolution structural information and elastic network theory, *J. Mol. Biol.*, *321*, 297 (2002).
11. Hinsen, K., The molecular modeling toolkit: A new approach to molecular simulations, *J. Comput. Chem.*, *21*, 79 (2000).
12. Kondrashov, D. A., Van Wynsberghe, A. W., Bannen, R. M., Cui, Q., and Phillips, G. N., Jr., Protein structural variation in computational models and crystallographic data, *Structure*, *15*, 169 (2007).
13. Lu, M., Poon, B., and Ma, J., A new method for coarse-grained elastic normal-mode analysis, *J. Chem. Theor. Comp.*, *2*, 464 (2006).
14. Lu, M. and Ma, J., The role of shape in determining molecular motions, *Biophys. J.*, *89*, 2395 (2005).
15. Doruker, P. and Jernigan, R. L., Functional motions can be extracted from on-lattice construction of protein structures, *Proteins*, *53*, 174 (2003).
16. Nicolay, S. and Sanejouand, Y. H., Functional modes of proteins are among the most robust, *Phys. Rev. Lett.*, *96* (2006).
17. Poon, B. K., Chen, X., Lu, M., Vyas, N. K., Quiocho, F. A., Wang, Q., and Ma, J., Normal mode refinement of anisotropic thermal parameters for a supramolecular complex at 3.42-Å crystallographic resolution, *Proc. Natl. Acad. Sci. U. S. A.*, *104*, 7869 (2007).
18. Chen, X., Poon, B. K., Dousis, A., Wang, Q., and Ma, J., Normal-mode refinement of anisotropic thermal parameters for potassium channel KcsA at 3.2 A crystallographic resolution, *Structure*, *15*, 955 (2007).
19. Wu, Y., Tian, X., Lu, M., Chen, M., Wang, Q., and Ma, J., Folding of small helical proteins assisted by small-angle X-ray scattering profiles, *Structure (Camb)*, *13*, 1587 (2005).
20. Zhang, Z., Shi, Y., and Liu, H., Molecular dynamics simulations of peptides and proteins with amplified collective motions, *Biophys. J.*, *84*, 3583 (2003).
21. Leitner, D. M., Energy flow in proteins, *Annu. Rev. Phys. Chem.*, *59*, 233 (2008).
22. Juanico, B., Sanejouand, Y. H., Piazza, F., and De Los Rios, P., Discrete breathers in nonlinear network models of proteins, *Phys. Rev. Lett.*, *99*, 238104 (2007).

23. Moritsugu, K., Miyashita, O., and Kidera, A., Vibrational energy transfer in a protein molecule, *Phys. Rev. Lett.*, *85*, 3970 (2000).
24. Piazza, F., De Los Rios, P., and Sanejouand, Y. H., Slow energy relaxation of macromolecules and nanoclusters in solution, *Phys. Rev. Lett.*, *94*, 145502 (2005).
25. Togashi, Y. and Mikhailov, A. S., Nonlinear relaxation dynamics in elastic networks and design principles of molecular machines, *Proc. Natl. Acad. Sci. U. S. A.*, *104*, 8697 (2007).
26. Reuveni, S., Granek, R., and Klafter, J., Proteins: coexistence of stability and flexibility, *Phys. Rev. Lett.*, *100*, 208101 (2008).
27. Chennubhotla, C. and Bahar, I., Signal propagation in proteins and relation to equilibrium fluctuations, *PLoS Comput. Biol.*, *3*, 1716 (2007).
28. Tama, F., Gadea, F. X., Marques, O., and Sanejouand, Y. H., Building-block approach for determining low-frequency normal modes of macromolecules, *Proteins*, *41*, 1 (2000).
29. Li, G. and Cui, Q., A coarse-grained normal mode approach for macromolecules: An efficient implementation and application to Ca(2+)-ATPase, *Biophys. J.*, *83*, 2457 (2002).
30. Ming, D. and Wall, M. E., Allostery in a coarse-grained model of protein dynamics, *Phys. Rev. Lett.*, *95* (2005).
31. Zhou, L. and Siegelbaum, S. A., Effects of surface water on protein dynamics studied by a novel coarse-grained normal mode approach, *Biophys. J.*, *94*, 3461 (2008).
32. MacKerell, A. D., Bashford Jr., D., Bellott, M., Dunbrack Jr., R. L., Evanseck, J. D., Field, M. J., Fischer, S., Gao, J., Guo, H., Ha, S., Joseph-McCarthy, D., Kuchnir, L., Kuczera, K., Lau, F. T. K., Mattos, C., Michnick, S., Ngo, T., Nguyen, D. T., Prodhom, B., Reiher, I. W. E., Roux, B., Schlenkrich, M., Smith, J. C., Stote, R., Straub, J., Watanabe, M., JWiorkiewicz-Kuczera, J., Yin, D., and Karplus, M., All-atom empirical potential for molecular modeling and dynamics studies of proteins, *J. Phys. Chem.*, *B102*, 3586 (1998).
33. Neria, E., Fischer, S., and Karplus, M., Simulation of activation free energies in molecular systems, *J. Chem. Phys.*, *105*, 1902 (1996).
34. Brooks, B. R., Bruccoleri, R. E., Olafson, B. D., States, D. J., Swaminathan, S., and Karplus, M., CHARMM: A program for macromolecular energy, minimization, and dynamics calculations, *J. Comput. Chem.*, *4*, 187 (1983).
35. Wang, J. M., Cieplak, P., and Kollman, P. A., How well does a restrained electrostatic potential (RESP) model perform in calculating conformational energies of organic and biological molecules?, *J. Comput. Chem.*, *21*, 1049 (2000).
36. Ponder, J. W. and Case, D. A., Force fields for protein simulations, *Adv. Protein Chem.*, *66*, 27 (2003).
37. Case, D. A., Cheatham, T. E., 3rd, Darden, T., Gohlke, H., Luo, R., Merz, K. M., Jr., Onufriev, A., Simmerling, C., Wang, B., and Woods, R. J., The Amber biomolecular simulation programs, *J. Comput. Chem.*, *26*, 1668 (2005).
38. Wang, J., Wolf, R. M., Caldwell, J. W., Kollman, P. A., and Case, D. A., Development and testing of a general amber force field, *J. Comput. Chem.*, *25*, 1157 (2004).
39. Lu, M. and Ma, J., A minimalist network model for coarse-grained normal mode analysis and its application to biomolecular x-ray crystallography, *Proc. Natl. Acad. Sci. U. S. A.*, *105*, 15358 (2008).
40. Lazaridis, T. and Karplus, M., Effective energy function for proteins in solution, *Proteins*, *35*, 133 (1999).
41. Higham, N. J., Computing a nearest symmetric positive semidefinite matrix, *Linear Algebra Applic.*, *103*, 103 (1988).
42. Eyal, E., Chennubhotla, C., Yang, L. W., and Bahar, I., Anisotropic fluctuations of amino acids in protein structures: Insights from X-ray crystallography and elastic network models, *Bioinformatics*, *23*, I175 (2007).

43. Diamond, R., On the use of normal modes in thermal parameters refinement: Theory and application to the bovine pancreatic trypsin inhibitor, *Acta Cryst.*, *A46*, 425 (1990).
44. Kidera, A. and Go, N., Refinement of protein dynamic structure: Normal mode refinement, *Proc. Natl. Acad. Sci. U. S. A.*, *87*, 3718 (1990).
45. Song, G. and Jernigan, R. L., vGNM: A better model for understanding the dynamics of proteins in crystals, *J. Mol. Biol.*, *369*, 880 (2007).
46. Ming, D. M. and Wall, M. E., Quantifying allosteric effects in proteins, *Proteins Struct. Function Bioinformatics*, *59*, 697 (2005).
47. Suhre, K. and Sanejouand, Y. H., ElNemo: A normal mode web server for protein movement analysis and the generation of templates for molecular replacement, *Nucleic Acids Res.*, *32*, W610 (2004).
48. Lyman, E., Pfaendtner, J., and Voth, G. A., Systematic multiscale parameterization of heterogeneous elastic network models of proteins, *Biophys. J.*, *95*, 4183 (2008).
49. Hinsen, K. and Kneller, G. R., A simplified force field for describing vibrational protein dynamics over the whole frequency range, *J. Chem. Phys.*, *111*, 10766 (1999).
50. Moritsugu, K. and Smith, J. C., Coarse-grained biomolecular simulation with REACH: Realistic extension algorithm via covariance Hessian, *Biophys. J.*, *93*, 3460 (2007).
51. Granek, R. and Klafter, J., Fractons in proteins: Can they lead to anomalously decaying time autocorrelations?, *Phys. Rev. Lett.*, *95*, 098106 (2005).
52. Lidar, D. A., Thirumalai, D., Elber, R., and Gerber, R. B., Fractal analysis of protein potential energy landscapes, *Phys. Rev. E*, *59*, 2231 (1999).

# 11 Heat Transport in Proteins

*David M. Leitner*

## CONTENTS

## 11.1 INTRODUCTION

Protein function is intimately connected to structure, dynamics, and energy flow, properties that are closely related to each other. Vibrational energy transport, for example, mediates the range of structures proteins maintain for function while serving as an environment for chemical reactions in cells. The anisotropic flow of energy, which is coupled to protein geometry, opens the possibility of energy transport channels between distant sites potentially involved in allostery. Elucidation of networks of channels has become a central goal of computational studies of energy flow in proteins [1–7]. An understanding of thermal flow in proteins is useful in addressing the possibility of conformational change in response to changes in temperature, either due to rapid heating [8] or cooling [9], the latter a particular concern in cryocrystallography.

Two extreme views of vibrational energy and heat flow may be useful in describing these properties in protein molecules. Proteins are large on the molecular scale and so it is tempting to ascribe macroscopic properties to them, for instance a coefficient of thermal conductivity to describe the flow of heat. However, proteins have of course a discrete set of vibrations, and a rather detailed description of energy flow among the vibrational states may be needed to characterize the transport of heat. In this respect, we might expect linear response predictions of heat flow to break down at the level of protein molecules. Because of their relatively large surface to volume ratio, we may furthermore need to consider coupling to the solvent environment [10,11]. Indeed, recent experiments and companion simulations carried out by Hamm, Stock, and their collaborators [12–14], summarized in Chapter 7, indicate that a complete description of thermal flow in proteins requires some combination of all of these factors. Thermal transport depends very much on how energy is introduced to the protein, in a sense on the thermal gradient.

In addition to the desire to elucidate heat flow in proteins and its connection to function, proteins may serve as prototypes for understanding heat flow in other molecular machines and nanoscale objects. With increasing focus on miniaturization of integrated circuits and other machinery, the need to control or dispose of high concentrations of excess heat plays a greater role in design strategies [15–17]. In this respect, there is a growing need to understand in general terms how heat is transported on the nanoscale [18–20]. Recent experimental studies by Dlott, Cahill, and coworkers [21,22] on heat flow in alkane chains that are of the same length scale as proteins indicate that heat flow may still be ballistic for systems of this size, violating the Fourier heat law, which holds on the macroscopic scale. Their results highlight the importance of the discrete vibrations of the molecule in transporting heat, and that thermal transport can depend sensitively on how the system is prepared. Segal, who with Nitzan and Hänggi predicted the trends observed in the alkane chain experiments [20], discusses theoretical methods for the calculation of thermal conduction in molecular chains in Chapter 12. The alkane chains apparently transport heat rather efficiently if there is a large thermal gradient, in part due to their periodic structure. As we shall see below, proteins are typically less adept at transporting heat when there is such a large thermal gradient. The higher frequency modes that are excited if the thermal gradient is large are spatially localized and transport heat inefficiently.

With the lessons of the recent experimental studies on heat flow in peptides and other molecules, we address in this chapter computational techniques for the study of thermal transport from two perspectives. One is the limit of a small thermal gradient, where we calculate thermal transport coefficients in the linear response regime. In this limit, it is useful to start with a harmonic protein and then add anharmonic corrections to the calculation of thermal transport coefficients. In the other limit, a large thermal gradient, proteins are heated by exciting high-frequency modes. In this case, the rate of heat transport may be limited by the transfer of energy from the high-frequency modes to the low-frequency, heat-carrying modes of the protein via anharmonic interactions. The contribution of anharmonicity to thermal transport can depend sensitively on protein rigidity and thus temperature. For example, proteins exhibit a glass-like transition from relatively rigid dynamics to more flexible dynamics [14,23] at a temperature that depends on the solvent [24], and this transition influences rates of energy transfer by anharmonic interactions in the protein [25].

As a means to bridge these limits, we describe in this chapter a method for computing frequency-resolved local energy diffusivities, which when connected together reveal pathways between protein residues where energy transport is particularly facile as a function of the vibrational frequencies of the modes that carry energy between them. A global network of energy transport channels in a protein can then be identified over select frequency ranges, and these networks are seen to vary widely with vibrational frequency.

We turn first to computation of thermal transport coefficients, which provides a description of heat flow in the linear response regime. We compute the coefficient of thermal conductivity, from which we obtain the thermal diffusivity that appears in Fourier's heat law. Starting with the kinetic theory of gases, the main focus of the computation of the thermal conductivity is the frequency-dependent energy diffusion coefficient, or mode diffusivity. In previous work, we computed this quantity by propagating wave packets filtered to contain only vibrational modes around a particular mode frequency [26]. This approach has the advantage that one can place the wave packets in a particular region of interest, for instance the core of the protein to avoid surface effects. Another approach, which we apply in this chapter, is via the heat current operator [27], and this method is detailed in Section 11.2.

The heat current operator can furthermore be broken down to reveal the inter-residue pathways that contribute most to thermal flow, i.e., local energy diffusivities, which taken together form energy transport channels in the protein that give rise to the highly anisotropic flow of energy in proteins [6,7,28]. In Chapter 6, Yamato building on his previous work [29] describes the identification of local energy conductivities based on the heat current operator. We shall provide in Section 11.3 a complementary approach, where we break down the heat current operator further in terms of vibrational frequencies. We then obtain the contribution of mode frequency to the network of energy transport channels. If an excitation involves a high-frequency mode, such as an amide vibration, we can spot the transport channels at this frequency, which are generally very different transport channels than if skeletal modes of, for example, $200\,\mathrm{cm}^{-1}$ are excited. The calculation of frequency-resolved local energy diffusivity is presented in Section 11.3.

Finally, bottlenecks to energy flow in proteins arise also during the transfer of energy from relatively high-frequency vibrations, to which the proteins are often excited in chemical reactions or spectroscopic experiments, down to the lower-frequency, heat-carrying vibrational modes of the protein [30]. We shall address the computation of energy transfer via anharmonic interactions in Section 11.4. We conclude with remarks on some remaining challenges and areas of future study in the final section.

## 11.2  SMALL THERMAL GRADIENT: THE HEAT CURRENT OPERATOR AND THERMAL TRANSPORT COEFFICIENTS

The coefficient of thermal conductivity, which relates the net energy flux to the thermal gradient, is given by

$$\kappa = \sum_{\alpha} C_{\alpha} D_{\alpha}, \tag{11.1}$$

$$C_{\alpha} = \frac{\hbar^2 \omega_{\alpha}^2}{V k_{\mathrm{B}} T^2} \frac{\exp(\hbar\omega_{\alpha}/k_{\mathrm{B}}T)}{[\exp(\hbar\omega_{\alpha}/k_{\mathrm{B}}T)-1]^2}. \tag{11.2}$$

where

the sum is over each mode, $\alpha$, of the system

$C_\alpha$ the heat capacity of mode $\alpha$ with frequency $\omega_\alpha$ per unit volume, which in Equation 11.2 is given in the harmonic approximation

$T$ is the temperature

$k_B$ is Boltzmann's constant

$V$ is the volume

$D_\alpha$ the frequency-dependent energy diffusion coefficient, or mode diffusivity

The relaxation of a thermal gradient is described by the thermal diffusivity, $D_T$, which is just the thermal conductivity over the heat capacity, $C$, or

$$D_T = \kappa/C. \qquad (11.3)$$

The mode diffusivity, $D_\alpha$, can be calculated by propagation of vibrational wave packets comprised of modes filtered around frequency $\omega_\alpha$ [26,31–33]. We have thus far carried out the calculation of the mode diffusivity for protein molecules this way to be able to focus on energy transport in the interior of the protein only, as detailed elsewhere [26], thereby avoiding surface effects. Alternatively, $D_\alpha$ can be expressed in terms of the heat current operator, $S$, in the harmonic approximation [27,34], and it is this approach that we focus on here. No distinction between thermal transport on the surface and in the interior of the protein is made. The mode diffusivity, calculated in terms of the heat current operator, can be further broken down into local contributions in the protein to yield the frequency-resolved local energy diffusivity, detailed in Section 11.3.

### 11.2.1 MODE DIFFUSIVITY IN THE HARMONIC APPROXIMATION

We calculate the mode diffusivity in terms of the heat current operator as follows [27,34]: The local energy density, $h(x)$, is obtained by summing over all atoms, $l$, in a region, $A$, so that

$$h(x) = \sum_{l \in A} h_l. \qquad (11.4)$$

The condition of local energy conservation is

$$\frac{\partial h(x)}{\partial t} + \nabla \cdot S(x) = 0. \qquad (11.5)$$

The total heat current operator is

$$S = \frac{1}{V} \int d^3x\, S(x). \qquad (11.6)$$

In the harmonic approximation, the heat current operator can be written in second quantized form as

$$S = \sum_{\alpha,\beta} S_{\alpha\beta} a_\alpha^\dagger a_\beta, \tag{11.7}$$

where $\alpha$ and $\beta$ are two modes of the protein. The coefficient $S_{\alpha\beta}$ can be expressed in terms of the Hessian matrix, H, and eigenmodes, $\mathbf{e}$, of the protein as [27]

$$S_{\alpha\beta} = \frac{i\hbar(\omega_\alpha + \omega_\beta)}{4V\sqrt{\omega_\alpha\omega_\beta}} \sum_{r,r'\in(x,y,z)} \sum_{l,l'\in AA'} e_l^\alpha H_{rr'}^{ll'}(\mathbf{R}_l - \mathbf{R}_{l'}) e_{l'}^\beta, \tag{11.8}$$

where $\mathbf{R}_l$ is the position of atom $l$ and $r$ is a coordinate ($x$, $y$, or $z$). The mode diffusivity is then given by [27,34]

$$D_\alpha = \frac{\pi V^2}{3\hbar^2\omega^2} \sum_{\beta\neq\alpha} |S_{\alpha\beta}|^2 \delta(\omega_\alpha - \omega_\beta). \tag{11.9}$$

For a practical calculation on a finite-size system, we substitute a Lorentzian of width $\eta$ for the delta function, which should be large enough to envelop at least a few vibrational modes.

Equation 11.9 can be computed in terms of the Hessian matrix, the eigenvalues, and eigenvectors of the protein or protein–solvent system, from which thermal transport coefficients can be directly computed with Equations 11.1 through 11.3. We shall illustrate this calculation with the protein myoglobin, an oxygen storage protein with 153 amino acids. In Figure 11.1 we plot the mode diffusivity we calculate for myoglobin for all vibrational modes to 1850 cm$^{-1}$, which includes the main band of vibrations and excludes only the CH, NH, and OH stretches above $\approx$3000 cm$^{-1}$. As with all computational examples in this chapter, we have adopted the force field parameters of the program MOIL developed by Elber and coworkers [35]. Also plotted in Figure 11.1 is the density of vibrational modes. The mode diffusivity has been calculated with Equation 11.9 using $\eta = 15$ cm$^{-1}$, which is large enough to include many modes in the averaging; results for $D_\alpha$ did not change significantly with larger $\eta$. The mode diffusivity is largest at low frequency and decreases to values below 10 Å$^2$ ps$^{-1}$ for frequencies above about 150 cm$^{-1}$. At this and higher frequencies, the vibrational modes are localized, as discussed further below. Because the vibrational modes of myoglobin above about 150 cm$^{-1}$ are exponentially localized in space, the mode diffusivity corresponding to these modes would be 0 if myoglobin were an infinite system [27]. We observe in Figure 11.1 that at these higher frequencies variation of the mode diffusivity apparently parallels the variation of the vibrational density of states, in contrast to the variation at low frequency.

Thermal transport coefficients can be computed in the harmonic approximation with the mode diffusivities plotted in Figure 11.1 using Equations 11.1 through 11.3.

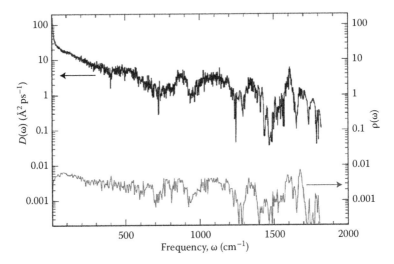

**FIGURE 11.1**  Mode diffusivity (black) and vibrational mode density (gray) computed for myoglobin. At low frequency, to about $100\text{--}150\,\text{cm}^{-1}$, where the vibrational modes are delocalized, the mode diffusivity is relatively large and becomes smaller with increasing frequency, in contrast to the vibrational mode density. Trends in the two quantities parallel one another at higher frequency, where the vibrational modes are localized.

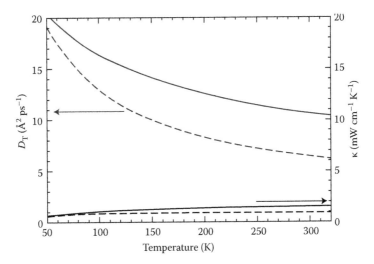

**FIGURE 11.2**  Thermal conductivity (below) and thermal diffusivity (above) computed for myoglobin as a function of temperature. Dashed curves indicate values obtained in the harmonic approximation, whereas solid curves correspond to calculations where anharmonicity is accounted for in the calculation of the thermal transport coefficient.

These are plotted for myoglobin in Figure 11.2, where we have used for the volumetric heat capacity the volume $2.1 \times 10^4\,\text{Å}^3$ used in previous calculations [26,30]. The results are similar to but about 20%–30% smaller than the thermal transport coefficients we computed in the harmonic approximation previously, where the mode diffusivity was obtained by propagating filtered wave packets in the interior of the protein [26]. Apparently the mode diffusivity for myoglobin computed without distinguishing between the surface of the protein and its interior, i.e., the calculation carried out here, is somewhat smaller than the value obtained by attempting to exclude transport on the surface in the previous calculation. We note that for this illustrative calculation, only one structure of myoglobin along a 100 ps simulation at 300 K was used, the same structure used in the previous calculation. For that calculation 10 different atoms were chosen as centers of the wave packets that were propagated in the normal mode representation to obtain the average mode diffusivity, and we observed that this average did not change much if we carried out the same procedure on alternative structures along the simulation. While we do not expect much variation, it would be useful to determine the sensitivity of the mode diffusivity plotted in Figure 11.1 to changes in structure along a molecular dynamics (MD) simulation.

Most vibrational modes of proteins are spatially localized [36,37], which as noted has a significant effect on the mode diffusivity. In the thermodynamic limit, the mode diffusivity vanishes when the eigenmodes are localized [27], and it is relatively small for a finite system such as a protein. To illustrate the extent to which the eigenmodes of a protein are localized, we plot the localization length for myoglobin in Figure 11.3. The localization length, $\xi$, which quantifies the exponential decay of the vibrational amplitude of atoms with increasing distance from the atom exhibiting the largest displacement in any mode, is obtained by fitting the eigenmodes to an exponential. The results in Figure 11.3 have been averaged over vibrational modes in 20 cm$^{-1}$ intervals. Also plotted in Figure 11.3 is the rate of vibrational energy transfer

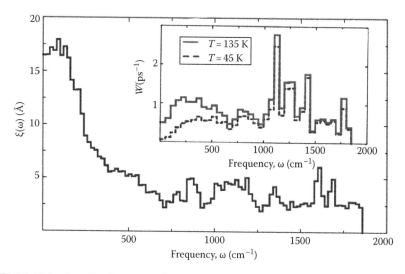

**FIGURE 11.3** Localization length, $\xi$, as a function of mode frequency for myoglobin. The results are averaged over 20 cm$^{-1}$ intervals. Inset shows the rate of vibrational energy transfer rate, $W$, computed with Equation 11.14 and averaged over 50 cm$^{-1}$ intervals for myoglobin at 45 K (dashed) and 135 K (solid).

due to anharmonic interactions, which we use in the next section to correct our estimates to the thermal transport coefficients in the harmonic limit.

## 11.2.2   ENHANCEMENT OF THERMAL TRANSPORT BY ANHARMONICITY

We have introduced the effect of anharmonic contributions to the mode diffusivity, and thus thermal transport, by including a term, $\kappa_{anh}(T)$, that arises from transfer of energy from anharmonic coupling [26,31],

$$\kappa(T) = \kappa_h(T) + \kappa_{anh}(T). \tag{11.10}$$

The harmonic contribution to the thermal conductivity is $\kappa_h(T)$, which we calculated in the previous section. The main purpose of $\kappa_{anh}(T)$ is to allow for localized modes, which in a protein such as myoglobin correspond to vibrational modes above 150 cm$^{-1}$, to contribute to thermal flow via anharmonic coupling to other vibrational modes. The contribution to the diffusion coefficient due to anharmonic coupling can be thought of as $D_\alpha \approx (1/\tau_\alpha)R^2$, where $1/\tau_\alpha$ is the transfer rate between modes due to anharmonic coupling and $R$ is the mean transfer distance for localized modes with frequency near $\omega_\alpha$. Similar corrections arising from anharmonic coupling were introduced earlier to estimate thermal conduction on percolation clusters, where there are also many localized vibrational modes [38].

We calculate the coefficient of thermal conductivity in the harmonic limit as

$$\kappa_h(T) \approx \sum_{\alpha=\text{deloc. mode}} C_\alpha D_\alpha \approx \int_0^{\omega_l} d\omega\, n(\omega)\, C(\omega)\, D_h(\omega), \tag{11.11}$$

where for myoglobin we use $\omega_l = 150$ cm$^{-1}$ as the upper limit, above which the normal modes are localized, as seen in Figure 11.2. Above this frequency the localization length becomes noticeably smaller than the protein radius with increasing frequency. In practice, making modest adjustments to $\omega_l$ has only a small effect on the thermal conductivity when anharmonicity is accounted for. Though the normal modes above 150 cm$^{-1}$ are localized, anharmonicity gives rise to finite lifetimes, which we have computed for myoglobin to be on the order of 1–10 ps for these modes, as illustrated in Figure 11.3 and discussed below. Energy localized to a normal mode is transferred to other modes on this timescale. To estimate the role of anharmonicity in thermal diffusion, we have adopted the following hopping model, which builds on earlier work on energy transport in amorphous systems [30,31,38].

If, near a given frequency, $\omega$, the localization length has a typical value $\xi(\omega)$, then energy can diffuse this distance before a transition to other modes takes place. If the anharmonic transition rate, $W(\omega)$, is sufficiently slow, then $\xi(\omega)$ is effectively the distance over which energy spreads in a time $W^{-1}(\omega)$. Energy that flows into a mode of similar frequency travels about the same length and spends about the same amount of time, $W^{-1}(\omega)$, in that mode. Since the modes that couple most strongly by cubic anharmonicity to the one at frequency $\omega$ have frequencies typically in the range $\approx\omega/3$ to $2\omega/3$, the localization lengths of such coupled modes are of the same order [30]. We take as a rough approximation to the mean free path the localization length, $\xi(\omega)$. The diffusion coefficient due to anharmonic transitions is then [26,31]

$$D_a(\omega) \approx \frac{1}{3}\xi^2(\omega)W(\omega),$$ (11.12)

where the subscript "a" indicates energy transfer due to anharmonic coupling of normal modes. The vibrational energy transfer rate, $W(\omega)$, is given by Equation 11.15. If, however, $W^{-1}(\omega)$ is sufficiently short, then a vibrational excitation will not have spread as far as $\xi(\omega)$ before a transition to other modes takes place. When anharmonic decay is rapid, transitions occur before the effects of localization influence diffusion of energy. In this case, $D(\omega)$ is given by $D_h(\omega)$.

We can estimate which diffusion coefficient to use in the regime of localized normal modes, $D_a(\omega)$ or $D_h(\omega)$, for $\omega > \omega_l \approx 150\,\text{cm}^{-1}$, by calculating the time for a vibrational excitation to diffuse a distance $\xi(\omega)$. This time, $t^*$, can be estimated as [26]

$$t^*(\omega) = \frac{1}{3}\xi^2(\omega)D_h^{-1}(\omega),$$ (11.13)

where for $D_h(\omega)$, we use Equation 11.9. Then for $\omega > 150\,\text{cm}^{-1}$, where normal modes of myoglobin are localized, we take the energy diffusion coefficient to be

$$D(\omega) = D_h(\omega), \quad \text{if } W^{-1}(\omega) \le t^*(\omega),$$ (11.14a)

$$D(\omega) = D_a(\omega), \quad \text{if } W^{-1}(\omega) > t^*(\omega),$$ (11.14b)

where $t^*(\omega)$ is calculated using Equation 11.13. Equation 11.14a represents the case where rapid anharmonic decay allows energy diffusion to occur without the restriction of localization, since mode lifetimes are shorter than the time it would take for energy to diffuse the length of the normal mode.

Calculation of $D_{anh}$ requires calculation of the rate of vibrational energy transfer due to anharmonic interactions, $W$. Computing $W$ in terms of the lowest-order anharmonic interactions has been carried out quantum mechanically by first-order perturbation theory for solid-state systems for some time. The energy transfer rate from mode $\alpha$, $W_\alpha$, can be written as the sum of terms that can be described as decay and collision, the former typically much larger except at low frequency where both terms are comparable. The anharmonic decay rate of vibrational mode $\alpha$ is then the sum of these two terms [39]:

$$W_\alpha^{\text{decay}} = \frac{\hbar\pi}{8\omega_\alpha}\sum_{\beta,\gamma}\frac{|\Phi_{\alpha\beta\gamma}|^2}{\omega_\beta\omega_\gamma}(1+n_\beta+n_\gamma)\delta(\omega_\alpha-\omega_\beta-\omega_\gamma),$$ (11.15a)

$$W_\alpha^{\text{coll}} = \frac{\hbar\pi}{4\omega_\alpha} \sum_{\beta,\gamma} \frac{|\Phi_{\alpha\beta\gamma}|^2}{\omega_\beta\omega_\gamma} (n_\beta - n_\gamma) \delta(\omega_\alpha + \omega_\beta - \omega_\gamma), \qquad (11.15b)$$

where $n_\alpha$ is the occupation number of mode $\alpha$, which at temperature $T$ we take to be $n_\alpha = (\exp(\hbar\omega_\alpha/k_B T) - 1)^{-1}$. The matrix elements $\Phi_{\alpha\beta\gamma}$ appear as the coefficients of the cubic terms in the expansion of the interatomic potential in normal coordinates.

In Figure 11.2, we plot $\kappa$ for myoglobin with both harmonic and anharmonic contributions. In the extended regime, i.e., vibrational frequencies of $\omega < 150\,\text{cm}^{-1}$, $D_\alpha$ appearing in Equation 11.1 is given by Equation 11.9. For $\omega > 150\,\text{cm}^{-1}$, we calculate $D_\alpha$ using the criterion of Equation 11.14. The localization lengths that enter into the calculation of $t^*$ in Equation 11.13 are those computed for myoglobin and plotted in Figure 11.3. The value of the thermal conductivity at 300 K, 1.6 mW cm$^{-1}$ K$^{-1}$, is almost double the value in the harmonic limit, highlighting the importance of anharmonicity toward enhancement of heat flow in proteins. The thermal conductivity that we have calculated for myoglobin is well below the thermal conductivity of $H_2O$ at 300 K, which is 6.1 mW cm$^{-1}$ K$^{-1}$ [32].

The thermal diffusivity, $D_T$, of myoglobin can be calculated in terms of $\kappa$ with Equation 11.3 and is also plotted in Figure 11.2. At 300 K we find the thermal diffusivity of myoglobin to be 11 Å$^2$ ps$^{-1}$, which can be compared with the value of 14 Å$^2$ ps$^{-1}$ for water [32,40]. We note also that the anharmonic contribution makes up almost half of the thermal diffusivity at 300 K. Our result is comparable to the value of the thermal diffusivity, 7 Å$^2$ ps$^{-1}$, estimated from a simulated cooling of the photosynthetic reaction center of *Rhodospseudomonas viridis* between 200 and 300 K by Tesch and Schulten [41].

Since the values of the thermal diffusivity for myoglobin and water are close, the sizable discrepancy between their thermal conductivities is largely due to differences in their respective heat capacities. At 300 K, the specific heat of water is 4.2 J K$^{-1}$ g$^{-1}$, whereas we calculate the specific heat of myoglobin to be 1.0 J K$^{-1}$ g$^{-1}$ at 300 K. This latter is reasonably close to measured specific heats for proteins. For example, the specific heat of lysozyme in dilute aqueous solution is 1.5 J K$^{-1}$ g$^{-1}$, and 1.3 J K$^{-1}$ g$^{-1}$ for the dry protein [42].

## 11.3  LOCAL ENERGY DIFFUSIVITIES AND ENERGY TRANSPORT CHANNELS

The previous section describes the computation of effective values of thermal transport coefficients for proteins. However, energy flows anisotropically in proteins, a consequence of protein geometry [28,43–45]. In this section, we discuss an approach to calculate local energy diffusivities in a protein as a function of vibrational frequency. Aggregates of regions of large diffusivities give rise to energy transport channels, which can depend sensitively on vibrational frequency. The wide range of local diffusivities manifest themselves in the anomalous subdiffusion of vibrational energy seen in simulations of energy flow in proteins [46]. Interestingly, the spread of relatively low-frequency vibrational energy (to roughly 100 cm$^{-1}$) with time in a protein mimics the energy flow on objects with a fractal dimension [46–48], as anticipated long ago by low-temperature spectroscopic studies of proteins [49], which appear to be well described by the vibrations of fractal objects [43,50]. In fact, the dimension of many globular proteins is quite close to a percolation cluster in 3-D [46–48,51], a property that appears to give rise to a network of channels in the protein along which energy flow is facile.

Specific channels for vibrational energy flow, and thus heat transport, have been elucidated computationally for heme proteins by Straub and coworkers [52–54] and Nagaoka and coworkers [6], and is discussed by them elsewhere in this book. More recently, Ishikura and Yamato have computed local conductivities in photoactive yellow protein [29], building on a procedure by Rabitz and coworkers [55] that focuses on the contributions of energy transport channels to the heat current operator. Continuing along this direction, our aim now is to explore the frequency dependence of these channels. We observe, for example, communication between the heme and specific residues of myoglobin to be relatively rapid for many vibrations below $300\,\mathrm{cm^{-1}}$, but substantially slower at higher frequency.

Local contributions to the mode diffusivity, $D_\alpha$, can be determined by the local contributions to the heat current operator, $\mathbf{S}_{\alpha\beta}$. We consider only those terms in the sum of Equation 11.8 for a particular residue pair $A$ and $A'$. By limiting the contribution to the heat current operator to only residues $A$ and $A'$, we compute $|\mathbf{S}_{\alpha\beta}|^2$, and thus $D_\alpha$, with Equation 11.9 to obtain the mode diffusivity within a local region spanned by $A$ and $A'$, i.e., a measure of the local diffusivity either within $A$, if $A' = A$, or between $A$ and $A'$, if $A \neq A'$. By connecting regions of large local diffusivities in the protein, we identify frequency-resolved channels along which energy transport is facile. We note that while the frequency-resolved local energy diffusivity is dimensionally a diffusion coefficient, vibrational dynamics on this local scale is generally not diffusive as there is no well established mean free path. Within residues, or between strongly communicating residues, the values that we obtain at frequencies below about $300\,\mathrm{cm^{-1}}$ can be well above $100\,\mathrm{\mathring{A}^2\,ps^{-1}}$, particularly at lower frequency.

In Figure 11.4, we plot the frequency-resolved local energy diffusivity at 50 and $300\,\mathrm{cm^{-1}}$. For the range of frequency displayed in Figure 11.4 the overall pattern of the regions where energy flow is rapid that is observed in this figure does not change dramatically in going from mode to mode around any particular frequency. There may be significant variation in pairs of residues between which energy transport is rapid in going from mode to mode, but from a coarser-grain perspective, the structural elements that are seen to communicate with each other do not vary much.

Below $100\,\mathrm{cm^{-1}}$, one finds for many modes, and illustrated by the $50\,\mathrm{cm^{-1}}$ mode in Figure 11.4, strong communication between the B and E helices, somewhat less so between A and E, and significant energy flow between the protein segment made up of helices A through C and the segment made up of G and H helices. There is also rapid energy flow between the G and H helices, and between the F and H helices. If we consider the heme and its communication with other residues of the protein, fastest transport is mainly to residues of helices C, E, and F. We also find that rapid energy flow between the heme and other parts of the protein occurs mainly at frequencies below $300\,\mathrm{cm^{-1}}$. At $300\,\mathrm{cm^{-1}}$, we still see significant energy flow between the heme and residues of helices C, E, and F, also G, but inspection

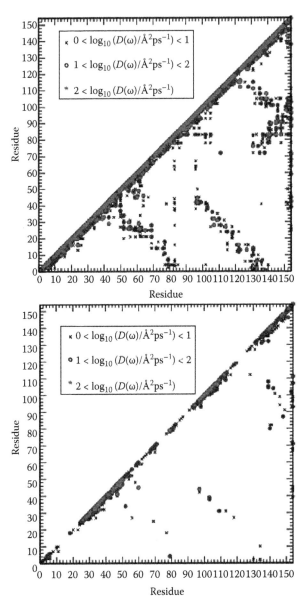

**FIGURE 11.4** Frequency-resolved local energy diffusivity within and between residues of myoglobin, where residue 154 is plotted as the heme. An average over 50 modes near 50 and 300 cm$^{-1}$ are plotted at the top and bottom, respectively. Gray stars indicate channels within or between residues where transport is fastest. At frequencies above about 300 cm$^{-1}$ communication between the heme and other parts of the protein is relatively slow.

of the local energy diffusivities for modes above $300\,cm^{-1}$ reveals that energy flow from the heme to other residues is carried slowly by these higher-frequency modes. For energy in higher-frequency vibrations of the heme to flow into the protein, it first decays anharmonically into modes of frequency below $\approx300\,cm^{-1}$, typically on the 1 to 10 ps timescale. Anharmonic decay of vibrational modes of proteins is discussed in the next section.

In Figure 11.5, we plot the local energy diffusivities between residues of myoglobin for two vibrations of the amide I band. The vibrations in this band reveal a wide range of signaling patterns, as illustrated by the two modes plotted in the figure. This trend of strong sensitivity of the local energy diffusivity to particular vibrational modes in a narrow range of mode frequency is representative of what one typically observes in going from mode to mode in a relatively small range of frequency for modes above about $400\,cm^{-1}$. In this case, one finds a wide variety of relatively local patterns of energy flow, though the amide I vibrations can in some cases span many residues, as illustrated by the vibrational mode at $1610\,cm^{-1}$. For this mode, which exhibits a relatively large mode diffusivity of $8\,\text{Å}^2\,ps^{-1}$, one finds facile energy transport between residues of the B and E helices, and in particular along the E helix, where an energy transport channel is observed close to the diagonal for residues along this helix. In contrast, for the mode at $1603\,cm^{-1}$, for which the mode diffusivity is a relatively small $2\,\text{Å}^2\,ps^{-1}$, we observe more limited energy transport, in this case along portions of the C and D helices and the F helix, with some communication between residues of the C and F helices.

## 11.4  LARGE THERMAL GRADIENT: BOTTLENECKS TO ENERGY TRANSFER FROM HIGH-FREQUENCY TO LOW-FREQUENCY VIBRATIONS

If vibrational states of a protein are expressed in terms of the populations of vibrational modes of the molecule, transitions among vibrational states arise from anharmonic interactions that give rise to Fermi resonances [56]. The role of Fermi resonances in energy flow in polyatomic molecules [56–58], including proteins [59,60], has been studied for some time, and we briefly summarize general features of vibrational energy flow among vibrational states of polyatomic molecules before turning specifically to proteins.

One relevant general feature of vibrational energy flow among vibrational states of a molecule is its local nature, a result of the predominantly low-order anharmonic coupling involved in Fermi resonances responsible for energy transfer [59–62]. It is often convenient to organize the vibrational states of a molecule into tiers, where each state couples to a relatively small subset of vibrational states by low-order Fermi resonance terms, and these in turn are coupled to another subset of states, and so on [57,63,64]. This local connectivity of states leads to interesting effects, in particular a localization transition determined by the local density of resonantly coupled states [57]. The localization transition in fact can depend sensitively on relatively high-order Fermi resonances in large molecules and is well below thermal energies in organic molecules as large as proteins [65].

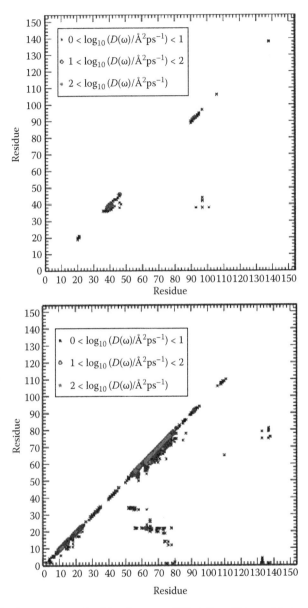

**FIGURE 11.5** Frequency-resolved local energy diffusivity within and between residues of myoglobin, where residue 154 is plotted as the heme. Two modes in the amide I region of the spectrum near 1600 cm$^{-1}$ are plotted, specifically at 1603 and 1610 cm$^{-1}$ at the top and bottom, respectively. Gray stars indicate channels within or between residues where transport is fastest. At these frequencies and essentially all above about 300 cm$^{-1}$ communication between the heme and other parts of the protein is relatively slow.

The rate of energy transfer in large molecules from a vibrational state of one tier to the next is largely determined by the low-order resonances, in contrast to the greater role played by high-order resonances in the localization transition [65,66]. To calculate the rate of energy transfer in a protein molecule, it is therefore reasonable to start with the lowest-order anharmonic interactions, i.e., the cubic terms. While incorporating only cubic anharmonicity is a significant approximation, albeit a practical one, it is particularly appropriate at low temperatures, and may well be appropriate for calculating the lifetimes of almost all vibrational modes up to at least ~200 K, which corresponds to a temperature below which the protein dynamics is relatively rigid [23,24]. At higher temperatures, the protein appears more flexible and anharmonic, and calculation of the lifetime using only cubic anharmonicity is a relatively crude approximation, particularly for the low-frequency modes corresponding to flexible, collective protein motions. Calculation of the vibrational mode lifetime in terms of the lowest-order anharmonic interactions has been carried out quantum mechanically by first-order perturbation theory for solid-state systems and glasses [67] for some time, and is given by Equation 11.15. Fujisaki and Straub have developed and applied a generalization that accounts for non-Markovian effects in vibrational energy transfer [68], which is discussed by Zhang and Straub in Chapter 9.

In Figure 11.3, we show the results of a calculation of the vibrational energy transfer rate for myoglobin as a function of frequency at temperatures 45 and 135 K. The rate of energy transfer has been computed for each mode of the protein to 2000 cm$^{-1}$, and the results plotted are average rates in 50 cm$^{-1}$ intervals. The rates plotted in Figure 11.3 were computed for myoglobin without water; computation of energy transfer rates of protein vibrations when water is present reduces the lifetime for most modes to frequencies of about 1000 cm$^{-1}$, but has only a small influence on the lifetime at higher frequency, as discussed in Refs. [28,69]. For example, there is almost no effect on the average rate of energy transfer from the well-studied amide I band, between 1600 and 1700 cm$^{-1}$, as seen in Figure 11.6, where here the energy transfer rate in solvated myoglobin is plotted. The rate of energy transfer from individual modes can be affected by solvent, which can tune resonances for energy transfer in the protein, but the average rate within the band changes very little with solvation. The lifetime of the amide I band of myoglobin is observed experimentally to be about 2 ps [25], whereas the time for energy transfer to the solvation layer is about 20 ps [70]. As shown in Figure 11.6, we similarly find a rate of ≈0.05 ps$^{-1}$ for energy transfer from the amide I band to water compared to a rate of ≈0.5 ps$^{-1}$ overall.

A noteworthy trend in the variation of the rate with temperature seen in Figure 11.3 is the general insensitivity of the computed energy transfer rates to temperature for vibrational modes of frequency above about 500 cm$^{-1}$. Such a weak temperature-dependence of the energy transfer rates of high-frequency modes of myoglobin have been observed in time-resolved spectroscopic studies. Pump–probe vibrational spectra of the amide I band measured at temperatures from 6 to 310 K reveal decay rates ranging only from ≈0.5 to 0.8 ps$^{-1}$ [25], where the rate holds at 0.5 ps$^{-1}$ up to the glass-like dynamical transition near 200 K. The computed lifetime of 2.0 ps plotted in Figure 11.3 matches closely that measured for temperatures up to about 200 K. At higher temperatures, the measured lifetime gradually decreases below 1.8 ps,

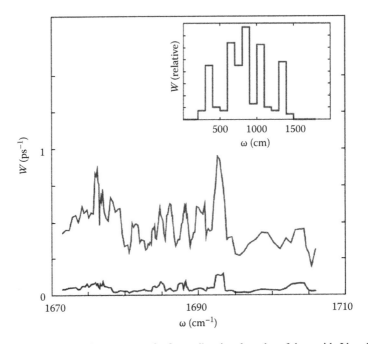

**FIGURE 11.6** The rate of energy transfer from vibrational modes of the amide I band (upper curve) and the rate of energy transfer directly to vibrations corresponding to displacement of surrounding water molecules (lower curve) at 135 K. The rate of the latter is about 1/20 the total rate of energy transfer, so that relatively little energy is directly transferred from the amide I vibration to the solvent by low-order anharmonic coupling. The inset shows the frequencies of the vibrational modes into which energy is transferred from the amide I by cubic anharmonicity.

reaching about 1.2 ps at 310 K [25], a change that likely corresponds to the more "anharmonic" dynamics of the protein at temperatures higher than 200 K. The near lack of temperature dependence of the amide I vibrational lifetime indicates that little energy flows directly into the low-frequency modes; if it did the rate would vary linearly with temperature (c.f. Equation 11.15). Time-resolved studies of other high-frequency vibrations reveal a similar trend. Pump–probe studies on myoglobin–CO indicate the decay of the CO stretch, about $1950 \, cm^{-1}$, is also essentially independent of temperature from about 6 to 310 K [24,71].

Low-frequency modes are abundant in proteins, as illustrated in Figure 11.1. The lack of temperature dependence indicates anisotropic flow of energy through the vibrational states, with energy transfer from higher-frequency modes to modes of intermediate frequency rather than into low-frequency modes. Consider, for example, the amide I band vibrations of myoglobin. In Figure 11.6, we plot the mode frequencies into which energy transfers predominantly from the amide I vibration. In this case, little energy flows directly into the solvent, as noted above. All of the vibrational modes into which energy flows, like the amide I, significantly overlap displacements of the C or O on the peptide bond, and all but the modes of frequency $\approx 300 \, cm^{-1}$ are fairly localized to displacements of atoms on the peptide bond.

In general, the flow of energy from high- to low-frequency modes of proteins is mediated by the anharmonic matrix elements that couple them. Given the high density of low-frequency modes, matrix elements coupling a given high-frequency mode to a pair of other modes must be typically small if the frequency of one of the pair is small to explain the observed temperature dependence. For this reason, we see in Figure 11.3 that the temperature dependence of the energy transfer rate for most localized modes is fairly small, such as the amide I vibrations.

The reason that the low-order anharmonic matrix elements coupling a high-frequency, localized mode and a low-frequency, heat carrying mode is typically small has been discussed in detail elsewhere [28,30]. We shall summarize the main points here. Consider cubic anharmonicity, the lowest-order anharmonicity, which gives rise typically to the largest anharmonic coupling matrix elements. For the matrix element arising from cubic anharmonicity coupling a triple of modes to be appreciable, the three modes must overlap in space. However, as mode frequency increases, the normal mode vibrations generally become more spatially localized, i.e., displacements are relatively large for a relatively small number of atoms in one part of the protein, as illustrated for myoglobin in Figure 11.3. As such, the higher-frequency vibrations of proteins have much in common with those of one-dimensional disordered systems, in contrast to the vibrations at much lower frequency, which, as discussed above, resemble those of a fractal object between two and three dimensions. One important consequence of strong localization of normal modes in one-dimensional disordered systems is that frequencies of normal modes whose localization centers overlap in space are generally very different [72]. This trend gives the appearance of "repulsion" of mode frequencies between pairs of nearby localized modes. Such mode repulsion has important consequences on the temperature dependence of the energy transfer rates in such systems. If energy in a high-frequency localized mode, $\alpha$, decays into a low-frequency mode of the protein, the rest of the energy must decay into a localized mode, $\beta$, whose frequency, $\omega_\beta$, is similar to $\omega_\alpha$. However, because of mode repulsion, higher-frequency modes of similar frequency do not overlap, and energy transfer to a localized mode with similar frequency and the remainder to a low-frequency mode occurs slowly due to the weak anharmonic interaction. The vibrational energy transfer rate from high-frequency modes is thus only weakly dependent on temperature.

The energy transfer rate in myoglobin is plotted in Figures 11.3 and 11.6 at temperatures below 200 K. These temperatures lie below the protein glass-like transition, or dynamical transition temperature, which is discussed by Tobias, Sengupta, and Tarek in Chapter 16. At temperatures below about 200 K, protein dynamics can be characterized as relatively rigid, vibrational dynamics [11,23,24], and we can approximate the dynamics reasonably well by limiting it to a single potential minimum with the inclusion of low-order anharmonicity. Indeed, it is for temperatures below 200 K that one observes essentially no temperature dependence of the amide I lifetime experimentally [25], whereas at higher temperature some dependence is observed, consistent with the expectation of considerably enhanced anharmonicity at these higher temperatures. A more complete description of vibrational energy transfer above the protein glass-like transition thus requires going beyond the lowest-order anharmonic contributions that we have thus far included in our calculations.

Though the lowest-order contributions still yield an energy transfer rate for the amide I vibration that is reasonably close to the observed rate at 300 K, higher-order terms are still needed to more faithfully reproduce trends that may appear at the dynamical transition and higher temperatures. In this respect, we can expect that if heat transport depends on the relaxation of energy from high-frequency modes into the lower-frequency heat-carrying modes of the protein, then the temperature dependence will likely be relatively weak for temperatures below the dynamical transition, but may be more dramatic at higher temperatures.

## 11.5  CONCLUDING REMARKS

There is a large and growing interest in understanding heat flow in molecular-scale machines, and computational tools aiming to address this problem are rapidly developing. One focus of this work has been on protein molecules, which function very efficiently in a relatively narrow temperature range. Recent experimental and computational work addressing energy flow in proteins point to the complexity and specificity of energy transport. Energy transport in proteins is highly anisotropic. While global properties such as thermal transport coefficients serve as useful guides to describing energy flow in the limit of modest thermal gradients, for a wider range of conditions information is needed about specific pathways or network of channels through which energy flows, either through space or through the vibrational states of the molecule.

In Chapter 6, Yamato describes an approach for computing local conductivities. In this chapter, we break down the local conductivities to examine their frequency dependence. This information is particularly useful if there is a nonthermal distribution of energy, for example, if a particular set of vibrational modes of the protein are photoexcited experimentally. Our approach is complementary to the recent frequency-dependent method developed to search for energy transport channels, pump–probe MD [4]. In this method, MD simulations are run while "pumping" an atom with a particular force, frequency, and direction. Other atoms are then "probed" as to their response, yielding frequency-dependent signaling in the protein. Our approach does not require excitation of particular atoms and instead provides a global mapping of the frequency-resolved local energy diffusivities, which, when residues in regions of large local diffusivities are connected together, yield the frequency-dependent energy transport channels.

Networks of energy transport channels depend sensitively on mode frequency, particularly at higher frequency. We have illustrated the calculation of frequency-resolved local energy diffusivities for myoglobin. We find energy transport to be particularly facile between the heme and other parts of the protein for frequencies below 300 cm$^{-1}$. Figure 11.4 reveals the network of residues among which energy flows rapidly for the thermally accessible vibrations. At higher frequency, where the vibrational modes are localized, the network of residues among which energy flows is quite sensitive to the excitation of particular modes, as illustrated in Figure 11.5 for two vibrations of the amide I band. The search for channels between distant sites on a protein molecule, for example, distant binding sites that may be involved in allostery, is currently an important goal in describing energy flow in proteins and connecting it to protein function [3,4,73,74]. The calculation of frequency-resolved local energy diffusivities and the global networks they form in a protein is one approach that can be usefully brought to bear on this problem. While our calculation has thus far been carried out in harmonic approximation, we expect to be able to include anharmonic contributions by adding low-order anharmonic corrections to the heat current operator [75].

We compute effective thermal transport coefficients for proteins using linear response theory and beginning in the harmonic approximation, with anharmonic contributions included as a correction. The correction can in fact be rather large, as we compute anharmonicity to nearly double the magnitude of the thermal conductivity and thermal diffusivity of myoglobin. We expect that anharmonicity will generally enhance thermal transport in proteins, in contrast, for example, to crystals, where anharmonicity leads to thermal resistance, since most of the harmonic modes of the protein are spatially localized and transport heat only inefficiently.

When the thermal gradient is large, energy transport is even more sensitive to anharmonicity, since in this case, many higher-frequency, localized modes of the protein that transport heat only inefficiently are excited initially. The rate of thermal transport may then be limited by the rate of anharmonic decay from the high to low frequency, heat-carrying modes. We have therefore presented a computational approach to calculate the rate of energy transfer via anharmonic interactions in a protein. We have illustrated the approach with the calculation of the energy transfer rate for all vibrational modes of myoglobin in the "main band" of vibrations, up to about $1850\,cm^{-1}$. For this illustrative calculation, we have included the lowest order anharmonic terms, which typically give the largest contribution to the rate of energy transfer and should yield a good approximation to the rate at low temperature, where the protein is relatively rigid. Indeed, upon comparing results of this calculation with measurements of the lifetime of the amide I band, we find that the calculation gives a close estimate of the energy transfer rate for all temperatures up to the protein dynamical transition of about 200 K, and gives a reasonable estimate at higher temperature, though our calculations predict a slower rise in the rate with increasing temperature above 200 K than is observed experimentally. At temperatures above the protein dynamical transition, the protein becomes more flexible and higher-order anharmonic interactions will be needed to more quantitatively estimate the rate of energy transfer.

Inclusion of anharmonic coupling beyond third order will also be required to capture qualitative trends in the variation of the energy transfer rate with temperature. When the protein is relatively rigid, at temperatures below the protein dynamical transition, the rate of energy transfer from high to low frequency, heat-carrying modes does not increase much with temperature. In this case cubic anharmonic coupling contributes mainly to the energy transfer rate, and, as discussed in the previous section, cubic coupling between a pair of high frequency, localized modes and a low-frequency, heat-carrying mode is typically weak. When the protein is more flexible, i.e., at temperatures above the protein dynamical transition, higher-order anharmonic interactions make a greater contribution to energy transfer and the restrictions on the interactions that arise from the mode repulsion between localized modes, described in the previous section, influence less the coupling between high and low frequency, heat-carrying modes. In this case, there can be more facile energy transfer to the heat-carrying modes and a more significant increase in the rate of heat transfer with temperature. A challenge to computational work on energy transfer in proteins is the incorporation of sufficiently high-order anharmonic interactions to capture this effect.

Finally, the variation of protein rigidity with temperature depends on solvent. This is just one way in which the protein environment mediates heat transport in the molecule. The nature of dynamic coupling between protein and its solvent environment and its role in protein function are currently the focus of much attention [10,76]. Zhang and Straub describe recent work [77] on the influence of solvent on energy flow in heme proteins in their chapter. Future computational work will undoubtedly play a major role in further clarifying how and the extent to which solvent modifies energy transport in proteins.

## ACKNOWLEDGMENT

Support for this work by the National Science Foundation (NSF CHE-0506020 and CHE-0910669) is gratefully acknowledged.

## REFERENCES

1. Formaneck MS, Ma L, Cui Q. 2006. Reconciling the "old" and "new" view of protein allostery: A molecular simulation study of chemotaxis Y protein (CheY). *Proteins: Struct. Func. Bioinform.* 63: 846–867.
2. Freire E. 1999. The propagation of binding interactions to remote sites in proteins: Analysis of the binding of the monoclonal antibody D1.3 to lysozyme. *Proc. Natl. Acad. Sci. U. S. A.* 96: 10118–10122.
3. Ota N, Agard DA. 2005. Intramolecular signaling pathways revealed by modeling anisotropic thermal diffusion. *J. Mol. Biol.* 351: 345–354.
4. Sharp K, Skinner JJ. 2006. Pump–probe molecular dynamics as a tool for studying protein motion and long range coupling. *Proteins: Struct. Func. Bioinform.* 65: 347–361.
5. Liu T, Whitten ST, Hilser VJ. 2006. Ensemble-based signatures of energy propagation in proteins: A new view of an old phenomenon. *Proteins Struct. Func. Bioinform.* 62: 728–738.
6. Takayanagi M, Okumura H, Nagaoka M. 2007. Anisotropic structural relaxation and its correlation with the excess energy diffusion in the incipient process of photodissociated MbCO: High-resolution analysis via ensemble perturbation method. *J. Phys. Chem. B* 111: 864–869.
7. Sagnella DE, Straub JE. 2001. Directed energy "funneling" mechanism for heme cooling following ligand photolysis or direct excitation in solvated carbonmonoxy myoglobin. *J. Phys. Chem. B* 105: 7057–7063.
8. Lampa-Pastirk S, Beck WF. 2006. Intramolecular vibrational preparation of unfolding transition state of ZnII-substituted cytochrome c. *J. Phys. Chem. B* 110: 22971–22974.
9. Halle B. 2004. Biomolecular cryocrystallography: Structural changes during flash cooling. *Proc. Natl. Acad. Sci. U. S. A.* 101: 4793–4798.
10. Leitner DM, Havenith M, Gruebele M. 2006. Biomolecule large amplitude motion and solvation dynamics: Modeling and probes from THz to X-rays. *Int. Rev. Phys. Chem.* 25: 553–582.

11. Tarek M, Tobais DJ. 2002. Role of protein–water hydrogen bond dynamics in the protein dynamical transition. *Phys. Rev. Lett.* 88: 138101, 1–4.

12. Botan V, Backus EHG, Pfister R, Moretto A, Crisma M, Toniolo C, Nguyen PH, Stock G, Hamm P. 2007. Energy transport in peptide helices. *Proc. Natl. Acad. Sci. U. S. A.* 104: 12749–12754.

13. Backus EHG, Nguyen PH, Botan V, Pfister R, Moretto A, Crisma M, Toniolo C, Stock G, Hamm P. 2008. Energy transport in peptide helices: A comparison between high- and low-energy excitations. *J. Phys. Chem. B* 112: 9091–9099.

14. Backus EHG, Nguyen PH, Botan V, Moretto A, Crisma M, Toniolo C, Zerbe O, Stock G, Hamm P. 2008. Structural flexibility of a helical peptide regulates vibrational energy transport properties. *J. Phys. Chem. B* 112: 15487–15492.

15. Cahill DG, Goodson K, Majumdar A. 2002. Thermometry and thermal transport in micro/nanoscale solid-state devices. *J. Heat Transfer* 124: 223–241.

16. Cahill DG, Ford WK, Goodson KE, Mahan GD, Majumdar A, Maris HJ, Merlin R, Phillpot SR. 2003. Nanoscale thermal transport. *J. Appl. Phys.* 93: 793–818.

17. Galperin M, Nitzan A, Ratner MA. 2007. Heat conduction in molecular transport junctions. *Phys. Rev. B* 75: 155312-1–15.

18. Buldum A, Leitner DM, Ciraci S. 1999. Thermal conduction through a molecule. *Europhys. Lett.* 47: 208–212.

19. Leitner DM, Wolynes PG. 2000. Heat flow through an insulating nanocrystal. *Phys. Rev. E* 61: 2902–2908.

20. Segal D, Nitzan A, Hänggi P. 2003. Thermal conductance through molecular wires. *J. Chem. Phys.* 119: 6840–6855.

21. Wang Z, Carter JA, Lagutchev A, Koh YK, Seong N-H, Cahill DG, Dlott DD. 2007. Ultrafast flash thermal conductance of molecular chains. *Science* 317: 787–790.

22. Wang Z, Cahill DG, Carter JA, Koh YK, Lagutchev A, Seong N-H, Dlott DD. 2008. Ultrafast dynamics of heat flow across molecules. *Chem. Phys.* 350: 31–44.

23. Frauenfelder H, Sligar SG, Wolynes PG. 1991. The energy landscapes and motions of proteins. *Science* 254: 1598–1603.

24. Fayer MD. 2001. Fast protein dynamics probed with infrared vibrational echo experiments. *Annu. Rev. Phys. Chem.* 52: 315–356.

25. Peterson KA, Rella CW, Engholm JR, Schwettman HA. 1999. Ultrafast vibrational dynamics of the myoglobin amide I band. *J. Phys. Chem. B* 103: 557–561.

26. Yu X, Leitner DM. 2005. Heat flow in proteins: Computation of thermal transport coefficients. *J. Chem. Phys.* 122: 054902-1–11.

27. Allen PB, Feldman JL. 1993. Thermal conductivity of disordered harmonic solids. *Phys. Rev. B* 48: 12581–12588.

28. Leitner DM. 2008. Energy flow in proteins. *Annu. Rev. Phys. Chem.* 59: 233–259.

29. Ishikura T, Yamato T. 2006. Energy transfer pathways relevant for long-range intramolecular signaling of photosensory protein revealed by microscopic energy conductivity analysis. *Chem. Phys. Lett.* 432: 533–537.

30. Yu X, Leitner DM. 2003. Vibrational energy transfer and heat conduction in a protein. *J. Phys. Chem. B* 107: 1698–1707.

31. Leitner DM. 2005. Heat transport in molecules and reaction kinetics: The role of quantum energy flow and localization. *Adv. Chem. Phys.* 130B: 205–256.

32. Yu X, Leitner DM. 2005. Thermal transport coefficients for liquid and glassy water computer from a harmonic aqueous glass. *J. Chem. Phys.* 123: 104503-1–10.

33. Yu X, Leitner DM. 2006. Thermal conductivity computed for vitreous silica and methyl-doped silica above the plateau. *Phys. Rev. B* 74: 184305-1–11.

34. Feldman JL, Kluge MD, Allen PB, Wooten F. 1993. Thermal conductivity and localization in glasses: Numerical study of a model amorphous silicon. *Phys. Rev. B* 48: 12589–12602.

35. Elber R, Roitberg A, Simmerling C, Goldstein R, Li H, Verkhivker G, Kaesar C, Zhang J, Ulitsky A. 1995. MOIL: A program for simulations of macromolecules. *Comp. Phys. Comm.* 91: 159–189.
36. Brooks CL, Karplus M, Pettitt BM. 1988. Proteins: A theoretical perspective of dynamics, structure and thermodynamics. *Adv. Chem. Phys.* 71: 1–150.
37. Nishikawa T, Go N. 1987. Normal modes of vibration in bovine pancreatic trypsin inhibitor and its mechanical property. *Proteins: Struct. Func. Genetics* 2: 308–329.
38. Nakayama T, Kousuke Y, Orbach RL. 1994. Dynamical properties of fractal networks: Scaling, numerical simulations, and physical realizations. *Rev. Mod. Phys.* 66: 381–443.
39. Maradudin AA, Fein AE. 1962. Scattering of neutrons by an anharmonic crystal. 128: 2589–2608.
40. Yu X, Leitner DM. 2004. Thermal transport in liquid and glassy water computed with normal modes. *Chem. Phys. Lett.* 398: 480–485.
41. Tesch M, Schulten K. 1990. A simulated cooling process for proteins. *Chem. Phys. Lett.* 169: 97–102.
42. Yang P-H, Rupley JA. 1979. Protein–water interactions. Heat capacity of the lysozyme–water system. *Biochemistry* 18: 2654–2661.
43. Elber R, Karplus M. 1986. Low-frequency modes in proteins: Use of the effective-medium approximation to interpret the fractal dimension observed in electron-spin relaxation measurements. *Phys. Rev. Lett.* 56: 394–397.
44. Granek R, Klafter J. 2005. Fractons in proteins: Can they lead to anomalously decaying time autocorrelations? *Phys. Rev. Lett.* 95: 098106-1–4.
45. Reuveni S, Granek R, Klafter J. 2008. Proteins: Coexistence of stability and flexibility. *Phys. Rev. Lett.* 100: 208101, 1–4.
46. Yu X, Leitner DM. 2003. Anomalous diffusion of vibrational energy in proteins. *J. Chem. Phys.* 119: 12673–12679.
47. Enright MB, Leitner DM. 2005. Mass fractal dimension and the compactness of proteins. *Phys. Rev. E* 71: 011912-1–9.
48. Enright MB, Yu X, Leitner DM. 2006. Hydration dependence of the mass fractal dimension and anomalous diffusion of vibrational energy in proteins. *Phys. Rev. E* 73: 051905-1–9.
49. Stapleton HJ, Allen JP, Flynn CP, Stinson DG, Kurtz SR. 1980. Fractal form of proteins. *Phys. Rev. Lett.* 45: 1456–1459.
50. Alexander S, Orbach R. 1982. Density of states of fractals: 'Fractons'. *J. Phys. Lett.* 43: L625–L631.
51. Liang J, Dill KA. 2001. Are proteins well-packed? *Biophys. J.* 81: 751–766.
52. Bu L, Straub JE. 2003. Vibrational energy relaxation of 'tailored' hemes in myoglobin following ligand photolysis supports energy funneling mechanism of heme 'cooling'. *J. Phys. Chem. B* 107: 10634–10639.
53. Bu L, Straub JE. 2003. Simulating vibrational energy flow in proteins: Relaxation rate and mechanism for heme cooling in cytochrome *c*. *J. Phys. Chem. B* 107: 12339–12345.
54. Sagnella DE, Straub JE, Thirumalai D. 2000. Timescales and pathways for kinetic energy relaxation in solvated proteins: Application to carbonmonoxy myoglobin. *J. Chem. Phys.* 113: 7702–7711.
55. Wang Q, Wong CF, Rabitz H. 1998. Simulating energy flow in biomolecules: Application to tuna cytochrome *c*. *Biophys. J.* 75: 60–69.
56. Uzer T. 1991. Theories of intramolecular vibrational energy transfer. *Phys. Rep.* 199: 73–146.
57. Logan DE, Wolynes PG. 1990. Quantum localization and energy flow in many-dimensional Fermi resonant systems. *J. Chem. Phys.* 93: 4994–5012.
58. Stuchebrukhov AA, Kuzmin MV, Bagratashvili VN, Letokhov VS. 1986. Threshold energy dependence of intramolecular vibrational relaxation in polyatomic molecules. *Chem. Phys.* 107: 429–443.

59. Leitner DM. 2001. Vibrational energy transfer in helices. *Phys. Rev. Lett.* 87: 188102-1–4.
60. Moritsugu K, Miyashita O, Kidera A. 2000. Vibrational energy transfer in a protein molecule. *Phys. Rev. Lett.* 85: 3970–3973.
61. Moritsugu K, Miyashita O, Kidera A. 2003. Temperature dependence of vibrational energy transfer in a protein molecule. *J. Phys. Chem. B* 107: 3309–3317.
62. Leitner DM. 2002. Anharmonic decay of vibrational states in helical peptides, coils and one-dimensional glasses. *J. Phys. Chem. A* 106: 10870–10876.
63. Sibert EL, Reinhardt WP, Hynes JT. 1984. Intramolecular vibrational relaxation and spectra of CH and CD overtones in benzene and perdeuterobenzene. *J. Chem. Phys.* 81: 1115–1134.
64. Stuchebrukhov AA, Marcus RA. 1993. Theoretical study of intramolecular vibrational relaxation of acetylenic CH vibration for $v = 1$ and 2 in large polyatomic molecules (CX3)3YCCH, where X = H or D and Y = C or Si. *J. Chem. Phys.* 98: 6044–6061.
65. Leitner DM, Wolynes PG. 1996. Vibrational relaxation and energy localization in polyatomics: Effects of high-order resonances on flow rates and the quantum ergodicity transition. *J. Chem. Phys.* 105: 11226–11236.
66. Leitner DM, Wolynes PG. 1997. Vibrational mixing and energy flow in polyatomic molecules: Quantitative prediction using local random matrix theory. *J. Phys. Chem. A* 101: 541–548.
67. Fabian J, Allen PB. 1996. Anharmonic decay of vibrational states in amorphous silicon. *Phys. Rev. Lett.* 77: 3839–3842.
68. Fujisaki H, Straub JE. 2005. Vibrational energy relaxation in proteins. *Proc. Natl. Acad. Sci. (USA)* 102: 7626–7631.
69. Yu X, Leitner DM. 2005. Anharmonic decay of vibrational states in proteins. In Q. Cui, I. Bahar, eds. *Normal Mode Analysis: Theory and Applications to Biological and Chemical Systems*, Taylor & Francis, Boca Raton, FL.
70. Austin RH, Xie A, Meer Lvd, Redlich B, Lingård P-A, Frauenfelder H, Fu D. 2005. Picosecond thermometer in the amide I band of myoglobin. *Phys. Rev. Lett.* 94: 128101, 1–4.
71. Rector KD, Rella CW, Hill JR, Kwok AS, Sligar SG, Chien EYT, Dlott DD, Fayer MD. 1997. Mutant and wild-type myoglobin–CO protein dynamics: Vibrational echo experiments. *J. Phys. Chem. B* 101: 1468–1475.
72. Leitner DM. 2001. Vibrational energy transfer and heat conduction in a one-dimensional glass. *Phys. Rev. B* 64: 094201-1–9.
73. Lockless SW, Ranganathan R. 1999. Evolutionarily conserved pathways of energetic connectivity in protein families. *Science* 286: 295–299.
74. Swain JF, Gierasch LM. 2006. The changing landscape of protein allostery. *Curr. Opin. Struct. Biol.* 16: 102–108.
75. Hardy RJ. 1963. Energy-flux operator for a lattice. *Phys. Rev.* 132: 168–177.
76. Frauenfelder H, Fenimore PW, Chan G, McMahon BH. 2006. Protein folding is slaved to solvent motions. *Proc. Natl. Acad. Sci. (USA)* 103: 15469–15472.
77. Zhang Y, Fujisaki H, Straub JE. 2007. Molecular dynamics study on the solvent dependent heme cooling following ligand photolysis in carbonmonoxymyoglobin *J. Phys. Chem. B* 111: 3243–3250.

# 12 Heat Transfer in Nanostructures

*Dvira Segal*

## CONTENTS

## 12.1  INTRODUCTION

What is the thermal conductivity of silicon nanowires, *n*-alkane single molecules, carbon nanotubes, or thin films? How does the conductivity depend on the nanowire dimension, nanotube chirality, molecular length and temperature, or the film thickness and disorder? More profoundly, what are the mechanisms of heat transfer at the nanoscale, in constrictions, at low temperatures? Recent experiments and theoretical studies have demonstrated that the thermal conductivity of nanolevel systems significantly differ from their macroscale analogs [1]. In macroscopic-continuum objects, heat flows diffusively, obeying the Fourier's law (1808) of heat conduction, $J = -K\nabla T$, $J$ is the current, $K$ is the thermal conductivity and $\nabla T$ is the temperature gradient across the structure. It is however obvious that at small scales, when the phonon mean free path is of the order of the device dimension, distinct transport mechanisms dominate the dynamics. In this context, one would like to understand the violation of the Fourier's law, derive complete set of necessary and sufficient conditions for its validity, and ultimately derive it from first principle arguments [2].

Thermal conductivity at the nanoscale may significantly deviate from the bulk material value from various reasons, of classical and quantum origin [1]. First, size effects are influential if the structure dimension is comparable (or smaller) to the phonon characteristic length. This may result in a ballistic motion and the failure of the Fourier's law, even when phonons are treated as particles. Moreover, if the phases of the phonon waves are fixed, for example when interfaces in the structure are flat, wave effects sustain, leading to phonon interference and diffraction effects.

More fundamentally, in the ballistic phonon regime at low enough temperatures, one-dimensional (1D) wires should manifest the quantization of the thermal conductance for the lowest energy modes [3,4]. Here $K = J/\Delta T$ is the thermal conductance with $\Delta T$ as the temperature difference. The fundamental quantum of thermal conductance is $\pi^2 k_B^2 T/3h$ where $k_B$ is the Boltzmann constant, $h$ is Planck's constant, and $T$ is the temperature. This value is universal, independent not only of the conducting material, but also of the particle statistics, i.e., the quantum conductance is the same for bosons and fermions [3].

Resolving classical and quantum transport mechanisms in various structures, for example in amorphous materials [5], proteins [6], nanotubes and nanowires [7], understanding the (in)validity of the Fourier's law in low dimensions [8], and exploring the quantization of the thermal conductance in molecular wires [3,9], are all topics of great theoretical interest. Likewise, these issues are very much the core of actual challenges in molecular electronics, information processing and computation, thermoelectricity, and nanoscale machinery. In molecular electronics, for example, engineering good thermal contacts and cooling of the junction are crucial for preventing device overheating, important for a stable operation mode [10–12]. Therefore, in conjunction with the endeavor to build good molecular-level electrical conductors, significant effort is invested in controlling and improving the thermal conductivity of the relevant structures. In contrast, in energy conversion and management sciences, we seek the opposite goal: In order to increase thermoelectric efficiency, it is advantageous to *reduce* the material thermal conductivity, keeping the resistivity and the Seebeck coefficient intact [13–15].

In this chapter, we discuss three theoretical-numerical approaches for studying heat flow in molecular systems. The different methods were developed in order to focus on various aspects of heat flow at the nanoscale, and are applicable at different parameters regimes: quantum, semiclassical and classical. Ultimately, developing a unified approach is of great interest [16,17].

The organization of the chapter is as follows. Section 12.2 presents our generic model and lists some of the main challenges. Sections 12.3 through 12.5 describe the theoretical approaches, followed by several applications. Section 12.6 discusses recent experimental results. Section 12.7 concludes, presenting some long-term goals.

## 12.2   MODEL AND TOPICS OF INTEREST

Our generic system consists a molecular junction, i.e., a molecular unit connecting two dielectrics (thermal reservoirs), $v = L, R$, of inverse temperatures $\beta_v = T_v^{-1}$. Henceforth we take the Boltzmann constant as $k_B = 1$. We neglect the electronic degrees of freedom, assuming the system is a perfect electrical insulator, and focus only on vibrational energy transfer between the solids.

The Hamiltonian of this model is the sum of the molecular Hamiltonian $H_M$, the Hamiltonian of the two dielectric reservoirs $H_B = H_L + H_R$ and the molecule-bath interaction $H_{MB}$,

$$H = H_M + H_B + H_{MB}. \tag{12.1}$$

The two baths are usually represented by sets of independent harmonic oscillators. For the molecular unit, we are mainly interested in 1-dimensional periodic chains made of $N$ units with anharmonic force fields. System-bath coupling is either harmonic or nonlinear and is typically taken to be weak. We will assume that only the end atoms of the chain, 1 and $N$, couple to the surfaces, L and R, respectively, and neglect direct interactions between the reservoirs.

For this generic setup, we focus on the following topics: (1) *Resolving transport mechanisms.* We study the role of harmonic/anharmonic internal molecular interactions on the heat current characteristics, as well as the effect of the molecular structure, system size, dimensionality, and the interaction strength with the solids. (2) *Proposing molecular level mechanical device.* We discuss the operation principle of various devices, e.g., a thermal rectifier and a heat pump. Such systems are of interest both fundamentally, manifesting nonlinear transport characteristics, and for practical applications.

In order to analyze these issues, we do the modeling within different levels. (a) We study quantum heat transfer in realistic chains, e.g., $n$-alkane systems, using the explicit molecular force field. This is done in the harmonic limit within the quantum generalized Langevin equation formalism [18]. (b) We study heat transfer using the quantum master equation approach [19], assuming for simplicity that a single mode dominates the phononic current across the junction. This approximation allows us to dissect the effect of nonlinear interactions on the current and to discuss the operation principles of simple devices: a thermal rectifier [20] and a molecular heat pump [21]. (c) We use classical Langevin-type molecular dynamics simulations in order to understand the onset of diffusional dynamics in nanosystems, and for clarifying the role of non-Markovian memory effects [22].

## 12.3 QUANTUM GENERALIZED LANGEVIN EQUATION: A FULLY HARMONIC MODEL

We present here the generalized Langevin equation approach [23] useful for studying quantum heat transfer through harmonic molecular chains. We show that a Landauer-type expression for the heat flux is obtained in this (fully harmonic) limit [18]. The formalism is useful for studying the thermal conductance of relatively short molecules with weak anharmonic interactions.

### 12.3.1 FORMALISM

Considering a realistic bath–molecule–bath junction, in the lowest order all modes in the system and bath are taken to be harmonic. To further simplify our presentation we assume a 1D model. The generalization to 3D systems is trivial [18]. The Hamiltonian (Equation 12.1) includes the following terms:

$$H_B + H_{MB} = \sum_{l \in L} \left\{ \frac{1}{2} m_l \omega_l^2 \left( x_l - \frac{g_{1,l} x_1}{m_l \omega_l^2} \right)^2 + \frac{p_l^2}{2m_l} \right\}$$

$$+ \sum_{r \in R} \left\{ \frac{1}{2} m_r \omega_r^2 \left( x_r - \frac{g_{N,r} x_N}{m_r \omega_r^2} \right)^2 + \frac{p_r^2}{2m_r} \right\}. \tag{12.2}$$

Here $x_q$, $p_q$, $m_q$, and $\omega_q$ ($q = l, r$) are coordinates, momenta, masses, and frequencies associated with the degrees of freedom of the reservoirs. $g_{1,l}$, $g_{N,r}$ are the coupling strengths of the L and R solids to the first (1) and last ($N$) atoms, respectively, of displacement $x_1$ and $x_N$. The molecule is represented by its $N$ independent collective harmonic modes,

$$H_M = \sum_{k=1}^{N} \left( \frac{1}{2} \omega_k^2 \bar{x}_k^2 + \frac{\bar{p}_k^2}{2} \right), \tag{12.3}$$

where $\bar{x}_k$ and $\bar{p}_k$ are the mass-weighted displacement and momentum associated with the $k$ normal mode. This normal mode representation can be obtained from the (local) atomic coordinate representation by diagonalizing the molecular Hessian matrix, $x = C\bar{x}$, assumed to be known for the structure analyzed. Next we transform the system–bath coupling to the normal mode representation:

$$H_B = \sum_l \left( \frac{1}{2} \omega_l^2 \bar{x}_l^2 + \frac{p_l^2}{2} \right) + \sum_r \left( \frac{1}{2} \omega_r^2 \bar{x}_r^2 + \frac{p_r^2}{2} \right), \tag{12.4}$$

$$H_{MB} = \sum_{l,k,k'} \frac{1}{2\omega_l^2} V_{l,k} V_{l,k'} \bar{x}_k \bar{x}_{k'} - \sum_{l,k} V_{l,k} \bar{x}_l \bar{x}_k + \sum_{r,k,k'} \frac{1}{2\omega_r^2} V_{r,k} V_{r,k'} \bar{x}_k \bar{x}_{k'} - \sum_{r,k} V_{r,k} \bar{x}_r \bar{x}_k,$$

where the transformed molecule–bath coupling constants are given by

$$V_{l,k} = V_l C_{1,k}; \quad V_l = \frac{g_{1,l}}{\sqrt{m_1 m_l}}$$

$$V_{r,k} = V_r C_{N,k}; \quad V_r = \frac{g_{N,r}}{\sqrt{m_N m_r}}. \tag{12.5}$$

Here $C_{1,k}$, $C_{N,k}$ are the matrix elements of the transformation matrix defined above, and $m_1$, $m_N$ are the masses of the end atoms. Following a standard procedure [18,23], it can be shown that in the long-time limit, the coordinates of the molecular modes obey the (classical) generalized Langevin equation ($v = L, R$),

$$\ddot{\tilde{x}}_k = -\omega_k^2 \tilde{x}_k + \sum_l V_{l,k} \tilde{x}_l + \sum_r V_{r,k} \tilde{x}_r - \sum_{v,k'} \int_{t_0 \to -\infty}^t \gamma_{k,k'}^v(t-\tau)\dot{\tilde{x}}_{k'}(\tau)d\tau, \quad (12.6)$$

where the reservoirs modes ($q = l, r$) evolve freely as

$$\tilde{x}_q(t) = \tilde{x}_q(t_0)\cos[\omega_q(t-t_0)] + \frac{\dot{\tilde{x}}_q(t_0)}{\omega_q}\sin[\omega_q(t-t_0)]. \quad (12.7)$$

The memory kernel of the $v$ bath satisfies

$$\gamma_{k,k'}^v(t) = \sum_{q \in v} \frac{V_{q,k} V_{q,k'}}{\omega_q^2}\cos(\omega_q t). \quad (12.8)$$

It can be easily shown that the ("random") forces induced by the $v$ reservoir, $F_v^{(k)} \equiv \sum_{q \in v} V_{q,k}\tilde{x}_q$ and the associated memory kernel are related by a fluctuation–dissipation relation,

$$\left\langle F_v^{(k)}(t)F_{v'}^{(k')}(0)\right\rangle = T_v\gamma_{k,k'}^v(t)\delta_{v,v'}. \quad (12.9)$$

It is notable that due to the system harmonicity, the bath forces act on the system additively, and the effect of each may be considered separately. We can therefore consider a version of Equation 12.6 with only one driving mode of frequency $\omega_0$.

We next write the classical equations of motion for the bath coordinates by including a small damping element $\eta$, enforcing a steady-state solution:

$$\ddot{\bar{x}}_l = -\omega_l^2 \bar{x}_l + \sum_k V_{l,k}\bar{x}_k - \eta\dot{\bar{x}}_l; \quad l \in L$$

$$\ddot{\bar{x}}_r = -\omega_r^2 \bar{x}_r + \sum_k V_{r,k}\bar{x}_k - \eta\dot{\bar{x}}_r; \quad r \in R \quad \eta \to 0+. \quad (12.10)$$

The long-time solution of Equations 12.6 through 12.10 is a steady state in which energy flows from a specific driving mode $\omega_0$ into the $\{l\}$ and $\{r\}$ modes through the molecular modes $\{k\}$. In particular, the steady-state heat flux channeled through, e.g., the mode $r$, is given by the rate of energy dissipation out of this mode:

$$J_{0 \to r} = \eta\left\langle \dot{\bar{x}}_r^2\right\rangle. \quad (12.11)$$

The total transmitted flux due to the driving force is given by $J_{0\to R} = \sum_r J_{0\to r}$. The reflected flux is $J_{0\to L} = \sum_l J_{0\to l}$. We should next sum over all driving frequencies, members of the L and R manifolds, where the net current is given by the difference of these terms.

We quantize the equations of motion (Equations 12.6 through 12.10), $\bar{x}_j(t) = \sqrt{\frac{1}{2\omega_j}}(a_j^\dagger(t) + a_j(t))$, $\bar{p}_j(t) = i\sqrt{\frac{\omega_j}{2}}(a_j^\dagger(t) - a_j(t))$, and transform it into frequency domain $(a_j^\dagger + a_j) = A_j e^{i\omega_0 t} + B_j e^{-i\omega_0 t}$. Here $j$ notates both molecular $(k)$ and bath modes $(l, r)$. This yields the Fourier-space Langevin equation $(B_j = A_j^\dagger)$

$$\left(\omega_k^2 - \omega_0^2 + i\omega_0\left[\gamma_{k,k'}^L(\omega_0) + \gamma_{k,k'}^R(\omega_0)\right]\right)A_k(\omega_0)$$

$$+ i\omega_0 \sum_{k\neq k'} \left[\gamma_{k,k'}^L(\omega_0) + \gamma_{k,k'}^R(\omega_0)\right]A_{k'}(\omega_0)\sqrt{\frac{\omega_k}{\omega_{k'}}} = \sqrt{\frac{\omega_k}{\omega_0}}V_{0,k}a_0^\dagger, \qquad (12.12)$$

providing the operators $A_k(\omega_0)$ in terms of the driving $a_0$. The thermal information enters here via the bath occupation factors $\langle a_0^\dagger a_0\rangle_v = [e^{\beta v\omega_0} - 1]^{-1}$.

The steady-state heat flux, independent of the position along the chain, is calculated by quantizing Equation 12.11. It can be rewritten in terms of the operators $A_k$ and $B_k$ whose steady-state solution is given in Equation 12.12 [18]. This yields the net current in the system, integrated over all driving frequencies,

$$J = \int \omega T(\omega)\left[n_L(\omega) - n_R(\omega)\right]d\omega, \qquad (12.13)$$

with the transmission coefficient

$$T(\omega) = \frac{2\omega^2}{\pi} \sum_{k,k'} \gamma_{k,k'}^R(\omega)\gamma_{k,k'}^L(\omega)\left[\bar{A}_k(\omega)\bar{A}_{k'}^\dagger(\omega) + \bar{A}_k^\dagger(\omega)\bar{A}_\kappa(\omega)\right]/2 \qquad (12.14)$$

and the Bose–Einstein occupation factors $n_v(\omega) = [e^{\omega/T_v} - 1]^{-1}$. The scalar function $\bar{A}_k(\omega_0)$ is defined through the relation $A_k(\omega_0) = \bar{A}_k(\omega_0)V_{0,k}a_0^\dagger\sqrt{\frac{\omega_k}{\omega_0}}$ [18].

Equation 12.13 is the thermal Landauer formula derived previously in different forms (see e.g., [3,9]). It is analogous to the Landauer result $J_{el} = \int T(\epsilon)\left[f_L(\epsilon) - f_R(\epsilon)\right]d\epsilon$, for the electrical current of a junction connecting two-electron reservoirs characterized by the Fermi distributions $f_v(\epsilon)$ and a transmission function $T(\epsilon)$. Note that our treatment could be easily generalized to describe 3D structures [18].

Concluding, given a molecule, i.e., a force field in the harmonic approximation, assuming harmonic solids and bilinear system–bath coupling, one can calculate the thermal conductance of the junction, $K = J/\Delta T$; $\Delta T = T_L - T_R$, by evaluating the energy-dependent transmission function $T(\omega)$ using Equation 12.14. This function is then multiplied by the energy carried per mode and by the phonons occupation factors. Integration over all driving frequencies (Equation 12.13) finally yields the total current $J$.

## 12.3.2 RESULTS: HEAT TRANSFER IN ALKANE CHAINS

We employ the above formalism and study the conductance of alkane chains of variable length (1–30 units). We have numerically verified that heat conduction in these systems is dominated by harmonic interactions [18]. A recent experiment confirmed this observation, demonstrating ballistic energy transfer in short hydrocarbon monolayers [24].

The information needed for any given molecular bridge is the normal mode spectrum of the molecular system and the corresponding transformation matrix C (see text after Equation 12.3). These were obtained using the HYPERCHEM 6 computer package [18]. The parameters that characterize the reservoirs are the Debye cut-off frequency $\omega_c$, which is taken in the range $\omega_c = 200$ to $800\,cm^{-1}$, and the temperatures $T_L$ and $T_R = T_L - \Delta T$ which are studied in the range $10$–$1000\,K$. $\Delta T$ itself is assumed to be small, typically $\Delta T = 10^{-3}\,K$, so $T$ represents the average temperature of the two reservoirs. We use a Debye model for describing the baths spectral properties:

$$\gamma(\omega) = ae^{-\omega/\omega_c}; \quad a = \frac{\pi}{4} \frac{g^2 N_B}{m_l m_B \omega_c^3}, \tag{12.15}$$

where
$N_B$ is the number of reservoirs modes
$m_B$ is the mass of the bath atoms
$g = g_{1,l} = g_{N,r}$

The two baths are assumed to be equivalent.

Figure 12.1 shows the thermal conductance as a function of chain length $N$ at several temperatures. While at high temperature, the conductance increases for short chains, then decays into (approximately) a constant, at low $T$ the conductance is increasing noncontinuously until saturation. This can be explained as a transition from a tunneling-type transport to a resonant-energy mode of transfer with increasing temperature. Specifically, variation of the chain length affects the molecular normal mode spectrum in two ways. First, the overall density of states is increased linearly. Second, the lower bound on this density is shifted to lower values. For example, for a pentane ($N=5$), the lowest vibrational frequency mode is $\omega=84\,cm^{-1}$, for decane ($N=10$), it is $\omega=28\,cm^{-1}$, while for $N=20$, it is $\omega=7\,cm^{-1}$. At low temperatures, the heat current is carried mostly by low-frequency phonons, and when the chain becomes longer, more molecular modes come into resonance with these incoming phonons. This causes an increase of the thermal flux at very low temperatures. In contrast, at high enough $T$ since all relevant modes are populated, after a short rise (reflecting the red shifting of the modes), the energy is carried through a resonant transmission mechanism, resulting in a size-independent current. The transition from tunneling to

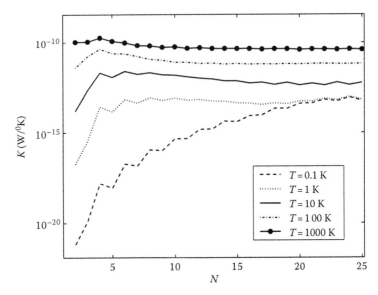

**FIGURE 12.1** Thermal conductance as a function of length for alkane chains, using $\omega_c = 400\,\text{cm}^{-1}$ and $a = 8000\,\text{cm}^{-1}$. $T = 0.1\,\text{K}$ (dashed). $T = 1\,\text{K}$ (dotted); $T = 10\,\text{K}$ (full); $T = 100\,\text{K}$ (dash-dotted); $T = 1000\,\text{K}$ (circles). (From Segal, D. et al., *J. Chem. Phys.* 119, 6840, 2003. With permission.)

resonant energy transmission can be directly detected by focusing on the conduction properties of subsets of modes [18]. This analysis shows that $K$ of the low-energy modes ($\omega < T$) is a constant, while the contribution of high-energy modes decays exponentially with distance.

The quantum Langevin equation formalism described here could be also employed for studying the conductance properties of disordered harmonic systems or chains combining impurities. In such systems, mode localization becomes a crucial factor [5,18].

We conclude that for short-intermediate chains, the principal factors that determine heat conduction are (a) the molecular vibrational spectrum, (b) the localization properties of the molecular modes, (c) the coupling of the system to the reservoirs, and (d) the cut-off frequency that characterizes the reservoirs spectral properties. For harmonic junctions, the heat conductance of long-enough chains at high temperatures essentially saturates or the thermal conductivity ($JN/\Delta T$) diverges. Extensions of this formalism to nonlinear molecular junctions are of great interest [16,17].

## 12.4 MASTER EQUATION FORMALISM: NONLINEAR INTERACTIONS

The Landauer expression for heat transfer (Equation 12.13) assumes the absence of inelastic scattering processes in the system, and the two opposite phonon flows of different temperatures are out of equilibrium with each other. This leads to an anomalous transport of heat, where (classically) the energy flux is proportional to the temperature difference, $T_L - T_R$, rather than to the temperature gradient $\nabla T$, as asserted by the Fourier's law of heat conductivity,

$$J = -\kappa \nabla T. \tag{12.16}$$

An outstanding problem in statistical physics is to find out the necessary and sufficient conditions for attaining the Fourier's law (Equation 12.16) in low dimensional systems. While classical systems were intensively investigated in the last 50 years, usually by means of computer simulations [2], there is also an ongoing effort in deriving the law using quantum principles [25,26]. These studies suggest that a crucial requirement for showing normal transport in 1D systems is that the molecular potential energy constitutes anharmonic interactions. Nonlinear interactions are also a tool for controlling heat flow in molecular junctions with potential technological applications, e.g., for building thermal diodes [20,27,28] and mechanical transistors [29,30].

In this section, we present a simple, analytically solved, model that yields a qualitative picture of heat conduction in anharmonic molecular chains. Our treatment is done at the level of the master equation for the vibrational states occupation, assuming dephasing processes die out at the relevant timescale for transport.

### 12.4.1 FORMALISM

In our minimal picture, the heat current flowing through a nano junction is dominated by a *single* vibrational mode of frequency $\omega_0$. The molecular part of the total Hamiltonian (Equation 12.1) is therefore given by

$$H_M = \sum_{n=0}^{N-1} E_n |n\rangle \langle n|; \ E_n = n\omega_0, \tag{12.17}$$

where $|n\rangle$ denotes the $n$ vibrational state. For a harmonic molecule, $N$ is taken up to infinity. Strong anharmonicity is enforced by limiting $n$ to 0,1, i.e., by considering a two-level system (TLS) as a model for the molecular vibration. One can think of the latter case as if only two bound states persist in the anharmonic (e.g., Morse) potential. We emphasize that in our model, anharmonicity is included only by truncating the spectrum of the single molecular mode. We do not include other phonon–phonon scattering processes, e.g., umklapp processes, that can lead to normal conductivity.

This simplified picture neglects, of course, detailed molecular properties, yet it contains two important elements that are crucial for resolving heat flow at the nanoscale: (1) Nonlinear effects may be easily included by considering an anharmonic-truncated molecular mode. In standard descriptions, anharmonic interactions are cumbersome to treat [17]. (2) Molecule–solid interactions and bath memory effects can be analyzed without being masked by the complicated molecular spectrum [22]. This treatment thus allows us to gain a qualitative understanding of heat conductance in single-mode junctions: its dependence on the internal interparticle potential, molecule–reservoirs contact interactions, and junctions spatial asymmetry.

The molecular mode is coupled either linearly or nonlinearly to the L and R thermal baths, represented by sets of independent harmonic oscillators,

$$H_B = H_L + H_R; \quad H_K = \sum_{j \in v} \omega_j a_j^\dagger a_j; \quad v = L, R. \tag{12.18}$$

$a_j^\dagger, a_j$ are boson creation and annihilation operators associated with the phonon modes of the harmonic baths. We use the following model for the molecule–reservoirs interaction:

$$H_{MB} = \sum_{n=1}^{N-1} \left( B|n-1\rangle\langle n| + B^\dagger|n\rangle\langle n-1| \right) \sqrt{n}, \tag{12.19}$$

where $B$ are bath operators. This model assumes that transitions between the molecular vibrational states occur due to the environment excitations. Note that in general this interaction does not need to be additive in the thermal baths, i.e., we may consider situations in which $B \neq B_L + B_R$ [19].

Under the assumption of *weak* system–bath interactions, going into the Markovian limit, the probabilities $P_n$ to occupy the $n$ state of the molecular oscillator satisfy the master equation [31]:

$$\dot{P}_n = -\left[ nk_d + (n+1)k_u \right] P_n + (n+1)k_d P_{n+1} + nk_u P_{n-1}, \tag{12.20}$$

where the occupations are normalized $\sum_n P_n = 1$, and $k_d$ and $k_u$ are the vibrational relaxation and excitation rates, respectively. In second-order perturbation theory, these rates are given by the Fermi golden rule, adjusted here to include two independent continua,

$$k_d = \int_{-\infty}^{\infty} d\tau e^{i\omega_0\tau} \left\langle B^\dagger(\tau)B(0) \right\rangle; \quad k_u = \int_{-\infty}^{\infty} d\tau e^{-i\omega_0\tau} \left\langle B(\tau)B^\dagger(0) \right\rangle. \tag{12.21}$$

In these expressions, the average is done over the baths thermal distributions, irrespective of the fact that it involves two distributions of different temperatures.

The nonlinear-strong coupling case is presented in details in Ref. [19]. In what follows, we discuss only the bilinear–weak coupling limit. Here the bath operator coupled to the system is $B = B_L + B_R$, where $B_v = \sum_{j \in v} \bar{\alpha}_j x_j, x_j = (2\omega_j)^{-1/2}(a_j^\dagger + a_j)$. The resulting rates (Equation 12.21) are additive in the L and R reservoirs, $k_d = k_L + k_R$, $k_u = k_L e^{-\beta_L \omega_0} + k_R e^{-\beta_R \omega_0}$, with

$$k_v = \Gamma_v(\omega_0)(1 + n_v(\omega_0)); \quad v = L, R. \tag{12.22}$$

Here $n_v(\omega) = (e^{\beta_v \omega} - 1)^{-1}$ is the Bose–Einstein distribution function, $\Gamma_v(\omega) = \frac{\pi}{2m\omega^2} \sum_{j \in v} \alpha_j^2 \delta(\omega - \omega_j)$ and $\alpha_j = \bar{\alpha}_j \sqrt{2m\omega_0}$ [20], where $m$ and $\omega_0$ are the mass and frequency of the molecular oscillator.

The heat conduction properties of this model are obtained from the steady-state solution of Equation 12.20 with the rates Equation 12.22. The steady-state heat flux calculated, e.g., at the right contact, is given by the sum

$$J = \omega_0 \sum_{n=1}^{N-1} n \left( k_R P_n - k_R P_{n-1} e^{-\beta_R \omega_0} \right), \qquad (12.23)$$

where positive sign indicates current going from left to right. In this expression, the first term denotes the energy flux going out of the molecular mode into the R reservoir. The second term provides the oppositely going current from the R reservoir into the system. The current could be equivalently calculated at the L contact. A first-principle derivation of Equation 12.23 is given in [32]. The current (Equation 12.23), calculated for harmonic ($N \rightarrow \infty$) and anharmonic ($n = 0, 1$) modes, is given exactly by [19]

$$J = \omega_0 (n_L - n_R) \frac{\Gamma_L \Gamma_R}{\Gamma_L + \Gamma_R} \frac{\omega_0}{T_M} f_{A,H}, \qquad (12.24)$$

where $T_M = \omega_0 [\Gamma_L (1 + 2n_L) + \Gamma_R (1 + 2n_R)]/2(\Gamma_L + \Gamma_R)$ is an effective molecular temperature. All the $\Gamma_\nu$ coefficients and Bose–Einstein terms $n_\nu$ are evaluated at the molecular frequency $\omega_0$. $f$ is an effective molecular occupation factor

$$f_{A,H} = \begin{cases} 1/2, & \text{anharmonic (TLS) case} \\ T_M/\omega_0, & \text{harmonic case}. \end{cases} \qquad (12.25)$$

In the harmonic limit, we thus retrieve the (resonant) Landauer type expression (Equation 12.13), $J = \omega_0 \Gamma_L \Gamma_R (\Gamma_L + \Gamma_R)^{-1} (n_L - n_R)$, while in the TLS model, the effective transmission function, $\mathcal{T} \propto 1/T_M$, is temperature dependent. This may potentially lead to nonlinear current–temperature characteristics as we discuss below.

The master equation formalism can be extended to include more than a single molecular oscillator by considering several localized vibrational sets [20]. Furthermore, it can be utilized for describing heat flow in a multiterminal junction [33], and for investigating electronic energy transfer between metals [34].

## 12.4.2 Results: Nanomechanical Devices

We explain here the operation principles of simple molecular devices, a thermal rectifier [20] and a heat pump [21]. First we present the heat current in the anharmonic (TLS) model. Figure 12.2 demonstrates that the current increases monotonically with $\Delta T$, then saturates at high temperature differences. It can be indeed shown that $\partial J/\partial \Delta T > 0$, which indicates that negative differential thermal conductance (NDTC), a decrease of $J$ with increasing $\Delta T$, is impossible in the present (bilinear coupling) case. As shown in Ref. [19], NDTC requires nonlinear system–bath interactions, resulting in an effective temperature-dependent molecule–bath coupling term.

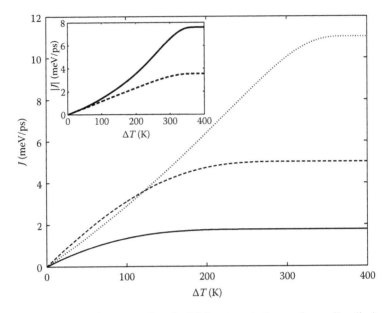

**FIGURE 12.2** Conduction properties of a TLS system in the weak coupling limit, generated by simulating 12.24. $\omega_0 = 150\,\text{meV}$ (full), $100\,\text{meV}$ (dashed), $25\,\text{meV}$ (dotted). $T_L = 400\,\text{K}$, $T_R = T_L - \Delta T$, $\Gamma_K = 1.2\,\text{meV}$. Inset: Rectifying behavior of this model, $\omega_0 = 25\,\text{meV}$, $\chi = 0.75$ and $T_L = 400\,\text{K}$, $T_R = T_L - \Delta T$ (full); $T_R = 400\,\text{K}$, $T_L = T_R - \Delta T$ (dashed).

*Thermal rectifier.* In analogy with the electric diode, a thermal rectifier is a device manifesting asymmetric heat conduction upon interchanging the reservoirs temperatures. Detailed theoretical studies concluded that the effect can be realized by combining some spatial asymmetry with nonlinear (anharmonic) interactions persisting either in the molecule, the solids, or the contacts [20,27,28,35]. A pioneering experiment has recently measured thermal rectification in an asymmetrically mass loaded carbon nanotube junction [36]. While the actual rectification ratios are typically low, of few percents, rectification is a topic of great fundamental interest, demonstrating nonlinear transport characteristics.

We can analyze this phenomenon in our model: While in the harmonic limit [(Equation 12.24) with $f = T_M/\omega_0$] the current is the same with opposite signs for forward and backward temperature biases, the TLS model, representing molecules incorporating anharmonic interactions, brings in thermal rectification, given that the system incorporates a spatial asymmetry. Defining the asymmetry parameter $\chi$ such that $\Gamma_L = \Gamma(1 - \chi)$; $\Gamma_R = \Gamma(1 + \chi)$ with $-1 \leq \chi \leq 1$ we obtain

$$\Delta J \equiv J(T_L = T_h; T_R = T_c) + J(T_L = T_c; T_R = T_h)$$

$$= \frac{\omega_0 \Gamma \chi (1 - \chi^2)(n_L - n_R)^2}{(1 + n_L + n_R)^2 - \chi^2 (n_L - n_R)^2}. \tag{12.26}$$

Here $T_c$ ($T_h$) relates to the cold (hot) bath. Equation 12.26 implies that for small $\Delta T$, $\Delta J$ increases as $\Delta T^2$ increases, and that the current is larger (in absolute value) when the cold bath is coupled more strongly to the molecular system. We exemplify this behavior at the inset of Figure 12.2.

*Heat pump.* By generalizing the master equation formalism to time-dependent situations, we can also consider a molecular machine that pumps heat against a thermal gradient [21]. The pumping action is achieved by applying an external force that periodically modulates the molecular levels. As we show below, since in our model the modulation affects both the internal temperature of the molecule, and the strength of its coupling to each reservoirs, this results in a net heat flow against the temperature bias.

For a two-level system, $H_M = \frac{\omega(t)}{2}\left(|1\rangle\langle1|-|0\rangle\langle0|\right)$. Here $\omega(t) = \omega_0 + F(t)$ with $\omega_0$ as a static frequency and $F(t) = F(t + 2\pi/\Omega)$ denoting a periodic modulation. Assuming bilinear interactions and weak coupling, going into the Markovian limit, the probabilities to occupy the states 0 and 1 satisfy the master equation (12.20) with time-dependent rates $(s = u, d)$, $k_{s,\nu}(t) = \sum_{n,m} \Re[\alpha_{n,m}e^{i(n-m)\Omega t}]k_{s,\nu}^{(n)}$, $k_{s,\nu}^{(n)}$ is given by Equation 12.21, calculated at the frequency $\omega_0 + n\Omega$. The coefficients $\alpha_{n,m}$ ($n$, $m$ are integers) depend on the specific modulation parameters [21], $\Re$ refers to the real part thereof.

An essential ingredient of our system is its spatial asymmetry. We build it into the pump by choosing reservoirs with difference cutoff frequencies, $\omega_c^L < \omega_c^R$, and by assuming that the TLS is asymmetrically connected to the solids, $\kappa_L > \kappa_R$. These parameters are defined through the relaxation rates (Equation 12.21), $\nu = L, R$

$$k_{d,\nu}^{(n)} = \begin{cases} \kappa_\nu, & \omega_0 + n\Omega < \omega_c^\nu \\ \kappa_\nu e^{-(\omega_0+n\Omega)/\omega_c^\nu}, & \omega_0 + n\Omega > \omega_c^\nu \end{cases} \tag{12.27}$$

For a schematic representation see Figure 12.3a and b. Panel Figure 12.3c shows a square shape driving signal and the ensuing TLS temperature. We find that during the wait time at the minimal energy spacing, the TLS heats from 150 K up to 200 K by accepting energy from both reservoirs 12.3d. Upon increasing the TLS energy, the effective TLS temperature becomes very high, higher than $T_h$. When the TLS reaches its maximal value, it becomes effectively decoupled from the cold reservoir and its temperature reduces from 600 to ~400 K by transferring energy to the right (hot) reservoir, completing the pumping cycle.

## 12.5 MOLECULAR DYNAMICS SIMULATIONS

Classical molecular dynamics (MD) simulations are a useful tool for elucidating the interplay between the molecular structure, solids properties, and interface–molecule interactions in determining the heat transport properties of nanometer systems [37,38]. Moreover, MD simulations were extensively used for testing the applicability of the Fourier's law in low dimensional systems [2] and for suggesting molecular level machines [29,30]. In standard-classical MD studies, one essentially disregards the electronic degrees of freedom, and considers an all-atom force-field for the atomic coordinates, with dynamics ruled by Newtonian equations of motion.

We mention here two of the multitude MD techniques developed for treating heat transfer in molecules [1,2,38]: (1) Nonequilibrium MD (NEMD) methods, where a

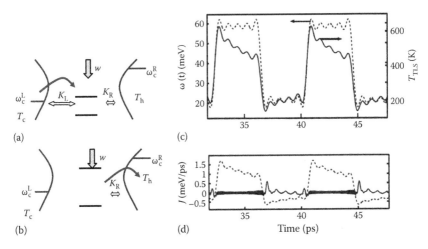

**FIGURE 12.3** (a) Schematic picture of the idealized pumping cycle. At low frequencies, the TLS is strongly coupled to the left (cold) reservoir. Thus when $T_{TLS} < T_c$ heat is transferred from the left bath to the TLS. (b) At high frequencies, the TLS is coupled only to the right (hot) reservoir, thus its internal energy is transmitted into the right bath when $T_{TLS} > T_h$. (c) The TLS energy spacing modulation (dashed line, left axis) and the resulting TLS temperature (full line, right). $T_c = 200$ K, $T_h = 300$ K, $\omega_0 = 40$ meV, $\omega_c^L = 6$ meV, $\omega_c^R = 250$ meV, $k_L = 2.5$ meV, $k_R = 0.1$ meV. (d) The resulting heat currents at the cold contact (full) and at the hot contact (dashed). The current is defined positive when flowing left to right.

heat source and a heat sink are built into the simulations; (2) The Green–Kubo (equilibrium) approach, where thermal fluctuations are used to compute the heat current using the fluctuation–dissipation relation. As a variant of the first (nonequilibrium) approach, one can impose a fixed temperature at the boundaries, where Nose–Hoover or Langevin thermostats [2] simulate the reservoirs.

### 12.5.1 FORMALISM: LANGEVIN DYNAMICS

We present here classical molecular dynamics simulations of steady-state heat transfer through 1D periodic chains coupled to non-Markovian reservoirs in order to elucidate the effect of the contacts on the junction thermal conductance.

We model the molecule as a chain of $N$ identical atoms. The end particles 1 and $N$ are connected to heat baths of temperatures $T_L$ and $T_R$, respectively. The dynamics is governed by the generalized Langevin equation

$$\ddot{x}_k(t) = -\frac{1}{m}\frac{\partial H_M}{\partial x_k}, \quad k = 2,3,\ldots,N-1$$

$$\ddot{x}_1(t) = -\frac{1}{m}\frac{\partial H_M}{\partial x_1} - \int_0^t dt'\gamma_L(t-t')\dot{x}_1(t') + \eta_L(t), \quad (12.28)$$

$$\ddot{x}_N(t) = -\frac{1}{m}\frac{\partial H_M}{\partial x_N} - \int_0^t dt'\gamma_R(t-t')\dot{x}_N(t') + \eta_R(t).$$

$x_k$ is the position of the $k$ particle of mass $m$, and $p_k$ (see Equation 12.31) is the particle momentum. $H_M$ is the internal molecular Hamiltonian. $\gamma_L$ and $\gamma_R$ are friction constants, and $\eta_L$ and $\eta_R$ are fluctuating forces that represent the effect of the thermal reservoirs. These terms are related through the fluctuation–dissipation relation

$$\langle \eta_v \rangle = 0; \quad \langle \eta_v(t)\eta_v(t') \rangle = \frac{T_v}{m}\gamma_v(t-t'), \quad v = L, R. \tag{12.29}$$

We consider here an exponentially correlated Ornstein–Uhlenbeck (O–U) noise [39]

$$\gamma_v(t-t') = \frac{\epsilon_v}{\tau_c^v}e^{-|t-t'|/\tau_c^v}, \tag{12.30}$$

with the intensity $\epsilon$ and a correlation time $\tau_c$. For short correlation times, the heat baths generate an uncorrelated (white) noise. The Fourier transform of the O–U function is a Lorentzian, $\gamma_v(\omega) \equiv \int e^{-i\omega t}\gamma_v(t)dt = \frac{2\epsilon_v}{1+(\omega\tau_c^v)^2}$.

A simple approach for implementing the O–U noise in numerical simulations is to introduce auxiliary dynamical variables $y_1(t)$ and $y_N(t)$ for the L and R baths, respectively [22,40]. The new equations of motion, for, e.g., the first particle are

$$\dot{x}_1(t) = \frac{p_1(t)}{m}$$

$$\ddot{x}_1(t) = -\frac{1}{m}\frac{\partial H_0}{\partial x_1} - y_1(t) + \eta_L(t)$$

$$\dot{y}_1(t) = -\frac{y_1(t)}{\tau_c^L} + \frac{\epsilon_L}{m\tau_c^L}p_1(t) \tag{12.31}$$

$$\dot{\eta}_L(t) = -\frac{\eta_L(t)}{\tau_c^L} + \frac{1}{\tau_c^L}\sqrt{\frac{2\epsilon_L k_B T_L}{m}}\mu_L(t),$$

where
$\mu_L(t)$ is a Gaussian white noise
$\langle \mu_L(t) \rangle = 0$ and $\langle \mu_L(t)\mu_L(t') \rangle = \delta(t-t')$

An equivalent set of equations exists for the $N$ particle, interacting with the R thermal bath. The coupled equations, Equation 12.28 for particles $2,\ldots,N-1$ and Equation 12.31 with its $N$ equivalent, are integrated using the fourth-order Runge–Kutta method to yield the positions and velocities of all particles. The heat flux can be calculated from the trajectory using [2]

$$J = \frac{1}{2(N-1)}\sum_{k=1}^{N-1}\left\langle (v_k + v_{k+1})F(x_{k+1} - x_k)\right\rangle, \tag{12.32}$$

where $F(r) = -\partial H_M(r)/\partial r$, $\upsilon_k = p_k/m$, and we average over time after steady state is reached.

Using this approach, we can address the following questions: (1) How does the heat current depend on the molecular size and force field? What is the role of anharmonic interactions? (2) What is the effect of the interface spectral properties? (3) How does the conductance depend on system–bath couplings? While our goal here is to discuss those issues within a phenomenological force field, detailed MD simulations can reasonably reproduce experimental data [38].

## 12.5.2  RESULTS

### 12.5.2.1  Heat Flux in Non-Markovian Chains

In this section, we elucidate on the role of the solids spectral properties. We specify first the molecular force field: We adopt a Morse function for the interatomic potential of dissociation energy $D$, width $\alpha$, and an equilibrium separation $x_{eq}$,

$$H_M = \sum_{k=1}^{N} \frac{p_k^2}{2m} + D \sum_{k=0}^{N-1} \left[ e^{-\alpha(x_{k+1} - x_k - x_{eq})} - 1 \right]^2.$$

(12.33)

The atoms indexed by 0 and $N+1$ are the left and right reservoirs atoms. We use $D = 367.8/\lambda^2$ kJ/mol, $\alpha = 1.875\lambda$ Å$^{-1}$, $x_{eq} = 1.54$ Å, and $m = 12$ g/mol. These numbers describe a c–c (carbon–carbon) stretching mode for $\lambda = 1$ [41]. Two sets of parameters are considered: In the first case, $\lambda = 0.01$, so as the potential energy is practically harmonic (H). We also take parameters where the anharmonic coefficient is large, $\lambda = 6$, referring to the later case as "anharmonic" (A).

We assume that the reservoirs have the same type of spectral function (O–U) with equal strength $\epsilon = \epsilon_v$ and noise correlation time $\tau_c = \tau_c^v$. Depending on the situation, we use integration time step $\Delta t = 10^{-3} - 10^{-4}$ $1/\omega$, where $\omega$ is the molecular frequency in the harmonic limit. We also take care of the required inequality $\Delta t \ll \tau_c$.

Figure 12.4 presents the heat current for harmonic ($J_H$) and anharmonic ($J_A$) chains calculated with different memory times $\tau_c$. Figure 12.4a shows the heat current in the Markovian limit. We find that the energy flux in harmonic systems does not depend on size, while it decays with distance for anharmonic chains in agreement with standard results [2]. When the noise correlation time is increased, an interesting behavior is observed (Figure 12.4b): While $J_H$ shows an initial increase for $N < 5$, then saturates to a good approximation, the anharmonic flux initial rise is followed by a decay for long-enough chains. As the memory time is further increased (Figure 12.4c), the current of an anharmonic system saturates, interestingly showing a counteracting effect between the molecular contribution to the heat current and the reservoir's spectral properties. For highly correlated reservoirs (Figure 12.4d), both harmonic and anharmonic currents are slightly enhanced with distance. In this case, the anharmonic junction conducts better than a fully harmonic system. This intricate behavior sustains over a broad range of parameters [22].

We explain next these observations. First we clarify why $J_H$ and $J_A$ increase with size for short chains for non-Markovian baths. As was shown in Ref. [18], the dominant heat conducting vibrational modes of n-alkane chains are shifted toward

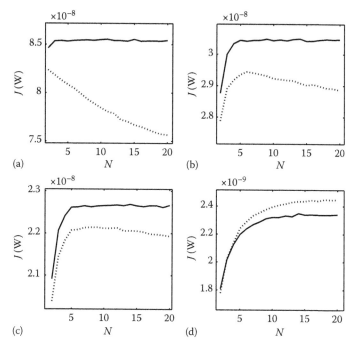

**FIGURE 12.4** Distance dependence of the heat current in non-Markovian systems for harmonic (full), and anharmonic (dotted) models. (a) Gaussian white noise; (b) O–U noise with $\tau_c = 8 \times 10^{-3}$ ps; (c) O–U noise with $\tau_c = 0.01$ ps; (d) O–U noise with $\tau_c = 0.04$ ps. $T_R = 300$ K, $T_L = 0$ K, $\epsilon = 50$ ps$^{-1}$ in all cases.

lower frequencies with increasing molecular size. Since within the O–U model $\gamma(\omega)$ (reflecting the system–bath coupling) is larger at lower frequencies, the current is enhanced with distance. Next we explain the intricate current–distance behavior of anharmonic systems. Scattering processes induced by anharmonic interactions are more influential in long chains. In Markovian systems, this results in the enhancement of the junction resistance, thus in the reduction of current with $N$. However, in non-Markovian systems, these scattering effects are actually beneficial for transferring energy from molecular modes, which are above the reservoirs' cutoff frequencies, into low-energy modes that overlap with the solids' vibrations. The interplay between these two effects leads to a rich behavior: If $\tau_c^{-1}$ is higher than the molecular frequencies, here of the order of 150 ps$^{-1}$, anharmonic effects lead to the reduction of current with size (see Figure 12.4a and b) In the opposite small cut-off limit (Figure 12.4d), $\tau_c^{-1} = 25$ 1/ps, harmonic systems can transfer only those modes that are in the reservoirs energy window, while anharmonic junctions better conduct by scattering high-energy modes into low frequencies. For $\tau_c \sim 0.01$ ps the two effects practically counteract and $J_A$ weakly depends on distance (Figure 12.4c).

### 12.5.2.2 Thermal Rectification

Thermal rectification (asymmetric energy conductance), was discussed in Section 12.4.2 within a quantum master equation approach. Here we display rectification

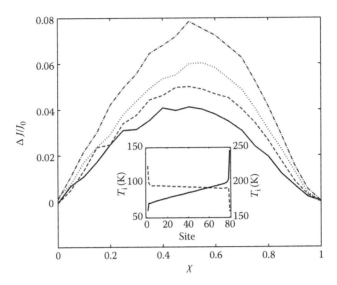

**FIGURE 12.5** Thermal rectification in anharmonic chains ($\lambda = 6$). Solid, dashed, dotted, dashed-dotted lines correspond to $N = 10, 20, 40$, and $80$, respectively, with $\epsilon = 25$ 1/ps, $T_h = 300$ K, $T_c = 0$ K. (Inset) The temperature profile for $N=80$, $\chi = 0.5$ case with $T_L = T_c$; $T_R = T_h$ (solid line), $T_L = T_h$; $T_R = T_c$ (dashed line, left y-axis).

in a classical system. Figure 12.5 exemplifies the effect in an anharmonic molecular chain, as described above, assuming white thermal reservoirs with $\epsilon_{R,L} = \epsilon(1 \pm \chi)$; $|\chi| \leq 1$ measures the asymmetry in the system. We plot $\Delta J \equiv J(T_L = T_h; T_R = T_c) + J(T_L = T_c;$ $T_R = T_h)$ over $J_0 \equiv |J(\chi) = 0|$, and find that rectification is enhanced with size due to the increasing importance of inelastic processes.

## 12.6 EXPERIMENTAL CHALLENGES AND RESULTS

The experimental effort in probing vibrational energy flow at the nanoscale is driven by miscellaneous challenges. First, there is a great interest in developing devices with high thermal conductivity. This is crucial in the field of molecular electronics, as overheating of nanoscale devices might cause structural instabilities undermining the junction integrity [10–12]. The thermal conductivities of nanostructures are often substantially reduced from the bulk material values due to phonon confinement effects. Carbon nanotubes, in contrast, were proven to be superior thermal conductors, and the thermal conductivity of a single-wall carbon nanotube was measured to be higher than 3000 W/m K at room temperature [42].

The second practical challenge in phononic transport rises from heat-to-electricity conversion technology. Here, in order to develop efficient thermoelectric materials, one seeks structures with ultralow heat conductivity in conjunction with high electrical conductivity and a high Seebeck coefficient. Ultralow conductivity was manifested in disordered-layered $WSe_2$ crystals [13], probably due to the localization of lattice vibrations. Si nanowires show enhanced thermoelectric efficiency compared to the bulk value. Here the reduction of the wire thermal conductivity probably stems from scattering events with the edges [14,15].

From a different perspective, theoretical research laying the groundwork for the Fourier's law in low dimensions has driven experimental efforts into elucidating transport mechanisms. Topics of particular interest are identifying a diffusive–ballistic transition in nanoscale systems [43], and building nonlinear conducting devices, e.g., an asymmetrically mass-loaded carbon nanotube, where heat might be carried by lattice solitons [36]. Intramolecular vibrational energy flow in bridged azulene–anthracene compounds could be explained by assuming ballistic energy transfer across the chain [44]. Likewise, self-assembled monolayers of hydrocarbon molecules display ballistic transport [24,45]; A micron-length individual carbon nanotube conducts heat ballistically without showing signatures of phonon–phonon scattering for temperatures up to 300 K [46].

Finally, minimization of mechanical devices, e.g., refrigerators and pumps, to the molecular scale is a topic of great interest for technologies such as chemical sensing, power generation, and energy conversion. A tunable thermal link [47], a nanotube-based thermal rectifier [36], and a phonon waveguide [48] are among the nanoscopic mechanical devices that were recently realized.

## 12.7  SUMMARY

We have presented here several theoretical tools for exploring mechanisms of heat flow in molecular junctions: a generalized Langevin equation approach, the master equation formalism, and classical MD simulations. Using these techniques, we have inferred that the thermal conductance of nanosystems results from an intricate interplay between the molecular structure, the contact properties, and the reservoir's spectral functions. Our minimal single-mode junction model clearly exposed the role of anharmonic interactions in bringing out nonlinear transport characteristics. This has lead us to propose unique thermal devices, a thermal rectifier and a heat pump.

Specific challenges to be pursued are the following: (1) Developing a general framework for describing quantum heat flow in molecules, incorporating anharmonic interactions and strong molecule-interface couplings [17]. (2) Understanding mechanisms for enhancement and reduction of heat transfer, important for molecular electronics and thermoelectric applications [13]. (3) Exploring dynamics while considering both electronic and nuclear degrees of freedom. Here we would like to reveal how does the interaction between the two components, electrons, and phonons, essentially affect charge transfer as well as energy transmission in complex systems [12].

Understanding heating issues in molecular systems is important both from the fundamental aspect, clarifying the connection between nonequilibrium statistical mechanics and its manifestations at the macroscopic scale, and for technological applications, for building molecular-level devices. Heating processes should also be explored for controlling molecular reactivity and for tuning thermal properties of materials. More broadly, exploration of the quantum properties of nanoelectromechanical structures might lead to a new paradigm for information processing and computation [49]. Integration of nanoscale devices with biological systems is another promising avenue for emerging biomedical applications such as chemical recognition, detection, and single-molecule manipulations in the living cell [50].

## REFERENCES

1. D. G. Cahill, et al., *J. App. Phys.* **93**, 793 (2003).
2. S. Lepri, R. Livi, and A. Politi, *Phys. Rep.* **377**, 1 (2003), and references therein.
3. L. G. C. Rego and G. Kirczenow, *Phys. Rev. Lett.* **81**, 232 (1998).
4. K. Schwab, E. A. Henriksen, J. M. Worlock, and M. L. Roukes, *Nature* **404**, 974 (2000).
5. P. B. Allen and J. L. Feldman, *Phys. Rev. B* **48**, 12581 (1993).
6. X. Yu and D. M. Leitner, *J. Phys. Chem. B* **107**, 1698 (2003); X. Yu and D. M. Leitner, *J. Chem. Phys.* **122**, 054902 (2005).
7. N. Mingo and D. A. Broido, *Phys. Rev. Lett.* **95**, 096105 (2005).
8. D. K. Campbell, P. Rosenau, and G. M. Zaslavsky, *Chaos* **15**, 015101 (2005).
9. A. Ozpineci and S. Ciraci, *Phys. Rev. B* **63**, 125415 (2001).
10. D. Segal and. Nitzan, *J. Chem. Phys.* **117**, 3915 (2002).
11. R. D'Agosta, N. Sai, and M. Di Ventra, *Nano Lett.* **6**, 2935 (2006).
12. M. Galperin, M. A. Ratner, and A. Nitzan, *J. Phys. Cond. Mat.* **19**, 103201 (2007).
13. C. Chiritescu, et al., *Science* **315**, 351 (2007).
14. A. I. Boukai, et al., *Nature* **451**, 168 (2008).
15. A. I. Hochbaum, et al., *Nature* **451**, 163 (2008).
16. N. Mingo, *Phys. Rev. B* **74**, 125402 (2006).
17. J.-S. Wang, J. Wang, and T. J. Lue, *Eur. Phys. J. B* **62**, 381 (2008).
18. D. Segal, A. Nitzan, and P. Hänggi, *J. Chem. Phys.* **119**, 6840 (2003).
19. D. Segal, *Phys. Rev. B* **73**, 205415 (2006).
20. D. Segal and A. Nitzan, *Phys. Rev. Lett.* **94**, 034301 (2005); *J. Chem. Phys.* **122**, 194704 (2005).
21. D. Segal and A. Nitzan, *Phys. Rev. E* **73**, 026109 (2006).
22. D. Segal, *J. Chem. Phys.* **128**, 224710 (2008).
23. E. Cortes, B. J. West, and K. Lindenberg, *J. Chem. Phys.* **82**, 2708 (1985).
24. Z. Wang, et al., *Science* **317**, 787 (2007).
25. M. Michel, G. Mahler, and J. Gemmer, *Phys. Rev. Lett.* **95**, 180602 (2005); M. Michel, J. Gemmer, and G. Mahler, *Phys. Rev. E* **73**, 016101 (2006).
26. L.-A. Wu and D. Segal, *Phys. Rev. E* **77**, 060101(R) (2008).
27. M. Terraneo, M. Peyrard, and G. Casati, *Phys. Rev. Lett.* **88**, 094302 (2002).
28. B. Li, L. Wang, and G. Casati, *Phys. Rev. Lett.* **93**, 184301 (2004).
29. B. Li, L. Wang, and G. Casati, *App. Phys. Lett.* **88**, 143501 (2006).
30. L. Wang and B. Li, *Phys. Rev. Lett.* **99**, 177208 (2007).
31. S. H. Lin, *J. Chem. Phys.* **61**, 3810 (1974).
32. L.-A. Wu and D. Segal, *J. Phys. A: Math. Theor.* **42**, 025302 (2009).
33. D. Segal, *Phys. Rev. E* **77**, 021103 (2008).
34. D. Segal, *Phys. Rev. Lett.* **100**, 105901 (2008).
35. G. Wu and B. Li, *Phys. Rev. B* **76**, 085424 (2007).
36. C. W. Chang, D. Okawa, A. Majumdar, and A. Zettl, *Science* **314**, 1121 (2006).
37. V. P. Carey, G. Chen, C. Grigoropoulos, M. Kaviany, and A. Majumdar, *Nanoscale Microscale Thermophys. Eng.* **12**, 1 (2008).
38. P. K. Schelling, S. R. Phillpot, and P. Keblinski, *Phys. Rev. B* **65**, 144306 (2002); H. A. Patel, S. Garde, and P. Keblinski, *Nano Lett.* **5**, 2225 (2005).
39. G. E. Uhlenbeck and L. S. Ornstein, *Phys. Rev.* **36**, 823 (1930).
40. J. Luczka, *Chaos* **15**, 026107 (2005).

41. S. Lifson and P. S. Stern, *J. Chem. Phys.* **77**, 4542 (1982).
42. P. Kim, L. Shi, A. Majumdar, and P. L. McEuen, *Phys. Rev. Lett.* **87**, 215502 (2001).
43. J. Shiomi and S. Maruyama, Jap. *J. Appl. Phys.* **47**, 2005 (2008).
44. D. Schwarzer, P. Kutne, C. Schröder, and J. Troe, *J. Chem. Phys.* **121**, 1754 (2004).
45. R. Y. Wang, R. A. Segalman, and A. Majumdar, *Appl. Phys. Lett.* **89**, 173113 (2006).
46. H.-Y. Chiu, et al., *Phys. Rev. Lett.* **95**, 226101 (2005).
47. C. W. Chang, et al., *Appl. Phys. Lett.* **90**, 193114 (2007).
48. C. W. Chang, D. Okawa, H. Garcia, A Majumdar, and A. Zettl, *Phys. Rev. Lett.* **99**, 045901 (2007).
49. K. C. Schwab and M. L. Roukes, *Phys. Today* **58**, 36 (2005).
50. J. L. Arlett, et al., *Lect. Notes Phys.* **711**, 241–270 (2007).

# Part IV

**Conformational Transitions and Reaction Path Searches in Proteins**

# 13 Tubes, Funnels, and Milestones

*Ron Elber, Krzysztof Kuczera, and Gouri S. Jas*

## CONTENTS

## 13.1 INTRODUCTION

A great promise of atomically detailed simulations is the generality of the model. Molecular dynamics (MD) simulations are used to study highly complex and diverse processes (such as energy flow and conformational transitions in proteins) with essentially the same mathematical model. We call this approach "the standard model of computational biophysics," or in short "the standard model." The classical equations of motion are solved with an empirical force field $U(\vec{X})$, which is transferable between different molecular systems. The vector $\vec{X}$ points to the positions of all the atoms in the molecular system. The number of atoms that are included in simulations (and the length of the vector $\vec{X}$) varies from a few hundreds to a submillion, making these simulations computationally challenging.

The potential energy $U(\vec{X})$ has a relatively fixed functional form [1] that is used by a large number of researchers. The standard form is based on a sum of covalent terms of bonds, angles, and torsions. It also includes noncovalent terms accounting for atomic hard cores, dispersion forces, and electrostatic interactions. A great deal of work was done on the design of the functional form of the potential energy function and on determining a suitable set of parameters. While much more remains to be optimized, current potentials allow for sound simulations of many systems. In this chapter, we take the molecular potential for granted and focus on other challenges that MD simulations face.

A significant challenge of MD is to expand the timescale of the simulations. The time is too short in many cases to address the phenomenon of interest. The problem is easy to understand considering the basic numerical algorithm of MD. A MD simulation computes a time series. We start from an initial coordinate and velocity vectors, say $X(0)$, and $V(0)$. The vector $X(0)$ is a crystallographic structure of a protein, or a predetermined molecular structure. The velocities are sampled from a known distribution (e.g., the Maxwell probability density,

$$p(v_i) = \left( \frac{m_i}{2\pi kT} \right)^{1/2} \exp\left( -\frac{m_i v_i^2}{2kT} \right)$$

where
 $v_i$ is one of the elements of the vector $V$
 $m_i$ is the mass of atom $i$
 $k$ is the Boltzmann constant
 $T$ the absolute temperature

Newton's equations of motion are integrated numerically. That is, the differential equations

$$M \frac{d^2 X}{dt^2} = -\nabla U(X)$$

are solved using the above initial conditions. An integration step with the velocity Verlet algorithm [2] is illustrated below:

$$X(t + \Delta t) = X(t) + V(t) \cdot \Delta t - \frac{M^{-1}}{2} \Delta t^2 \cdot \nabla U(X(t))$$

$$V(t + \Delta t) = V(t) - \frac{M^{-1}}{2} \Delta t \cdot (\nabla U(X(t)) + \nabla U(X(t + \Delta t)))$$

The integration step is repeated until the time of interest is reached. To maintain numerical stability the time step $\Delta t$ must be smaller than the fastest motion in the system. Bond vibrations and hard-core atomic collisions have periods of femtoseconds and picoseconds. A typical time step that is used to integrate the equations of motion is therefore a femtosecond ($10^{-15}$ s). The number of steps accessible to a typical computational setup is about $10^8$, which makes it possible to reach timescales of hundreds of nanoseconds or $10^{-7}$ s in biological molecules. A considerable body of research in numerical analysis aims at increasing this time step and extending the overall timescale. So far, these efforts were not able to increase the computational efficiency by more than a factor of 2 or 3.

   Why should we care about longer timescales? We should care because many interesting processes in molecular biophysics require timescales significantly longer than the fraction of a microsecond approachable by routine MD calculations. For example, the conversion of chemical energy to mechanical energy in the protein myosin requires about 10 ms [3]. To reach the timescale relevant to the energy conversion, $10^{13}$ steps (each step 1 fs) are needed. At present, there is no computer system, algorithm, or software capable of computing a millisecond trajectory of the standard model in a reasonable running time (less than several months). We call this challenge "overcoming the temporal barrier."

Moreover, even if we were able to produce a trajectory of a relevant timescale, a single trajectory of time length $t$ is insufficient to study an experiment of the same time ($t$). Each of the large number of molecules in the experiment (different initial conditions of the differential equations in the computational model) reacts differently and contributes to a distribution of observed quantities. A distribution cannot be captured by a single trajectory executing one transition.

To compute significant samples and averages of time-dependent phenomena, a collection (ensemble) of trajectories with time lengths comparable to the timescale of the experiment are needed. The requirement for a significant sample makes the calculation even more challenging compared to our estimate above for a single trajectory. A typical (and actually quite minimal) sample would be of 100 trajectories, making the complete calculation 100 times more expensive. It is therefore no wonder that theories and approximate procedures were developed to allow simulations of the standard model at much longer timescales.

We note that the temporal barrier is also connected to potential design. Deficiencies of the potential are detected nowadays when experimental measurements are compared to novel long-time calculations. These computations exploit advances in algorithms, software, and hardware to explore new time domains. The availability of simulation times not accessible before suggests new opportunities to calibrate force fields. Another connection of temporal barriers and potentials is via coarse-grained modeling. The timescales are extended by further reduction and simplification of the standard model (e.g., grouping a number of atoms into a point mass). However, these coarse-grained approaches are not as widely used as the standard form and the approximations are not always controllable. In this chapter, we remain with the standard model.

A key idea of theories and algorithms to extend timescales is the use of short time trajectories to construct a probabilistic model of rate computed at much longer timescales [4–8]. From a physical perspective, long-time dynamics can come in many flavors; however, it is typically a mix of two kinds of fundamental processes: (1) diffusive and (2) activated dynamics. One class of slow processes typical to chemical reactions (activated dynamics) was the focus of theoreticians in the last half a century or so [9].

Activated processes are a special case of long time trajectories in which a significant free energy barrier separates the states of reactants and products (Figure 13.1). The large barrier dictates one type of transitions that are rare and fast. If we take pictures of the system at some fixed time intervals, then most of the time we do not observe special activity or progress. The system is observed again and again in the reactant state. If we wait long enough, we are likely to suddenly observe it in the product state without observing the actual transition event. The transition time in activated processes is much shorter than the waiting time. It is difficult to estimate this time by the sampling procedure described above since the time interval must be very short to observe the transition. The experimental time, which is an average over many trajectories, is a function of both the transition and the waiting times. Ironically, it is similar in magnitude to the much longer waiting time during which nothing happens. Nevertheless, one's curse can be a another's blessing.

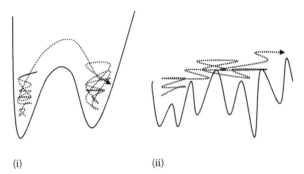

(i)                                                      (ii)

**FIGURE 13.1**   A schematic representation of (i) an activated and (ii) a diffusive energy landscapes and their corresponding trajectories. The diffusive energy landscape is typical for downhill processes that follow protein activation (e.g., in the case of allosteric transition in which local perturbation fundamentally changes the shape of the underlining energy surface).

The observation that the transition time of the trajectory is extremely short, and the existence of a single dominant barrier, can be exploited to design algorithms for the calculation of rate constants of these rare (but rapid) processes. The calculations are based on statistical properties of the transitional (short) trajectories and their weights in the complete ensemble of pathways. In the calculations of rate constants, it is typically assumed that the kinetics follows an exponential law. Let $P(t)$ be the probability that the system remains in the reactant state after time $t$. Exponential kinetics means that the time evolution of the reactant probability is $P(t) = \exp(-\kappa t)$ where $\kappa$ is a rate constant (with units of the inverse of time). The initial condition is $P(0) = 1$. This is considered the simplest behavior and is frequently observed experimentally in activated processes.

The dynamics can be complex in biophysics. The energy landscape is rough in the sense that it includes large number of barriers and local minima. Studies of minima and saddles of energy surfaces of peptides in vacuum illustrate this property [10,11]. For example, in protein activation, these barriers are rarely high and the motions are only moderately activated. By moderately activated we mean that the energy difference between the barrier and the minimum $\Delta U = U_{barrier} - U_{minimum}$, is of order of a few $kT$. As a result the Boltzmann factor $\exp(-\Delta U/kT)$ is not exceptionally small. In this case, there is no reason to believe that the kinetics is exponential. Indeed a number of studies suggest significant deviation from the simplest behavior.

Power law ($t^\alpha$) and stretched exponential ($\exp(-\lambda t^\beta) 0 < \beta < 1$) were used in the past to describe the nonexponential kinetics of protein relaxation [12]. From a pictorial viewpoint, this type of reaction is fundamentally different from activated dynamics. The system progresses from reactant to product less abruptly (Figure 13.1). Different time slices show progress between the two states and the system moves continuously toward the product. The complex energy surfaces of protein folding and of conformational transitions frequently make simple exponential relaxation the exception rather than the rule. This is particularly relevant in biology in which conformational transition are initiated by a local structural change that impact the underlining energy surface and make it less activated and more accessible to room temperature processes.

Computational methods that focus on short time transitional trajectories between states separated by large barriers may be inappropriate for many biophysical processes. Perhaps the most distinct property of milestoning, the approach we discuss in the present chapter, is that it can handle with comparable efficiency mixed processes. It provides significant speedup factors that vary (depending on the system investigated) from about 10 [13] to a factor of 1000 [14] and to 1 million [15]. It is based on physically motivated approximations that can be tested and made more accurate (at the expense of smaller computational speedup) [13]. Interestingly, while milestoning can be used for a wide range of dynamics (Newton, Langevin, etc.), for Brownian dynamics, milestoning can be made a mathematically exact procedure [16].

In the next section, we describe the theory and algorithm of milestoning. We then continue to describe an application to the folding kinetics of a helical heteropeptide peptide.

## 13.2 MILESTONING

### 13.2.1 THEORY

We describe below the approach of milestoning. The milestoning theory follows from an intuitive expression in which we capture the characteristics of the microscopic dynamics in a non-Markovian kernel and then solve the dynamics within the assumption of local equilibrium or constant flux.

We partition the space that includes the reactants and products by hypersurfaces that are denoted by $s$ and are called "milestones" (Figure 13.2). The hypersurfaces $s = 0$ and $s = L$ are the reactants and products respectively. The other $L - 1$ hypersurfaces $s = 1, \ldots, L - 1$ are intermediates that we used to speed up the calculation. The choice of the hypersurfaces is critical and corresponds to a correct selection of an order parameter or a reaction coordinate. In the example given in this chapter, we compute the reaction coordinate and describe how to make the calculations less sensitive to the choice of the milestones.

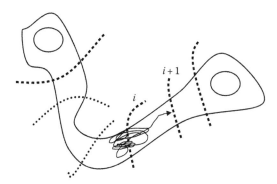

**FIGURE 13.2** A schematic representation of milestoning surfaces on a two-dimensional energy surface (dotted lines). Also shown a single trajectory initiated at milestone $i$ and terminated on milestone $i + 1$.

The probability of being at milestone $s$ at time $t$ is $P_s(t)$. Microscopically, "being at a milestone" means that the last milestone passed by the trajectory is $s$ (and not that the trajectory is located exactly at the hypersurface $s$). As discussed below, the trajectory terminates when it touches either milestone $s - 1$ or $s + 1$. It therefore goes without saying that the trajectory is always between milestones $s - 1$ and $s + 1$. The "uncertainty" in the position of the system illustrates the spatial coarse-graining of milestoning.

To compute $P_s(t)$, we estimate from atomically detailed simulations the transition probability density $K_{s,s'}(\tau)$. It is the probability of making a transition from state $s$ into state $s'$ after an "incubation" or "waiting" time $\tau$ in the state $s$. In Section 13.2.2, we describe how $K_{s,s'}(\tau)$ is computed from short time trajectories. The incubation time captures memory effects in which the transition probability between $s$ and $s'$ depends on the time spent already in $s$.

The transition probability density $K_{s,s'}(\tau)$ is normalized as follows: the integral $P_{s,s'} = \int_0^\infty K_{s,s'}(\tau)d\tau$ is the probability that a trajectory that is initiated at milestone $s$ arrives at milestone $s'$ before any other milestone $s''(s'' \neq s', s)$. At the long time limit, the trajectory must arrive to another milestone, and we have $\sum_{s'} p_{s,s'} = \sum_{s'} \int_0^\infty K_{s,s'}(\tau)d\tau = 1$. Due to the potentially complex dependence on time of the incubation the spatial transition can be non-Markovian $s$ space. If and only if the transition probability density follows

$$K_{s,s'}(\tau) = \frac{p_{s,s'}}{\kappa_{s,s'}} \exp(-\kappa_{s,s'}\tau)$$

the process is Markovian. Formally the transition can be made Markovian if the state is defined as the pair $(s, \tau)$ [8].

The transition probability density $K_{s,s'}(\tau)$ is the only microscopically derived function that we need for milestoning. Note that we assume that $K_{s,s'}(\tau)$ is independent of the absolute time. This assumption is not valid in systems that strongly deviate from equilibrium or from a stationary state. Recently Vanden Eijnden et al. have shown that milestoning can be made mathematically exact if the microscopic dynamics is Brownian and the hypersurfaces are committers [16].

To facilitate accurate calculation for more general dynamics we restrict our attention to systems that satisfy the following requirements.

*Condition* (i): Trajectories that arrive at milestone $s'$ are distributed in the hypersurface according to a known time-independent distribution $\rho_{s'}$.

*Condition* (ii): Let the transition times between two nearby milestones $s$ and $s + 1$ be $\tau_{s,s+1}$ and $\tau_{s+1,s}$. For all $s$, the transition times $\tau_{s+1,s}$ and $\tau_{s,s+1}$ are statistically independent.

If the system on arrival to the nearby milestone had sufficient time to equilibrate and the equilibrium distribution is known (e.g., $\rho_s \propto \exp\left[-\beta U(\vec{X})\right](\vec{X} \in s)$) then condition (i) is satisfied. Condition (ii) is not necessary if we are only after the first

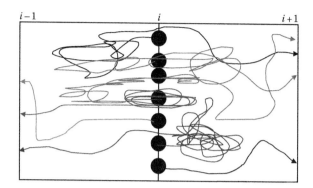

**FIGURE 13.3** **(See color insert following page 172.)** A schematic representation of sampling terminating trajectories between milestones $i$ and $i \pm 1$. The milestones are planes. Each trajectory is drawn with a different color and initiated from a lack circle at milestone $i$. The probability of termination at milestone $i+1$ is 4/7 and to terminate at $i$ is 3/7.

moment of the overall first passage time [16]. In the present study, our statistics are not very high and the first moment is all we can get reliably.

The calculation of $K_{s,s'}(\tau)$ is done by sampling trajectories between a pair of hypersurfaces $s$ and $s'$ (Figure 13.3). The trajectories are initiated at according to the stationary distribution $\rho_s$ and their termination times at $s'$ are recorded. These distributions are binned to estimate the probability that the system transitions between $s$ and $s'$ after incubation time $\tau$.

We argued and illustrated [13,14] that it is much easier to compute the above matrix than to perform the complete simulations from reactants to products. Arguments in support of the expected speedup are presented in Section 13.2.2. This matrix is finally used in a probabilistic non-Markovian framework to obtain the overall kinetics of the system.

To demonstrate how the transition probability density $K_{s,s'}(\tau)$ can be used to compute $P_s(t)$—the probability of being at milestone $s$ at time $t$—it is convenient to define another function $Q_s(t)$. It is the probability density that a trajectory will make a transition into $s$ at time $t$. The normalization of $Q_s(t)$ is a little trickier compared to $K_{s,s'}(\tau)$ and is done with respect to trajectories. The set of events are trajectories of time periods between $t$ and $t + dt$. Each of these trajectories pass (as a function of time) through milestones and is said to be in $s$ if the last milestone it passes is $s$. The time spent at $s$ (incubation time $\tau$) is set to zero when the trajectory passes $s$ and the milestone it passes earlier was not $s$, (say $s'$ ($s' \neq s$)). A fraction of the trajectories of time $t + dt$ transition between $s$ and $s'$ at the time interval $[t, t + dt]$. This fraction is denoted by $Q_s(t)dt$. It is explicitly given by a ratio of path integrals

$$Q_s(t)dt = \frac{\sum_{s' \neq s} \int \delta(y(X(t)) - s') DX(t) \int \delta(y(X(t + dt)) - s) DX(t + dt)}{\int \sum_{s'} \delta(y(X(t + dt)) - s') DX(t + dt)}$$

The expression $DX(t)$ denotes a volume element in path space of time length $t$. The function $y(X(t)) = z$ maps the trajectory, $X(t)$, into a corresponding milestoning index $z$, which is the last milestone crossed by the trajectory before or at $t$. The time interval $dt$ is small allowing one crossing between $t$ and $t + dt$ (at time zero the system is at the reactant ($s = 0$)). The above equation reads as follows: For a transition into $s$ to occur between $t$ and $t + dt$, it is necessary for the system to be at $s$ at time $t + dt$ and at $s'$ (which is different from $s$) at time $t$. The denominator is a summation over all trajectories regardless of the milestone they end up at time $t + dt$. For clarity, we kept the summation over $s'$ of the delta functions in the denominator. This sum is simply one and can be omitted. Only the path integral is necessary for the normalization to work.

With $K_{s,s'}(\tau)$ and the definition of $Q_s(t)$ known, Equation 13.1, which is the central equation of milestoning, simply balances transition probabilities. The system is initiated at zero time and starting probabilities are injected into the milestones. At later times, we consider transitions between the states. To make a transition into $s$ from one of the nearby milestones $s'$, it is necessary to transition first to $s'$, wait (or incubate) at $s'$ for time $\tau$, and then transition into $s$. The probability density of making a transition into $s'$ at time $t - \tau$ is $Q_{s'}(t - \tau)$. The probability density of making a transition from $s'$ to $s$ after waiting time $\tau$ is $K_{s',s}(\tau)$. Finally, a summation over all states $s'$ that are directly connected to $s$, and over all incubation times $\tau$ from 0 to $t$ gives the equation

$$Q_s(t) = \delta(t - 0^+)P_s(0) + \int_0^t Q_{s'}(t - \tau)K_{s',s}(\tau) \cdot d\tau \tag{13.1}$$

It is a vector-matrix equation in $s$ space and an integral equation in time. The unknown is the vector of functions $Q_s(t)$. This equation is called continuous time random walk (CTRW) and was used in phenomenological modeling of transport [17]. Equation 13.1 is closed and can be solved provided that $K_{s',s}(\tau)$ is known. Our contribution is to show how detailed microscopic dynamics is used to compute $K_{s',s}(\tau)$ or its moments (see below).

Our interest focuses on $P_s(t)$, the probability of being at $s$ at time $t$. With $Q_s(t)$ determined from Equation 13.1 we write $P_s(t)$ as the integral of the probability density to make a transition into $s$ at an earlier time $t'$ and to remain at $s$ (avoid transitions to other states $s'$) until time $t$. Summation over all channels $s'$ gives

$$P_s(t) = \int_0^t Q_s(t')\left[1 - \sum_{s'} \int_0^{t-t'} K_{s';s}(\tau) \cdot d\tau\right] dt' \tag{13.2}$$

The explicit numerical solution of Equations 13.1 and 13.2 is now a topic in applied mathematics and was obtained with different approaches [8,13,16,18]. The most obvious one is to solve the integral equation (Equation 13.1) by small time steps. Since the number of degrees of freedom was greatly reduced and the functions considered are much smoother in time compared to MD, the computational effort of solving the integral equation is still negligible compared with the calculations of the trajectories:

$$Q_s(0) = P_s(0)/\Delta t$$

$$Q_s(\Delta t) = Q_s(0) + \sum_{s'} K_{s,s'}(\Delta t) \cdot Q_{s'}(0) \cdot \Delta t$$

(13.3)

$$Q_s(2\Delta t) = Q_s(0) + \sum_{s'} [K_{s,s'}(\Delta t) \cdot Q_{s'}(\Delta t) + K_{s,s'}(2 \cdot \Delta t) \cdot Q_{s'}(0)]$$

. . .

The finite range of the memory in time (or the decay of the function $K_{s',s}(\tau)$) makes Equation 13.3 accessible to direct numerical calculations.

Another solution is based on Laplace transforms of Equations 13.1 and 13.2 and algebraic manipulation of the transforms [18]. Below we quote one analytical result. Define the off diagonal matrix $(\hat{K})_{s,s'}(t) = K_{s,s'}(t)$ $s \neq s'$, the time integral $\langle f \rangle = \int_0^\infty f(t)dt$, and an average over an ensemble of trajectories $\bar{f}$. A useful measure of the kinetic properties of the system is the overall mean first passage time. It is defined as the mean time required for trajectories initiated at the reactant to reach the product for the first time. The ensemble average of the overall first passage time $\bar{\tau}$ is given by

$$\bar{\tau} = I \cdot \langle \tau \cdot \hat{K}_{s,s'}(\tau) \rangle \left[ I - \langle \hat{K}_{s,s'}(\tau) \rangle \right]^{-1} \cdot \varepsilon_i$$

(13.4)

where

$I$ is the identity matrix

$\varepsilon_i$ is a unit vector in the direction of the initial milestone $i$

The last milestone is set to be absorbing. It is interesting to comment that the last formula can be computed without explicit knowledge of the probability density $K_{s,s'}(\tau)$. We only need the first moments of $\tau$ (the nominator) and the zero moments of the distributions (the denominator), which we can obtain with direct sampling. The zero moment of $K_{ij}(\tau)$ is the probability that a trajectory that starts at milestone $i$ ends at milestone $j$.

Equation 13.4 is useful since the calculations of moments are easier than estimates of the distributions (the statistics required for accurate estimates of the moments are much smaller). We emphasize that the first passage time is not the inverse of a rate constant. The use of a rate constant assumes that the time evolution is exponential. The mean first passage time is a well-defined measure of the progress of the reaction with no reference to exponential relaxation. The inverse of the mean of the first passage time corresponds to a rate constant only if the distribution of first passage times is exponential. This is true not only for the overall mean first passage time but also for the local first passage times computed as the first moments of the distributions $K_{s,s'}(\tau)$.

## 13.2.2 Algorithm

It is important to emphasize that the theory described in the previous section is "dynamic free." We do not assume a particular model of the dynamics as Langevin,

Brownian, or Newtonian mechanics. All reasonable forms of dynamics (dynamics that approach equilibrium) can be used to compute the transition probability density matrix or its moments. We compute the transition matrix locally as discussed below and we do not globally fit parameters to a particular coarse-grained dynamical model.

What is the gain in studying the kinetic with milestones instead of using straight-forward MD? This question has been discussed extensively [13]. Practical gains were illustrated in the examples of a conformational transition in alanine dipeptide and in the allosteric transition of a dimeric hemoglobin [13,14]. The larger the system, the more significant was the gain. Recently, a millisecond process in myosin was studied with milestoning using nanosecond trajectories (a factor of million speedup) [15].

The dominant gain comes from exponential enrichment of rare trajectories. Consider three sequential milestones $s - 1$, $s$, and $s + 1$. We wish to reach milestone $s + 1$ starting from $s-1$, a process that is rare. Let the number of trajectories that are required to sample one transitional event starting from $s - 1$ and reaching $s$ be $n$, and from $s$ to $s + 1$ be $n'$. To sample a transitional event from $s - 1$ to $s + 1$ with straightforward MD we need $n \times n'$ trajectories. In contrast, in milestoning, we need only $n + n'$ trajectories [13]. The gain is exponential. We typically employ about 100 milestones and 100 trajectories (per milestone) for a problem. The speed in this case for a monotonically activated process is the difference between $100^{100}$ and $100 \cdot 100$.

Another speedup factor is for barrierless processes. Consider a reaction in which the system diffuses freely for a length $L$. The time to reach the end of the line using straightforward trajectories is proportional to $L^2$. In milestoning, we chop the complete length $L$ to (say) $N$ pieces. The time to diffuse through one piece is $(L/N)^2$. There are $N$ pieces and therefore the time to reach the destination in milestoning is $(L/N)^2 \cdot N = L^2/N$. We obtain a speedup proportional to the number of milestones.

In the next section, we discuss a concrete example. We consider progress along a reaction coordinate measured by passing hypersurfaces (milestones) orthogonal to it. In Figure 13.3, we sketch a terminating trajectory between milestones $s$ and $s - 1$. An ensemble of such trajectories is used to compute moments of the local first passage times (between $s$ and $s \pm 1$) and the overall mean first passage time according to Equation 13.4.

## 13.3   EXAMPLE: FOLDING OF A SOLVATED HELICAL PEPTIDE

### 13.3.1   INTRODUCTION

We consider the structural transition of the peptide Ac-WAAAH(AAARA)$_3$ A-NH$_2$ from a helix to an unfolded state. This transition was studied extensively experimentally [19,20]. The peptide is rich in alanine residues with three more positively charged amino acids. Alanine has a significant tendency to form a helix and the charged residues such as arginine help to dissolve the peptide in buffer solution. Residues of the same charge seek an arrangement in space that will maximize the distance between the charges and a helical configuration is one way of maximizing charge–charge separation. Hence, the peptide is designed to prefer a helical conformation.

An acetyl and an amide group block the N and C termini of the peptide chain. The tryptophan residue is added as a probe to collect time-resolved fluorescence signal under nanosecond T-jump spectroscopy, allowing measurement of coil to helix transition. The experimentally estimated relaxation time at room temperature for this transition is about 300 ns. The inverse of the experimental relaxation time is the sum of two rate constants, from the unfolded to the folded state and back. The equilibrium constant of this transition is about 1, which indicates that the forward and the backward rates are almost the same. The experimental first passage time from the folded to the unfolded state (which we estimate computationally in this chapter) is therefore 600 ns. This timescale seems achievable within the standard model and atomically detailed simulations. However, one should keep in mind that an ensemble of trajectories is required to study kinetics. The calculation of kinetics will be at least 100 times more expensive than the calculation of a single trajectory and therefore difficult to do with the usual standard model.

A widely used mechanistic picture of protein folding (and helix formation is a subprocess in protein folding) is that of a funnel (Figure 13.4) [21]. The unfolded state is rich in multiple unstructured conformations and the diversity of structures is decreasing profoundly as we approach the unique three-dimensional structure of the folded state. A simple illustration is the formation of hydrogen bonds. There are a lot more options of forming the first hydrogen bond between backbone amides than after the first few were made. The significant changes in entropy make direct simulations with MD difficult since the characterization of the unfolded state requires a large computational effort.

A significant help in the studies of folding is the use of collective order parameters to follow the thermodynamics of the reaction. For example, we ask what is the

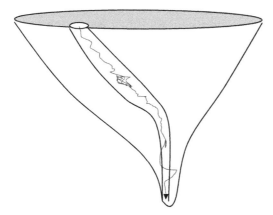

**FIGURE 13.4** A schematic view of a funnel and a tube. The funnel is drawn with the assumption that the underlining initial structure is made unstable by an abrupt change in the initial condition and the energy surface, e.g., pH. The upper part is the unfolded state and the bottom corresponds to a folded configuration. Also shown a folding trajectory (dotted line) that passes through the tube. Note that the width of the tube does not necessarily correspond to the funnel width.

free energy for a fixed value of the order parameter or a reaction coordinate? Order parameters that were used in past simulations include the radius of gyration, the fraction of native contacts, and the fraction of native hydrogen bonds [22–24]. These collective variables were proven useful in the studies of thermodynamic and equilibrium properties. The definition of the free energy of the reduced set of variable (called here $\Gamma$) is straightforward. Let $F(\Gamma)$ be the free energy expressed in terms of the collective variables. It is exactly and directly derived from the microscopic model with coordinate $X$ as

$$F(\Gamma) = -\frac{1}{\beta}\log\left[\int \exp\left[-\beta U(X)\right]\delta\big(\gamma(X)-\Gamma\big)dX\right]$$

where $\gamma(X)$ is a mapping of the Cartesian coordinates to the collective variables. Unfortunately and in contrast to the above definition of the free energy landscape, there is no such nicely close formula for the study of the kinetics.

A transformation to a set of collective variables in kinetics is likely to satisfy similar conditions as those we discussed for milestoning. For example, we require that the times of relaxation to equilibrium for degrees of freedom that are not collective must be shorter compared to typical times for transitions along the collective variables (milestones). Consider, for example, the radius of gyration as a collective variable. In some cases, a single bond rotation can rapidly change the radius of gyration. In other cases, cooperative (and slow) rotations of multiple bonds are required to switch between two conformations with the same radius of gyration. Hence, the timescale of a conformational transition within the hypersurface of a poorly selected milestone can be exceptionally long. Collective variables for thermodynamics are not necessarily good collective variables for kinetics. The last should be chosen with relaxation times in mind, and it is not even obvious that they exist (e.g., the spectrum of relaxation times can be continuous with no separation of timescales).

It is possible to "design" the milestones to satisfy the timescale requirement. In contrast to the usual definition of reaction coordinates, which is continuous, milestones are discrete and spatially separated. Therefore the user can control the transition times, moving the hypersurfaces away from each other. Of course, larger separation makes the milestoning procedure less efficient. A compromise must be made between efficiency and accuracy.

It is clear that the funnel shape with a vast number of chain conformations in the unfolded state makes a difficult sampling target for calculations of the dynamics. We therefore prefer to work in a tube, which is more accessible to simulations. Hence, we do not require absolute equilibrium in the direction perpendicular to the reaction coordinate. We are satisfied with a local equilibrium in a tube that follows the milestones (sampling is always constrained in time and space and it is assumed that it is appropriate for the problem at hand) (Figure 13.4). Obviously, this picture is not a complete view of the funnel. Nevertheless, it allows for the sampling of a group of reactive trajectories of special properties and of interest.

The tube models a concrete folding mechanism by following the center of the tube and the fluctuations near that center. This concrete mechanism can be examined with respect to the overall funnel picture. For example, if the rate in the tube is exceptionally slow compared to the experimental rate, it is unlikely that the sampled tube is important. Computationally and conceptually, we should partition the funnel to tubes and analyze them one by one. We note that all the tubes meet at the folded state. Therefore the weight of each tube can be measured by overlapping the equilibrium flux into the reactant with the flux from a tube. The calculation of weights of tubes is a topic for future research.

We define the center of the tube with the help of (1) a guiding trajectory, (2) a steepest descent path (SDP) calculation, (3) and a flexible choice of collective variables.

The guiding trajectory for a helix unfolding pathway was generated using a previous replica-exchange MD (REMD) simulations of the blocked WH21 peptide [25]. The simulations used 8 replicas with temperatures spanning 280–450 K and a generalized Born/surface area (GB/SA) continuum solvation model and the CHARMM program [26]. In a 30 ns long REMD trajectory starting from an $\alpha$-helical structure, conformational equilibrium was reached after 15 ns. For the unfolding transition, this simulation predicted a melting temperature $T_m$ of 330–350 K, unfolding enthalpy $\Delta H = -10$ kcal/mol and entropy of $\Delta S = -30$ cal/ (mol K). These values were in reasonable agreement with the experimentally observed values of $T_m = 296$ K, $\Delta H = -12$ kcal/mol and entropy of $\Delta S = -40$ cal/ (mol K) [25]. Our path generation procedure employed the sample of 37,500 structures from the 367 K replica, selected to lie above the calculated melting temperature, in order to assure presence of folded and unfolded states as well as a wide range of intermediate conformers. In this sample, we calculated the root-mean-square deviations (RMSD) of backbone atoms from an ideal helix structure and the number of $\alpha$-helical hydrogen bonds (NAHB) for each structure. For the subset of 3901 structures with NAHB = 8, we performed hierarchical clustering based on backbone RMSD using the MMTSB tool kit [27]. At the highest level, there were two clusters, with populations of 1255 and 2646. A randomly chosen structure from the most highly populated cluster was chosen as a starting point. For this structure, closest neighbors in backbone RMSD were found with NAHB = 7 and NAHB = 9. The search for structures differing by 1–2 hydrogen bonds and backbone RMSD below 2.5 Å was continued until a chain of 16 structures (with NAHB = 0,1,2,3,4, 5,6,7,8,9,10,11,12,14,16, and 17) was generated, spanning the range from NAHB = 17 (corresponding to the $\alpha$-helix) to NAHB = 0 (fully unfolded state). We did not count the hydrogen bonds of the blocking groups. The backbone RMSD between the neighbors in the chain varied from 0.7 to 2.5 Å.

All the calculations described below were done with the program MOIL [28]. Moil is a freely available molecular simulation package with a focus on reaction path and rate calculations. Structures selected from the trajectory described in the previous paragraph were fed into MOIL. They were refined into a detailed SDP from a helix to a "coil" conformation with the action formulation of Olender and Elber [29]. The SDP was computed in a reduced subspace of coarse variables, namely the positions of the $C_\alpha$-s of the peptide. We are making the plausible assumption that side chain motions equilibrate more rapidly than backbone degrees of freedom.

The numerically computed SDP is presented by a set of structures equally spaced along the path $-\{X_i\}_{i=1}^{N}$, where $X_i$ is a Cartesian coordinate vector that includes the positions of the $C_\alpha$ atoms only. We denote the reaction coordinate by a scalar function $q$, which is the arc-length of the path $dq = \sqrt{dX^t \cdot dX}$ and at the position of the reactant $X_R$, we have $q(X_R) = 0$. $q(X_i)$ is therefore the position along the reaction coordinate and $\nabla q(X_i)$ is the slope of the reaction coordinate at $i$. A unit vector in the direction of $\nabla q(X_i)$ is approximated by the finite difference

$$\frac{\nabla q(X_i)}{\left|\nabla q(X_i)\right|} \simeq \frac{\left(X_{i+1} - X_i\right)}{\left|X_{i+1} - X_i\right|} \tag{13.5}$$

There are 50 structures in the SDP. A few snapshots at different positions along the SDP are shown in Figure 13.5. The unfolding along this tube starts at the N terminal. It flashes on and off at the C terminal and finally progresses through the rest of the chain.

The path was prepared using an implicit solvent model. (The replica exchange simulations and the action optimization use a generalized Born model to mimic solvation.) To better model kinetics, the milestoning simulations were done with explicit water molecules (TIP3P [30]). Each of the structures of the SDP was first solvated in a $41.8 \times 41.8 \times 41.8 \,\text{Å}^3$ box of water and neutralized by adding four $Cl^-$ counterions. The water structures and counterions were equilibrated around the frozen peptides for 40 ps. Particle mesh Ewald method [31] accounted for long-range electrostatic forces and the distance cutoff for Lennard–Jones interactions was 8 Å. Individual water molecules were kept rigid with the matrix form of SHAKE [32,33]. The time step was 0.5 fs.

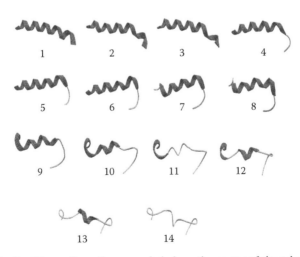

**FIGURE 13.5** Peptide configurations sampled along the center of the tube. There are 50 structures along the tube and we sampled only a few for clarity. The plots were prepared by the ZMOIL program (Authors: Avijit Ghosh, Baohua Wang, and Thomas Blom).

The 50 equilibrated solvent structures around the frozen peptide were used to initiate a sampling of configurations in the hypersurfaces perpendicular to the reaction coordinate. We approximate the general hypersurfaces by planes. A plane is defined by a point in it ($X_i$) and by the normal to the plane ($\nabla q(X_i)$) estimated by the finite difference formula of Equation 13.5. MD simulations at room temperature were conducted and constrained to the hyperplanes by linear constraints as discussed in [13]. Trajectories of 50 ps in the hyperplanes were used to sample and save configurations of each 0.5 ps and to probe the width of the tube. The 100 configurations at each milestone initiate constant energy trajectories between the milestones that were terminated and their time recorded upon touching nearby hypersurfaces (Figure 13.3).

Some milestones have many trajectories that terminate too quickly. Their termination time was shorter than 200 fs typically required for velocity de-correlation. We remove these milestones (say milestone $i$) and compute instead transitional trajectories from the nearby milestones $i - 1$ and $i + 1$. Moreover, in a few milestones termination is found primarily or only in one direction. In this case, we added an intermediate milestone in the direction that was difficult to sample and repeated the calculation. For example, if the sampling between milestones $i$ and $i + 1$ is difficult, we add a new milestone half the way between $i$ and $i + 1$ (say $i + 0.5$) and compute the transition between $i$ and $i + 0.5$, and between $i + 0.5$ and $i + 1$.

This above refinement procedure, which includes deletions and insertions of milestones, is not computationally expensive. The milestones with trajectories that are too short are computed rapidly and are not a significant computational burden. Moreover, the number of milestones that we need to add is typically small. Only one milestone (the first) requires the addition of intermediates in the present study. Nevertheless, while the refinement is not expensive computationally, it is costly in human time. Significant human time is required to examine the current data, to decide on the proper action, and set up the additional simulations. We plan to automate a significant part of this refinement procedure. For the current simulation, the refinement procedure drastically reduced the number of milestones that we finally used 26 milestones.

Even counting the runs before the refinement the calculations provide significant saving. We need to compute $50 \times 100 = 5000$ trajectories to estimate the overall rate. This may seem substantial but since the length of the trajectories between the milestones rarely exceeds tens of picoseconds (even with the reduced set of 26 milestones), the aggregated time of simulation over all trajectories is $10 \times 5000 = 50{,}000$ ps $= 50$ ns. This timescale is even shorter than the overall timescale of the folding process probed by a single trajectory, and with 50 ns worth of trajectories we obtain a sound description of the kinetics in a tube.

A sample of first passage times for milestone 9 is shown in Figure 13.6. The indices of the trajectories are given from left to right and the termination (first passage) times along the horizontal axis. For every trajectory, we put one first passage time bar (the termination times are in femtoseconds). For convenience, we denote times of termination at the milestone 10 as positive and times of termination at milestone 8 as negative. Note that the statistics that we sample is insufficient to generate accurate distribution $K_{89}(\tau)$, however, computing the first moments is easier to do. None of the trajectories presented in this plot is shorter than 200 fs, which is the rough correlation time we established for the velocity.

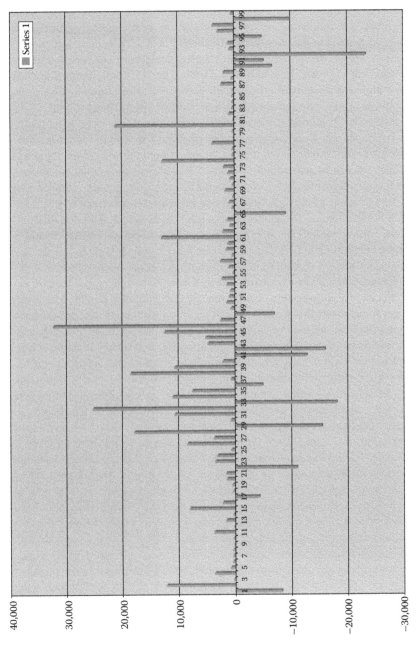

**FIGURE 13.6** The first passage times sampled from milestone 9. The positive number are from terminating trajectories that made it to milestone 10 and the negative times from the trajectories that terminate on milestone 8. A total of 99 events are shown as sticks. The statistic is insufficient to obtain accurate estimate of the density but allows for sound calculation of the first moments. The times are in femtoseconds.

The average first passage time from the first milestone to the last was $2.8\,\mu s$ computed with Equation 13.4. This timescale is about a factor of 5 larger than the experimental number (600 ns), however, considering that the sample includes only one tube from an ensemble of potential tubes, our estimate is not unreasonable. Because we did not sample the complete funnel, we are unable in this case to assess our results against experiment. Therefore, it is probably immature to reach a conclusion on the quality of the force field given the deviation from the empirical measurements [19,20,25].

It is of interest to examine mechanistic questions, probing the progress of the unfolding process along the unfolding pathway. We compute the sampling of a few critical torsions, making it possible to evaluate the shape of the tube as we progress from the reactant (folded helix) to the product (unstructured chain). We consider a torsion defined by four sequential $C_\alpha$. To reduce complexity, we show only the five torsions with sequential nonoverlapping atoms. For example, the first torsion was constructed from the alpha carbons of the first four residues, the second torsion from the fifth to the eight $C_\alpha$-s and so on to the fifth torsion. The last alpha carbon of the peptide with 21 amino acids was not used in this analysis. The torsion values were computed for all the conformations sampled in the in the 50 milestones (even the milestones that were excluded from the calculations of kinetics were added to better characterize the tube). The results are displayed in Figure 13.7 in five panels (each panel probes the progress of one of the torsions). It is striking that for both the reactants and the products the distribution of the torsions is narrow and is less than $100°$ at each of the ends of the path.

One interpretation of an unfolded state, keeping in mind the tube picture, is that the tube at the unfolded state becomes considerably broader allowing rapid exchange between alternative unfolded structures (no significant energy barrier between the unfolded conformations). This is however not a picture followed in the computed reaction coordinate. The final state is confined to the same extent as the initial folded state. A search for broad unfolded state will probably pick up a hyperplane near milestone 30. A calculation of the transition time from milestone 1 (helical configuration) to an ensemble of disparate structures (milestone 30) provides a mean first passage time of 132 ns, which is faster by a factor of about 5 compared to experiment. This small calculation illustrates the significant error bars that we still have in estimating transition times of complex systems. Studies of the shapes of concrete tubes are likely to provide significant insight into the nature of the unfolded state.

## 13.4   CONCLUDING COMMENTS

We have outlined in this chapter one of the most significant challenges to the standard model of molecular biophysics, the timescale limitation of MD. This limitation makes it exceedingly difficult to study complex and moderately complex processes in molecular biophysics. We describe milestoning, a procedure that combines theory, algorithm, and software development to study the kinetics of complex biochemical processes. Of course, the characterization of a folding funnel still presents a significant challenge, and further studies of tubes their individual contributions and possible connections are of significant interest for mechanisms and timescales. Additional future applications

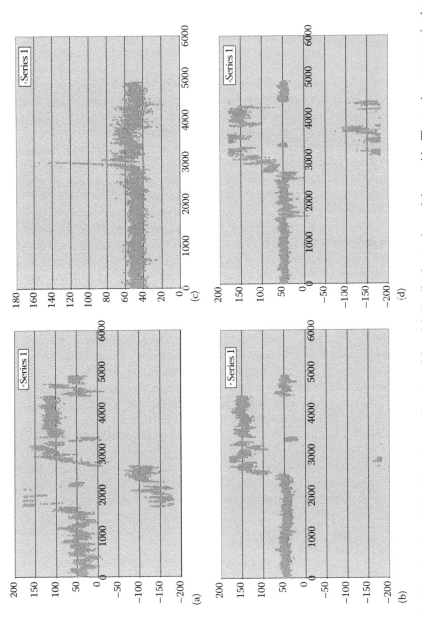

**FIGURE 13.7** Measuring the width of the tube as a function of five global collective torsions of the peptide. The torsions are examined sequentially. (a) shows torsion of the first four $C_\alpha$ atoms and (b) includes the next four alpha carbons and so on. The vertical axis is the sampling in the milestones in degrees. The sampled points are displayed as a function of the trajectory indices. The first 100 trajectories are from the first milestone, the second 100 are from the second milestone, and so on. Note the significant large fluctuations in *all* torsions around milestone 30. A total of 5000 trajectories are summarized in this plot.

will include the investigations of more complex systems such as myosin, GroEl, and other biological motors. Of particular interest is the understanding of allosteric systems in which free energy profile and the widths of the tubes (entropy) may play a significant role in driving the reaction to a desired direction [34].

On the methodological side, we expect to further develop estimates of errors and automated refinement protocols for milestoning. Milestoning in phase space (as opposed to coordinate space) will be investigated. The ability of extracting useful information for long time behavior from short time flux through milestone interfaces is playing a useful role in extending the scope of the standard model to new time domains.

## REFERENCES

1. Schlick, T., *Molecular Modeling and Simulation: An Interdisciplinary Guide.* Mathematical Biology, 2002, New York: Springer Verlag.
2. Verlet, L., Computer "Experiments" on classical fluids. I. Thermodynamical properties of Lennard–Jones molecules. *Physical Review*, 1967, **159**(1): 98–103.
3. Malnasi-Csizmadia, A., et al., Selective perturbation of the myosin recovery stroke by point mutations at the base of the lever arm affects ATP hydrolysis and phosphate release. *Journal of Biological Chemistry*, 2007, **282**(24): 17658–17664.
4. Dellago, C., P.G. Bolhuis, and P.L. Geissler, Transition path sampling. *Advances in Chemical Physics*, 2002, **123**: 1–78.
5. Ren, W., et al., Transition pathways in complex systems: Application of the finite-temperature string method to the alanine dipeptide. *Journal of Chemical Physics*, 2005, **123**(13): 13419.
6. Hummer, G. and I.G. Kevrekidis, Coarse molecular dynamics of a peptide fragment: Free energy, kinetics, and long-time dynamics computations. *Journal of Chemical Physics*, 2003, **118**(23): 10762–10773.
7. Voter, A.F., F. Montalenti, and T.C. Germann, Extending the time scale in atomistic simulation of materials. *Annual Review of Materials Research*, 2002, **32**: 321–346.
8. Faradjian, A.K. and R. Elber, Computing time scales from reaction coordinates by milestoning. *Journal of Chemical Physics*, 2004, **120**(23): 10880–10889.
9. Truhlar, D.G., B.C. Garrett, and S.J. Klippenstein, Current status of transition-state theory. *Journal of Physical Chemistry*, 1996, **100**(31): 12771–12800.
10. Czerminski, R. and R. Elber, Reaction path study of conformational transitions in flexible systems—Applications to peptides. *Journal of Chemical Physics*, 1990, **92**(9): 5580–5601.
11. Mortenson, P.N., D.A. Evans, and D.J. Wales, Energy landscapes of model polyalanines. *Journal of Chemical Physics*, 2002, **117**(3): 1363–1376.
12. Frauenfelder, H., et al., The role of structure, energy landscape, dynamics, and allostery in the enzymatic function of myoglobin. *Proceedings of the National Academy of Sciences of the United States of America*, 2001, **98**(5): 2370–2374.
13. West, A.M.A., R. Elber, and D. Shalloway, Extending molecular dynamics time scales with milestoning: Example of complex kinetics in a solvated peptide. *Journal of Chemical Physics*, 2007, **126**(14): 1451104.
14. Elber, R., A milestoning study of the kinetics of an allosteric transition: Atomically detailed simulations of deoxy Scapharca hemoglobin. *Biophysical Journal*, 2007, **92**(9): L85-L87.
15. West, A. and R. Elber, The recovery strike in myosin: Atomically detailed simulations with milestoning (in progress).

16. Vanden Eijnden, E., et al., On the assumption underlying milestoning. *Journal of Chemical Physics*, 2008, **129**(17), 174102.

17. Kenkre, V.M. and R.S. Knox, Generalized master equation theory of excitation transfer. *Physical Review B*, 1974, 5279–5290.

18. Shalloway, D. and A.K. Faradjian, Efficient computation of the first passage time distribution of the generalized master equation by steady-state relaxation. *Journal of Chemical Physics*, 2006, **124**(5): 054112.

19. Jas, G.S., W.A. Eaton, and J. Hofrichter, Effect of viscosity on the kinetics of alpha-helix and beta-hairpin formation. *Journal of Physical Chemistry B*, 2001, **105**(1): 261–272.

20. Thompson, P.A., et al., The helix-coil kinetics of a heteropeptide. *Journal of Physical Chemistry B*, 2000, **104**(2): 378–389.

21. Bryngelson, J.D., et al., Funnels, pathways, and the energy landscape of protein-folding—A synthesis. *Proteins-Structure Function and Genetics*, 1995, **21**(3): 167–195.

22. Boczko, E.M. and C.L. Brooks, First-principles calculation of the folding free-energy of a 3-helix bundle protein. *Science*, 1995, **269**(5222): 393–396.

23. Ghosh, A., R. Elber, and H.A. Scheraga, An atomically detailed study of the folding pathways of protein A with the stochastic difference equation. *Proceedings of the National Academy of Sciences of the United States of America*, 2002, **99**(16): 10394–10398.

24. Cardenas, A.E. and R. Elber, Kinetics of cytochrome C folding: Atomically detailed simulations. *Proteins: Structure Function and Genetics*, 2003, **51**(2): 245–257.

25. Jas, G.S. and K. Kuczera, Equilibrium structure and folding of a helix-forming peptide: Circular dichroism measurements and replica-exchange molecular dynamics simulations. *Biophysical Journal*, 2004, **87**(6): 3786–3798.

26. Brooks, B.R., et al., CHARMM—a program for macromolecular energy, minimization, and dynamics calculations. *Journal of Computational Chemistry*, 1983, **4**(2): 187–217.

27. Feig, M., J. Karanicolas, and C.L. Brooks, *MMTSB Tool Set*, ed. NIH Research Resource. 2001, The Scripps Research Institute, La Jolla, CA.

28. Elber, R., et al., Moil a program for simulations of macromolecules. *Computer Physics Communications*, 1995, **91**(1–3): 159–189.

29. Olender, R. and R. Elber, Yet another look at the steepest descent path. *Theochem-Journal of Molecular Structure*, 1997, **398**: 63–71.

30. Jorgensen, W.L., et al., Comparison of simple potential functions for simulating liquid water. *Journal of Chemical Physics*, 1983, **79**(2): 926–935.

31. Essmann, U., et al., A smooth particle mesh Ewald method. *Journal of Chemical Physics*, 1995, **103**(19): 8577–8593.

32. Ryckaert, J.P., G. Ciccotti, and H.J.C. Berendsen, Numerical integration of Cartesian equations of motion of a system with constraints—Molecular dynamics of *n*-alkanes. *Journal of Computational Physics*, 1977, **23**(3): 327–341.

33. Weinbach, Y. and R. Elber, Revisiting and parallelizing SHAKE. *Journal of Computational Physics*, 2005, **209**(1): 193–206.

34. Xu, C.Y., D. Tobi, and I. Bahar, Allosteric changes in protein structure computed by a simple mechanical model: Hemoglobin T ↔ R2 transition. *Journal of Molecular Biology*, 2003, **333**(1): 153–168.

# 14 Pathways and Rates for Structural Transformations of Peptides and Proteins

*David J. Wales, Joanne M. Carr, Mey Khalili,*
*Vanessa K. de Souza, Birgit Strodel,*
*and Chris S. Whittleston*

## CONTENTS

## 14.1 INTRODUCTION: TRANSITION NETWORKS AND GRAPHS

Understanding the structure and function of biomolecules requires insight into both thermodynamic and kinetic properties. Unfortunately, many of the dynamical processes of interest occur too slowly for standard molecular dynamics (MD) simulations to gather meaningful statistics. This problem is not confined to biomolecular systems, and the development of methods to treat such "rare events" is currently an active field of research.[1–31] If the kinetic system can be represented in terms of linear rate equations between a set of $M$ states, then the complete spectrum of $M$ relaxation timescales can be obtained in principle by solving a memoryless master equation.[32,33] This approach was used in the last century for a number of studies involving atomic and molecular clusters, as well as biomolecules.[34–53] These applications range from analysis of model energy landscapes to coarse-grained and atomistic models. In the present chapter, the focus is upon networks composed of stationary points of the potential energy surface (PES), particularly local minima and the transition states that connect them (Section 14.2.1). The first studies of biomolecules employing such techniques include applications to the tetrapeptide IAN[53] and the fk506 binding protein.[54] Berry and coworkers subsequently analyzed sequences of connected local minima for a coarse-grained model,[55] which has since been used in a number of other studies.[56–59]

More recent work has focused either on building up transition networks from geometry optimization more systematically, or on alternative approaches that use dynamical methods to construct the networks.[18,60–69] For smaller systems, it may be possible to capture the pathways of interest without specifically seeking transition states that improve the overall rate constants.[70–72] However, for larger biomolecules some sort of directed sampling is likely to be needed. The discrete path sampling approach described in Section 14.2.2 is designed to locate dynamically relevant stationary points that determine the rate constant for interconverting two given end points, which may be local minima or groups of minima.[6,13] Algorithms from graph theory and network analysis are used in sampling and analyzing such stationary point databases, and can also be used for constructing free energies.[73] For example, a bracketing procedure, which successively refines upper and lower bounds on selected energy barriers,[74] has been applied to analyze the dynamics of the Ras p21 molecular switch.[75] Applications of stationary point databases to biomolecules have generally employed implicit solvent models to avoid the complication of trying to distinguish or group together conformations that include solvent degrees of freedom. The use of implicit solvent can produce a qualitative change in the energy landscape, since a different secondary structure may be favored compared to a gas phase description. Assuming that the chosen implicit solvent description produces a faithful description of the polypeptide, the main difference compared to explicit solvent would be an ensemble average over solvent configurations. If these configurations were included explicitly, then every local minimum would split into a spectrum of minima, differing principally in the solvent degrees of freedom. Ideally, the implicit solvent treatment should produce energies for local minima and transition states that reproduce the average values for explicit solvent. Hence this component of the solvent "friction" is included, i.e., shifts in the energy of different configurations, but barriers corresponding to changes in specific internal solvent degrees of freedom are not.

Transition networks have also been constructed for biomolecules directly from MD data,[18,60–69] where they are sometimes referred to as Markov state models. In order to gather sufficient statistics, the transitions between states must be observable on an MD timescale. Alternative methods exist, not based on MD, where connections between conformations are inferred based on distance criteria. The edge weights of the corresponding graphs are then based on the energies of the two geometries that are assumed to be connected,[76,77] rather than calculated barriers or rate constants.

Disconnectivity graphs provide a convenient visualization of transition or weighted graphs,[41,56,78,79] and have been generalized to use free energy[73,80] and transition probabilities from kinetic networks.[69] These graphs can be analyzed in terms of basic network properties, such as the number of edges (transition states) on paths between different minima, and the connectivity. For example, in scale-free networks,[81] the probability distribution for the number of connections follows a power law, rather than the binomial distribution expected for a random graph. Scale-free characteristics have been been identified in networks of stationary points constructed for atomic clusters, where low-lying minima on the PES tend to be highly connected and act as hubs.[82] Disconnectivity graphs with well-defined, highly connected global minima observed in biological systems may exhibit scale-free properties for the same reason. For example, a network constructed from MD snapshots for the three-stranded β-sheet peptide Beta3s has been characterized as scale-free.[60] Transition state ensembles were constructed by identifying states with high connectivity and low statistical weight, and analyzed by calculating the folding commitment probability, $P^{fold}$ (Section 14.2.2). Network properties were compared with the corresponding free energy surfaces in subsequent work.[83]

## 14.2  METHODS

### 14.2.1  Geometry Optimization

If we wish to analyze global kinetics in a coarse-grained picture of local minima connected by transition states, then it is essential to have efficient tools for locating such stationary points. Here we employ the geometrical definition of a transition state as a stationary point of the PES with a single negative Hessian eigenvalue.[84] In the absence of branch points,[79,85] each transition state links two minima via the two steepest-descent paths defined by the Hessian eigenvector corresponding to the unique negative Hessian eigenvalue. Hence we characterize such connections by a minimum–transition state–minimum triplet, and define a discrete path between two minima, $\min_A$ and $\min_B$, in terms of overlapping triplets: $\min_A$–$ts_1$–$\min_1$–$ts_2$–$\min_2$–$\cdots$–$\min_B$. The number of steps in a discrete path is defined as the number of transition states.

All the minimization procedures in the present work employed a slightly modified version of the limited-memory Broyden–Fletcher–Goldfarb–Shanno (LBFGS) algorithm,[86,87] as implemented in the OPTIM code.[88] Double-ended searches for connections between specified local minima were performed using the doubly-nudged[89] elastic band[53,90–93] (DNEB) algorithm. The band consists of a series of image structures, which are optimized simultaneously using the LBFGS approach until local maxima become well defined. These local maxima are then taken as candidates for further refinement using hybrid eigenvector-following (EF),[94,95] where a selected direction is searched uphill in steps that alternate with LBFGS minimization projected onto the tangent space to prevent interference with the uphill step. The dimer method, described by Henkelman and Jónsson[91] employs a similar approach, and a variety of other methods use analogous ideas.[96–98] In our applications, the step length and uphill direction are usually specified by characterizing the smallest nonzero Hessian eigenvalue and the corresponding eigenvector. An analytic Hessian can be used, together with complete or partial diagonalization, or a gradient-only formulation can be employed by defining the Rayleigh–Ritz ratio[99]

$$\lambda(\mathbf{x}) = \frac{\mathbf{x}^{\mathrm{T}}\mathbf{H}\mathbf{x}}{|\mathbf{x}|^2}, \tag{14.1}$$

where
   $\mathbf{x}$ is a displacement from the current configuration, $\mathbf{X}$
   superscript T denotes the transpose
   $\mathbf{H}$ is the Hessian matrix at $\mathbf{X}$

Using the central difference approximation for the second derivatives of $V$

$$\lambda(\mathbf{x}) \approx \frac{V(\mathbf{X}+\mathbf{x})+V(\mathbf{X}-\mathbf{x})-2V(\mathbf{X})}{|\mathbf{x}|^2}, \tag{14.2}$$

where $V(\mathbf{X})$ is the potential energy at configuration $\mathbf{X}$ and $|\mathbf{x}| \ll 1$, the gradient can be written

$$\frac{\partial\lambda(\mathbf{x})}{\partial\mathbf{x}} = \frac{\nabla V(\mathbf{X}+\mathbf{x})-\nabla V(\mathbf{X}-\mathbf{x})-2\lambda(\mathbf{x})\mathbf{x}}{|\mathbf{x}|^2}. \tag{14.3}$$

$\lambda(\mathbf{x})$ is minimized using the LBFGS algorithm with overall translation and rotation projected out to prevent the search from converging to one of the corresponding eigenvectors.[94,95]

Two modifications of the Rayleigh–Ritz procedure were found useful in the present work. Expanding the gradients in Equation 14.3 to first-order in a Taylor series about $\mathbf{X}$ gives

$$\frac{\partial\lambda(\mathbf{x})}{\partial x_\alpha} = \frac{2}{|\mathbf{x}|^2}\left(\sum_\beta H_{\alpha\beta}x_\beta - \lambda(\mathbf{x})x_\alpha\right), \tag{14.4}$$

where $H_{\alpha\beta} = \partial^2 V(\mathbf{X})/\partial X_\alpha\partial X_\beta$. Hence the dot product of the gradient with the step can be written

$$\sum_\alpha x_\alpha\frac{\partial\lambda(\mathbf{x})}{\partial x_\alpha} = \frac{2}{|\mathbf{x}|^2}\left\{\sum_{\alpha\beta} x_\alpha H_{\alpha\beta}x_\beta - \lambda(\mathbf{x})|\mathbf{x}|^2\right\} = 2\{\lambda(\mathbf{x})-\lambda(\mathbf{x})\} = 0. \tag{14.5}$$

The step should therefore be perpendicular to the gradient, and it was found that orthogonalizing $\mathbf{x}$ to $\partial\lambda(\mathbf{x})/\partial\mathbf{x}$ improved the convergence of the minimization.

We have also considered an alternative formulation of the estimate for $\lambda(\mathbf{x})$ using

$$\lambda(\mathbf{x}) \approx \frac{\{\nabla V(\mathbf{X}+\mathbf{x})-\nabla V(\mathbf{X}-\mathbf{x})\}\cdot\mathbf{x}}{2|\mathbf{x}|^2}. \tag{14.6}$$

The truncation error in $\lambda(\mathbf{x})$ for Equations 14.2 and 14.6 has the same leading term except for an additional factor of two for Equation 14.6. However, there is a significant difference in the roundoff error from the two formulations. For systems with numerically large values of $V(\mathbf{X})$, we have sometimes encountered a serious loss of precision for Equation 14.2 due to roundoff error. Switching to Equation 14.6 generally produced a better estimate for $\lambda(\mathbf{x})$ in such cases, especially when the magnitude of the gradient is small compared to the energy itself.

Employing an internal coordinate representation can reduce the number of steps required for convergence of geometry optimizations with anisotropic potentials.[100] However, there is also a significant overhead associated with the coordinate transformation, and in the applications discussed below, we only employed internal coordinates for initial interpolations between minima. Interpolating in this way can help to avoid stationary points with high energies or unphysical geometries that are sometimes supported by empirical biomolecular force fields. The main advantage of geometry optimization is that it enables us to overcome high barriers easily, but it can also lead us into irrelevant regions of configuration space. For example, we systematically reject pathways that involve *cis–trans* isomerization of peptide bonds (except for proline residues), which might otherwise be added to the database.

Complete discrete paths between distant minima are unlikely to be found in a single connection attempt. For a given pair of minima, we therefore consider successive DNEB/hybrid EF searches up to a maximum number of cycles, terminating if a connected path is found. Each of these connection attempts generates a separate local stationary point database that is subsequently merged with the global database. Unless the connection attempt succeeds in one cycle, there is therefore another local decision to be made about which pair of minima to consider next. The number of possible pairs grows as the square of the number of minima found, and hence a systematic way to select minima that are most likely to produce a successful connection is needed. This problem is solved using a missing connection algorithm[101] that treats the known minima in the local database as nodes in a complete graph, where an edge exists between every pair of nodes. We define edge weights for this graph as zero where direct connections via transition states have already been found, and as a monotonically increasing function of the minimized Euclidean distance where no connection exists. The Dijkstra algorithm[102] is then used to identify the shortest path between the two end points in terms of the sum of edge weights, and new DNEB/hybrid EF searches are performed for all the missing connections on this path. This approach can speed up pathway searches between distant minima by several orders of magnitude.[101]

## 14.2.2 Discrete Path Sampling

Discrete path sampling (DPS) is a framework for harvesting connected stationary point databases of minima and transition states to represent the kinetics for a specified structural transition.[6,13,103] The two end points, denoted $A$ and $B$, can be single potential energy minima or sets of minima, which can be defined by order parameters, clustering schemes, or regrouping based upon rates or free energy barriers (Section 14.2.3). The set of minima outside $A$ and $B$ is denoted as intervening, $I$. If the intervening minima can be treated within the steady-state approximation, where the rate of change of the occupation probability $p_i(t)$ can be neglected for $i \in I$, then phenomenological two-state rate constants can be formulated as

$$k_{BA}^{SS} = \frac{1}{p_A^{eq}} \sum_{b \leftarrow a} \frac{k_{bi_1} k_{i_1 i_2} \cdots k_{i_n a} p_a^{eq}}{\sum_{\alpha_1} k_{\alpha_1 i_1} \sum_{\alpha_2} k_{\alpha_2 i_2} \cdots \sum_{\alpha_n} k_{\alpha_n i_n}}, \qquad (14.7)$$

where

$k_{ij}$ is the net rate constant from minimum $j$ to minimum $i$

$p_a^{eq}$ is the equilibrium occupation probability of minimum $a \in A$

$p_A^{eq}$ is the equilibrium occupation probability of the $A$ group of minima, with

$$p_A^{eq} = \sum_{a \in A} p_a^{eq}$$

The corresponding formula for $k_{AB}^{SS}$ is obtained by exchanging $A$ and $B$ minima. This expression also assumes that the minima within each of the $A$ and $B$ sets are in local equilibrium on the timescale corresponding to the $A \leftrightarrow B$ transitions of interest. A more compact formula can be obtained as[103,104]

$$k_{BA}^{SS} = \frac{1}{p_A^{eq}} \sum_{b \leftarrow a} P_{bi_1} P_{i_1 i_2} \cdots P_{i_{n-1} i_n} P_{i_n a} p_a^{eq} \tau_a^{-1} = \frac{1}{p_A^{eq}} \sum_{a \in A} \frac{C_a^B p_a^{eq}}{\tau_a}, \qquad (14.8)$$

where $\tau_i = 1 / \sum_\alpha k_{\alpha i}$ is the mean waiting time for any direct transition to occur from minimum $i$, and $P_{ji} = k_{ji} \tau_i$ is the branching probability for the $j \leftarrow i$ transition. $C_a^B$ is the probability that a random walk starting from $a$ will encounter a $B$ minimum before it returns to the $A$ state, and corresponds to a "committor" or "splitting" probability.[105] The $P_\alpha^{fold}$ index, defined as the probability that a protein will fold before unfolding, starting from some initial condition $\alpha$,[61,106,107] is another example of a committor probability. We have evaluated these quantities using the first-step analysis:[108]

$$C_\alpha^B = \sum_\beta C_\beta^B P_{\beta\alpha}, \qquad (14.9)$$

which can be solved using successive overrelaxation[109] within a compressed row storage scheme[110] for sparse matrices.

Analogous expressions for the phenomenological rate constants can be derived[103] that relax the assumptions of steady-state kinetics ($k_{BA}^{NSS}$) for the $I$ minima and of local equilibrium in the $A$ and $B$ states ($k_{BA}^{KMC}$)

$$k_{BA}^{NSS} = \frac{1}{P_A^{eq}} \sum_{a \in A} \frac{C_a^B P_a^{eq}}{t_a}, \quad k_{BA}^{KMC} = \frac{1}{p_A^{eq}} \sum_{a \in A} \frac{p_a^{eq}}{\mathcal{T}_{Ba}}, \qquad (14.10)$$

where

$t_a \geq \tau_a$ is the average waiting time for transitions between minimum $a$ and any minimum in the $A$ or $B$ sets

$\mathcal{T}_{Ba}$ is the mean first passage time between minimum $a$ and any $B$ minimum, with $\mathcal{T}_{Ba} \geq t_a$

The superscripts NSS and KMC stand for nonsteady-state and kinetic Monte Carlo, respectively. The latter label is used to highlight the correspondence between $k_{BA}^{KMC}$ and the rate constant that would be obtained using a kinetic Monte Carlo approach[111-115] within the state space of local minima. The same information could be obtained in principle by solving the corresponding master equation directly, but we have found that the latter approach is not practical for large state spaces and slow $A \leftrightarrow B$ dynamics.[103,104,116] Instead, our preferred approach for extracting overall rate constants now employs a graph transformation algorithm. This method avoids the numerical problems of matrix diagonalization in direct solutions of the master equation, and can treat cases with large overall barriers compared to the thermal energy, $k_B T$, much faster than kinetic Monte Carlo.[103,104,116] Here $k_B$ is the Boltzmann constant and $T$ is the temperature. The disconnectivity graphs presented below provide a convenient way to visualize the transition network. However, all the rate constant calculations that we present are equivalent to solving the corresponding master equation (possibly with regrouping of states, see Section 14.2.3). Hence this work is a direct descendent of the master equation treatment for a tetrapeptide presented by Czerminski and Elber in 1990.[53]

In essence, the DPS approach reduces the problem of global kinetics to a discrete space of stationary points. Phenomenological rate constants can then be extracted under the assumption of Markovian dynamics within this space, which requires that the system has time to equilibrate between transitions and lose any memory of how it reached the current minimum. The Markovian assumption is therefore an essential part of the framework. However, we can regroup the stationary points into states whose members are separated by low barriers so that the Markov property is likely to be better obeyed between the groups (Section 14.2.3).

For both individual stationary points and regrouped states, we must employ a consistent theory for the densities of states and minimum-to-minimum rate constants that appear in the formulas above. In principle, these quantities could be determined more accurately using methods that involve explicit dynamics.[1,10,11] In practice, the applications presented in Section 14.3 employ a harmonic normal mode approximation to obtain a local density of states for every stationary point, along with harmonic transition state theory[117-122] for the rate constants. This choice introduces additional systematic errors in addition to uncertainties associated with convergence of the stationary point database and the underlying Markov assumption for the chosen state space. The Markov assumption itself is likely to be least reliable for barriers that are small or comparable to $k_B T$. However, if the rate-determining steps are associated with larger barriers, then we do not expect the overall rate constants to depend significantly on the fast processes associated with crossing low barriers. If all the barriers in question are low, then it may be more efficient to calculate rates using a direct dynamical method, such as transition path sampling.[11,123,124]

The accuracy of a Markov model for a given state space can be evaluated by comparing the timescale for trajectories to lose memory of their history with the lifetimes of the states of interest. For example, Altis et al. concluded that the timescale for memory loss of a few hundred picoseconds, which they calculated from explicit solvent simulations of $Ala_7$, was too long to permit a Markov description of the chosen state space.[125] Chodera et al. were unable to formulate a Markovian picture for the trpzip2 hairpin, but identified a suitable separation of timescales for $Ala_2$.[126] Regrouping procedures for stationary point databases, which should produce a state space with a better separation between lifetimes and memory loss timescales, are discussed in Section 14.3.

The overall strengths and weaknesses of the DPS approach are generally complementary to those of methods founded on explicit dynamics.[11,123,124] For example, basing path searches on geometry optimization enables us to overcome barriers on virtually any energy scale. However, some of these barriers may correspond to unphysical regions of configuration space, such as *cis–trans* isomerization of peptide bonds, and therefore need to be systematically removed. The assumption of Markovian dynamics in the analysis of stationary point databases requires the existence of a state space with a suitable separation of timescales between equilibration within the groups and interconversion between them. In contrast, the use of harmonic densities of states to derive partition functions and minimum-to-minimum rate constants is not intrinsic to the DPS framework, but seems appropriate in view of the errors associated with the empirical potentials employed for the applications in Section 14.3. Since geometry optimization provides no analog to the importance sampling step in transition path sampling,[11,123,124] our stationary point databases are expanded by adding results from every search, screening out only unphysical geometries. We must therefore remove artifical kinetic frustration that may appear due to undersampled regions of configuration space, where lower barriers exist, but have not yet been located.

In the first DPS applications to atomic and molecular clusters,[6,13,127] met-enkephalin,[128] and the "GB1" peptide,[129] stationary points were added to the existing database by considering new connection attempts between minima belonging to particular discrete paths. A fixed number of attempts was considered for a specified number of discrete paths that made the largest contribution to $k_{BA}^{SS}$ or $k_{AB}^{SS}$. Subsequent work has considered different approaches that select pairs of minima from the current database for new connection attempts. Most of these methods apply Dijkstra's shortest-path algorithm[102] to extract the discrete path that makes the largest contribution to $k_{BA}^{SS}$ or $k_{AB}^{SS}$. [Note that this is a different application of the Dijkstra procedure from the missing connection algorithm used to construct initial paths[101] (Section 14.2.1).] In this graph theoretical representation the local minima are nodes of the graph, and each transition state corresponds to two directed edges with weights $W_{\alpha\beta} = -\ln P_{\alpha\beta}$ for the transition from $\beta$ to $\alpha$. The Dijkstra algorithm identifies the discrete path between given minima from the reactant and product sets, e.g., $a \in A$ and $b \in B$, with the smallest value for the sum of edge weights. This sum therefore gives the largest value for the product of branching probabilities from $a$ to $b$, and we then include the conditional probability $p_a^{eq}/P_A^{eq}$ to compare the contributions to $k_{BA}^{SS}$. We refer to the discrete path with the largest contribution as the "fastest" path, although it actually includes the conditional probability factor and does not include any recrossings between minima. These recrossings were accounted for in the early DPS applications where perturbations of particular discrete paths were considered. However, in subsequent work, where we account for the whole stationary point database in making new connection attempts, the Dijkstra algorithm provides a more convenient way to identify likely kinetic bottlenecks. Similar network analysis algorithms were used by Krivov and Karplus[73] in their construction of free energy disconnectivity graphs. We have also employed the recursive enumeration algorithm[130] to obtain additional discrete paths in the order that they contribute to $k_{BA}^{SS}$.[131]

The "fastest" discrete path in the database changes as new stationary points are added. One scheme that we have found useful for the location of missing connections chooses minima from the current fastest path that are separated by a minimum number of transition states and by a minimal Euclidean distance after optimal alignment.[132] We refer to this scheme as SHORTCUT, while an alternative SHORTCUT BARRIER procedure selects minima on different sides of the largest barrier on the path. We have programmed both schemes so that they can operate on either potential energy minima or groups of minima (with associated free energies) obtained via a regrouping scheme (Section 14.2.3). Related graph theoretical methods for refining stationary point databases have also been used successfully in biological applications,[74,75] as well as condensed matter systems.[133]

Both SHORTCUT schemes generally improve the fastest discrete path in the database. However, they can also introduce artificial kinetic traps via minima that are located during the searches, but for which the escape pathways with the lowest barriers have not been found. To remove such features, we have introduced the FREEPAIRS procedure, where minima are selected for connection attempts if they belong to groups with similar free energies separated by a large free energy barrier.[134] We consider the ratio of the barrier height to the free energy difference to quantify this frustration, and select local minima from such groups for connection attempts based upon their minimized Euclidean distance. Hence this approach employs a measure that is related to the Z-score,[135] defined as the normalized difference between the energy of the native structure and the average energy of a representative set of non-native structures. It is also related to the measure of frustration defined by the folding temperature divided by the glass transition temperature.[136]

## 14.2.3 GROUPING SCHEMES FOR STATIONARY POINT DATABASES

Various schemes have been considered for coarse-graining stationary point databases.[79,80,128,129,134,137] Pruning schemes, e.g., recursively removing minima with only one connection, have also been used to reduce the dimension of the problem for kinetic analysis. The motivation and justification for grouping local potential energy minima together is that sets of minima linked by low barrier rearrangements are likely to be in local equilibrium on the timescale of interest. Regrouping can be based on the values of suitable order parameters, such as structural similarity, or using kinetic or thermodynamic criteria.[69,79,80,128,129,134,137,138] In the following applications, we regroup recursively according to a threshold for the free energy barriers between the current groups[134]:

1. Assign each minimum under consideration to its own free energy group. Calculate the free energy of each group and of the intergroup transition state ensembles.
2. For each intergroup transition state ensemble in turn, merge the two directly connected groups together if their forward and backward free energy barriers both lie below a fixed threshold value, $\Delta F_{barrier}$.

3. Calculate the free energy of each changed group, and of the changed inter-group transition state ensembles.
4. Go to step 2, or exit if the overall group membership did not change during step 2.

This scheme is designed to generate groups of minima that are in local equilibrium on a timescale determined by the choice of the barrier height, $\Delta F_{\text{barrier}}$. In the applications described in Section 14.3, the free energies are all calculated from densities of states obtained using harmonic normal mode analysis for each stationary point. Here we employ the superposition approach[79,139–145] and write the total partition function for groups of minima as the sum of individual partition functions, $Z_\alpha(T)$, for temperature $T$. In principle, this is a rigorous decomposition, since an integral over configuration space can be broken down into the sum of integrals over the basins of attraction[146] of local minima. The free energy of minimum $j$ is then written as $F_j(T) = -k_B T \ln Z_j(T)$. The free energy of a transition state connecting minima $j$ and $l$ is similarly defined as $F_{lj}^\dagger(T) = -k_B T \ln Z_{lj}^\dagger(T)$, where $Z_{lj}^\dagger(T)$ does not include the reactive mode corresponding to the negative Hessian eigenvalue. The occupation probability and free energy of a group of minima, $J$, are calculated as

$$p_J^{\text{eq}}(T) = \sum_{j \in J} p_j^{\text{eq}}(T) \quad \text{and} \quad F_J(T) \equiv -k_B T \ln \sum_{j \in J} Z_j(T) \equiv -k_B T \ln Z_J(T), \quad (14.11)$$

and the free energy of the group of transition states that directly connect minima in group $J$ to minima in group $L$ is

$$F_{LJ}^\dagger(T) = -k_B T \ln \sum_{l \leftarrow j} Z_{lj}^\dagger(T) \equiv -k_B T \ln Z_{LJ}^\dagger(T). \quad (14.12)$$

The corresponding intergroup rate constant from $J$ to $L$, $k_{LJ}$, is then[137,147]

$$k_{LJ}(T) = \sum_{l \leftarrow j} \frac{p_j^{\text{eq}}(T)}{p_J^{\text{eq}}(T)} k_{lj}(T) = \sum_{l \leftarrow j} \frac{Z_j(T)}{Z_J(T)} \frac{k_B T}{h} \frac{Z_{lj}^\dagger(T)}{Z_j(T)}$$

$$= \frac{k_B T}{h} \frac{Z_{LJ}^\dagger(T)}{Z_J(T)} = \frac{k_B T}{h} e^{-\left[F_{LJ}^\dagger(T) - F_J(T)\right]/k_B T}. \quad (14.13)$$

## 14.3 APPLICATIONS

### 14.3.1 FOLDING THE THREE-STRANDED β-SHEET PEPTIDE BETA3S

Beta3s is a 20-residue peptide specifically designed to adopt a three-stranded anti-parallel β-sheet conformation with type II′ hairpin turns at Gly-6–Ser-7 and Gly-14–Ser-15.[148] The rapid folding observed experimentally (an upper bound of between 4 and 14 μs at 283 K from 1D-$^1$H nuclear magnetic resonance (NMR) spectra[148]) has made it an attractive system for computer simulation.[60,69,83,134,149–159] Results have been reported for both parallel tempering[150] and MD studies.[152,154,156] The MD results revealed a dominant folding pathway where contacts between the second and third (middle and C-terminal) strands form first, starting close to the hairpin turn, followed by the docking of the first (N-terminal) strand onto the existing hairpin. The same order of events was deduced from a free energy surface constructed for Betanova,[160] which is also a three-stranded β-sheet. An average folding time of 85 ns was calculated at 330 K using a set of independent MD runs and the CHARMM19/SASA force field.[156] Conformations from these simulations were used in a network analysis for this system,[60,83] as discussed in Section 14.1. Φ-Values[161] have been obtained from multiple MD simulations at 330 K for both the original peptide and single-point mutants.[158] The results suggest that for most mutants, the C-terminal hairpin forms first, as for the wild-type. However, mutations that destabilize strands two and three significantly may favor the path in which the N-terminal hairpin forms first.[158]

DPS calculations were conducted for Beta3s using the CHARMM19 united-atom potential[162] and the EEF1 implicit solvent model.[163] Small modifications of the potential were made to eliminate discontinuities in the EEF1 terms[164] and ensure that side-chain rotamers have identical energies and geometries.[165] To obtain suitable end points representative of the folded and unfolded states, we first ran constant temperature MD simulations, saving configurations at regular intervals for structural analysis. The initial local minimum corresponding to the folded state contained a complete set of native contacts, as defined in Ref. 152, while an extended configuration containing no native contacts was chosen as the initial unfolded endpoint. An initial discrete path connecting the two minima was obtained using our Dijkstra missing-connection algorithm[101] (Section 14.2.1). For conformational changes involving around 10 or more transition states on the PES, the initial discrete path generated in this way is not expected to be kinetically relevant. The final database considered for kinetic analysis was generated by a series of SHORTCUT and FREEPAIRS runs (Section 14.2.2), and contained around 500,000 stationary points.[134] A folding rate constant of around $3.3 \times 10^6$ s$^{-1}$ at 298 K was extracted from this database after regrouping using a threshold of $\Delta F_{\text{barrier}} = 5$ kcal mol$^{-1}$. The 250 "fastest" discrete paths (in terms of contributions to $k_{BA}^{SS}$) were obtained using the recursive enumeration algorithm,[131] and analyzed in terms of mechanistic details. These paths correspond to initial formation of the C-terminal hairpin, starting at the turn and proceeding via a zipper-like mechanism. The N-terminal strand then docks against the first hairpin. Configurations from this path are shown in Figure 14.1, together with the total potential energy (including the EEF1 solvation contribution) and the nonbonding contribution (from the sum of the electrostatic and van der Waals terms). The total energy is seen to correlate closely with the nonbonded energy.

The disconnectivity graph illustrated in Figure 14.2 provides a simplified view of the landscape, where only the stationary points involved in the 250 "fastest" discrete paths are included. Aside from minor differences in detail, the corresponding pathways are very similar, and all correspond to the mechanism described above, with initial C-terminal hairpin formation. The red and blue branches of this graph correspond to minima that are visited in either the fastest or the slowest of the 250 fastest paths, respectively, and appear mostly in an intermediate energy range, between formation of the two hairpins. Overall, this graph exhibits little frustration, since barriers

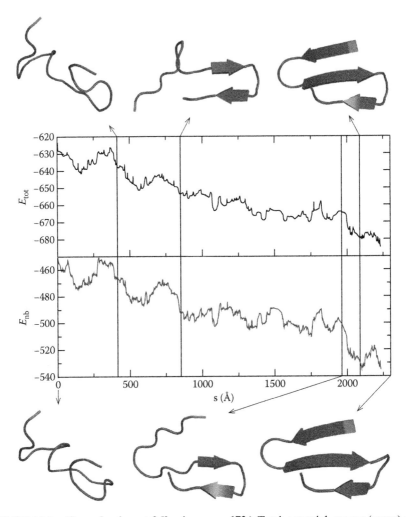

**FIGURE 14.1**  (**See color insert following page 172.**) Total potential energy (upper) and nonbonding component (lower) as a function of the integrated path length, $s$, for a selected folding pathway in Beta3s.[134] The energy is in units of kcal mol$^{-1}$. Vertical lines indicate roughly where formation of the the two hairpins starts and finishes. These four conformations are illustrated, together with the two endpoints. The C-termini are colored red and the N-termini are green. The C-terminal hairpin forms first and then the N-terminal strand docks against it to complete the β-sheet.

between minima of similar energy are usually small. We associate this structure with the "palm tree" pattern that corresponds to efficient structure-seeking behavior.[78]

A free energy disconnectivity graph for Beta3s is shown in Figure 14.3, constructed from the groups obtained using the regrouping scheme described in Section 14.2.3 with $\Delta F_{\text{barrier}} = 5$ kcal mol$^{-1}$.[134] This graph also exhibits a relatively low degree of frustration, although some of the downhill barriers from higher energy groups may be as large as 10 kcal mol$^{-1}$. The barriers to folding from groups with comparable free energies to the denatured minimum found from MD simulations are more like 8 kcal mol$^{-1}$

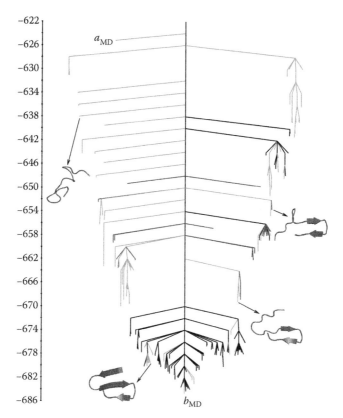

**FIGURE 14.2** **(See color insert following page 172.)** Disconnectivity graph for Beta3s based on the stationary points that appear in the 250 "fastest" discrete paths (potential energy in kcal mol$^{-1}$). The denatured and folded end points are labelled $a_{MD}$ and $b_{MD}$, respectively. The branches are colored according to whether the corresponding local minima appear in the fastest or slowest of these paths, emphasising that only minor differences in the folding pathway occur within this set. Green branches lead to minima present on both the fastest and slowest paths in the set, red branches correspond to minima on the fastest path, but not the slowest, and blue branches correspond to minima on the slowest path, but not the fastest. The intervening configurations from the fastest path illustrated in Figure 14.1 are also included here.

at 298 K. This value seems to be reasonably consistent with the barriers inferred from MD simulations corresponding to a higher temperature (330 K) reported by Krivov et al.[69] These 20 µs trajectories included around 100 folding/unfolding events, and the corresponding free energy profiles generally include a number of barriers. This complexity is consistent with the structure we observe in the above disconnectivity graphs, and with previous work.[70–73,80,128,129,137,166,167] The number of distinct free energy minima decreases rapidly as we increase the coarse-graining threshold $\Delta F_{barrier}$, but there are still many distinct states for a threshold of 5 kcal mol$^{-1}$ at 298 K.

## 14.3.2 CONFORMATIONAL CHANGE IN NITROGEN REGULATORY PROTEIN C

Nitrogen regulatory protein C (NtrC) belongs to the bacterial "two-component system" signal transduction pathway.[168–170] Phosphorylation at aspartate-54 plays a key

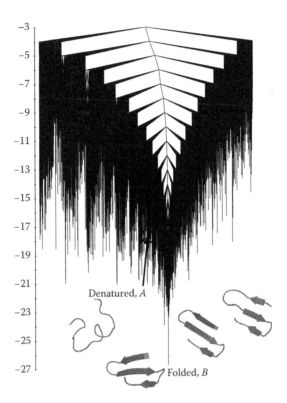

**FIGURE 14.3**  Free energy disconnectivity graph for the grouped Beta3s database calculated at 298 K.[134] The two minima corresponding to the original folded and unfolded states are shown, along with two minima belonging to the next two groups above the folded state. The energy is in kcal mol[-1].

role in activating the protein and is associated with a relatively large conformational change.[171–174] MD simulations have revealed significant structural fluctuations, particularly for the inactive form.[175] Pathways and rate constants have been calculated using DPS for the conformational change corresponding to activation in the unphosphorylated protein.[167] Initial conformations were taken from NMR structures for the 124-residue N-terminal receiver domain of NtrC in protein data bank entries 1NTR and 1KRW, which correspond to the inactive and active forms, respectively. We found that the active conformation unfolded after 1 ns of MD simulation for the CHARMM19[162]/EEF1[163] potential. Further short MD runs were then performed starting from low-lying minima located in a preliminary DPS run. Comparing conformations from active and inactive states, it was found that the distance between residues in a particular loop and helix appeared to separate the two regions of configuration space most effectively. Clustering local minima on this basis led to initial end points for a subsequent DPS calculation (Figure 14.4).

The free energy disconnectivity graph corresponding to $\Delta F_{\text{barrier}} = 5$ kcal mol[-1] (Section 14.2.3) at 298 K is shown in Figure 14.5. Branches corresponding to inactive

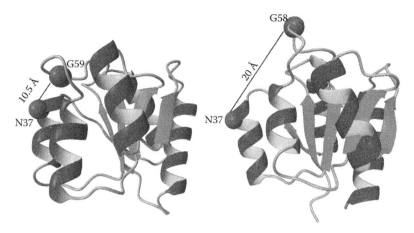

**FIGURE 14.4**   Initial end points for NtrC corresponding to the active or closed state (left) and inactive or open state (right). The distance between the $C_\alpha$ atoms of residues N37 and G59 was employed to define the two states.

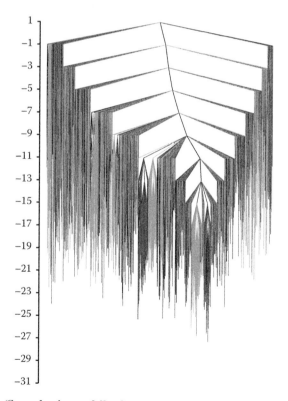

**FIGURE 14.5   (See color insert following page 172.)** Free energy disconnectivity graph for NtrC at 298 K for regrouping threshold $\Delta F_{barrier} = 5$ kcal mol$^{-1}$. The energy is in kcal mol$^{-1}$. The branches corresponding to groups of structures classified as inactive (open) and active (closed) are colored green and blue, respectively.[167]

and active conformations are colored green and purple, respectively. Branches of the same color cluster together, as in the region around the global minimum, when the barriers between similar conformations are smaller than the barriers for interconverting the active and inactive states. For the lowest two sets of active and inactive structures, the calculated transition rate is about $200 \, s^{-1}$.

The estimated rate constant lies in the range expected for large conformational transitions. However, the uncertainty due to all the approximations involved is likely to be at least an order of magnitude. The predicted sequence of events corresponding to the dominant pathways probably provides more appropriate features for comparison with experiment, especially since phosphorylation may reduce the barrier between active and inactive conformations, and increase the rate constants. Stationary points from the "fastest" path between the lowest-energy active and inactive conformations are shown in Figure 14.6. This sequence suggests cooperative motion of the $\alpha$-helices, which "slide" past one another before the active site loop changes to the active (closed) conformation. The alternative view in Figure 14.6B also suggests that there is a correlated change in orientation, and these changes appear to agree with experimental observations.[172] Two residues (R3 and K46) play a key role in changing the $\alpha3$–$\beta3$ loop from the inactive (open) to the active (closed) conformations.[167] Residues Y94 and Y101 also interact significantly in the active conformation via ring stacking, and the protein becomes "stiffer," in agreement with experiment.

### 14.3.3 INTERCONVERSION BETWEEN DIMERS OF AN AMYLOIDOGENIC PEPTIDE

Protein misfolding and aggregation is an active field of contemporary research due to the implication of misfolding in serious diseases.[176–178] Misfolded amyloid aggregates exhibit a generic cross $\beta$-sheet structure, with $\beta$-strands perpendicular to the long fiber axis, and backbone hydrogen-bonds stabilizing the sheets along the direction of the fiber.[179–183] The polar heptapeptide from the N-terminal prion-determining region of the yeast prion protein Sup35, GNNQQNY, has become a popular target for simulation studies[184–188] since the publication of a microcrystal structure for the corresponding amyloid.[189] The full Sup35 protein contains 685 residues, yet the peptide (residues 7 to 13) exhibits similar amyloidogenic properties.[190]

In the GNNQQNY microcrystal peptide monomers align to give in-register parallel $\beta$-sheets. Two sheets associate via a dry interface involving interdigitation of side-chains in a "steric zipper".[189,191] MD simulations suggest that this zipper motif is maintained when GNNQQNY protofibrils twist in thermal equilibrium.[188] Application of a more coarse-grained model suggested that GNNQQNY has intrinsic $\beta$-propensity,[192] and would form fibrils via addition of monomers to an existing cross-$\beta$ structure, rather than by random aggregation followed by a large-scale reorganization. Our results for free energy surfaces of the GNNQQNY dimer and DPS analysis of pathways between distinct $\beta$ structures support this picture.[166] These calculations again employed the CHARMM19[162]/EEF1[193] force field, using the implementation of replica exchange[194] from the MMTSB tool set[195] and weighted histogram analysis to construct free energy surfaces.[196] The peptides were confined using a spherical container,[197] giving a concentration of 10 mM, which is about an order of magnitude larger than experiment.

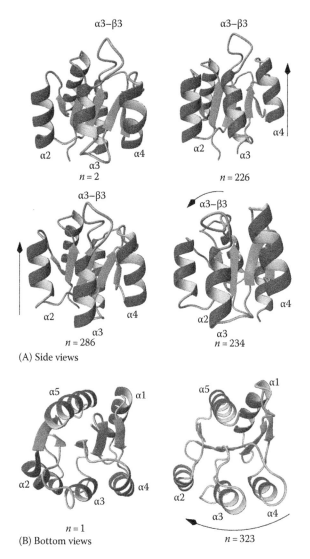

**FIGURE 14.6** Alternative views of conformations selected from the "fastest" pathway between active and inactive conformations of NtrC.[167] $n$ corresponds to the number of stationary points along the discrete path. (A) The helices move cooperatively before loop $\alpha 3-\beta 3$ closes. (B) In this view, the reorientation of the helices in the two forms is clear.

Simulations for the GNNQQNY monomer with the same force field indicate that coil structures are favored, in agreement with previous work.[186,190] Rapid interconversions between different conformations are observed, with characteristic timescales varying from microseconds to nanoseconds.[166] The free energy surface for the dimer in Figure 14.7 supports minima for several different $\beta$-sheet structures. One of these, the in-register parallel (IP) conformation, is similar to the geometry in the microcrystal,

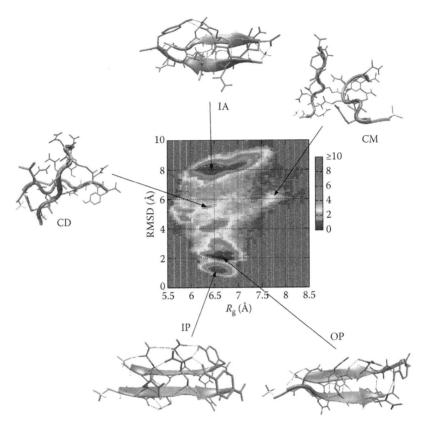

**FIGURE 14.7** **(See color insert following page 172.)** Free energy surface calculated at 298 K for the GNNQQNY dimer in terms of the radius of gyration, $R_g$, and $C_\alpha$-RMSD from the amyloid β-sheet.[166] The free energy (kcal mol$^{-1}$) is indicated by shading according to the scale on the right. Representative structures are indicated for five of the minima.

aside from a twist. The others can be characterized as an in-register antiparallel (IA) β-sheet, and a parallel structure where the two strands are off-register by one residue (OP). A detailed analysis of the different energetic contributions in the CHARMM19/ EEF1 potential for these structures indicates that they are stabilized by both electrostatic and hydrophobic interactions, which include π–π stacking.[166]

Rate constants were estimated using DPS for various interconversion processes. The three dimers IP, OP, and IA were all found to be stable against dissociation at 298 K, with association rate constants about 10 orders of magnitude larger than for dissociation. The interconversion between in-register and out-of-register parallel β structures can be described as a "reptation move." This process requires a significant reorganization of structure, although the $C_\alpha$-RMSD between the IP and OP geometries is only 2.4 Å. The calculated rate constants correspond to mean first-passage times of order 20 days and 15 h for IP → OP and OP → IP, respectively,[166] and the barrier is comparable to values calculated for analogous processes in protein G[198] and Aβ$_{16-22}$.[199–201] Experimentally, a timescale of about 8 h was reported for the

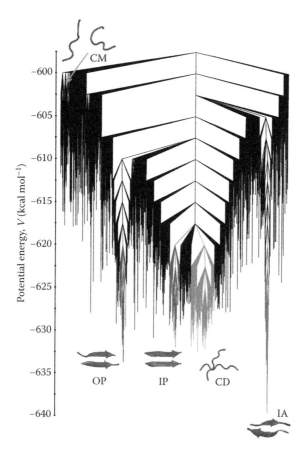

**FIGURE 14.8** (**See color insert following page 172.**) Potential energy disconnectivity graph for GNNQQNY.[166] Branches are colored for minima with $C_\alpha$-RMSD < 1 Å from the lowest minima in the IP, OP, IA, and CD sets and $V$ < −620 kcal mol⁻¹. CD structures correspond to conformations where the two peptides are roughly orthogonal and the CM structures correspond to loosely associated dimers.

growth of amyloid fibrils for the $PrP_{109-122}$ peptide at 308 K.[202] The corresponding mean first-passage times calculated for the GNNQQNY dimer at this temperature are 5.5 days and 3.5 h for IP → OP and OP → IP, respectively. At physiological temperatures, our calculations indicate that reptation moves may produce faster alignment of parallel β-sheet aggregates than a dissociation/reassociation mechanism. The calculated timescales for interconversions between parallel and antiparallel conformations are comparable to the values quoted above for reptation moves.[166]

Comparing disconnectivity graphs based on potential energy (Figure 14.8) and free energy (Figure 14.9), we find that the antiparallel sheet is favored by potential energy and the in-register parallel sheet is favored by entropy. The stronger intermolecular interaction for IA clearly reduces the conformational flexibility, and hence the entropy. The lowest free energy groups for IP and IA correspond to competing morphologies at 298 K, in agreement with the surface shown in Figure 14.7.

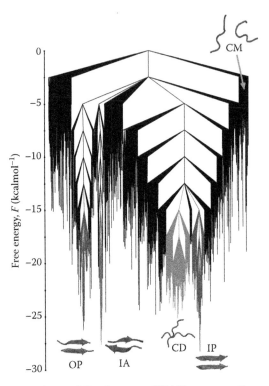

**FIGURE 14.9** (**See color insert following page 172.**) Free energy disconnectivity graph for the dimeric phase of GNNQQNY at 298 K.[166] Branches are colored for free energy groups containing minima of the PES with $C_\alpha$-RMSD < 1 Å from the lowest minima in the IP, OP, IA, and CD sets and $F < -15$ kcal mol$^{-1}$.

## 14.4 CONCLUSIONS

The examples discussed in Section 14.3 show how geometry optimization tools, combined with statistical rate theory, can be employed to access experimental timescales corresponding to folding, conformational changes associated with function, and amyloid formation. Most of the computer time used in such calculations is spent on finding transition states on the potential energy surface. These algorithms have been tested quite extensively, and it does not seem likely that much improvement will be possible beyond the DNEB/hybrid EF approach described in Section 14.2.1, or related schemes.

For pathways involving very large changes in structure, which may require discrete paths containing more than 100 transition states on the PES, finding an appropriate initial path can present a problem. The Dijkstra missing connection algorithm[101] (Section 14.2.1) enables such paths to be found in an automated fashion, but large barriers are likely to be present due to the lengthy interpolations involved. Locating kinetically relevant paths from such starting points may require weeks or months of computer time, and even when searches are based on geometry optimization tools,

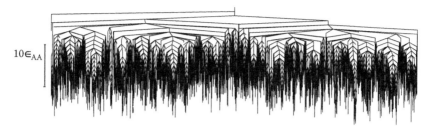

$10\in_{AA}$

**FIGURE 14.10**  Disconnectivity graph calculated for the minima sampled[203] over a locally ergodic time interval in a binary Lennard–Jones system of 60 particles modeled with periodic boundary conditions at a number density of 1.3 and $k_BT/\in_{AA} = 0.96$. Here, $\in_{AA}$ is the pair well depth between atoms of type A.

there is no guarantee that the best paths will be found. Future efforts are therefore likely to focus on improving the initial pathway, and converging the stationary point databases in terms of the calculated rate constants.

We conclude by noting that disconnectivity graphs provide a direct visualization of frustration in the potential or free energy landscape, in terms of low-lying minima that are separated by high barriers. As expected, the graphs that we have illustrated for Beta3s and NtrC exhibit only limited frustration compared to the landscape of a model bulk glass former illustrated in Figure 14.10.[204] This disconnectivity graph was constructed from local minima that were sampled during a locally ergodic MD trajectory.[203,205,206] It can be described as a large number of separate "palm tree" features,[78] with barriers that exceed about $30k_BT$ at the glass transition temperature for interconverting the lowest minima in different subtrees. This is an example of what we would consider a "rough" or "rugged" energy landscape. Sometimes the same description has been applied to the energy landscapes of proteins, in the sense that there are a large number of conformations. However, the number of local minima is expected to grow exponentially with system size,[142,207] and is not a useful indication of the structure-seeking[40,208] properties of the landscape. We therefore prefer to associate rough or rugged energy landscapes with disconnectivity graphs such as Figure 14.10, which exhibit the frustration we would expect for a good glass former.

We finally note that the average number of transition states per local minimum is expected to scale linearly with the number of atoms.[207] Obtaining a useful visualization of the potential energy surface for a many-body system is therefore very difficult, and projections onto one or two dimensions can be rather misleading.[31,72,125,138,209] In particular, conformations that are well separated in configuration space may overlap in the low-dimensional projection, distorting or removing barriers.[31] Disconnectivity graphs avoid this problem, and can be coarse-grained in a self-consistent fashion without the need to specify any structural order parameters (Section 14.2.3).

## ACKNOWLEDGMENTS

We gratefully acknowledge funding from the EPSRC and BBSRC, and helpful comments on the original manuscript from Dr. Jason Green.

## REFERENCES

1. E. A. Carter, G. Ciccotti, J. T. Hynes, and R. Kapral, *Chem. Phys. Lett.* **156**, 472 (1989).
2. A. F. Voter, *J. Chem. Phys.* **106**, 4665 (1997).
3. A. F. Voter, *Phys. Rev. Lett.* **78**, 3908 (1997).
4. A. F. Voter, *Phys. Rev. B* **57**, 13985 (1998).
5. D. Passerone and M. Parrinello, *Phys. Rev. Lett.* **87**, 108302 (2001).
6. D. J. Wales, *Mol. Phys.* **100**, 3285 (2002).
7. W. E, W. Ren, and E. Vanden-Eijnden, *Phys. Rev. B* **66**, 052301 (2002).
8. A. Laio and M. Parrinello, *Proc. Nat. Acad. Sci. U. S. A.* **99**, 12562 (2002).
9. T. S. van Erp, D. Moroni, and P. G. Bolhuis, *J. Chem. Phys.* **118**, 7762 (2003).
10. C. Dellago, P. Bolhuis, and P. L. Geissler, *Adv. Chem. Phys.* **123**, 1 (2002).
11. P. G. Bolhuis, D. Chandler, C. Dellago, and P. L. Geissler, *Annu. Rev. Phys. Chem.* **53**, 291 (2002).
12. R. A. Miron and K. A. Fichthorn, *J. Chem. Phys.* **119**, 6210 (2003).
13. D. J. Wales, *Mol. Phys.* **102**, 891 (2004).
14. D. M. Zuckerman, *J. Phys. Chem. B* **108**, 5127 (2004).
15. G. Hummer, *J. Chem. Phys.* **120**, 516 (2004).
16. A. K. Faradjian and R. Elber, *J. Chem. Phys.* **120**, 10880 (2004).
17. W. Ren, E. Vanden-Eijnden, P. Maragakis, and W. E, *J. Chem. Phys.* **123**, 134109 (2005).
18. S. Sriraman, I. G. Kevrekidis, and G. Hummer, *J. Phys. Chem. B* **109**, 6479 (2005).
19. J. E. Basner and S. D. Schwartz, *J. Am. Chem. Soc.* **127**, 13822 (2005).
20. K. N. Kudin and R. Car, *J. Chem. Phys.* **122**, 114108 (2005).
21. R. J. Allen, D. Frenkel, and P. R. ten Wolde, *J. Chem. Phys.* **124**, 024102 (2006).
22. R. J. Allen, D. Frenkel, and P. R. ten Wolde, *J. Chem. Phys.* **124**, 194111 (2006).
23. J. Hu, A. Ma, and A. R. Dinner, *J. Chem. Phys.* **125**, 114101 (2006).
24. S. L. Quaytman and S. D. Schwartz, *Proc. Natl. Acad. Sci. U. S. A.* **104**, 12253 (2007).
25. L. Maragliano and E. Vanden-Eijnden, *Chem. Phys. Lett.* **426**, 168 (2006).
26. I. V. Khavrutskii, K. Arora, and C. L. Brooks, *J. Chem. Phys.* **125**, 174108 (2006).
27. K. E. Becker and K. A. Fichthorn, *J. Chem. Phys.* **125**, 184706 (2006).
28. C. A. F. de Oliveira, D. Hamelberg, and J. A. McCammon, *J. Chem. Phys.* **127**, 175105 (2007).
29. A. M. A. West, R. Elber, and D. Shalloway, *J. Chem. Phys.* **126**, 145104 (2007).
30. T. F. Miller and C. Predescu, *J. Chem. Phys.* **126**, 144102 (2007).
31. F. Noé and S. Fischer, *Curr. Opin. Struct. Biol.* **18**, 154 (2008).
32. N. G. van Kampen, *Stochastic Processes in Physics and Chemistry*, North-Holland, Amsterdam, the Netherlands (1981).
33. R. E. Kunz, *Dynamics of First-Order Phase Transitions*, Deutsch, Thun Switzerland (1995).
34. B. A. Huberman and M. Kerszberg, *J. Phys. A: Math. Gen.* **18**, L331 (1985).
35. P. E. Leopold, M. Montal, and J. N. Onuchic, *Proc. Natl. Acad. Sci. U. S. A.* **89**, 8721 (1992).
36. R. S. Berry and R. Breitengraser-Kunz, *Phys. Rev. Lett.* **74**, 3951 (1995).
37. R. E. Kunz and R. S. Berry, *J. Chem. Phys.* **103**, 1904 (1995).
38. B. Kunz, R. S. Berry, and T. Astakhova, *Surf. Rev. Lett.* **3**, 307 (1996).
39. J. Wang, J. N. Onuchic, and P. G. Wolynes, *Phys. Rev. Lett.* **76**, 4861 (1996).
40. J. P. K. Doye and D. J. Wales, *J. Chem. Phys.* **105**, 8428 (1996).
41. O. M. Becker and M. Karplus, *J. Chem. Phys.* **106**, 1495 (1997).
42. R. E. Kunz, P. Blaudeck, K. H. Hoffmann, and R. S. Berry, *J. Chem. Phys.* **108**, 2576 (1998).
43. T. R. Walsh and D. J. Wales, *J. Chem. Phys.* **109**, 6691 (1998).
44. M. Cieplak, M. Henkel, J. Karbowski, and J. R. Banavar, *Phys. Rev. Lett.* **80**, 3654 (1998).
45. K. D. Ball and R. S. Berry, *J. Chem. Phys.* **109**, 8541 (1998).
46. K. D. Ball and R. S. Berry, *J. Chem. Phys.* **109**, 8557 (1998).

47. N. P. Kopsias and D. N. Theodorou, *J. Chem. Phys.* **109**, 8573 (1998).

48. K. M. Westerberg and C. A. Floudas, *J. Chem. Phys.* **110**, 9259 (1999).

49. K. D. Ball and R. S. Berry, *J. Chem. Phys.* **111**, 2060 (1999).

50. M. A. Miller, J. P. K. Doye, and D. J. Wales, *J. Chem. Phys.* **110**, 328 (1999).

51. J. P. K. Doye and D. J. Wales, *Phys. Rev. B* **59**, 2292 (1999).

52. M. A. Miller, J. P. K. Doye, and D. J. Wales, *Phys. Rev. E* **60**, 3701 (1999).

53. R. Czerminski and R. Elber, *J. Chem. Phys.* **92**, 5580 (1990).

54. S. Fischer, S. Michnick, and M. Karplus, *Biochemistry* **32**, 13830 (1993).

55. R. S. Berry, N. Elmaci, J. P. Rose, and B. Vekhter, *Proc. Natl. Acad. Sci. U. S. A.* **94**, 9520 (1997).

56. M. A. Miller and D. J. Wales, *J. Chem. Phys.* **111**, 6610 (1999).

57. D. J. Wales and P. E. J. Dewsbury, *J. Chem. Phys.* **121**, 10284 (2004).

58. T. Komatsuzaki, K. Hoshino, Y. Matsunaga, G. J. Rylance, R. L. Johnston, and D. J. Wales, *J. Chem. Phys.* **122**, 084714 (2005).

59. J. Kim and T. Keyes, *J. Phys. Chem. B* **111**, 2647 (2007).

60. F. Rao and A. Caflisch, *J. Mol. Biol.* **342**, 299 (2004).

61. N. Singhal, C. D. Snow, and V. S. Pande, *J. Chem. Phys.* **121**, 415 (2004).

62. C. Schultheis, T. Hirschberger, H. Carstens, and P. Tavan, *J. Chem. Theor. Comput.* **1**, 515 (2005).

63. S. Park and V. S. Pande, *J. Chem. Phys.* **124**, 054118 (2006).

64. I. Horenko, E. Dittmer, A. Fischer, and C. Schütte, *Multiscale Model. Sim.* **5**, 802 (2006).

65. F. Noé, I. Horenko, C. Schütte, and J. C. Smith, *J. Chem. Phys.* **126**, 155102 (2007).

66. N. S. Hinrichs and V. S. Pande, *J. Chem. Phys.* **126**, 244101 (2007).

67. J. D. Chodera, K. A. Dill, N. Singhal, V. S. Pande, W. C. Swope, and J. W. Pitera, *J. Chem. Phys.* **126**, 155101 (2007).

68. N.-V. Buchete and G. Hummer, *J. Phys. Chem. B* **112**, 6057 (2008).

69. S. V. Krivov, S. Muff, A. Caflisch, and M. Karplus, *J. Phys. Chem. B* **112**, 8701 (2008).

70. P. N. Mortenson and D. J. Wales, *J. Chem. Phys.* **114**, 6443 (2001).

71. P. N. Mortenson, D. A. Evans, and D. J. Wales, *J. Chem. Phys.* **117**, 1363 (2002).

72. S. V. Krivov and M. Karplus, *Proc. Natl. Acad. Sci. U. S. A.* **101**, 14766 (2004).

73. S. V. Krivov and M. Karplus, *J. Chem. Phys.* **117**, 10894 (2002).

74. F. Noé, M. Oswald, G. Reinelt, S. Fischer, and J. C. Smith, *Multiscale Model Sim.* **5**, 393 (2006).

75. F. Noé, D. Krachtus, J. C. Smith, and S. Fischer, *J. Chem. Theory Comput.* **2**, 840 (2006).

76. N. M. Amato and G. Song, *J. Comput. Biol.* **8**, 149 (2002).

77. D.-W. Li, H. Yang, L. Han, and S. Huo, *Biophys. J.* **94**, 1622 (2008).

78. D. J. Wales, M. A. Miller, and T. R. Walsh, *Nature* **394**, 758 (1998).

79. D. J. Wales, *Energy Landscapes*, Cambridge University Press, Cambridge, U.K. (2003).

80. D. A. Evans and D. J. Wales, *J. Chem. Phys.* **118**, 3891 (2003).

81. A.-L. Barabási and R. Albert, *Science* **286**, 509 (1999).

82. J. P. K. Doye, *Phys. Rev. Lett.* **88**, 238701 (2002).

83. D. Gfeller, D. M. de Lachapelle, P. De Los Rios, G. Caldarelli, and F. Rao, *Phys. Rev. E* **76**, 026113 (2007).

84. J. N. Murrell and K. J. Laidler, *Trans. Faraday. Soc.* **64**, 371 (1968).

85. P. Valtazanos and K. Ruedenburg, *Theor. Chim. Acta* **69**, 281 (1986).

86. J. Nocedal, *Math. Comput.* **35**, 773 (1980).

87. D. Liu and J. Nocedal, *Math. Prog.* **45**, 503 (1989).

88. D. J. Wales, *OPTIM: A program for optimising geometries and calculating pathways*, http://www-wales.ch.cam.ac.uk/software.html.

89. S. A. Trygubenko and D. J. Wales, *J. Chem. Phys.* **120**, 2082 (2004).

90. R. Elber and M. Karplus, *Chem. Phys. Lett.* **139**, 375 (1987).

91. G. Henkelman and H. Jónsson, *J. Chem. Phys.* **111**, 7010 (1999).

92.  G. Henkelman, B. P. Uberuaga, and H. Jónsson, *J. Chem. Phys.* **113**, 9901 (2000).
93.  G. Henkelman and H. Jónsson, *J. Chem. Phys.* **113**, 9978 (2000).
94.  L. J. Munro and D. J. Wales, *Phys. Rev. B* **59**, 3969 (1999).
95.  Y. Kumeda, L. J. Munro, and D. J. Wales, *Chem. Phys. Lett.* **341**, 185 (2001).
96.  S. Fischer and M. Karplus, *Chem. Phys. Lett.* **194**, 252 (1992).
97.  W. Quapp and D. Heidrich, *J. Mol. Struct. (Theochem)* **585**, 105 (2002).
98.  M.-R. Yun, N. Mousseau, and P. Derreumaux, *J. Chem. Phys.* **126**, 105101 (2007).
99.  F. B. Hildebrand, *Methods of Applied Mathematics*, Dover, New York (1992).
100. E. F. Koslover and D. J. Wales, *J. Chem. Phys.* **127**, 234105 (2007).
101. J. M. Carr, S. A. Trygubenko, and D. J. Wales, *J. Chem. Phys.* **122**, 234903 (2005).
102. E. W. Dijkstra, *Numerische Math.* **1**, 269 (1959).
103. D. J. Wales, *Int. Rev. Phys. Chem.* **25**, 237 (2006).
104. S. A. Trygubenko, and D. J. Wales, *Mol. Phys.* **104**, 1497 (2006).
105. L. Onsager, *Phys. Rev.* **54**, 554 (1938).
106. R. Du, V. S. Pande, A. Y. Grosberg, T. Tanaka, and E. I. Shakhnovich, *J. Chem. Phys.* **108**, 334 (1998).
107. C. D. Snow, E. J. Sorin, Y. M. Rhee, and V. S. Pande, *Annu. Rev. Biophys. Biomol. Struct.* **34**, 43 (2004).
108. H. M. Taylor and S. Karlin, *An Introduction to Stochastic Modeling*, Academic Press, Orlando, FL (1984).
109. W. Press, B. Flannery, S. Teukolsky, and W. Vetterling, *Numerical Recipes*, Cambridge University Press, Cambridge, U.K. (1986).
110. Z. Bai, J. Demmel, J. Dongarra, A. Ruhe, and H. van der Vorst (eds.), *Templates for the Solution of Algebraic Eigenvalue Problems: a Practical Guide*, SIAM, Philadelphia, PA (2000).
111. A. B. Bortz, M. H. Kalos, and J. L. Lebowitz, *J. Comput. Phys.* **17**, 10 (1975).
112. D. A. Reed and G. Ehrlich, *Surf. Sci.* **105**, 603 (1981).
113. A. F. Voter, *Phys. Rev. B* **34**, 6819 (1986).
114. K. A. Fichthorn and W. H. Weinberg, *J. Chem. Phys.* **95**, 1090 (1991).
115. A. F. Voter, *Radiation Effects in Solids* (pp. 1–22), Springer-Verlag, Berlin (2005).
116. S. A. Trygubenko and D. J. Wales, *J. Chem. Phys.* **124**, 234110 (2006).
117. H. Pelzer and E. Wigner, *Z. Phys. Chem.* **B15**, 445 (1932).
118. H. Eyring, *Chem. Rev.* **17**, 65 (1935).
119. M. G. Evans and M. Polanyi, *Trans. Faraday Soc.* **31**, 875 (1935).
120. H. Eyring, *J. Chem. Phys.* **3**, 107 (1935).
121. M. G. Evans and M. Polanyi, *Trans. Farady Soc.* **33**, 448 (1937).
122. W. F. K. Wynne-Jones and H. Eyring, *J. Chem. Phys.* **3**, 492 (1935).
123. C. Dellago, P. G. Bolhuis, F. S. Csajka, and D. Chandler, *J. Chem. Phys.* **108**, 1964 (1998).
124. C. Dellago, P. G. Bolhuis, and D. Chandler, *J. Chem. Phys.* **110**, 6617 (1999).
125. A. Altis, M. Otten, P. H. Nguyen, R. Hegger, and G. Stock, *J. Chem. Phys.* **128**, 245102 (2008).
126. J. D. Chodera, S. Singhal, V. S. Pande, K. A. Dill, and W. C. Swope, *J. Chem. Phys.* **126**, 155101 (2007).
127. F. Calvo, F. Spiegelman, and D. J. Wales, *J. Chem. Phys.* **118**, 8754 (2003).
128. D. A. Evans and D. J. Wales, *J. Chem. Phys.* **119**, 9947 (2003).
129. D. A. Evans and D. J. Wales, *J. Chem. Phys.* **121**, 1080 (2004).
130. V. M. Jiménez and A. Marzal, in Algorithm Engineering: 3rd International Workshop, WAE'99, London, UK, July 1999. J. S. Vitter, and C. D. Zaroliagis (eds.), (vol. 1668, p. 1529). Springer, Berlin, Heidelberg (1999).
131. J. M. Carr and D. J. Wales, The Energy Landscape as a Computational Tool, in *Latest Advances in Atomic Cluster Collisions: Structure and Dynamics from the Nuclear to the Biological Scale*. J.-P. Connerade, and A. Solov'yov (eds.). 321-330, Imperial College Press, London (2008).

132. S. K. Kearsley, *Acta Cryst. A* **45**, 208 (1989).
133. G. C. Boulougouis and D. N. Theodorou, *J. Chem. Phys.* **127**, 084903 (2007).
134. J. M. Carr and D. J. Wales, *J. Phys. Chem. B* **112**, 8760 (2008).
135. A. Godzik, A. Koliński, and J. Skolnick, *J. Comput. Aid. Mol. Des.* **7**, 397 (1993).
136. J. D. Bryngelson and P. G. Wolynes, *Proc. Natl. Acad. Sci. U. S. A.* **84**, 7524 (1987).
137. J. M. Carr and D. J. Wales, *J. Chem. Phys.* **123**, 234901 (2005).
138. S. Muff and A. Caflisch, *Proteins: Struct. Func. Bioinf.* **70**, 1185 (2008).
139. D. J. McGinty, *J. Chem. Phys.* **55**, 580 (1971).
140. J. J. Burton, *J. Chem. Phys.* **56**, 3133 (1972).
141. M. R. Hoare, *Adv. Chem. Phys.* **40**, 49 (1979).
142. F. H. Stillinger and T. A. Weber, *Science* **225**, 983 (1984).
143. G. Franke, E. R. Hilf, and and P. Borrmann, *J. Chem. Phys.* **98**, 3496 (1993).
144. D. J. Wales, *Mol. Phys.* **78**, 151 (1993).
145. D. J. Wales, J. P. K. Doye, M. A. Miller, P. N. Mortenson, and T. R. Walsh, *Adv. Chem. Phys.* **115**, 1 (2000).
146. P. G. Mezey, *Theo. Chim. Acta* **58**, 309 (1981).
147. J. S. Shaffer and A. K. Chakraborty, *Macromolecules* **26**, 1120 (1993).
148. E. de Alba, J. Santoro, M. Rico, and M. A. Jiménez, *Protein Sci.* **8**, 854 (1999).
149. H. W. Wang and S. S. Sung, *J. Am. Chem. Soc.* **122**, 1999 (2000).
150. S. Mohanty and U. H. E. Hansmann, *Biophys. J.* **91**, 3573 (2006).
151. S. Mohanty and U. H. E. Hansmann, *Phys. Rev. E* **76**, 012901 (2007).
152. P. Ferrara and A. Caflisch, *Proc. Natl. Acad. Sci. U. S. A.* **97**, 10780 (2000).
153. P. Ferrara and A. Caflisch, *J. Mol. Biol.* **306**, 837 (2001).
154. A. Cavalli, P. Ferrara, and A. Caflisch, *Proteins Struct., Func. Gen.* **47**, 305 (2002).
155. R. Davis, C. M. Dobson, and M. Vendruscolo, *J. Chem. Phys.* **117**, 9510 (2002).
156. A. Cavalli, U. Haberthür, E. Paci, and A. Caflisch, *Prot. Sci.* **12**, 1801 (2003).
157. F. Rao and A. Caflisch, *J. Chem. Phys.* **119**, 4035 (2003).
158. G. Settanni, F. Rao, and A. Caflisch, *Proc. Natl. Acad. Sci. U. S. A.* **102**, 628 (2005).
159. F. Rao, G. Settanni, E. Guarnera, and A. Caflisch, *J. Chem. Phys.* **122**, 184901 (2005).
160. B. D. Bursulaya and C. L. Brooks III, *J. Am. Chem. Soc.* **121**, 9947 (1999).
161. A. R. Fersht, *Structure and Mechanism in Protein Science*, W. H. Freeman and Company, New York (1999).
162. B. R. Brooks, R. E. Bruccoleri, B. D. Olafson, D. J. States, S. Swaminathan, and M. Karplus, *J. Comp. Chem.* **4**, 187 (1983).
163. T. Lazaridis, and M. Karplus, *Proteins Struct., Func. Gen.* **35**, 133 (1999).
164. J. D. Bloom, Computer Simulations of Protein Aggregation, Master's thesis, University of Cambridge (2002).
165. http://www-wales.ch.cam.ac.uk/~sat39/charmm/rotamers/rotamers_in_CHARMM19.pdf.
166. B. Strodel, C. S. Whittleston, and D. J. Wales, *J. Am. Chem. Soc.* **129**, 16005 (2007).
167. M. Khalili and D. J. Wales, *J. Phys. Chem. B* **112**, 2456 (2008).
168. J. S. Parkinson and E. C. Kofoid, *Annu. Rev. Genet.* **26**, 71 (1992).
169. T. Mizuno, T. Kaneko, and S. Tabata, *DNA Res.* **3**, 407 (1996).
170. T. Mizuno, *DNA Res.* **4**, 161 (1997).
171. I. Hwang, T. Thorgeirsson, J. Lee, S. Kustu, and Y. Shin, *Proc. Natl. Acad. Sci. U. S. A.* **96**, 4880 (1999).
172. D. Kern, B. F. Volkman, P. Luginbuhl, M. J. Nohaile, S. Kustu, and D. E. Wemmer, *Nature* **402**, 894 (1999).
173. B. F. Volkman, D. Lipson, D. E. Wemmer, and D. Kern, *Science* **291**, 2429 (2001).
174. C. A. Hastings, S.-Y. Lee, H. S. Cho, D. Yan, S. Kustu, and D. E. Wemmer, *Biochemistry* **42**, 9081 (2003).
175. X. Hu and Y. Wang, *J. Biomol. Struct. Dyn.* **23**, 509 (2006).
176. D. J. Selkoe, *Nature* **426**, 900 (2003).
177. F. Chiti and C. M. Dobson, *Annu. Rev. Biochem.* **75**, 333 (2006).

178. V. N. Uversky and A. L. Fink, *Protein Misfolding, Aggregation and Conformational Disease*, Springer, Berlin (2006).

179. J. D. Sipe and A. S. Cohen, *J. Struct. Biol.* **130**, 88 (2000).

180. E. D. Eanes and G. G. Glenner, *J. Histochem. Cytochem.* **16**, 673 (1968).

181. A. J. Geddes, K. D. Parker, A. D. Atkins, and E. Beighton, *J. Mol. Biol.* **32**, 343 (1968).

182. M. Sunde, L. C. Serpell, M. Bartlam, P. E. Fraser, M. B. Pepys, and C. C. F. Blake, *J. Mol. Biol.* **273**, 729 (1997).

183. O. S. Makin and L. C. Serpell, *Fib. Diffr. Rev.* **12**, 29 (2004).

184. J. Gsponer, U. Haberthür, and A. Caflisch, *Proc. Natl. Acad. Sci. U. S. A.* **100**, 5154 (2003).

185. M. Cecchini, F. Rao, M. Seeber, and A. Caflisch, *J. Chem. Phys.* **121**, 10748 (2004).

186. J. Lipfert, J. Franklin, F. Wu, and S. Doniach, *J. Mol. Biol.* **349**, 648 (2005).

187. J. Zheng, B. Ma, C.-J. Tsai, and R. Nussinov, *Biophys. J.* **91**, 824 (2006).

188. L. Esposito, C. Pedone, and L. Vitagliano, *Proc. Natl. Acad. Sci. U. S. A.* **103**, 11533 (2006).

189. R. Nelson, M. R. Sawaya, M. Balbirnie, A. O. Madsen, C. Riekel, R. Grothe, and D. Eisenberg, *Nature* **435**, 773 (2005).

190. M. Balbirnie, R. Grothe, and D. S. Eisenberg, *Proc. Natl. Acad. Sci. U. S. A.* **98**, 2375 (2001).

191. M. R. Sawaya, S. Sambashivan, R. Nelson, M. I. Ivanova, S. A. Sievers, M. I. Apostol, M. J. Thompson, M. Balbirnie, J. J. Wiltzius, H. T. McFarlane, A. Ø. Madsen, C. Riekel, and D. Eisenberg, *Nature* **447**, 453 (2007).

192. R. Pellarin and A. Caflisch, *J. Mol. Biol.* **360**, 882 (2006).

193. T. Lazaridis and M. Karplus, *Proteins Struct. Func. Gen.* **35**, 133 (1999).

194. Y. Sugita and Y. Okamoto, *Chem. Phys. Lett.* **314**, 141 (1999).

195. M. Feig, J. Karanicolas, and C. L. Brooks, *J. Mol. Graph. Mod.* **22**, 377 (2004).

196. S. Kumar, D. Bouzida, R. H. Swendsen, P. A. Kollman, and J. M. Rosenberg, *J. Comp. Chem.* **13**, 1011 (1992).

197. D. K. Klimov, D. Newfield, and D. Thirumalai, *Proc. Natl. Acad. Sci. U. S. A.* **99**, 8019 (2002).

198. G. Wei, P. Derreumaux, and N. Mousseau, *J. Chem. Phys.* **119**, 6403 (2003).

199. S. Santini, G. Wei, N. Mousseau, and P. Derreumaux, *Structure.* **12**, 1245 (2004).

200. S. Santini, N. Mousseau, and P. Derreumaux, *J. Am. Chem. Soc.* **126**, 11509 (2004).

201. P. Derreumaux and N. Mousseau, *J. Chem. Phys.* **126**, 025101 (2007).

202. S. A. Petty and S. M. Decatur, *Proc. Natl. Acad. Sci. U. S. A.* **102**, 14272 (2005).

203. V. K. de Souza and D. J. Wales, *J. Chem. Phys.* **123**, 134504 (2005).

204. F. Calvo, T. V. Bogdan, V. K. de Souza, and D. J. Wales, *J. Chem. Phys.* **127**, 044508 (2007).

205. V. K. de Souza and D. J. Wales, *Phys. Rev. Lett.* **96**, 057802 (2006).

206. V. K. de Souza and D. J. Wales, *Phys. Rev. B* **74**, 134202 (2006).

207. D. J. Wales and J. P. K. Doye, *J. Chem. Phys.* **119**, 12409 (2003).

208. K. D. Ball, R. S. Berry, R. E. Kunz, F. Y. Li, A. Proykova, and D. J. Wales, *Science* **271**, 963 (1996).

209. S. V. Krivov and M. Karplus, *J. Phys. Chem. B* **110**, 12689 (2006).

# 15 Energy Flow and Allostery in an Ensemble

*Vincent J. Hilser and Steven T. Whitten*

## CONTENTS

## 15.1 INTRODUCTION

The ability of proteins to couple the binding of a ligand at one site of a macromolecule or macromolecular complex to the regulation of another nonoverlapping site is known as allostery. This functional property of proteins was first observed more than 100 years ago [1], and has been studied intensively since [2–5]. Early efforts to understand allostery focused on the development of phenomenological models to explain the observed coupling between sites located in symmetrically arranged monomers of homo-oligomeric systems [6,7], in which oxygen binding in hemoglobin received the most attention [3,8–10]. Although the different models initially provided no physical insight into the origins of the allosteric coupling energies, the subsequent availability of structures for various ligated states of hemoglobin [11] led to a seemingly important connection. Namely that the physical basis of allosteric control could be reconciled in the context of the atomic-level changes in the bonds that are made or broken in conversion between the various states of the protein or the quaternary structures of the oligomer [11–13]. Indeed, numerous protein systems have been observed to behave in a manner consistent with a structural–mechanical basis of allosteric control between coupled sites [14,15], suggesting that an exclusively structural representation provides a useful framework for understanding allostery, albeit qualitatively. What has not been established, however, is that such a framework provides a suitable set of organizing principles within which an understanding of allosteric phenomena in one system can be ascertained from data obtained in another. In short, will a structure-based description of allostery that focuses on identifying and characterizing pathways between allosteric sites provide a quantitative, predictive model?

One potential pitfall in describing allostery in strictly structural terms is the observation that proteins are not the static structures usually used to depict them. Instead, proteins exist as ensembles of interconverting conformational states [16], and these fluctuations appear to be relevant to function [17,18]. Furthermore, the structural description of allostery is also challenged by the observation that thousands of different proteins are likely to be intrinsically disordered (ID), or have ID domains, under physiological conditions [19–26]. These ID domains are found in disproportionally higher amounts in cell signaling proteins and transcription factors [23], suggesting an important regulatory role. A strictly structural description of allostery that includes ID domains would be difficult to conceive, for obvious reasons. In spite of these observations, there has been comparatively little progress in the development of a framework within which to understand the complex interplay between protein structure, conformational fluctuations, ID domains, and the observed allosteric coupling between sites.

In this chapter, we develop a hierarchical model of proteins that allows us to investigate the phenomena of allostery as it pertains to an ensemble of conformations that includes partially unfolded states. We show that such an ensemble representation of allostery provides unique insights into the ground rules governing site-to-site communication in macromolecular systems, rules that would be difficult if not impossible to glean from a purely structural view of allostery. We will show that rather than representing a special case of allosteric control, the ground rules obtained from the ensemble model of allosteric control establish the fundamental relationships between conformational equilibria, binding energy and allosteric coupling, and provide a set of organizing principles within which allostery can be understood in quantitative terms. In addition to showing that allostery is related in complex ways to the conformational equilibria in the ensemble, the model illuminates fundamental shortcomings of other structural models that attempt to assign allosteric free energies to pathways that connect the coupled sites by a network of intramolecular interactions between atoms. Instead, the ensemble model reveals how the binding energy can be distributed throughout the structure, such that even surface-exposed positions in proteins can have a profound effect on allosteric coupling.

## 15.2  THE ENSEMBLE ALLOSTERIC MODEL

### 15.2.1  Theory

Regulatory proteins are often multidomain oligomers that segregate the binding sites for each ligand into different structural domains [27]. Monomeric proteins have also been observed to possess allosteric properties [28], and it is well established that proteins in general have a modular structure [29]. Accordingly, an allosteric protein system can be represented, to a first approximation, as a group of interacting domains, the simplest of which is the two-domain protein in which a single binding

Allosteric model

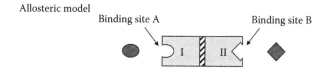

Thermodynamic parameters

| | | | | |
|---|---|---|---|---|
| N | $\supset$ I ▨ II $\langle$ | 0 | 1 | $P_N = \dfrac{1}{Q}$ |
| 1 | $\supset$ I ▨ | $\Delta G_{II} + \Delta g_{int}$ | $K_{II}\varphi_{int}$ | $P_1 = \dfrac{K_{II}\varphi_{int}}{Q}$ |
| 2 | ▨ II $\langle$ | $\Delta G_{I} + \Delta g_{int}$ | $K_{I}\varphi_{int}$ | $P_2 = \dfrac{K_{I}\varphi_{int}}{Q}$ |
| U | | $\Delta G_{I} + \Delta G_{II} + \Delta g_{int}$ | $K_{I}K_{II}\varphi_{int}$ | $P_U = \dfrac{K_{I}K_{II}\varphi_{int}}{Q}$ |

**FIGURE 15.1** Model for allosteric coupling. In the top panel, shown is a schematic representation of a test model for allosteric coupling. The model consists of a hypothetical protein that can bind two different ligands, A and B, one in each domain. Each domain can be folded or unfolded, resulting in four possible states: N, 1, 2, and U. Each domain is modeled such that it can bind its putative ligand only when it is folded. The thermodynamic parameters that describe this model energetically are given in the lower panel.

site resides in each domain (Figure 15.1). This simplified description of a protein containing two binding sites will be used to establish a quantitative model of allostery as described previously [30]. To explore the contributions of intrinsic disorder and conformational fluctuations in allostery, each domain in the model can be independently folded or unfolded. Thus, as opposed to assuming that the entire protein cooperatively unfolds as a unit, the protein is divided into an arbitrary number of units that unfold cooperatively. For the simplest case involving just two domains, there are four possible states (i.e., N, 1, 2, and U) that represent all combinations of folding and unfolding each domain. In this model, the energy of each state, which is presented relative to the N state in the figure, is composed of the free energy of unfolding each domain plus the energy of breaking the interactions between them ($\Delta g_{int}$). For this system, the partition function, $Q$, is the sum of the statistical weights of all states in the ensemble

$$Q = 1 + K_{II}\varphi_{int} + K_{I}\varphi_{int} + K_{I}K_{II}\varphi_{int}, \tag{15.1}$$

where
$$K_{II} = \exp(-\Delta G_{II}/RT)$$
$$K_{I} = \exp(-\Delta G_{I}/RT)$$
$$\varphi_{int} = \exp(-\Delta g_{int}/RT)$$

The probability of each state is simply the statistical weight normalized to the partition function.

As will be shown below, allostery between the binding sites results for two reasons: (1) the binding of ligand within a domain affects the stability of that domain (i.e., value of $\Delta G$) and (2) the two domains interact with each other (i.e., $\Delta g_{int} \neq 0$). Because of the interaction energy between the domains, each domain can "sense" the folding status (or a change in the folding status) of the other. We note that in our simple hierarchical model there are no assumptions regarding the physical origins of the energies. In fact, $\Delta g_{int}$ (and even $\Delta G$) can arise from any number of sources. For pedagogical purposes however, associating $\Delta g_{int}$ with a physically meaningful origin may add clarity. As an example, for cases where $\Delta g_{int}$ is positive, it is energetically unfavorable to break the interaction. Such a situation would exist with two complementary hydrophobic surfaces, wherein it would be energetically more favorable to interact with each other than with solvent. For situations where $\Delta g_{int}$ is negative, on the other hand, exactly the opposite effect would be observed; interaction of the surfaces with solvent would be more favorable than the interaction with each other. Such a situation might exist for the same hydrophobic surfaces described above, but at low temperatures [31]. We note however, and will demonstrate below, that the physical basis for the interaction between domains is not relevant, only the magnitude of the value. In other words, whether the interface between two domains is stabilized by hydrogen bonds, hydrophobic interactions, salt bridges, or a combination of each, is immaterial to the ability of the protein to propagate the thermodynamic effects from one domain or region to another.

To explore the extent of energetic coupling between the domains, a perturbation to the system is applied. In principle, this perturbation can be in the form of a mutation, the ionization or chemical modification of a residue, or, as in our allosteric model, the binding of a ligand. For the current case, we are interested in determining how the binding of ligand A to domain I can influence the ability of the protein to bind ligand B in domain II. The binding of ligand by ID proteins is usually associated with folding of the disordered domain [19], indicating that the affinity for ligand of the folded conformation is typically greater than the affinity for ligand of the disordered conformation. To capture this observation, the model is constructed such that each domain can bind its putative ligand only when it is folded. Thus, state 1 can only bind ligand A, state 2 can only bind ligand B, state N can bind both ligands, and state U can bind neither ligand.

Because states N and 1 are able to bind ligand A, the partition function (Equation 15.1) in the presence of ligand A becomes

$$Q = Z_{\text{Lig,A}}\left(1 + K_{II}\varphi_{int}\right) + K_I\varphi_{int} + K_I K_{II}\varphi_{int}, \tag{15.2}$$

where $Z_{\text{Lig,A}} = 1 + K_{a,A}[A]$, and $K_{a,A}$ is the intrinsic association constant of domain I for ligand A (we note that at concentrations of $[A] \ll 1/K_{a,A}$, Equation 15.2 reduces to Equation 15.1). As Equation 15.2 reveals, adding ligand A to the system results in a redistribution of the ensemble probabilities. This is evident in Figure 15.2A, where the state probabilities are plotted in response to increasing concentrations of ligand A, which favors population of states N and 1 at the expense of states 2 and U. The

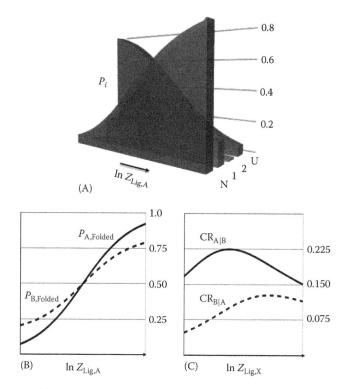

(A)

(B)   $\ln Z_{\text{Lig,A}}$

(C)   $\ln Z_{\text{Lig,X}}$

**FIGURE 15.2** Allosteric coupling in the model was evaluated by the addition of ligand. (A) Increasing concentrations of ligand A, shown by $\ln Z_{\text{Lig,A}}$ (where $Z_{\text{Lig,A}} = 1 + K_{a,A}$ [A]), affects the state probabilities, $P_i$, by preferentially stabilizing those states competent to bind ligand A, states N and 1. (B) Because state N is also competent to bind ligand B, increasing concentrations of ligand A not only increased the summed probability of states competent to bind ligand A ($P_{\text{A,Folded}}$), but also the summed probability of states competent to bind ligand B ($P_{\text{B,Folded}}$). (C) The allosteric coupling response (CR, see Equation 15.5) is directionally dependent. As can be seen in this figure, the sensitivity of site B to ligand binding at site A, $CR_{\text{B|A}}$, was not equal to the sensitivity of site A to ligand binding at site B, $CR_{\text{A|B}}$.

question with regard to allostery is, what effect did the binding of ligand A (to states N and 1) have on the probability of the states that can bind ligand B (states N and 2)? This question can be resolved analytically by examining the expression for the combined probability of states N and 2 ($P_{\text{B,Folded}}$) both without and with ligand A.

$$\text{Without ligand A: } P_{\text{B,Folded}} = \frac{1 + K_I\varphi_{\text{int}}}{1 + K_{II}\varphi_{\text{int}} + K_I\varphi_{\text{int}} + K_I K_{II}\varphi_{\text{int}}}, \quad (15.3a)$$

$$\text{With ligand A: } P_{\text{B,Folded}} = \frac{Z_{\text{Lig,A}} + K_I\varphi_{\text{int}}}{Z_{\text{Lig,A}}\left(1 + K_{II}\varphi_{\text{int}}\right) + K_I\varphi_{\text{int}} + K_I K_{II}\varphi_{\text{int}}}, \quad (15.3b)$$

As the expressions show, the effect of $Z_{\text{Lig,A}}$ on the probability of states in which the binding site for B is folded (and therefore competent to bind ligand B) will depend on the magnitudes of the statistical weights of the individual states, which in turn depend on the intrinsic stabilities of each domain, $\Delta G_{\text{I}}$ and $\Delta G_{\text{II}}$, and the interaction energy between them, $\Delta g_{\text{int}}$. This is demonstrated in Figure 15.2B with an arbitrary set of parameter values for $\Delta G_{\text{I}}$, $\Delta G_{\text{II}}$, and $\Delta g_{\text{int}}$. For this example, the free energy of each domain, $\Delta G_{\text{I}}$ and $\Delta G_{\text{II}}$, without ligand was modeled as $-2.5$ and $-1.0$ kcal/mol, respectively, which implies that both domains are essentially unfolded in the absence of ligand (see Figure 15.2A). A favorable interaction free energy between the domains ($\Delta g_{\text{int}} = 2$ kcal/mol) was modeled and results in a summed probability for the states that bind ligand B as modest under these conditions (i.e., $P_{\text{B,Folded}} \sim 0.20$). Thus, domain I is almost never folded, and domain II is folded only 20% of the time. The addition of ligand A, however, stabilizes each state that binds ligand A by the amount

$$\Delta g_{\text{Lig,A}} = -RT \cdot \ln Z_{\text{Lig,A}} = -RT \cdot \ln(1 + K_{a,\text{A}}[\text{A}]), \tag{15.4}$$

and the probabilities of each state are redistributed (Figure 15.2A). For this example, the stabilization of states that bind ligand A results in a substantial shift in the probability of state N, which is also competent to bind ligand B. The physical basis of this coupling between the two sites will be discussed in detail next. But first, it is helpful to define a metric of the observed coupling between the two sites in our model. To quantify the observed coupling, we define the allosteric coupling response (CR) as the degree to which the probability of states that can bind ligand B is affected for a given perturbation to states that can bind ligand A

$$\text{CR}_{\text{B|A}} = \frac{\Delta P_{\text{B,Folded}}}{\Delta \ln Z_{\text{Lig,A}}} = \frac{\left| P_{\text{B,Folded}}(\text{A}) - P_{\text{B,Folded}}(\text{A}=0) \right|}{\ln Z_{\text{Lig,A}}}. \tag{15.5}$$

In general, the CR (Equation 15.5) is a measure of the sensitivity of site B to the binding of ligand at site A. As stated above, this model could equally represent other types of perturbations at site A, such as mutation or chemical modifications. Also, it needs to be noted that the sensitivity of site A to perturbations at site B does not necessarily equal the sensitivity of site B to perturbations at A. This is shown in Figure 15.2C. For comparison to the first simulation (i.e., the effect at site B due to ligand binding at site A), the reciprocal simulation was performed, where ligand binds at site B. For this simulation, identical free energy parameters were used ($\Delta G_{\text{I}} = -2.5$ kcal/mol, $\Delta G_{\text{II}} = -1.0$ kcal/mol, $\Delta g_{\text{int}} = 2$ kcal/mol). As can be seen in this figure, the magnitude of the allosteric coupling between two sites is directionally dependent, and thus sensitive to the precise energetic balance in the protein (i.e., the stabilities of the domains and their interaction energies). To determine which parameters maximize the allosteric coupling between sites, an unbiased search of parameter space was performed by systematically exploring all possible combinations of values for $\Delta G_{\text{I}}$, $\Delta G_{\text{II}}$, and $\Delta g_{\text{int}}$.

## 15.2.2 LIGAND-INDUCED POPULATION SHIFTS IN THE ENSEMBLE ALLOSTERIC MODEL

The classical view of allosteric coupling is that two sites are coupled through a network of interactions that extend throughout the protein and connect the two sites—in essence, that there is an energetic or mechanical pathway linking the sites [32]. If this is the case, it might be expected that site-to-site coupling would be maximized when a well-defined pathway of stable, folded structure connects the two sites. Paradoxically, such a conclusion is not borne out in the current analysis. Instead, an inverse relationship between allosteric coupling potential and the stability within the molecule was observed, and this relationship provides insight into the ground rules governing site-to-site communication.

Inspection of the parameter space that maximizes the allosteric coupling response (CR) reveals that there are distinct regions of energetic space within which different combinations of energetic values can facilitate linked behavior. This is shown in Figure 15.3A, which was generated by calculating the sensitivity of site B for bound ligand at site A. Evident in the simulation was that there were two nodes representing optimized coupling for the two domain protein. The origin of this behavior became apparent when the stabilities from Figure 15.3A were converted to the summed probability of states that can bind ligand A and B, which is presented in Figure 15.3B. Several observations are apparent in the data. First and most important, allosteric

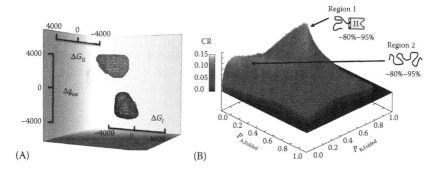

**FIGURE 15.3** **(See color insert following page 172.)** Allosteric coupling is optimized in proteins that contain ID. (A) The parameter combinations of $\Delta G_I$, $\Delta G_{II}$, and $\Delta g_{int}$ that generate a CR ≥ 0.10. A wide range of parameter combinations was sufficient to elicit a high CR values. The absence of points in the middle region of the 3D plot indicates that although a wide range of parameter values can combine to produce a high CR, significant interaction energy (i.e., for this case $|\Delta g_{int}| \geq 1$ kcal/mol) was a prerequisite to coupling. All energies are presented in calories per mole. (B) Plot of the summed probability of those states of the model that are competent to bind ligand A ($P_{A,Folded} = P_N + P_I$) vs. the summed probability of those states of the model that are competent to bind ligand B ($P_{B,Folded} = P_N + P_2$), shown as a function of CR (color-coded as indicated). The maximum CRs were observed in two regions. In region 1, domain I is unfolded and domain II is folded. Binding of ligand A folds domain I, but because of unfavorable domain coupling, domain II unfolds. In region 2, both domains are unfolded. Binding of ligand A folds domain 1, and as a result of favorable domain coupling, domain II folds.

coupling was maximized when the domains containing one or both binding sites were ID in the absence of ligand, a result that is consistent with the prevalence of ID segments in certain signaling proteins [23].

Second, although the individual energetic parameters that optimized coupling varied considerably (i.e., a unique set of parameter values for $\Delta G_I$, $\Delta G_{II}$, and $\Delta g_{int}$ was not necessary), in all cases significant interaction energy was required. This point is discussed in more detail below where a thermodynamic basis for allosteric coupling is presented. We note, however, that because all possible conditions were tested, the results obtained were not predetermined by the specifics of the simulation. In fact, varying the degree of complexity of the model (e.g., introducing additional domains) did not affect the results. The extension of our allosteric model to three domains is presented elsewhere [30] where it was demonstrated that for all multidomain systems that were modeled, allosteric coupling between sites was optimized when one or a number of domains were ID, indicating that the principles described here are not artifacts of the simplicity of the two-domain assumption and are extendable to other multidomain systems.

### 15.2.3 Thermodynamic Basis for Coupling

Considering the classical view of allostery, i.e., site-to-site coupling should be maximized when a well-defined pathway of stable, folded structure connects the two sites, the results presented in Figure 15.3 are somewhat surprising. Rather than being a continuous function of stability, optimum allosteric coupling was observed in the ensemble model when the equilibrium was poised in either of two regions of thermodynamic parameter space. More explicitly, coupling between sites was optimized when the ensemble was dominated by the state in which just domain I was unfolded (Region 1 in Figure 15.3B), or when the ensemble was dominated by the state in which both domains were unfolded (Region 2 in Figure 15.3B). The origin of the bimodal response is that the two sites can be either positively or negatively coupled.

In the case of positive coupling (Region 2), the effect of ligand is as described in Figure 15.2. Namely, the equilibrium was modeled such that the unfolded state dominates the ensemble probabilities. Upon adding ligand A, those states with domain I folded were preferentially stabilized. Since domains I and II were positively coupled, then the energy of breaking the interaction between them was positive (unfavorable), and states with only one domain unfolded were highly improbable. As a result, stabilizing the binding site for ligand A had the effect of also stabilizing the binding site for ligand B, simply because states where both domains were folded were more probable.

In the case of negative coupling (Region 1), the principles are the same but the effect is opposite. When negative coupling was modeled, the interaction energy, $\Delta g_{int}$, was negative, meaning that it was energetically unfavorable to have both domains folded simultaneously. States with only one domain unfolded thus dominated the ensemble probabilities. In this case, stabilization of domain I, via the binding of ligand A, resulted in a destabilization of domain II, which caused a decrease in affinity for ligand B. In either case, the results of this analysis are clear. Proteins containing ID domains were more able to propagate the effects of binding through different domains if the binding was coupled to the folding of the molecule.

Although the multimodal behavior in the distribution of CR values (Figure 15.3) was due to the existence of negative and positive coupling, it is important to note that the magnitude of the perturbation induced by ligand A, (i.e., $\Delta g_{Lig,A}$), also contributed to determining where the equilibrium was poised to elicit the maximum response. As shown in Figure 15.4, in the limit where the system must respond to only

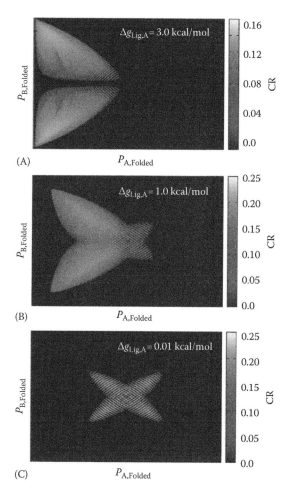

**FIGURE 15.4 (See color insert following page 172.)** The effect of binding energy on the optimum distribution for allosteric coupling. Each plot shows the summed probability of those states of the model that are competent to bind ligand A ($P_{A,Folded} = P_N + P_1$) versus the summed probability of those states of the model that are competent to bind ligand B ($P_{B,Folded} = P_N + P_2$), color-coded as a function of CR (as indicated in the plots). Shown is the dependence of the calculated CR values on the binding free energies ($\Delta g_{Lig,A}$) modeled as (A) 3.0 kcal/mol, (B) 1.0 kcal/mol, and (C) 0.01 kcal/mol. For clarity in the plots, CR values less than 0.10, 0.15, and 0.20 were zeroed to highlight the maxima for $\Delta g_{Lig,A} = 3.0$, 1.0, and 0.01, respectively.

minute changes in the fraction of molecules that are bound to site A (e.g., $K_a[A] = 0.017$, which according to Equation 15.4 gives $\Delta g_{\text{Lig,A}} \sim 0.01$ kcal/mol), the maximum response was observed when each of the domains were unfolded 50% of the time. This scenario is likely to be rare, as allosteric effectors are usually bound tightly and are molecules which have been selected by nature to act as effectors because they vary in concentration as a result of cellular or environmental changes. Nonetheless, even in cases when modest changes in energy are expected from ligand A (e.g., $K_a[A] = 4.4$, which according to Equation 15.4 gives $\Delta g_{\text{Lig,A}} \sim 1.0$ kcal/mol), the equilibrium that will produce the optimum response involved a significant fraction of molecules wherein one or both domains were unfolded.

In any case, Figure 15.4 indicates that where the equilibrium is poised prior to the addition of ligand A will depend on how much binding energy (Equation 15.4) is available to the system to elicit the desired signal. If the binding affinity for the effector ligand is low and/or the anticipated change in concentration of ligand is small relative to $K_a$, the ensemble will have a higher fraction of states that are structured. If, on the other hand, the binding affinity for the effector ligand is high and/or the anticipated change in effector concentration is large, the ensemble will be dominated by states that are partially or fully disordered. It is noteworthy that such a continuum in relative structure has been observed in disordered proteins [25].

### 15.2.4 UNDERSTANDING ENERGY FLOW IN AN ENSEMBLE

The model of allostery described above can be used to highlight the relationship, or lack thereof, between the observed coupling and any apparent pathway between the linked sites. Shown in Figure 15.5A are the results obtained from the range of stability and interaction parameters used in simulating the allosteric CR of Figure 15.3.

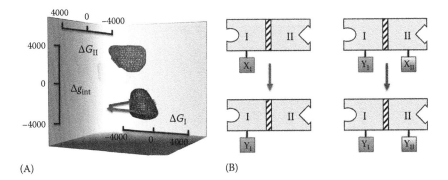

(A)                                                                                 (B)

**FIGURE 15.5** Mutational effects on allostery in a two-domain protein. (A) Reproduced from Figure 15.3A are the parameter combinations of $\Delta G_I$, $\Delta G_{II}$, and $\Delta g_{\text{int}}$ that generate a CR $\geq 0.10$. Mutations that destabilize domain I, decreasing $\Delta G_I$ (shown by the blue arrow), can eliminate or greatly reduce the allosteric coupling between sites A and B. A compensating mutation that destabilizes domain II, decreasing $\Delta G_{II}$ (shown by the green arrow), can restore the allosteric coupling between the sites that was eliminated by the mutation in domain I. (B) The mutational scheme of panel A, represented in terms of the two-domain model protein.

As noted, a broad spectrum of parameter values was sufficient to result in an observed coupling between the two sites. To highlight the difficulty in assigning importance to a pathway connecting the sites, we introduce a set of hypothetical experiments, which are represented schematically in Figure 15.5B. In the first experiment, a mutation at a surface exposed site in domain I (i.e., $X_I$ to $Y_I$) is made, which destabilizes that domain. An example of such a mutation would be a change from Ala to Gly at a surface-exposed position. Such a mutation would not change the structure of the folded state [33], but would nonetheless destabilize domain I, as well as the protein as a whole, by increasing the number of available conformations when unfolded.

Because of the surface mutation in domain I, which is destabilizing to this domain, the allosteric CR of the system decreases. This decrease in CR is represented in Figure 15.5A by the blue arrow. We note that such a decrease in response occurs even in the absence of a structural perturbation to the folded structure. In effect, there are, by definition, no changes to the network of interactions, except that the stability alters the probability and thus the fraction of time that those interactions are intact. Further undermining the notion that quantitative importance can be assigned only to residues within a connectivity network, an additional, compensating surface mutation can be made to the system to restore the coupling. In this second case (Figure 15.5B), a destabilizing mutation is introduced into domain II (i.e., $X_{II}$ to $Y_{II}$), which pushes the coupling energy back toward the original value, and is represented in Figure 15.5A by the green arrow. These hypothetical experiments show that a perturbation in one domain, which affects coupling, can be compensated by perturbations in other domains. Most importantly, however, these results reveal that such compensation does not require changes in the ground state structure. Instead the compensating change involved modulating the equilibria of the ensemble. In other words, the conformational fluctuations around the average structures play a significant role in the coupling between sites of the protein (a fact which is now well-known [34,35,38,65]), and that an understanding of the allosteric coupling cannot be inferred solely from, or attributed only to, the interactions that connect the two coupled sites.

### 15.2.5 BUILDING A STRUCTURE-BASED LOCAL UNFOLDING MODEL OF PROTEINS: COREX/BEST

We note that the model presented thus far is a simple construct that does not infer any structural interpretations. In order to see how such a model can relate to real proteins, it is useful to use the high-resolution structure as the starting point for the development of a structure-based approach. Such an approach is justified by the observation that conformational fluctuations around the canonical native structure of proteins are modulated during allosteric coupling [34–39]. In the limit that these conformational fluctuations resemble local folding/unfolding transitions [35,39], our laboratory has developed over the past decade a structural-thermodynamic model of the protein ensemble, called COREX/BEST [40], which models the fluctuations as local two-state folding/unfolding transitions. Because the stability of each state is calculated by COREX/BEST (as described below), and thus the intrinsic stabilities of isolated segments as well as the interaction energies between them can be calculated, COREX/BEST provides a means of investigating the relationship between structure and allosteric coupling, and thus understanding allostery in terms of the impact that perturbations have on the ensemble through changes in local stability.

Briefly, the COREX/BEST algorithm can be used to generate a large (>10$^6$ states) conformational ensemble based upon a high-resolution structure [40]. The individual conformations are simulated by treating contiguous groups of residues as being either folded or unfolded. By systematically varying the boundaries of the folding units and by sampling all possible combinations of multiple groups being unfolding simultaneously, an exhaustive enumeration of conformations can be generated (Figure 15.6). The statistical weight of each state ($K_i = \exp(-\Delta G_i/RT)$) is determined from the calculated Gibbs energy, which has been calibrated previously and tested extensively [41–43]. The probability of each state $i$ is determined by

$$P_i = \frac{K_i}{\sum_i K_i},\qquad(15.6)$$

where the summation is over all states in the ensemble.

Within the context of an ensemble-based description of equilibria, allostery is manifested in the relative probabilities of the different ensemble states [44,45].

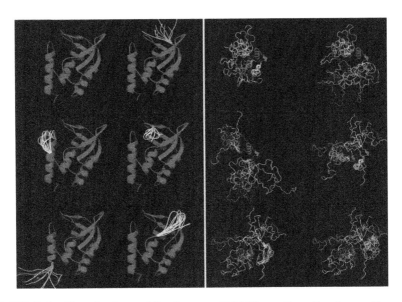

**FIGURE 15.6** (See color insert following page 172.) The conformational ensemble as calculated by the COREX/BEST algorithm. Shown is a sample of 12 (out of more than 10$^6$) of the more probable conformational states calculated by COREX/BEST for SNase. Ensemble states on the left were chosen among those that retained most of the native structure, whereas those on the right were chosen among the mostly unfolded states. Segments of the protein colored red indicate regions that would be folded; regions colored yellow would be unfolded. To illustrate the concept that the unfolded segments freely sample accessible conformational space, these regions were modeled in the figure as multiple flexible loops.

For regions of the protein that are positively coupled, the probability of states in which both regions are folded or unfolded is greater than the probability of states in which only one is folded. For regions that are negatively coupled, the probability of states in which only one region is folded is greater that the probability of states in which both are folded or unfolded. For regions that are uncoupled, the folding status of one is independent of the folding status of the other. As demonstrated previously, in proteins such as staphylococcal nuclease (SNase) [46], dihydrofolate reductase (DHFR) [35], and eglin c [45], a connecting network of intramolecular interactions is not required for allosteric coupling between distant regions of the protein structure.

By introducing a perturbation to the ensemble, as already shown above with the simple model, the allosteric linkage between different regions of the protein can be simulated. This perturbation to the ensemble can be introduced in the form of a ligand-specific interaction [35,39,45,46], a position-specific structural change in the protein, such as mutation [39,45], an environmental change such as temperature [47,48], pH [46,49], or chemical denaturation [50], or a purely computational perturbation modeled as a site-specific energetic stabilization of a residue or segment of the protein [44]. As an example, the mechanism of communication between the two binding sites in DHFR, which are known to be allosterically coupled [51], was investigated by simulating ligand binding in the COREX/BEST model [35]. This is shown schematically in Figure 15.7A, where those ensemble states that are competent to bind folate were preferentially stabilized. This selective stabilization of binding competent states results in a redistribution of the ensemble, the effects of which can be quantitated at positions distant to the folate site (Figure 15.7B). Three important results were observed in this simulation. First, the loop containing residues 63–69 had a strong negative coupling with folate binding, which was also observed experimentally [52–54]. Second, this loop is more than 15 Å from the folate binding site, and the residues surrounding it show no connectivity to the folate site. This important result indicates that residues can be energetically coupled in the absence of a visible connectivity pathway, similar to what was observed with the general ensemble allosteric model described above. Thirdly, and equally important, the results contradict the classical view that binding "freezes out" protein conformations, resulting in a decrease in motion or dynamics [55]. The experimental observation that binding can increase dynamics in some systems [53,54,56–60] emphasizes the importance of entropic contributions to allosteric coupling between sites.

Similarly, the effect of proton binding on allosteric coupling and global unfolding can be modeled using this ensemble-based method [46,49]. By modifying the probability of each ensemble state (Equation 15.6) to account for proton-binding energies

$$P(\mathrm{pH})_i = \frac{K_i \cdot \prod_j \left(1 + 10^{(pK_{a,i,j} - \mathrm{pH})}\right)}{\sum_i \left(K_i \cdot \prod_j \left(1 + 10^{(pK_{a,i,j} - \mathrm{pH})}\right)\right)}, \qquad (15.7)$$

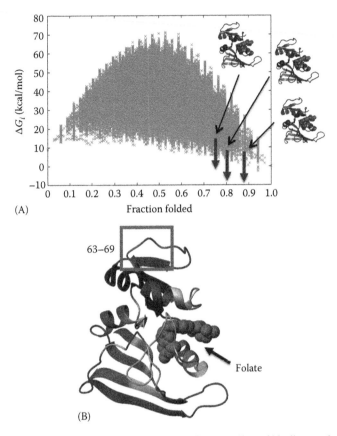

(A)

(B)

**FIGURE 15.7** **(See color insert following page 172.)** The effect of binding on the ensemble. (A) The theory of linked functions explains that the effect of increasing concentration of ligand is to stabilize those states of the ensemble that have greater affinity for that ligand [63,64]. To simulate folate binding, those ensemble states of DHFR that were competent to bind folate, which was approximated as those states in which the folate binding site remained folded, were preferentially stabilized relative to all other states of the ensemble [35]. This is represented graphically in the figure by the blue arrows. Each green point in the plot represents one state in the DHFR ensemble. Here, the fraction of residues folded in each state is shown as a function of the calculated state stability, $\Delta G_i$. (B) Redistribution of the ensemble states in response to folate binding had the effect of stabilizing some regions of DHFR while destabilizing (promoting unfolding) other regions. The cartoon structure of DHFR was color-coded such that regions colored red were stabilized the most due to folate binding, green represents modest stability increases, whereas blue is the least. Regions colored purple indicate a negative effect; that folate binding decreased the stability of the protein in these regions, and is highlighted in the figure for the loop containing residues 63–69.

where $pK_{a,i,j}$ is the $pK_a$ value of proton titration of residue $j$ in state $i$, the effect of pH on the stability of SNase was simulated and is presented in Figure 15.8. The residues implicated in the ensemble simulation as determining the pH-dependent stability of SNase were not clustered in any obvious way in the high-resolution

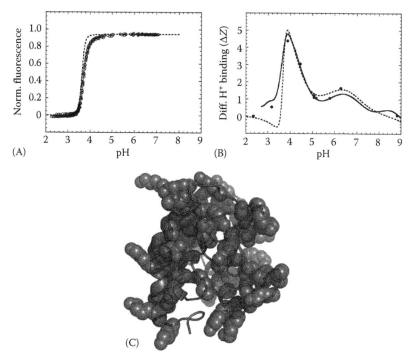

**FIGURE 15.8** (**See color insert following page 172.**) The effect of proton (H$^+$) binding on the ensemble. (A) The experimental unfolding of SNase (open circles) as monitored by pH titration of the intrinsic fluorescence of Trp-140 [46]. The stippled line overlaid on the experimental data represents the calculated unfolding of the SNase ensemble as determined by $\langle \text{Fraction\_Native} \rangle = \sum_i \text{Fraction\_Native}_i \cdot P(\text{pH})_i$, where $P(\text{pH})_i$ is given by Equation 15.7. The *Fraction Native* value of each state $i$ was calculated as the number of residues folded divided by the total number of residues of SNase. The change in fluorescence due to the acid unfolding of SNase was normalized to a value of 1 for direct comparison to the ensemble simulation. (B) The experimental difference in H$^+$ binding between GdnHCl-unfolded and native SNase was determined by the continuous (solid line) and batch (solid circles) potentiometric techniques [46]. The calculated H$^+$ binding difference (stippled line) was determined by $\langle Z \rangle = \sum_i Z_i \cdot P(\text{pH})_i$, where $Z_i$ was calculated as the number of H$^+$ bound to each ensemble state $i$ as a function of pH. (C) The positions of the residues that can titrate H$^+$ in SNase are shown by the semitransparent space-fill representation of the side chain atoms. Of the 61 titratable groups (59 residues + the N- and C-termini) in SNase, only 18 (colored red) were calculated to contribute to its pH-dependent stability in the ensemble simulations [46,49]. Interestingly, those 18 titratable groups were not clustered in any obvious manner in the native structure of SNase and are intermixed among the other titratable groups that apparently do not play a dominant role in its pH-dependent stability.

native structure and, most surprisingly, were intermixed among titratable groups that contribute negligibly to its pH-dependent stability (Figure 15.8C). This result suggests that the pH-linked contributions to stability of this protein are not generated through a well-defined network of intramolecular electrostatic interactions.

Instead, the common characteristic of the titratable groups that were predicted to govern pH-dependent stability is that they were all in folded regions of the protein in the majority of the highly probable states of the ensemble under native conditions. In other words, the hierarchy of states in the ensemble played the dominant role in determining the pH sensitivity of SNase, more so than any connected network of charged groups observed in the high-resolution structure. Experimental verification of the residues responsible for the acid-induced unfolding of SNase was presented elsewhere [46] and was shown to be consistent with the ensemble calculations.

## 15.3 CONCLUSIONS

The ability to understand the determinants of allostery in proteins is the cornerstone to a quantitative description of functional biology. In this regard, the results presented here provide a general quantitative rationale for the observation that many regulatory proteins have ID regions; intrinsic disorder can maximize the ability to allosterically couple sites. More importantly, the results reveal the general thermodynamic ground rules for site-to-site coupling, wherein the ability to propagate the effects of binding are determined not necessarily by a mechanical pathway linking the two sites, but by the energetic balance within the states of an ensemble. The significance of this result with regard to allosteric mechanisms cannot be overstated. As Figure 15.3A reveals, the parameter combinations that produce optimal conditions for site-to-site coupling are highly degenerate, meaning that the stability of any one domain is not critical to the coupling. Changes in stability in one domain (or region) can be compensated by changes to another domain, or to changes in the interactions between domains. Indeed, the results indicate that the precise stabilization mechanism is not a determinant of the coupling at all. In effect, the sites can be coupled without the requirement of a connecting network of specific intramolecular interactions. The relatively degenerate requirements for coupling described here appear to undermine the view that allostery is a precisely evolved property of proteins that relies on a specific mechanical pathway. Indeed, these results suggest that site-to-site coupling in proteins can be robustly encoded [35,39] and amenable to significant sequence divergence, a result that is also borne out in the low sequence similarities observed for the disordered regions within specific classes of proteins [27].

As discussed in Section 15.1, numerous protein systems have been observed to behave in a manner consistent with a structural-mechanical basis of allosteric control between coupled sites, in which the sites are connected structurally by a network of interactions or "pathways" [14,15]. In these systems, allostery typically has been presented in terms of the Monod–Wyman–Changeux (MWC) [8] and Koshland–Nemethy–Filmer (KNF) [9] models, both of which are special cases of a more general model [61]. Briefly, the MWC model describes allostery as originating from the equilibria between two macroscopic states, each of which can bind ligand with different affinities. In this model, the equilibrium is driven toward the high-affinity state by mass action. The KNF model, however, relies on an "induced fit" mechanism, wherein binding is facilitated by only one form. In general, the notion of a pathway for allosteric control arises intuitively from the KNF model: binding of the allosteric ligand is viewed as inducing a series of conformational changes that begin at the binding site and propagate toward the affected site. In effect, the pathway is seen as being mechanical, with work being produced through the structural distortion that is initiated upon binding.

Although structural differences are likely to accompany allosteric coupling [11–15], such changes are not obligatory [34], and as the ensemble allosteric model (as well as COREX/BEST) indicate, significant effects can occur that will not be manifested as changes in the ground state structure. Consequently, any structural changes that are observed cannot be assumed to constitute the entire set determinants of coupling in a given system. The question then becomes, how does one add clarity to our understanding of allosteric control given that the determinants will be both structural and thermodynamic in nature? The answer to this question is likely in the use of parallel modeling strategies. Clearly, it would be most beneficial to have an all-atom representation [13,14] of the allosteric system so that all of the conformational states can be quantitatively sampled. Unfortunately, as the ensemble allosteric model indicates, capturing allostery quantitatively requires knowing all of the states in the system, and these are likely to include conformational excursions that are locally unfolded [35,38], making a complete enumeration computationally inaccessible through standard all-atom approaches. Thus, a complementary strategy to all atom modeling approaches would be an ensemble model such as COREX/BEST, which captures the breadth of the distribution.

The COREX/BEST analyses of DHFR and SNase suggests that many solution properties of proteins can be described quantitatively without having to go past the level of average properties of ensemble states and into the description of individual atoms and bonds. In this respect, the ensemble view of proteins constitutes a level of complexity intermediate between the unsophisticated treatment of proteins as strictly static structures and the daunting complexity of their treatment with explicit "all-atom" representations of each accessible conformation. An especially appealing feature of the ensemble-based approach described here is that it offers a computationally tractable method of apparently sufficient resolution to study and unify disparate biological phenomena [35,39,44–50], reconcile the role of conformational fluctuations in modulating functionally important processes [62], and understand proteins in terms of its cooperative substructure [47,48]. In effect, the ensemble based approach compromises atomic resolution for accuracy in the determination of energetics associated with the breadth of the distribution, whereas all-atom models compromise inclusivity of states in order to obtain a more detailed structural description of a subset of relevant conformations.

## ACKNOWLEDGMENT

This work was supported by grants from the NSF (MCB-0446050), NIH (GM-63747), and the Welch Foundation (H-1461).

## REFERENCES

1. Bohr, C., Hasselbach, K. A., Krogh, A. 1904. Über einen in biologischen Beziehung wichtigen Einfluss, den die Kohlen-sauerspannung des Blutes auf dessen Sauerstoffbindung übt. *Skand. Arch. Physiol.* **15**, 401–412.
2. Adair, G. S. 1925. A critical study of the direct method of measuring osmotic pressure of hemoglobin. *Proc. R. Soc. London Ser. A* **108A**, 627–637.
3. Pauling, L. 1935. The oxygen equilibrium of hemoglobin and its structural interpretation. *Proc. Natl. Acad. Sci. U. S. A.* **21**, 186–191.
4. Eaton, W. A., Henry, E. R., Hofrichter, J., Mozzarelli, A. 1999. Is cooperative oxygen binding by hemoglobin really understood? *Nat. Struct. Biol.* **6**, 351–358.
5. Cui, Q., Karplus, M. 2008. Allostery and cooperativity revisited. *Protein Sci.* **17**, 1295–1307.
6. Monod, J., Jacob, F. 1961. Teleonomic mechanisms in cellular metabolism, growth, and differentiation. *Cold Spring Harb. Symp. Quant. Biol.* **26**, 389–401.
7. Changeux, J. P. 1961. The feedback control mechanisms of biosynthetic L-threonine diaminase by L-isoleucine. *Cold Spring Harb. Symp. Quant. Biol.* **26**, 313–318.
8. Monod, J., Wyman, J., Changeux, J. P. 1965. On the nature of allosteric transitions: A plausible model. *J. Mol. Biol.* **12**, 88–118.
9. Koshland, D. E., Nemethy, G., Filmer, D. 1966. Comparison of experimental binding data and theoretical models in proteins containing subunits. *Biochemistry* **5**, 365–385.
10. Bonaventura, J., Bonaventura, C., Giardina, B., Antonini, E., Brunori, M., Wyman, J. 1972. Partial restoration of normal functional properties in carboxypeptidase A-digested hemoglobin. *Proc. Nat. Acad. Sci. U. S. A.* **69**, 2174–2178.
11. Perutz, M. F. 1970. Stereochemistry of cooperative effects in haemoglobin. *Nature* **228**, 726–739.
12. Perutz, M. F., Wilkinson, A. J., Paoli, M., Dodson, G. G. 1998. The stereochemistry of the cooperative effects in hemoglobin revisited. *Ann. Rev. Biophys. Biomol. Struct.* **27**, 1–34.
13. Szabo, A., Karplus, M. 1972. A mathematical model for structure–function relations in hemoglobin. *J. Mol. Biol.* **72**, 163–197.
14. Bray, D., Duke, T. 2004. Conformational spread: the propagation of allosteric states in large multiprotein complexes. *Annu. Rev. Biophys. Biomol. Struct.* **33**, 53–73.
15. Changeux, J. P., Edelstein, S. J. 2005. Allosteric mechanisms of signal transduction. *Science* **308**, 1424–1428.
16. Englander, S. W. 2000. Protein folding intermediates and pathways studied by hydrogen exchange. *Annu. Rev. Biophys. Biomol. Struct.* **29**, 213–238.
17. Lu, H., Xun, L., Xie, X. S. 1998. Single-molecule enzymatic dynamics. *Science* **282**, 1877–1882.
18. Yang, H., Luo, G., Karnchanaphanurach, P., Louie, T. M., Rech, I., Cova, S., Xun, L., Xie, X. S. 2003. Protein conformational dynamics probed by single-molecule electron transfer. *Science* **302**, 262–266.
19. Wright, P. E., Dyson, H. J. 1999. Intrinsically unstructured proteins: Re-assessing the protein structure–function paradigm. *J. Mol. Biol.* **293**, 321–231.
20. Uversky, V. N. 2002. Natively unfolded proteins: A point where biology waits for physics. *Protein Sci.* **11**, 739–756.
21. Uversky, V. N., Oldfield, C. J., Dunker, A. K. 2005. Showing your ID: Intrinsic disorder as an ID for recognition, regulation and cell signaling. *J. Mol. Recognit.* **18**, 343–384.
22. Dunker, A. K., Lawson, J. D., Brown, C. J., Williams, R. M., Romero, P., Oh, J. S., Oldfield, C. J., Campen, A. M., Ratliff, C. M., Hipps, K. W., Ausio, J., Nissen, M. S., Reeves, R., Kang, C., Kissinger, C. R., Bailey, R. W., Griswold, M. D., Chiu, W., Garner, E. C., Obradovic, Z. 2001. Intrinsically disordered protein. *J. Mol. Graph. Model.* **19**, 26–59.

23. Liu, J., Perumal, N. B., Oldfield, C. J., Su, E. W., Uversky, V. N., Dunker, A. K. 2006. Intrinsic disorder in transcription factors. *Biochemistry* **45**, 6873–6888.

24. Fink, A. L. 2005. Natively unfolded proteins. *Curr. Opin. Struct. Biol.* **15**, 35–41.

25. Dunker, A. K., Cortese, M. S., Romero, P., Iakoucheva, L. M., Uversky, V. N. 2005. Flexible nets. The roles of intrinsic disorder in protein interaction networks. *FEBS Lett.* **272**, 5129–5148.

26. Tompa, P. 2005. The interplay between structure and function in intrinsically unstructured proteins. *FEBS Lett.* **579**, 3346–3354.

27. Kumar, R., Thompson, E. B. 2005. Gene regulation by the glucocorticoid receptor: structure:function relationship. *J. Steroid Biochem. Mol. Biol.* **94**, 383–394.

28. Di Cera, E. 1998. Site-specific analysis of mutational effects in proteins. *Adv. Protein Chem.* **51**, 59–119.

29. Brändén, C. I., Tooze, J. 1991. *An Introduction to Protein Structure* (Garland, New York).

30. Hilser, V. J., Thompson, E. B. 2007. Intrinsic disorder as a mechanism to optimize allosteric coupling in proteins. *Proc. Nat. Acad. Sci. U. S. A.* **104**, 8311–8315.

31. Baldwin, R. L. 1986. Temperature dependence of the hydrophobic interaction in protein folding. *Proc. Nat. Acad. Sci. U. S. A.* **83**, 8069–8072.

32. Lockless, S. W., Ranganathan, R. 1999. Evolutionarily conserved pathways of energetic connectivity in protein families. *Science* **286**, 295–299.

33. Blaber, M., Zhang, X. J., Lindstrom, J. L., Pepiot, S. D., Baase, W. A., Matthews, B. W. 1994. Determination of {alpha}-helix propensity within the context of a folded protein, sites 44 and 131 in bacteriophage T4 lysozyme. *J. Mol. Biol.* **235**, 600–624.

34. Cooper, A., Dryden, D. T. F. 1982. Allostery without conformational change. *Eur. Biophys. J.* **11**, 103–109.

35. Pan, H., Lee, J. C., Hilser, V. J. 2000. Binding sites in Escherichia coli dihydrofolate reductase communicate by modulating the conformational ensemble. *Proc. Nat. Acad. Sci. U. S. A.* **97**, 12020–12025.

36. Fuentes, E. J., Der, C. J., Lee, A. L. 2004. Ligand-dependent dynamics and intramolecular signaling in a PDZ domain. *J. Mol. Biol.* **335**, 1105–1115.

37. Igumenova, T. I., Frederick, K. K., Wand, A. J. 2006. Characterization of the fast dynamics of protein amino acid side chains using NMR relaxation in solution. *Chem. Rev.* **106**, 1672–1699.

38. Popovych, N., Sun, S., Ebright, R. H., Kalodimos, C. G. 2006. *Nat. Struct. Mol. Biol.* **13**, 831–838.

39. Liu, T., Whitten, S. T., Hilser, V. J. 2006. Ensemble-based signatures of energy propagation in proteins: A new view of an old phenomenon. *Protein Struct. Funct. Bioinform.* **62**, 728–738.

40. Hilser, V. J, Freire, E. 1996. Structure-based calculations of the equilibrium folding pathway of proteins. Correlation with hydrogen exchange protection factors. *J. Mol. Biol.* **262**, 756–772.

41. D'Aquino, J. A., Gomez, J., Hilser, V. J., Lee, K. H., Amzel, L. M., Freire, E. 1996. The magnitude of the backbone conformational entropy change in protein folding. *Proteins* **25**, 143–156.

42. Gomez, J., Hilser, V. J., Xie, D., Freire, E. 1995. The heat capacity of proteins. *Proteins* **22**, 404–412.

43. Hilser, V. J., Gomez, J., Freire, E. 1996. The enthalpy change in protein folding and binding: refinement of parameters for structure-based calculations. *Proteins* **26**, 123–133.

44. Hilser, V. J., Dowdy, D., Oas, T. G., Freire, E 1998. The structural distribution of cooperative interactions in proteins: Analysis of the native state ensemble. *Proc. Natl. Acad. Sci. U. S. A.* **95**, 9903–9908.

45. Liu, T., Whitten, S. T., Hilser, V. J. 2007. Functional residues serve a dominant role in mediating the cooperativity of the protein ensemble. *Proc. Natl. Acad. Sci. U. S. A.* **104**, 4347–4352.

46. Whitten, S. T., García-Moreno, E. B., Hilser, V. J. 2005. Local conformational fluctuations can modulate the coupling between proton binding and global structural transitions in proteins. *Proc. Natl. Acad. Sci. U. S. A.* **102**, 4282–4287.

47. Babu, C. R., Hilser, V. J., Wand, A. J. 2004. Direct access to the cooperative substructure of proteins and the protein ensemble via cold denaturation. *Nat. Struct. Mol. Biol.* **11**, 352–357.

48. Whitten, S. T., Kurtz, A. J., Pometun, M. S., Wand, A. J., Hilser, V. J. 2006. Revealing the nature of the native state ensemble through cold denaturation. *Biochemistry* **45**, 10163–10174.

49. Whitten, S. T., García-Moreno, B. E., Hilser, V. J. 2008. Ligand effects on the protein ensemble: unifying the descriptions of ligand binding, local conformational fluctuations, and protein stability. *Methods Cell Biol.* **84**, 871–891.

50. Wooll, J. O., Wrabl, J. O., Hilser, V. J. 2000. Ensemble modulation as an origin of denaturant-independent hydrogen exchange in proteins. *J. Mol. Biol.* **301**, 247–256.

51. Fierke, C. A., Johnson, K. A., Benkovic, S. J. 1987. Construction and evaluation of the kinetic scheme associated with dihydrofolate reductase from *Escherichia coli*. *Biochemistry* **26**, 4085–4092.

52. Ohmae, E., Iriyama, K., Ichihara, S., Gekko, K. 1996. Effects of point mutations at the flexible loop glycine-67 of Escherichia coli dihydrofolate reductase on its stability and function. *J. Biochem.* **119**, 703–710.

53. Bolin, J. T., Filman, D. J., Matthews, D. A., Hamlin, R. C., Kraut, J. 1982. Crystal structures of *Escherichia coli* and *Lactobacillus casei* dihydrofolate reductase refined at 1.7 A resolution. I. General features and binding of methotrexate. *J. Biol. Chem.* **257**, 13650–13662.

54. Bystroff, C., Kraut, J. 1991. Crystal structure of unliganded *Escherichia coli* dihydrofolate reductase. Ligand-induced conformational changes and cooperativity in binding. *Biochemistry* **30**, 2227–2239.

55. Froloff, N., Windemuth, A., Honig, B. 1977. On the calculation of binding free energies using continuum methods: application to MHC class I protein–peptide interactions. *Protein Sci.* **6**, 1293–1301.

56. Akke, M., Skelton, N. J., Kördel, J., Palmer, A. G., Chazin, W. J. 1993. Effects of ion binding on the backbone dynamics of calbindin D9k determined by $^{15}$N NMR relaxation. *Biochemistry* **32**, 9832–9844.

57. Olejniczak, E. T., Zhou, M. M., Fesik, S. W. 1997. Changes in the NMR-derived motional parameters of the insulin receptor substrate 1 phosphotyrosine binding domain upon binding to an interleukin 4 receptor phosphopeptide. *Biochemistry* **36**, 4118–4124.

58. Yu, L., Zhu, C. X., Tse-Dinh, Y. C., Fesik, S. W. 1996. Backbone dynamics of the C-terminal domain of Escherichia coli Topoisomerase I in the absence and presence of single-stranded DNA. *Biochemistry* **35**, 9661–9666.

59. Stivers, J. T., Abeygunawardana, C., Mildvan, A. S. 1996. 15N NMR relaxation studies of free and inhibitor-bound 4-oxalocrotonate tautomerase: Backbone dynamics and entropy changes of an enzyme upon inhibitor binding. *Biochemistry* **35**, 16036–16047.

60. Zidek, L., Novotny, M. V., Stone, M. J. 1999. Increased protein backbone conformational entropy upon hydrophobic ligand binding. *Nat. Struct. Biol.* **6**, 1118–1121.

61. Wyman, J. 1972. On Allosteric Models. *Curr. Top. Cell. Regul.* **6**, 207–223.

62. Ferreon, J. C., Volk, D. E., Luxon, B. A., Gorenstein, D. G., Hilser, V. J. 2003. Solution structure, dynamics, and thermodynamics of the native state ensemble of the Sem-5 C-terminal SH3 domain. *Biochemistry* **42**, 5582–5591.

63. Wyman, J., Jr. 1948. Heme proteins. *Adv. Protein Chem.* **4**, 407–531.

64. Wyman, J., Jr. 1964. Linked functions and reciprocal effects in hemoglobin: A second look. *Adv. Protein Chem.* **19**, 223–286.

65. Daily, M. D., Gray, J. J. 2007. Local motions in a benchmark of allosteric proteins. *Proteins* **67**, 385–399.

# 16 Molecular Dynamics Simulation Studies of Coupled Protein and Water Dynamics

*Douglas J. Tobias, Neelanjana Sengupta, and Mounir Tarek*

## CONTENTS

## 16.1   INTRODUCTION

Proteins exhibit a broad spectrum of motions over a wide range of length scales, and spanning many decades of time. Under physiological conditions in their native states, proteins undergo localized, fast conformational fluctuations (picosecond to nanosecond timescales) around a well-defined three-dimensional structure. The folding and function of proteins generally involves much slower, large-scale transitions (more than microsecond timescales) between stable conformations. In the energy landscape picture of protein dynamics, protein motions on such disparate time scales are connected [1].

The energy landscape of proteins appears to be organized into tiers of conformational substates, classified by the height of the barriers that separate the substates [2]. The lowest tier consists of a vast number of substates that interconvert on picosecond to nanosecond timescales, while the highest tier consists of a small number of substates that interconvert slowly via large-scale conformational changes. The ubiquitous fast fluctuations are generally considered to promote the slower, rare events that are directly associated with protein folding and function. Although the interplay between the fast dynamics and slower, functionally relevant motions has been elucidated recently for an enzyme by a combination of nuclear magnetic resonance (NMR) experiments and molecular dynamics (MD) simulations [3], a detailed description of the energy flow between protein degrees of freedom fluctuating over a wide range of length and timescales is generally not available.

Proteins do not work in isolation, and it goes without saying that the solvent environment plays an important role in processes involving energy flow in proteins. In addition to comprising a major contribution to the relative thermodynamic stability of different protein conformations, the solvent environment has a major influence on protein dynamics. Indeed, the concept of "slaving" has been invoked to discuss the control of protein motion by bulk solvent dynamical properties, such as viscosity and dielectric relaxation rates [2,4].

The focus of this chapter is on the influence of the solvent environment on fast (picosecond to nanosecond) protein dynamics. Although the majority of the results presented herein are from our published MD simulation studies, we will mention selected results from the vast experimental literature, as well as simulation studies carried out by other groups, to develop a more complete picture. Following a brief description of the systems considered and simulation methodology, we will begin with a discussion of the coupling of protein and solvent dynamics by examining the influence of proteins on single-particle and collective water dynamics. This discussion will include an examination of the subtle changes in water dynamics that accompany the transition from the native to compact, denatured ("molten globule" [MG]) states. Next, we will introduce the use of investigations of the temperature and solvent dependence of protein dynamics as a paradigm for unraveling protein–solvent dynamical couplings in soluble and membrane proteins. We will place particular emphasis on the "dynamical transition" (also referred to as the glass transition) from liquid-like protein dynamics at room temperature to glass-like below the transition temperature, which is at ~200 K for soluble proteins. Specific connections between the dynamical transition and energy flow in proteins are discussed elsewhere in this volume (see Chapters 7 and 11). In the final section, we will discuss how this paradigm has exposed the key role of protein–solvent hydrogen bond dynamics in the solvent control of fast protein dynamics in the case of soluble proteins, and we will discuss important differences between soluble and membrane proteins with regard to protein–solvent dynamical coupling.

## 16.2  METHODS

We will illustrate the coupling of protein and water dynamics using results of MD simulations of several systems, including native and MG states of human α-lactalbumin (HαLA) in aqueous solution [5], ribonuclease A (RNase) in dry and hydrated powders and glycerol solution [6,7], maltose-binding protein (MBP) in a hydrated powder [8], and bacteriorhodopsin (BR) in purple membrane (PM) stacks [9,10]. Our choice of specific systems has generally been made based on the availability of experimental data to which the simulation results can be closely compared. We have accordingly set up the systems so that the simulations are very similar to the experiments, both in terms of sample composition and thermodynamic state points (i.e., temperature and pressure).

We will first give a brief summary of the setup and simulation of each system. All of the simulations reported here employed the CHARMM22 force field [11] for the proteins, the smooth particle mesh Ewald method [12] for the calculation of the electrostatic energies and forces (with electroneutrality imposed by addition of sodium cations or chloride anions), and multiple time-step algorithms to integrate the equations of motion [13,14] with holonomic constraints on the lengths of bonds involving hydrogen atoms [15,16]. Additional details may be found in the accompanying citations.

### 16.2.1  Molecular Dynamics Simulations of Native and Molten Globule States of α-Lactalbumin in Solution

The simulation of the native state of HαLA was initiated from a crystal structure obtained from the Protein Data Bank (entry 1A4V [17]). The stable, experimentally well-characterized low pH MG "A" state was modeled as an ensemble of 15 partially unfolded conformers. To create the MG conformers, the calcium ion was removed from the crystal structure, the ASP, GLU, and HIS side chains and the C-terminal carboxyl group were protonated, and the protein was partially unfolded by running a MD trajectory at elevated temperature in an implicit (GB/SA [18]) solvent. Fifteen different configurations were extracted from this unfolding trajectory and used to initiate MD simulations in the explicit solvent. The radii of gyration and amount of secondary structure present in the MG conformers were in good agreement with experimental data on the A state of HαLA. The native state and 15 MG conformers were placed in periodic boxes of SPC/E water and equilibrated at a constant temperature of 296 K and a constant pressure of 1 atm until the box volume fluctuated around a stable average. Production runs of each system were subsequently carried out for 400 ps under conditions of constant volume and energy for analysis of the water dynamics. Additional details of the HαLA simulations may be found in Ref. [5].

### 16.2.2    MOLECULAR DYNAMICS SIMULATIONS OF RIBONUCLEASE A IN DRY
AND HYDRATED POWDERS AND GLYCEROL SOLUTION

Our simulations of RNase model powder samples were set up starting with a crystal structure (PDB entry 7RSA [19]). The hydrated powder was modeled as a single unit cell of the monoclinic crystal containing two protein molecules. The empty space in the cell was filled with water molecules until the unit cell parameters remained close to the experimental values when the system was relaxed with a constant temperature and pressure MD simulation employing periodic boundary conditions with anisotropic cell fluctuations permitted. The final hydration level, $h = 0.57$ (g $D_2O$ per g protein), is characteristic of hydrated powder samples employed in neutron scattering studies of protein and hydration water dynamics. The "dry" powder model, which consisted of eight protein molecules, was generated by replicating the unit cell four times using lattice translations ($2a \times 2b \times c$). After removing the water molecules, the system was annealed by MD simulation at constant temperature and pressure with periodic boundary conditions to produce disordered protein configurations. The system was then hydrated to $h = 0.05$, which is roughly the amount of water in "dry" powder samples employed in neutron scattering experiments. Production runs of the dry and hydrated powder models were carried out over a range of temperature at a constant pressure of 1 atm.

To model the glycerol solution, a single RNase molecule was placed in a cubic box of 1467 glycerol molecules. A series of MD simulations of the resulting system were carried out with periodic boundary conditions over a range of temperature at a constant pressure of 1 atm. Additional details on the RNase simulations may be found in Refs. [6,7].

### 16.2.3    MOLECULAR DYNAMICS SIMULATIONS OF MALTOSE-BINDING
PROTEIN IN A HYDRATED POWDER

A model for a hydrated powder of MBP was constructed based on a crystal structure (PDB entry 1JW4 [20]). The model contained four protein molecules generated from four unit cells (a $2a \times 2b \times c$ lattice) of the triclinic crystal with the water molecules removed. A large box of water molecules was overlaid on the protein supercell, and all but the 3460 water molecules that were closest to the protein molecules were removed to give $h = 0.43$, corresponding to samples used in neutron scattering experiments carried out in conjunction with the simulations [8]. A constant pressure and temperature MD simulation at 1 atm and 300 K was used to allow the cell to collapse and anneal the protein–protein and protein–water contacts. A series of production runs were performed at constant pressure over a range of temperature.

### 16.2.4    MOLECULAR DYNAMICS SIMULATIONS OF BACTERIORHODOPSIN
IN PURPLE MEMBRANE

Our PM system was derived from an all-atom model constructed as described by Baudry et al. [21]. The hexagonal unit cell consists of a single BR trimer, 28 archeal lipids, and 1924 water molecules. The cell was replicated using periodic boundary conditions to produce a membrane stack with an interlamellar spacing of roughly 58 Å. The CHARMM22 force field for nucleic acids was used to model the sugar moieties in the glycolipids, and the CHARMM27 force field [22] was used for the remainder of the lipids. The TIP3P model [23] was used for water in the PM simulations. The simulations reported below were carried out for 5 ns following several nanoseconds of equilibration (sufficient to converge the lattice parameters and BR root-mean squared deviation from the initial structure) at a series of temperatures and constant pressure. The hexagonal symmetry of the unit cell was maintained during the simulations, and the temperature dependence of the in-plane lattice parameter was in good agreement with neutron diffraction data [10].

## 16.3  APPLICATIONS

### 16.3.1  PROTEIN EFFECTS ON SINGLE-PARTICLE WATER DYNAMICS

MD simulations, by virtue of the fact that they provide space–time trajectories of all the atoms in the simulated system, enable a quantitative characterization of the differences in the behavior of bulk and protein hydration water. To this end, it is useful to introduce a convenient measure of what distinguishes protein hydration water from bulk water. Any single parameter is bound to be somewhat arbitrary and restrictive given the variation in the sensitivity of experimentally measurable and theoretically calculable quantities to the details of protein–water and water–water interactions. In spite of this caveat, a distance criterion can be reasonably well justified based on experimental and simulation data. Radial solvent density profiles extracted from x-ray diffraction data show that the extent of the perturbation of the structure of water in protein crystals, judging from deviations from the bulk water density, is 4–5 Å [24]. The range of perturbation of water structure is similar in protein solutions, according to small angle x-ray and neutron scattering measurements [25,26]. MD simulation studies have shown that the change in water dynamics is by far most pronounced in the first solvation shell, as defined by distance criteria, or more sophisticated measures such as Voronoi tessellation [27]. When water rotational dynamical quantities computed from simulations are compared with corresponding parameters extracted from magnetic resonance relaxation measurements, good agreement is obtained when a distance of ~4 Å is used to define the range of perturbation of water dynamics in the vicinity of a protein [5]. We have therefore used a distance cutoff of 4 Å as the operational definition of the protein hydration shell in the analyses presented in this chapter.

MD simulations and experiments clearly show that the "single particle" motion of water molecules next to a protein surface is different than in the bulk. Here, "single particle" refers to measures of the average behavior of individual water molecules, as opposed to coherent behavior of collections of water molecules, which will be discussed in more detail below. The perturbation of the translational and rotational mobility of protein hydration water (defined using the 4 Å distance criterion) is depicted in Figure 16.1a and b, respectively. We will discuss the data for the native (N) state first, and subsequently compare the native and MG states. In bulk water, after an initial rapid (~2 ps) rise corresponding to ballistic motion,

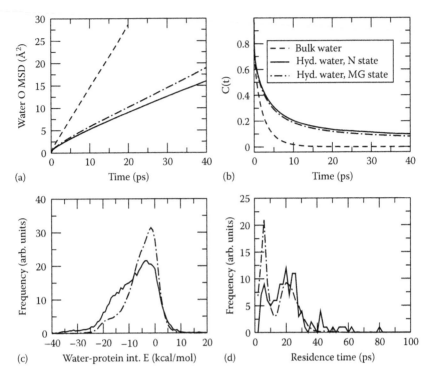

**FIGURE 16.1** Dynamics and energetics of the solvation shell (defined as water with O atoms within 4 Å of protein heavy atoms) in MD simulations of native and model molten globule conformers of HαLA in aqueous solution [5]. Data for the native conformation are plotted as solid curves. Data for the molten globule were averaged over simulations of 15 individual conformers, and are plotted as dot-dashed curves. Panels (a) and (b) include data for bulk water, plotted as dashed curves. (a) Water oxygen mean-squared displacements. (b) Water rotational correlation functions, $C(t) = <P_2[\cos\theta(t)]>$, where $\cos\theta(t) = \mathbf{u}(0) \cdot \mathbf{u}(t)$, $\mathbf{u}(t)$ is a unit vector along the water dipole moment at time $t$, and $P_2(x)$ is the second Legendre polynomial of argument $x$. (c) Histogram of interaction energies of individual water molecules in the hydration shell with the protein. (d) Histograms of water residence times within the hydration shell.

the mean-squared displacement (MSD) increases linearly in time, as expected for normal diffusion (Brownian motion) in liquids (Figure 16.1a). In contrast, the MSD of hydration water near the N state of HαLA in solution is sublinear, i.e., $MSD(t) \sim t^{\alpha}$, where $\alpha \sim 0.7$, and is indicative of anomalous diffusion. Fourier inversion of neutron scattering measurements of water dynamics in protein powders in which the hydration level is sufficiently low ($h \sim 0.3$ g $H_2O$ per g protein) that all of the water molecules can be considered in contact with the protein, yielded a water MSD exhibiting a sublinear time dependence with $\alpha = 0.4$ [28]. This is somewhat lower than the value of 0.7 we have computed for hydration water near a protein in solution, and the discrepancy could presumably be due to additional confinement effects on water in powder versus solution samples. Nonetheless, both simulation and experiment agree that the translational mobility of protein hydration water is suppressed relative to bulk water. Simulations further suggest that water translational mobility near peptides and proteins in solution is anisotropic, with bulk-like diffusion parallel to the peptide–protein surface, and significantly suppressed mobility perpendicular to the surface [29,30].

The data plotted in Figure 16.1b reveal corresponding modifications of the rotational mobility of water next to a protein in solution compared to bulk water. The second-rank rotational correlation functions shown in Figure 16.1b are well described by stretched exponential functions, $C(t) = \exp(-t/\tau)^\beta$, with $\beta = 0.7$ for SPC/E bulk water and $\beta = 0.4$ for water in the hydration shell of the native state of H$\alpha$LA. Deviations from $\beta = 1$ are indicative of departures from rotational Brownian motion. Our simulations suggest that rotational diffusion is substantially more anomalous in protein hydration water compared to bulk water. The rotational mobility of protein hydration water is also significantly retarded. The average rotational correlation time of water next to the N state of H$\alpha$LA, 9.5 ps, is substantially greater that the value of 1.9 ps computed for bulk SPC/E water. The ratio of the correlation times of H$\alpha$LA hydration water to bulk water, ~5, is in good agreement with the range of values, 4.9–5.4, extracted from $^{17}O$ nuclear magnetic resonance dispersion (NMRD) measurements on solutions of several soluble proteins, including bovine $\alpha$LA [31].

The retardation of water translational and rotational mobility that has been firmly established by numerous in vitro and in silico studies has recently been confirmed to be present also in vivo. This is not a foregone conclusion, given the high concentration of biomolecules in cellular compartments, which introduces complexities due to crowding and chemical heterogeneity [32]. Nonetheless, both neutron scattering and NMRD measurements have shown that the majority of water molecules in both prokaryotic and eukaryotic cells exhibit bulk-like translational and rotational mobility [33–35]. The vast majority of the remainder, which can be assigned as closely associated with biomacromolecules, displays a level of dynamical retardation that is typical of that established by in vitro and in silico investigations.

The sublinear time evolution of the MSD and the pronounced stretched exponential relaxation of the rotational correlation function of protein hydration water are signatures of anomalous translational and rotational diffusion. Anomalies in diffusive motion can be attributed to spatial and/or temporal heterogeneity, the former being connected to the roughness of the protein surface and the latter to a distribution of relaxation times of water dynamics at different locations on the protein surface [36]. Simulation studies have shown that both types of heterogeneities contribute to the anomalous character of protein hydration water dynamics. This is not too surprising, given that simple inspection of protein surfaces reveals a high degree of roughness and chemical heterogeneity on the length scale of a water molecule [37].

Although protein–water interaction energies are not dynamical quantities, they do highlight the existence of a broad range of different sites for water molecules on the surface of a protein. This is exemplified by the interaction energy distribution for the N state of H$\alpha$LA, plotted in Figure 16.1c, which is broad, ranging from −40 kcal/mol to +10 kcal/mol, with a mean of −8.2 kcal/mol. The distribution appears to be bimodal, with peaks at roughly −12 and −4 kcal/mol, representing two classes of water molecules with more and less energetically favorable interactions, respectively, with the protein.

Water residence times on the protein surface are not directly measurable experimentally, but can be defined as the relaxation time of time correlation functions of the population of the hydration shell [5]. Due to differences in the definition from one investigation to another, the values reported in the literature exhibit considerable variability. Nonetheless, heterogeneity in water dynamics near the protein surface is clearly manifested in distributions of water residence times. The distribution we have constructed for the N state of HαLA in solution from the residence time of water next to each residue is plotted up to 100 ps in Figure 16.1d. The distribution is very broad, ranging from 2.6 to 241 ps, but highly skewed toward shorter residence times. The mean residence time of 23 ps is about 2.5 times longer than the rotational correlation time for hydration water.

Several investigations have sought to correlate water residence times on protein surfaces with the type of functional group displayed on the surface [38]. There is considerable variability in the results reported, and this reflects, to some extent, differences in the definition of residence time and solvation shell. It is also quite likely that it is a manifestation of a true lack of correlation. The average residence times that we have computed for various functional groups in the N state of HαLA are listed in Table 16.1. The residence times are longest (~70 ps) next to anionic (ionized acidic) groups, and shortest (<10 ps) for cationic (basic) and nonpolar (methyl and methylene) groups, with intermediate values (12–25 ps) for neutral polar groups. This ordering is similar to the results of a previous analysis of crambin [39], but substantially different from similar analyses of BPTI [40] and azurin [38]. The large variation in the results of such analyses suggest that other factors, beside chemical functionality, such as surface shape and water site accessibility, are important in determining the extent to which interaction with a protein affects water dynamics. Indeed, an analysis of the azurin simulation demonstrated a convincing inverse correlation between solvent accessible surface area (SASA) and water residence time, with the longest times observed for sites with SASA between 1 and 16 Å$^2$, regardless of the chemical functionality of the site [38].

### TABLE 16.1
### Average Residence Times of Water Molecules in the Vicinity of Specific Chemical Functional Groups from an MD Simulation of Native HαLA in Solution

| Functional Group | Average Residence Time (ps) |
|---|---|
| Side chain COO$^-$ in ASP and GLU | 68.7 |
| Side chain OH in SER, THR, and TYR | 25.4 |
| Side chain NH$_2$ in ASN and GLN | 18.8 |
| Backbone NH | 14.7 |
| Backbone CO | 11.9 |
| Side chain NH$_3^+$ from LYS | 8.8 |
| Side chain CH$_2$ and CH$_3$ | 5.7 |

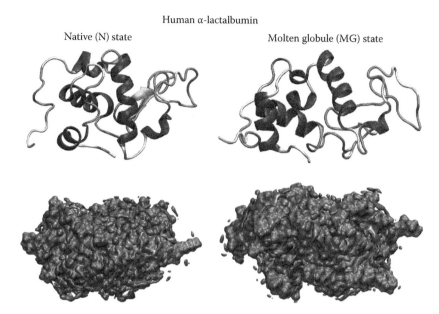

**FIGURE 16.2    (See color insert following page 172.)** Molecular graphics images of the native conformation and a selected molten globule conformer from MD simulations of HαLA in aqueous solution [5]. The top two images are cartoon representations of the protein backbone, with α-helices represented by purple ribbons, β-sheets by yellow ribbons, and reverse turns and random coil by cyan and white tubes. The MG conformer is less compact and contains less secondary structure than the native structure. The bottom two images depict the protein molecules in magenta colored surface representation, and water isodensity surfaces, contoured at ~1.5 times the bulk density, in gray. The solvation of the MG conformer is more diffuse, with less accumulation of water at high density in the nooks and crannies, than that of the native structure.

Inspection of protein hydration water density distributions, which may be readily computed from MD trajectories, reveals that there are regions of enhanced water density (relative to bulk) [5,37]. The water isodensity surface computed from our MD simulation of the N state of HαLA in solution, shown in Figure 16.2 (left panel), reveals that water tends to accumulate at relatively high density in nooks and crannies on the rough protein surface. Thus, it appears that there are water "binding sites" on the surface of a protein, where the probability of finding a water molecule is relatively high. Such sites also appear to coincide with restricted mobility, according to the analysis of Luise et al. [38], even though the thermodynamic (density) and dynamical (residence times) quantities are generally complementary, and not necessarily correlated [41].

## 16.3.2  DIFFERENCES IN HYDRATION OF NATIVE AND MOLTEN GLOBULE STATES OF SOLUBLE PROTEINS

Hydrophobic hydration and energetically favorable interactions between water molecules and polar groups on protein surfaces provide major contributions to the thermodynamic stability of the native state of a protein. The mobility of protein hydration water has also been suggested to be involved in the kinetics of protein folding [42]. Thus, characterization of the differences in the solvation structure, energetics, and dynamics between native and denatured states is a prerequisite to a detailed understanding of protein stability and the mechanism of protein folding. Although there is ample evidence of residual structure in unfolded proteins [43], it is clear that the folding of a protein is accompanied by dehydration of the polypeptide chain as secondary structural elements grow and pack together to form the tertiary structure. A question whose answer is of considerable fundamental interest is: how much does the number and mobility of water molecules hydrating a protein change during protein folding?

This question has been addressed in a number of experimental and theoretical investigations for the case of the so-called MG states [44,45]. Many proteins pass though subcompact kinetically stable intermediates as they move from unfolded states to the native state. MG states, which are stable, compact denatured states of certain proteins that can be prepared at low pH (e.g., α-lactalbumin and carbonic anhydrase II) or by removing a cofactor (such as the heme in myoglobin), are regarded as equilibrium analogs of folding intermediates. MG states are ensembles of conformers characterized by modest changes in compactness (e.g., ~10%–20% increase in radii of gyration) compared to the corresponding native states, with retention of the majority of secondary structure, and substantial disruption of tertiary contacts. A variety of experimental measurements (reviewed in Ref. [31]) have suggested that MG states are extensively solvent-penetrated, so that the number of water molecules hydrating proteins in the MG state is significantly greater than in the N state (e.g., the MG state of α-lactalbumin has been estimated to be >50% more hydrated than the N state). In contrast, NMRD measurements indicate only a small increase in the number and a slight reduction in the dynamical retardation of water molecules hydrating the MG state compared to the N state [31]. This conclusion is based on the observation that the product of parameters, $N_s\rho_s$, extracted from the NMRD data is very similar in the N and MG states. Here $N_s$ is the number of hydration water molecules, $\rho_s = <\tau_s>/\tau_b - 1$ is a dynamical retardation factor, with $\tau_s$ and $\tau_b$ the average rotational correlation time of protein hydration and bulk water molecules, respectively, The constancy of $N_s\rho_s$ could arise because neither $N_s$ nor $\rho_s$ change, or because $N_s$ increases while $\rho_s$ decreases, or vice versa. Although it is not possible to separately determine $N_s$ and $\rho_s$ from experimental data on the MG state, it has been argued that only small differences in these parameter between N and MG states are sensible [46].

We have recently used MD simulations to explore the differences in the structure, energetics, and dynamics of protein hydration water in N and MG states of HαLA. The MG state was modeled as an ensemble of 15 conformers prepared as summarized above in Section 16.2 and described in detail in Ref. [5]. The root-mean squared deviation (RMSD) of the backbone heavy atom positions from the initial crystal structure ranged from 2.8 to 7.6 Å, which is significantly greater than the 1.1 Å RMSD of the N state simulation. Radii of gyration demonstrated the loss of compaction of the MG conformers ($R_g$ = 15.2–17.8 Å) compared to the N state ($R_g$ = 14.3 Å). The increase in size is accompanied by a modest increase in the number of water molecules hydrating the MG state ($N_s = 546$, averaged over the 15 conformers) versus the N state ($N_s = 504$). Molecular graphics images of a selected MG conformer and its hydration shell are compared to corresponding images of the N state in Figure 16.2. It is evident from the water isodensity surfaces displayed in Figure 16.2 that the hydration shell is more diffuse and there is a loss of water "binding sites" (regions of relatively high water density/probability) in the MG conformer. The changes in the structure of the hydration shell is accompanied by a weakening of the protein–water interactions, as indicated by the shift of the interaction energy distribution toward more positive values in the MG versus N state (Figure 16.1c).

The data plotted in Figure 16.1a through d show that there are corresponding changes in the mobility of water molecules in the first hydration shell (defined here as water molecules with their O atoms within 4 Å of any protein heavy atom). Figure 16.1a shows that the translational diffusion is anomalous (MSD$(t) \sim t^\alpha$, $\alpha < 1$) for hydration water in both the N and MG states, but the translational mobility is enhanced in the MG state ($\alpha \sim 0.8$) versus the N state ($\alpha \sim 0.7$). Given that there are fewer water molecules found on the surface of the nooks and crannies on the surfaces of the MG conformers than in the N state, it is not surprising that the distribution of water residence times shifts to lower values (Figure 16.1d). The mean residence time of water in the hydration shell of the MG conformers, 14 ps, is substantially shorter than that of the N state (~23 ps). In addition to the shift, the distribution of residence times averaged over MG conformers is clearly bimodal, with one peak near the average residence time for the N state and another at a much shorter time (~5 ps).

Figure 16.1b shows that rotational diffusion is anomalous for water molecules hydrating both the MG and N states. In addition, it is evident from the slightly faster decay of the rotational correlation function that the rotational mobility is slightly greater for water hydrating the MG conformers compared to the N state. The average rotational correlation times computed from the correlation functions are $<\tau_s> = 9.5$ ps for the N state and $<\tau_s> = 8.4$ ps for the MG state. The values of $<\tau_s>$ and $N_s$ computed from the simulations enable a comparison with experimental data via the quantity $N_s\rho_s$ from NMRD data. We obtain the values $N_s\rho_s = 2.0 \times 10^3$ for the N state and $N_s\rho_s = (1.9 \pm 0.2) \times 10^3$ averaged over the MG conformers of HαLA, which compare favorably with the corresponding values, $(1.8 \pm 0.2) \times 10^3$ and $(2.0 \pm 0.2) \times 10^3$, obtained from NMRD measurements on bovine α-lactalbumin [31]. Thus, the MD simulations reproduce the experimental finding that $N_s\rho_s$ is essentially the same in the N and MG states. According to the simulations, the constancy of $N_s\rho_s$ results from the compensation of a slight increase in the number ($N_s$) by a slight decrease in the retardation ($\rho_s$) of water molecules in the hydration shell of the MG state versus the N state.

## 16.3.3 COLLECTIVE DYNAMICS OF PROTEIN HYDRATION WATER

MD simulations and normal mode analyses have revealed a large number of low-frequency vibrational modes in biopolymers [47–50]. The lowest frequency modes correspond to the largest amplitudes, and generally involve displacements of a large fraction of the atoms in the molecule. These "collective" modes contribute to functionally relevant protein dynamics [51,52]. Measurements (and simulations) of single-particle motion contain contributions from both true single-particle dynamics (e.g., Brownian motion of a tagged particle) and collective motion (e.g., a hinge-bending mode in an enzyme).

The number of experimental techniques available for isolating collective motions is limited. Terahertz spectroscopy, which probes a combination of single-particle and collective dipolar fluctuations on the picosecond timescale, is currently being applied to investigate coupled protein and solvent dynamics radial [53,54]. Inelastic x-ray scattering (IXS) [55] and coherent neutron scattering (CNS) [56], which measure the coherent dynamic structure factor, $S_c(Q,E)$, and neutron spin-echo spectroscopy (NSE) [52], which measures the coherent density correlation function, $I_c(Q,t)$, have emerged as potentially promising techniques for measuring collective density fluctuations in biomolecular systems on the picosecond to nanosecond timescales and angstrom to nanometer length scales. Here, $Q$ and $E$ are the wavevector transfer and energy of a collective excitation, respectively, and $S_c(Q,E)$ is the time–energy Fourier transform of $I_c(Q,t)$. NSE measurements have been used to detect and characterize overdamped, large-amplitude (~70 Å) correlated domain motion in *taq* DNA polymerase [52].

For collective density excitations in simple liquids, in the finite $Q$ region describable by generalized hydrodynamics, $S_c(Q,E)$ reduces to one Rayleigh (heat) mode at $E = 0$ and two Brillouin lines at $\pm E_s$ for each collective mode [57]. For bulk liquid water, inelastic neutron and x-ray scattering experiments [58] have revealed the existence of two Brillouin modes, as predicted by the pioneering MD simulations of Rahman and Stillinger [59]. The two modes, which are commonly referred to as ordinary and fast sound modes, propagate at 1500 and 3200 m/s, respectively, in bulk water at 278 K.

As far as we are aware, there are no published experimental studies of the coupling of collective density fluctuations in proteins and their hydration water. To fill an obvious void, we have analyzed the collective density fluctuations of water molecules in the RNase crystal at 150 and 300 K [60]. Essentially all of the water molecules in the crystal are in contact with the protein, and may be regarded as protein hydration water. In Figure 16.3, we show a contour plot of the longitudinal current spectrum for the water molecules, $C_L(Q,E) = (E^2/Q^2)S_c(Q,E)$, which enhances features present in $S_c(Q,E)$, at 300 K. The corresponding spectrum at 150 K (not shown) is almost identical to the 300 K spectrum. The low-energy Brillouin mode, indicated by a peak at $E \sim 10$ meV, is nondispersive, i.e., has a nearly constant energy, while the high energy Brillouin mode is clearly dispersive, over the $Q$ range considered here. From the limiting (low $Q$) slope of $E_s(Q)$ of the high frequency, dispersive mode, we obtain a value of ~3800 m/s for the speed of propagation of the fast sound mode in protein hydration water at 300 K, and we find that the speed increases with decreasing temperature [60]. Our calculations predict a somewhat higher speed of propagation for hydration water than is measured experimentally for bulk water. In the absence of experimental data on protein hydration water, it is not clear if the discrepancy is real, or it reflects a deficiency in the description of protein–water and/or water–interactions in the force fields employed in our simulations.

Our results for the collective behavior for protein hydration water on length scales of angstroms length scales and timescales up to a few picoseconds are remarkable for two reasons. First, the coherent dynamic structure factors we have computed for

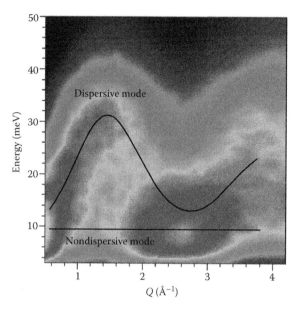

**FIGURE 16.3**    **(See color insert following page 172.)** Contour plots of the longitudinal current spectra, $C_L(Q,E) = (E^2/Q^2)S(Q,E)$, computed from coherent dynamic structure factors for protein hydration water in the RNase crystal at 300 K [60]. The black lines trace the maxima of the two Brillouin side peaks in the spectra, one of which is dispersive (i.e., the excitation energy is wavevector transfer/length scale dependent), and the other nondispersive.

protein hydration water are very similar to those reported for bulk water. Second, the coherent spectra are hardly affected by a large change in temperature (150 to 300 K). In contrast, as was demonstrated above, various measures of single-particle dynamics of water on similar length and timescales are significantly different for protein hydration water versus bulk water, and, as will be shown below, single-particle dynamics change qualitatively between 150 and 300 K.

## 16.3.4   DYNAMICAL TRANSITION AS A PROBE OF PROTEIN–SOLVENT DYNAMICAL COUPLING

Much of what is known about the coupling of protein and solvent motion has come from investigations of the temperature dependence of protein and solvent dynamics, and protein function. For example, the concept of solvent slaving of protein dynamics and function has emerged from comparing the temperature dependence of the rates of protein processes with solvent dynamical properties such as viscosity and dielectric relaxation rates [2,4]. While such studies have covered protein motions spanning many decades of time, we will restrict our attention here to protein motion occurring on the picosecond to nanosecond timescales, which includes atomic fluctuations, side chain and loop librations and conformational transitions, and limited rigid body-like displacements of secondary structural elements, with amplitudes of up to a few angstroms.

When measured over a wide range of temperature (e.g., from ~100 K to room temperature), the amplitudes of atomic fluctuations in dehydrated proteins increase linearly with temperature, as expected for a harmonic solid. In hydrated soluble proteins, the amplitudes increase linearly with temperature, as in dehydrated proteins, until around 200 K, when there is a change of slope, indicating the onset of additional motion not present in the dehydrated protein. The change of slope is commonly referred to as a "dynamical transition," although it is generally gradual, and not sharp, as expected for an abrupt transition such as a phase transition. In any event, the so-called dynamical transition has been detected by several experimental techniques (Mössbauer spectroscopy [61], neutron scattering [62], and Debye–Waller factors from x-ray crystallography [63]) in many soluble proteins with different sizes and folds. Neutron scattering data illustrating the transition in hydrated lysozyme, and lack of a transition in the dry protein, are shown in Figure 16.4a. The dynamical transition has also been reproduced in numerous MD simulation studies, an example of which is shown in Figure 16.4b. The transition has also been observed in a membrane protein (albeit at a higher temperature than in soluble proteins) [64], and in a tRNA [65], and it thus appears to be a generic phenomenon in biological macromolecules.

The functional relevance of the additional anharmonic and diffusive motion that is present at high temperature, in addition to the harmonic solid-like motion that is always present, is a subject of ongoing discussion. There are a few cases where a clear correlation between the dynamical transition and the onset of protein function

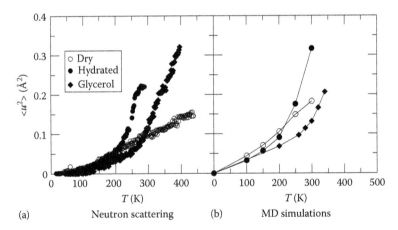

(a)        Neutron scattering        (b)        MD simulations

**FIGURE 16.4** (a) Mean-squared fluctuations (MSFs) of H atoms in lysozyme on a time scale of tens of ps determined from elastic incoherent neutron scattering [70]. The displacements increase linearly with temperature (as in a harmonic solid) below 200 K, and a dynamical transition (change in slope of MSFs vs. temperature) is evident at 200 K in a hydrated powder (filled circles), but not in a dry powder (open circles). In a mixture of water and glycerol (filled diamonds), the transition temperature is increased, and the amplitudes in the harmonic regime are reduced. (b) Temperature dependence of the MSFs of protein heavy atoms computed from MD simulations of RNase A in different environments: dry powder (open circles), hydrated powder (filled circles), glycerol (filled circles) [7]. The MSFs were computed as averages over 100 ps blocks of the trajectories. The solvent dependence of the dynamical transitions is accurately described by MD simulations.

has been drawn. For example, the increased flexibility above the transition has been shown to be necessary for CO escape from myoglobin [66] and substrate binding in RNase [67]. Moreover, it has been shown that the ability of BR to complete its photocycle is correlated with the onset of the additional motion that accompanies the dynamical transition in PM [68]. On the other hand, there are examples of certain enzymes for which no abrupt change in activity is observed when the proteins exhibit dynamical transitions in cryosolvents [69].

The solvent dependence of the dynamical transition is a clear demonstration of the influence of the solvent environment on the dynamics of soluble proteins. The transition is suppressed in dehydrated proteins, and the temperature at which the change of slope occurs, as well as the extent of additional motion at higher temperature, can be altered by changing solvent properties, such as viscosity [66,70], as exemplified by the neutron scattering and MD simulation data shown in Figure 16.4. Increasing solvent viscosity tends to raise the transition temperature and decrease the amplitudes at a given temperature above the transition. Simulation studies have shown that anharmonic protein motion above the dynamical transition temperature can be suppressed by holding the surrounding at low temperature [71,72] or by inhibiting water translational diffusion using harmonic restraints [73]. Simulations have also shown that the liquid-like motion of soluble proteins at room temperature is pronounced on the solvent-exposed protein surface, and the motion becomes more solid-like in the interior of the protein [74,75].

Simultaneous measurement of changes in both protein and solvent dynamics with temperature has provided insight into the nature of protein–solvent dynamical coupling. In the case of soluble proteins, it appears that the protein dynamical transition coincides with changes in water mobility. For example, time-dependent MSDs of protein hydration water extracted from neutron scattering data as a function of temperature have revealed the onset of water translational diffusion at the dynamical transition temperature of myoglobin [76]. More recently, a concerted neutron scattering and MD simulation study demonstrated the coincidence of dynamical transitions in the soluble protein MBP and its hydration water, as may be seen in Figure 16.5a. Neutron scattering and NMR experiments, as well as MD simulations, have demonstrated a super-Arrhenius to Arrhenius transition in the diffusivity of protein hydration water, signaling a fragile to strong dynamic crossover, at the dynamical transition temperature [77]. The dynamical crossover, along with concomitant changes in thermodynamic (e.g., specific heat) and structural (e.g., radial distribution function, static structure factor, tetrahedral order parameter), have led to the proposition that the protein dynamical transition is triggered by changes in hydration water that occur along a path that crosses the Widom line, and hence may be connected with the hypothetical liquid–liquid phase transition in supercooled water [78].

The correlation of the dynamical transition of soluble proteins and changes in properties of the hydration shell is clear from a growing body of experimental and theoretical data, although the details of the physical mechanism for the connection vary considerably from account to account. Below we will discuss a microscopic mechanism, which we have developed based on our simulation studies, for the triggering of the protein dynamical transition by changes in the dynamics of

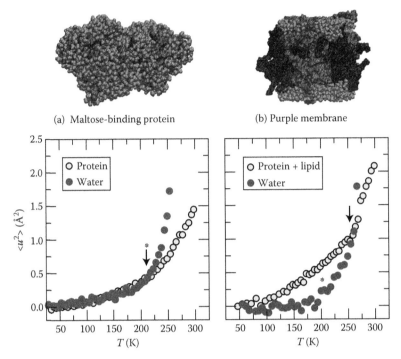

FIGURE 16.5  (See color insert following page 172.) Mean-squared fluctuations of nonexchangeable H atoms in protein/lipid (filled circles) and water molecules (open circles) in: (a) the soluble protein, maltose-binding protein (MBP) [64], and (b) the protein BR and its surrounding lipids in purple membranes (PM) [9]. The MSFs were measured by neutron scattering experiments that probe motions on up to nanosecond time scales. In each system, protein/lipid motion and water motion were measured separately by employing selective deuteration. Approximate locations of dynamical transitions in the protein/lipid data are indicated by arrows, and in the water data by asterisks. The accompanying molecular graphics images are from MD simulations, carried out in concert with the scattering experiments, of a hydrated powder model of MBP and a unit cell of PM [9,10]. The images depict water molecules in cyan, protein molecules in gray, and lipid molecules in black.

protein–water hydrogen bonds. However, we would first like to point out some important differences between soluble and membrane proteins with regard to their dynamical coupling to their surroundings.

## 16.3.5   DYNAMICAL TRANSITIONS IN PURPLE MEMBRANE

A full understanding of protein–water dynamical coupling in a cellular context requires a characterization of the influence of water motion on the dynamics of membrane proteins. The situation is more complicated for membrane proteins, which are not completely immersed in aqueous solution as are soluble proteins (or their complexes), but rather a large fraction of their surface is contained within the chemically heterogeneous environment of a lipid bilayer. The chemical nature of a lipid bilayer changes drastically, from the highly hydrophobic acyl chains in the middle to the highly polar, hydrated headgroups on the exterior, over a length scale of ~2 nm [79]. Lipid bilayer structure and dynamics are sensitive to both temperature and the level of hydration [80]. Hence, it is clear that a detailed description of the influence of water on membrane protein dynamics will require the unraveling of protein–water, lipid–water, and protein–lipid dynamical couplings.

Some progress along these lines has been made for the membrane protein BR. Data from the most recent neutron spectroscopic study of the dynamical transition, in which selective deuteration was employed to separately measure the PM and water dynamics, is shown in Figure 16.5b [9]. The MSFs of PM (protein and lipid) on the nanosecond timescale show two changes of slope: one at around 120 K attributed to methyl group rotations [81] and a second signaling a dynamical transition at ~250 K, which is higher than the transition temperature for soluble proteins. The water hydrating PMs undergoes a dynamical transition at ~200 K (Figure 16.5b), which is the same as in a hydrated powder of MBP (Figure 16.5a), but much lower than the temperature at which the proteins and lipids undergo their dynamical transition. The temperature dependence of the lamellar spacing of PM stacks suggested that the water transition at ~200 K is associated with the onset of water translational mobility [82], and this has been confirmed by MD simulations [9,10].

The solvent dependence of the proton pumping activity of BR suggests that biological function is correlated with the higher temperature (~250 K) transition, which is suppressed by dehydration [83,84]. In addition to a threshold level of hydration, experiments have also demonstrated that the lipid matrix is crucial in promoting the full range of mobility required for proper BR function. In particular, delipidation of PM has been show to lead to a significant reduction in internal dynamics and a corresponding attenuation of photocycle activity [83,84].

The results summarized here establish that the additional motion that accompanies the ~250 K dynamical transition of PM is required for proper BR function, and that the transition is hydration- and lipid-dependent. They also show that, in contrast to the case of soluble proteins, the PM transition is not coincident with the hydration water transition. Thus, the water and protein–lipid dynamics do not appear to be directly coupled as for soluble proteins. It should be clear from this summary that much remains to be done to develop a picture of membrane protein dynamics that is consonant with our current understanding of soluble protein dynamics. We now proceed to a comparison of protein–water dynamical coupling in the soluble protein MBP and the membrane protein BR, in the context of dynamical transitions.

## 16.3.6 Role of Protein–Solvent Hydrogen-Bond Dynamics in Dynamical Transitions

An important role for hydrogen bond dynamics in the dynamical transition was proposed by Doster and Settles [85], based on an interpretation of the transition as taking place between two levels in the hierarchy of conformational substates: a glassy state, in which the protein side chains are locked by rigid hydrogen bonds, and a liquid-like state in which the hydrogen bonds fluctuate on the picosecond timescale. Inspired by this proposition, we examined the dynamics of protein–water hydrogen bonds in the RNase crystal, taking care to distinguish between two types of processes, a fast process (with characteristic time, $\tau_{HB}$) of hydrogen bond breaking and forming due to vibration/rotation of water molecules without translational displacement, and a slower process (with characteristic time, $\tau_R$) that involves rearrangement of the hydrogen bond network via water translational displacements [73]. We found that $\tau_{HB}$ was ~1 ps and showed little variation with temperature, while $\tau_R$ was on the order of 10 ps at 300 K, and rapidly increased to several nanoseconds as the temperature was lowered and approached the dynamical transition temperature. Based on these observations, we proposed that the structural relaxation of the protein that occurs at temperatures above the dynamical transition temperature is triggered by structural relaxation of the protein–water hydrogen bond network, which, in turn, is associated with the onset of water translational diffusion.

The correlation of the dynamical transition of soluble proteins with the rapid increase in the rate of protein–solvent hydrogen bond fluctuations and the onset of solvent translational diffusion appears to be general. Results from a more recent simulation study of a hydrated powder of MBP that reinforces our earlier observations are reported in Figure 16.6 [8]. The data show that the onset of a rapid increase with temperature of $1/\tau_R$ (akin to a rate of relaxation of the protein–water hydrogen bond network) and the value of the water oxygen mean-squared displacement at 100 ps (a measure of diffusivity) are coincident with the change of slope in the protein hydrogen atom mean-squared displacements that signals the protein dynamical transition at 240 K. We have also recently shown that a similar connection between protein–solvent hydrogen bond dynamics, solvent diffusion, and the dynamical transition exists for RNase in glycerol [7]. The hydrogen bond network relaxation, and hence the dynamical transition, occur at higher temperature in glycerol, as expected from the greater viscosity compared to water.

Analysis of hydrogen bond fluctuations also provides some insight into the differences in protein–water dynamical coupling in soluble and membrane proteins. Results from our simulations of PMs are shown in Figure 16.7. The plot of the MSFs of protein/lipid nonexchangeable H atoms (corresponding to incoherent neutron scattering measurements) versus temperature (Figure 16.7a) displays a change of slope at 250 K, and the water diffusivity, measured as the value of the water O MSD at 30 ps, begins to increase rapidly with temperature at ~200 K (Figure 16.7c). These results are consistent with the neutron scattering data on PM shown in Figure 16.5b, and confirm the lack of coincidence in changes in protein/lipid and water dynamics, in contrast to the case of soluble proteins such as MBP (Figure 16.6). The temperature dependence of protein–water and lipid–water hydrogen bond dynamics in PM is also different from protein–lipid hydrogen bond dynamics in soluble proteins: the data show that the sharp increase in the hydrogen bond relaxation rate with temperature begins at a temperature

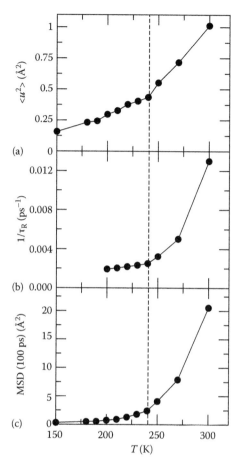

**FIGURE 16.6** Temperature dependence of protein and water dynamical properties from MD simulations of a hydrated powder of MBP [8]. (a) MSFs of protein nonexchangeable H atoms averaged over 1 ns blocks of the trajectories. (b) Temperature dependence of the inverse of the correlation times, $\tau_R$, of the protein-water hydrogen bond correlation functions [73]. (c) Value of the mean-squared displacement of water O atoms at $t = 30$ ps.

between the transitions in protein/lipid dynamics and water diffusivity (Figure 16.7b) [10]. Moreover, it is evident from Figure 16.7b that the increase in hydrogen bond relaxation rates with temperature is substantially greater for lipid–water versus protein–water hydrogen bonds. This observation underscores the potential importance of protein–lipid dynamical coupling in membranes. With regard to the dynamical transition in PM, we have hypothesized that the dynamical transition in the protein is triggered by a melting in the acyl chains in the lipids that is, in turn, enabled by a breakup of the interactions between lipid polar groups by enhanced mobility in their solvent shells [10].

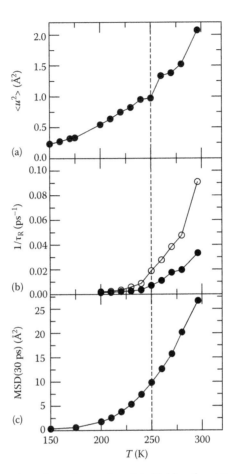

**FIGURE 16.7** Temperature dependence of protein/lipid and water dynamical properties from MD simulations of purple membrane [9,10]. (a) MSFs of protein/lipid nonexchangeable H atoms averaged over 5 ns blocks of the trajectories. (b) Temperature dependence of the inverse of the correlation times, $\tau_R$, of the protein–water (filled circles) and lipid–water (open circles) hydrogen bond correlation functions. (c) Value of the mean-squared displacement of water O atoms at $t = 100$ ps.

## 16.4   CONCLUDING REMARKS AND OUTLOOK

This chapter has examined several aspects of the coupling of protein and water dynamics from the point of view of MD simulations and complementary experimental measurements. The picture that has emerged from a large collection of results is that the hydration layer is limited in scope to roughly one layer of water, and that water next to a protein generally experiences retardation in single-particle dynamics (e.g., translational and rotational diffusion), typically by roughly a factor of 5. In contrast, the limited data presently available suggest that the collective density fluctuations of protein hydration water are remarkably similar to bulk water.

Compared to natively folded proteins, compact denatured states ("MGs") experience a modest increase in the number of water molecules in the hydration layer, and a slightly smaller perturbation of hydration water dynamics. Soluble protein–water dynamical coupling has been elucidated by simultaneous examination of transitions in protein and water dynamics as a function of temperature. Hydrated proteins at room temperature exhibit liquid-like motion on the subnanosecond timescale and behave like glasses at low temperature. The dynamical (or glass) transition between the low-temperature glassy state and room-temperature liquid-like state plays an important role in energy flow processes in proteins (see Ref. [86] and Chapters 7 and 11).

The liquid-like motion of proteins above the dynamical transition temperature appears to be facilitated by relaxation of the protein–water hydrogen bond network on a timescale of tens of picoseconds. Increasing solvent viscosity (e.g., by addition of glycerol) slows the hydrogen bond dynamics and suppresses diffusive motions in proteins. The tight coupling of protein and water dynamics that has been observed for soluble proteins does not appear to exist in membrane proteins, which are embedded in the amphipathic milieu of a lipid bilayer, and are only partially exposed to the aqueous solution on the exterior of the membrane. Rather, preliminary data suggest that membrane protein dynamics are more tightly coupled to lipid conformational fluctuations, which, in turn, may be more directly coupled to water dynamics [9,10].

The coupling of protein and water dynamics continues to be a very active area of research. Recent studies have raised a number of new questions that will need to be answered by a combination of experimental and theoretical investigations in the future. For example, the limited range of water perturbation by a protein that appeared to be well established has recently been called into question by tetrahertz spectroscopic measurements, which seem to imply that the solvation shell extends ~10Å from the protein surface [54]. This underscores the need for a better understanding of the underlying dynamics giving rise to the spectroscopic observable, as well as the need to develop a global description of hydration water dynamics that is consistent with a large body of experimental data that probe a variety of excitations over a range of timescales (e.g., neutron scattering, NMR, fluorescence stokes shift, THz and dielectric spectroscopies), and some of which are contradictory at first glance. For example, fluorescence Stokes shift (FSS) experiments [87] have been interpreted in terms of a long-time relaxation process in hydration water that is not present in direct measurements of single particle dynamics such as NMRD [88], and the fragile to strong crossover in hydration water that has been detected by neutron scattering and NMR, and has been proposed to be the trigger of the protein dynamical transition, is not observed in dielectric spectroscopic measurements [89]. Theory and simulation will certainly play a role in resolving some of the discrepancies (see, e.g., Refs. [88,90,91] concerning the interpretation of FSS data).

At the moment, very little is known about the coupling of collective density fluctuations in proteins and their hydration water (and lipids, in the case of membrane proteins). New instruments recently developed for coherent neutron scattering, combined with selective sample deuteration, may fill this gap soon, and MD simulations are expected to be helpful in the interpretation of the data [60,92].

Tetrahertz spectroscopy, coupled with MD simulations, can also be expected to provide insight into coupled protein and water collective dynamics [53,54].

The primary motivation for studying coupled protein and water dynamics is to gain insight into the role of the solvent environment in biological function. While it is clear that anharmonic and diffusive motion on picosecond to nanosecond timescales are tightly coupled to the dynamics of the solvent shell, the connection of these nonspecific fast fluctuations to directed energy flow into specific coordinates involved with particular functions is not clear, except in a few cases (e.g., in myoglobin, and to a lesser extent, BR). The situation is even less clear in the case of membrane proteins, where water-coupled lipid dynamics are expected to be relevant. Biological membranes contain multiple lipid and protein components, and are laterally heterogeneous [93]. Variations in local lipid composition have been implicated as a mechanism for the control of cellular signaling processes by membrane proteins [94,95]. Presumably, changes in hydration-dependent membrane properties could modulate protein function by altering the relative stability of different conformational states, and/or by modifying protein dynamics. Investigation of protein–lipid–water dynamical couplings in membranes is clearly a fertile ground for future research.

## ACKNOWLEDGMENTS

This work was supported by the National Science Foundation (grant CHE-0750175 to D. J. T.). We thank Martin Weik, Kathleen Wood, Joe Zaccaï, and Frank Gabel for fruitful collaborations and discussions, Emad Tajkhorshid for providing the coordinates for the PM model, and Jessica Morgan for help setting up and performing the MD simulations of PM. D. J. T. thanks the Université Joseph Fourier for a visiting professorship in 2007, during which some of the work reported here was carried out.

## REFERENCES

1. Henzler-Wildman KA, Kern D: Dynamic personalities of proteins. *Nature* 2007, **450**:964–972.
2. Fenimore PW, Frauenfelder H, McMahon B, Parak FG: Slaving: Solvent fluctuations dominate protein dynamics and function. *Proc. Natl. Acad. Sci. U. S. A.* 2002, **99**:16047–16051.
3. Henzler-Wildman KA, Lei M, Thai V, Kerns SJ, Karplus M, Kern D: A hierarchy of timescales in protein dynamics is linked to enzyme catalysis. *Nature* 2007, **450**:838–844.
4. Frauenfelder H, Fenimore PW, McMahon BH: Hydration, slaving and protein function. *Biophys. Chem.* 2002, **98**:35–48.
5. Sengupta N, Jaud S, Tobias DJ: Hydration dynamics in a partially denatured ensemble of the globular protein human alpha-lactalbumin investigated with molecular dynamics simulations. *Biophys. J.* 2008, **95**:5257–5267.
6. Tarek M, Tobias DJ: The dynamics of protein hydration water: A quantitative comparison of molecular dynamics simulations and incoherent neutron scattering experiments. *Biophys. J.* 2000, **79**:3244–3257.
7. Tarek M, Tobias DJ: The role of protein–solvent hydrogen bond dynamics in the structural relaxation of a protein in glycerol versus water. *Eur. Biophys. J.* 2008, **37**:701–709.

8. Wood K, Frölich A, Paciaroni A, Moulin M, Härtlein M, Zaccaï G, Tobias DJ, Weik M: Coincidence of dynamical transitions in a soluble protein and its hydration-water: Direct measurements by neutron scattering and MD simulations. *J. Am. Chem. Soc.* 2008, **130**:4586–4587.

9. Wood K, Plazanet M, Gabel FK, B., Oesterhelt D, Tobias DJ, Zaccai G, Weik M: Coupling of protein and hydration-water dynamics in biological membranes. *Proc. Natl. Acad. Sci. U. S. A.* 2007, **104**:18049–18054.

10. Tobias DJ, Sengupta N, Tarek M: Hydration dynamics of purple membranes. *Faraday Discuss.* 2009, **141**:99–116.

11. MacKerell Jr. AD, Bashford D, Bellott M, Dunbrack Jr. RL, Evanseck J, Field MJ, Fischer S, Gao J, Guo H, Ha S, et al.: All-atom empirical potential for molecular modeling and dynamics studies of proteins. *J. Phys. Chem. B* 1998, **102**:3586–3616.

12. Essmann U, Perera L, Berkowitz ML, Darden T, Pedersen LG: A smooth particle mesh Ewald method. *J. Chem. Phys.* 1995, **103**:8577–8593.

13. Grubmüller H, Heller H, Windemuth A, Schulten K: Generalized Verlet algorithm for efficient molecular dynamics simulation with long-range interactions. *Mol. Simul.* 1991, **6**:121–142.

14. Martyna GJ, Tuckerman ME, Tobias DJ, Klein ML: Explicit reversible integrators for extended systems dynamics. *Molec. Phys.* 1996, **87**:1117–1157.

15. Ryckaert J-P, Ciccotti G, Berendsen HJC: Numerical integration of the Cartesian equations of motion of a system with constraints: Molecular dynamics of $n$-alkanes. *J. Comp. Phys.* 1977, **23**:327–341.

16. Andersen HC: Rattle: A velocity version of the SHAKE algorithm for molecular dynamics calculations. *J. Comp. Phys.* 1983, **52**:24–34.

17. Acharya KR, Ren J, Stuart DI, Phillips DC, Fenna RE: Crystal structure of human alpha-lactalbumin at 1.7Å resolution. *J. Mol. Biol.* 1991, **221**:571–581.

18. Qiu D, Shenkin PS, Hollinger FP, Still WC: The GB/SA continuum model for solvation. A fast analytical method for the calculation of approximate Born radii. *J. Phys. Chem. A* 1997, **101**:3005–3014.

19. Wlodawer A, Svensson LA, Sjolin L, Gilliland G: Structure of phosphate free ribonuclease A refined at 1.26Å resolution. *Biochemistry* 1988, **27**:2705–2717.

20. Duan XQ, Quiocho FA: Structural evidence for a dominant role of nonpolar interactions in the binding of a transport/chemosensory receptor to its highly polar ligands. *Biochemistry* 2002, **41**:706–712.

21. Baudry J, Tajkhorshid E, Molnar F, Phillips J, Schulten K: Molecular dynamics study of bacteriorhodopsin and the purple membrane. *J. Phys. Chem. B* 2001, **105**:905–918.

22. Feller SE, MacKerrell Jr. AD: An improved empirical potential function for molecular simulations of phospholipids. *J. Phys. Chem. B* 2000, **104**:7510–7515.

23. Jorgensen WL, Chandrasekhar J, Madura JD, Impey RW, Klein ML: Comparison of simple potential functions for simulating liquid water. *J. Chem. Phys.* 1983, **79**:926–935.

24. Burling FT, Weis WI, Flaherty KM, Brünger AT: Direct observation of protein solvation and discrete disorder with experimental crystallographic phase. *Science* 1996, **217**:72–77.

25. Svergun DI, Richard S, Koch MHJ, Sayers Z, Zaccai G: Protein hydration in solution: Experimental observation by X-ray and neutron scattering. *Proc. Natl. Acad. Sci. U. S. A.* 1998, **95**:2267–2272.

26. Merzel F, Smith JC: Is the first hydration shell of lysozyme of higher density than bulk water? *Proc. Natl. Acad. Sci. U. S. A.* 2002, **99**:5378–5383.

27. Abseher R, Schreiber H, Steinhauser O: The influence of a protein on water dynamics in its vicinity investigated by molecular dynamics simulation. *Proteins: Struct., Funct., Genetics* 1996, **25**:366–378.

28. Settles M, Doster W: Anomalous diffusion of adsorbed water: A neutron scattering study of hydrated myoglobin. *Faraday Discuss.* 1996, **103**:269–279.

29. Makarov VA, Feig M, Andrews BK, Pettitt BM: Diffusion of solvent around biomolecular solutes. A molecular dynamics simulation study. *Biophys. J.* 1998, **75**:150–158.

30. Massi F, Straub JE: Structural and dynamical analysis of the hydration of the Alzheimer's beta-peptide. *J. Comp. Chem.* 2002, **24**:143–153.

31. Denisov DP, Jonsson BH, Halle B: Hydration of denatured and molten globule proteins. *Nat. Struct. Biol.* 1999, **6**:253–260.

32. Ball P: Water as a biomolecule. *Chem. Phys. Chem.* 2008, **9**:2677–2685.

33. Jasnin M, Moulin M, Haertlein M, Zaccai G, Tehei M: Down to atomic-scale intracellular water dynamics. *EMBO Rep.* 2008, **6**:543–547.

34. Stadler AM, Embs JP, Digel I, Artmann GM, Unruh T, Büldt G, Zaccai G: Cytoplasmic water and hydration layer dynamics in human red blood cells. *J. Am. Chem. Soc.* 2008, **130**:16852–16853.

35. Persson E, Halle B: Cell water dynamics on multiple time scales. *Proc. Natl. Acad. Sci. U. S. A.* 2008, **105**:6266–6271.

36. Pizzitutti F, Marchi M, Sterpone F, Rossky PJ: How protein surfaces induce anomalous dynamics of hydration water. *J. Phys. Chem. B* 2007, **111**:7584–7590.

37. Tobias DJ, Kuo IW, Razmara A, Tarek M: Protein hydration water. In *Water in Confining Geometries.* Edited by Devlin JP, Buch V: Springer-Verlag, Berlin; 2003, pp. 213–225.

38. Luise A, Falconi M, Desideri A: Molecular dynamics simulations of solvated azurin: correlation between surface solvent accessibility and water residence times. *Proteins: Struct. Funct. Genetics* 2000, **39**:56–67.

39. Garcia AE, Stiller L: Computation of the mean residence time of water in the hydration shells of biomolecules. *J. Comp. Chem.* 1993, **14**:1396–1406.

40. Brunne RM, Liepinsh E, Otting G, Wuthrich K, van Gunsteren WF: Hydration of proteins. A comparison of experimental residence times of water molecules solvating the bovine pancreatic trypsin inhibitor with theoretical model calculations. *J. Mol. Biol.* 1993, **231**:1040–1048.

41. Halle B: Protein hydration dynamics in solution: a critical survey. *Phil. Trans. R. Soc. Lond. B* 2004, **359**:1207–1224.

42. Levy Y, Onuchic JN: Water mediation in protein folding and molecular recognition. *Annu. Rev. Biophys. Biomol. Struct.* 2006, **35**:389–415.

43. Plaxco KW, Gross M: Unfolded, yes, but random? Never! *Nat. Struct. Biol.* 2001, **8**:659–660.

44. Kuwajima K: The molten globule state as a clue for understanding the folding and cooperativity of globular-protein structure. *Proteins: Struct. Funct. Genet.* 1989, **6**:87–103.

45. Ptitsyn OB: Molten globule and protein folding. *Adv. Protein Chem.* 1995, **47**:83–229.

46. Halle B, Denisov VP, Modig K, Davidoc M: Protein conformational transitions as seen from the solvent: magnetic relaxation dispersion studies of water, co-solvent, and denaturant interactions with nonnative proteins. In *Protein Folding Handbook.* Edited by Buchner J, Kiefhaber T: Wiley-VCH, Weiheim; 2005, pp. 201–242.

47. Go N: Shape of the conformational energy surface near the global minimum and low-frequency vibrations in the native conformation of globular proteins. *Biopolymers* 1978, **17**:1373–1379.

48. Brooks BR, Karplus M: Normal modes for specific motions of macromolecules: application to the hinge-bending mode of lysozyme. *Proc. Natl. Acad. Sci. U. S. A.* 1985, **82**:4995–4999.

49. Levitt M, Sander C, Stern PS: Protein normal-mode dynamics: trypsin inhibitor, crambin, ribonuclease, and lysozyme. *J. Mol. Biol.* 1985, **181**.

50. Smith J, Cusack S, Tidor B, Karplus M: Inelastic neutron scattering analysis of low-frequency motions in proteins: Harmonic and damped harmonic models of bovine pancreatic trypsin inhibitor. *J. Chem. Phys.* 1990, **93**:2974–2991.

51. Eisenmesser EZ, Millet O, Labeikovsky W, Korzhnev DM, Wolf-Watz M, Bosco DA, Skalicky JJ, Kay LE, Kern D: Intrinsic dynamics of an enzyme underlies catalysis. *Nature* 2005, **438**:117–121.

52. Bu Z, Biehl R, Monkenbusch M, Richter D, Callaway DJE: Coupled protein domain motion in Taq polymerase revealed by neutron spin-echo spectroscopy. *Proc. Natl. Acad. Sci. U. S. A.* 2005, **102**:17646–17651.

53. Zhang C, Durbin SM: Hydration-induced far-infrared absorption increase in myoglobin. *J. Phys. Chem. B* 2006, **110**:23607–23613.

54. Ebbinghaus S, Kim SJ, Keyden M, Heugen U, Gruebele M, Leitner DM, Havenith M: An extended dynamical hydration shell around proteins. *Proc. Natl. Acad. Sci. U.S.A.* 2007, **104**:20749–20752.

55. Chen SH, Liao CY, Huang HW, Weiss TM, Bellissent-Funel MC, Sette F: Collective dynamics in fully hydrated phospholipid bilayers studied by inelastic X-ray scattering. *Phys. Rev. Lett.* 2001, **86**:740–743.

56. Rheinstädter MC, Ollinger C, Fragneto G, Demmel F, Salditt T: Collective dynamics of lipid membranes studied by inelastic neutron scattering. *Phys. Rev. Lett.* 2004, **93**:108107.

57. Boon JP, Yip S: *Molecular Hydrodynamics*. Dover, New York; 1991.

58. Ruocco G, Sette F: The high-frequency dynamics of liquid water. *J. Phys. Cond. Matter* 1999, **11**:259–293.

59. Rahman A, Stillinger FH: Propagation of sound in water. A molecular-dynamics study. *Phys. Rev. A* 1974, **10**:368–378.

60. Tarek M, Tobias DJ: Single-particle and collective dynamics of protein hydration water: A molecular dynamics study. *Phys. Rev. Lett.* 2002, **89**:275501.

61. Parak F: Physical aspects of protein dynamics. *Rep. Prog. Phys.* 2003, **66**:103–129.

62. Doster W, Cusack S, Petry W: Dynamical transition of myoglobin revealed by inelastic neutron scattering. *Nature* 1989, **337**:754–756.

63. Tilton Jr. RF, Dewan JC, Petsko GA: Effect of temperature on protein structure and dynamics: x-ray crystallographic studies of the protein ribonuclease-A at nine different temperatures from 98 to 320 K. *Biochemistry* 1992, **31**:2469–2481.

64. Wood K, Plazanet M, Gabel F, Kessler B, Oesterhelt D, Zaccai G, Weik M: Dynamics of hydration water in deuterated purple membranes explored by neutron scattering. *Eur. Biophys. J.* 2008, **37**:619–626.

65. Caliskan G, Briber RM, Thirumalai D, Garcia-Sakai V, Woodson SA, Sokolov AP: Dynamic transition in tRNA is solvent induced. *J. Am. Chem. Soc.* 2006, **128**:32–33.

66. Lichtenegger H, Doster W, Kleinert T, Birk A, Sepiol B, Vogl G: Heme–solvent coupling: A Mössbauer study of myoglobin in sucrose. *Biophys. J.* 1999, **76**:414–422.

67. Rasmussen BF, Stock AM, Ringe D, Petsko GA: Crystalline ribonuclease A loses function below the dynamical transition at 220 K. *Nature* 1992, **357**:423–424.

68. Ferrand M, Dianoux AJ, Petry W, G. Zaccai G: Thermal motions and function of bacteriorhodopsin in purple membranes: effects of temperature and hydration studied by neutron scattering. *Proc. Natl. Acad. Sci. U. S. A.* 1993, **90**:9668–9672.

69. Daniel RM, Dunn RV, Finney JL, Smith JC: The role of dynamics in enzyme activity. *Annu. Rev. Biophys. Biomol. Struct.* 2003, **32**:69–92.

70. Tsai AM, Neumann DA, Bell LN: Molecular dynamics of solid-state lysozyme as affected by glycerol and water: A neutron scattering study. *Biophys. J.* 2000, **79**:2728–2732.

71. Vitkup D, Ringe D, Petsko GA, Karplus M: Solvent mobility and the protein 'glass' transition. *Nat. Struct. Biol.* 2000, **7**:34–38.

72. Tournier AL, Xu J, Smith JC: Translational water dynamics drives the protein glass transition. *Biophys. J.* 2003, **85**:1871–1875.

73. Tarek M, Tobias DJ: Role of protein–water hydrogen bond dynamics in the protein dynamical transition. *Phys. Rev. Lett.* 2002, **88**:138101.

74. Zhou YQ, Vitkuo D, Karplus M: Native proteins are surface-molten solids: Application of the lindemann criterion for the solid versus liquid state. *J. Mol. Biol.* 1999, **285**:1371–1375.

75. Dellerue S, Petrescu AJ, Smith JC, Bellissent-Funel MC: Radially softening diffusive motions in a globular protein. *Biophys. J.* 2001, **81**:1666–1676.

76. Doster W, Settles M: Protein–water displacement distributions. *Biochim. Biophys. Acta* 2005, **1749**:173–186.

77. Lagi M, Chu X, Kim C, Mallamace F, Baglioni P, Chen S-H: The low-temperature dynamic crossover phenomenon in protein hydration water: simulations vs. experiments. *J. Phys. Chem. B* 2008, **112**:1571–1575.

78. Kumar P, Yan Z, Xu L, Mazza MG, Buldryev SV, Chen S-H, Sastry S, Stanley HE: Glass transition in biomolecules and the liquid–liquid critical point of water. *Phys. Rev. Lett.* 2006:177802.

79. Wiener MC, White SH: Structure of a fluid dioleoylphosphatidylcholine bilayer determined by joint refinement of X-ray and neutron diffraction data. III. Complete structure. *Biophys. J.* 1992, **61**:434–447.

80. Small DM: *The Physical Chemistry of Lipids: From Alkanes to Phospholipids.* Plenum Press, New York; 1986.

81. Roh JH, Novikov VN, Gregory RB, Curtis JE, Chowdhuri Z, Sokolov AP: Onset of anharmonicity in protein dynamics. *Phys. Rev. Lett.* 2005, **95**:038101.

82. Weik M, Lehnert U, Zaccai G: Liquid-like water confined in stacks of biological membranes at 200 K and its relation to protein dynamics. *Biophys. J.* 2005, **89**:3639–3646.

83. Fitter J, Verclas SAW, Lechner RE, Seelert H, Dencher NA: Function and picosecond dynamics of bacteriorhodopsin in purple membrane at different lipidation and hydration. *FEBS Lett.* 1998, **433**:321–325.

84. Fitter J, Lechner RE, Dencher NA: Interactions of hydration water and biological membranes studied by neutron scattering. *J. Phys. Chem. B* 1999, **103**:8036–8050.

85. Doster W, Settles M: The dynamical transition in proteins: The role of hydrogen bonds. In *Hydration Processes in Biology: Experimental and Theoretical Approaches,* vol. 305. Edited by Bellissent-Funel M-C: IOS Press, Amsterdam, the Netherlands; 1999, pp. 177–191.

86. Backus EHG, Nguyen PH, Botan V, Moretto A, Crisma A, Toniolo C, Zerbe C, Stock G, Hamm P: Structural flexibility of a helical peptide regulates vibrational energy transport properties. *J. Phys. Chem. B* 2008, **112**:15487–15492.

87. Pal SK, Zewail AH: Dynamics of water in biological recognition. *Chem. Rev.* 2004, **104**:2099–2124.

88. Nilsson L, Halle B: Molecular origin of time-dependent fluorescence shifts in proteins. *Proc. Natl. Acad. Sci. U. S. A.* 2005, **102**:13867–13872.

89. Khodadadi S, Pawlus S, Sokolov AP: Influence of hydration on protein dynamics: Combining dielectric and neutron spectroscopy Data. *J. Phys. Chem. B* 2008, **112**: 14273–14280.

90. Li T, Hassanali AA, Kao Y-T, Zhong D, Singer SJ: Hydration dynamics and time scales of coupled water–protein fluctuations. *J. Am. Chem. Soc.* 2007, **129**:3376–3382.

91. Golosov AA, Karplus M: Probing polar solvation dynamics in proteins: A molecular dynamics simulation analysis. *J. Phys. Chem. B* 2007, **111**:1482–1490.

92. Tarek M, Tobias DJ, Chen S-H, Klein ML: Short wavelength collective dynamics in phospholipid bilayers: A molecular dynamics study. *Phys. Rev. Lett.* 2001, **87**:238101.

93. Jacobson K, Sheets ED, Simon R: Revisiting the fluid mosaic model of membranes. *Science* 1995, **268**:1441–1442.

94. Pike LJ: Lipid rafts: bringing order to chaos. *J. Lipid Res.* 2003, **44**:655–667.

95. McIntosh TJ, Simon SA: Roles of bilayer material properties in function and distribution of membrane proteins. *Annu. Rev. Biophys. Biomol. Struct.* 2006, **35**:177–198.

# Index

Printed and bound by CPI Group (UK) Ltd, Croydon, CR0 4YY

23/10/2024

01778266-0003